Rolf Herken (ed.)

The Universal Turing Machine
A Half-Century Survey

Second Edition

SPRINGER-VERLAG

Wien   New York

Computerkultur, edited by Rolf Herken, Volume II

Copy Editor: Danny Lee Lewis

© 1994 and 1995 Springer-Verlag/Wien
Printed in Slovenia

Typeset with TEX: Lewis & Leins, Berlin
Printing and binding: Tiskarna Ljudske pravice, 61104 Ljubljana

Printed on acid-free and chlorine-free bleached paper

With 29 Figures

ISSN 0946-9613
ISBN 3-211-82637-8 Springer-Verlag Wien New York
ISBN 3-211-82628-9 I. Aufl. Springer-Verlag Wien New York

Contents

Contents

List of Contributors

Michael A. Arbib
Computer Science Department, University of Southern California, Los Angeles, California, U.S.A.

Michael J. Beeson
Department of Mathematics and Computer Science, San Jose State University, San Jose, California, U.S.A.

Charles H. Bennett
IBM Thomas J. Watson Research Center, Yorktown Heights, New York, U.S.A.

Allen H. Brady
Department of Mathematics, University of Nevada Reno, Reno, Nevada, U.S.A.

Gregory J. Chaitin
IBM Thomas J. Watson Research Center, Yorktown Heights, New York, U.S.A.

Michael Conrad
Department of Computer Science and Biological Sciences, Wayne State University, Detroit, Michigan, U.S.A.

Elias Dahlhaus
Abt. Informatik III, Universität Bonn, Bonn, Germany

Martin Davis
Courant Institute of Mathematical Sciences, New York University, New York, New York, U.S.A.

Solomon Feferman
Department of Mathematics, Stanford University, Stanford, California, U.S.A.

Jens Erik Fenstad
Department of Mathematics, University of Oslo, Oslo, Norway

David Finkelstein
School of Physics, Georgia Institute of Technology, Atlanta, Georgia, U.S.A.

Robin Gandy
Mathematical Institute, University of Oxford, Oxford, England, U.K.

Oded Goldreich
Computer Science Department, Technion, Israel Institute of Technology, Technion City, Haifa, Israel

Yuri Gurevich
Department of Electrical Engineering and Computer Science, University of Michigan, Ann Arbor, Michigan, U.S.A.

Brosl Hasslacher
Los Alamos National Laboratory, Los Alamos, New Mexico, U.S.A.

Andrew Hodges
Mathematical Institute, University of Oxford, Oxford, England, U.K.

Stephen C. Kleene
Department of Mathematics, University of Wisconsin, Madison, Wisconsin, U.S.A.

Moshe Koppel
Department of Mathematics, Bar-Ilan University, Ramat Gan, Israel

Johann A. Makowsky
Computer Science Department, Technion, Israel Institute of Technology, Technion City, Haifa, Israel

Donald Michie
The Turing Institute, Glasgow, Scotland, U.K.

Roger Penrose
Mathematical Institute, University of Oxford, Oxford, England, U.K.

Robert Rosen
Department of Physiology and Biophysics, Dalhousie University, Halifax, Nova Scotia, Canada

Helmut Schnelle
Sprachwissenschaftliches Institut, Ruhr-Universität Bochum, Bochum, West Germany

Uwe Schöning
Seminar für Informatik, EWH Koblenz, Koblenz, West Germany

John C. Shepherdson
School of Mathematics, University of Bristol, Bristol, England, U.K.

Boris A. Trakhtenbrot
Raymond and Beverly Sackler Faculty of Exact Sciences, Department of Computer Science, Tel Aviv University, Tel Aviv, Israel

Oswald Wiener
Dawson City, Yukon, Canada

Preface

The papers contained in this volume were written on the occasion of the fiftieth anniversary of the publication of Alan Turing's paper "On Computable Numbers, with an Application to the Entscheidungsproblem" in 1937.

Turing's paper contained what may now be called Turing's Thesis, namely that every 'effective' computation can be programmed on a Turing machine. Furthermore it proved the unsolvability of the halting problem and of the decision problem for first order logic, and it presented the invention of the *universal* Turing machine. It is the publication of this idea that will presumably be acknowledged as marking sub specie aeternitatis the beginning of the "computer age".

As to the importance of Turing's notion of computability, Gödel's remark is worth quoting, that "with this concept one has for the first time succeded in giving an absolute definition of an interesting epistemological notion, i.e., one not depending on the formalism chosen"[1]. But not only did Turing introduce the most persuasive and influential concept of a machine model of effective computability a half century ago, he also anticipated in his work the diversity of topics brought together in this volume under the central notion of computability.[2]

Many of the papers that were written for this volume are not confined to mathematical logic or computer science proper. Today we witness the growing importance of the theory of computation in other areas of research – in mathematics as well as in physics, biology, cognitive science and artificial intelligence.

A striking example for this can be seen in the recent applications of computational complexity theory in physics, ranging from studies of computational properties emerging from the collective behavior of simple computing elements to the study of possible generalized recursion theoretic transcendations of the limits of Turing

---

1 Kurt Gödel, Remarks before the Princeton bicentennial conference on problems in mathematics–1946–, in: *The Undecidable,* ed. Martin Davis, p. 84. New York: Raven Press (1965)

2 See e.g., Hodges, this volume

computability in physical systems, and their implications for the nature of physical theory, arising in connection with the Church-Turing Thesis. With the ongoing work on various complexity measures based on distinguishability from randomness and the closely related notion of an observer-dependent computational complexity theory, the development that was initiated fifty years ago – combined with research in artificial intelligence (again initiated by Turing) – might even result in a natural setting for the understanding of intelligence.

This book is divided into two parts. Taken together they make the book fairly self-contained. In general, the level of technicality has been kept restricted – most of the contributions are intended to be read by the nonspecialist, too. Some acquaintance with Turing's work will of course be helpful. To a greater extent this can be achieved by reading Part I of the book.

Part I consists of five papers – among them fundamental contributions by Kleene, Gandy, and Feferman – that are directly concerned with Turing's work, its relation to the work of Hilbert, Gödel, Church, Kleene, Post, and others, and its consequences. Hodges' essay can also be read as an introduction to this volume. A general reader could then continue with the essay by Davis, in which Turing's role in the development of the modern computer is described.

Part II contains twenty three papers surveying contemporary research. Most of them contain recent or new results by their authors.

The table of contents and the abstracts or first pages of the contributions will enable the reader to individually plan a path through the volume.

THE STORY OF THE BOOK

In 1985, during studies for my doctoral dissertation, it occurred to me that the emerging picture of an interplay or even synthesis of ideas from mathematical logic, computer science, and physics, with an additional influx from epistemology, might be adequately reflected in a volume dedicated to Turing.

As it happened to be the case that no such project was in progress, I made a proposal to the German publishers Kammerer & Unverzagt to publish such a book in collaboration with a well-known, international publishing house.

In the beginning of 1986, I wrote personal invitational letters to a number of renowned researchers in the areas of research which were to be represented in the volume. The response by those who had been invited to contribute was extraordinarily positive. Due to the fact that many of them had already made a commitment, the project was well under way when in Autumn 1986 Oxford University Press obtained the rights to publish and distribute the book worldwide, with the exception of the German speaking countries.

While some of the final versions of the contributions had been sent to me by the end of 1986, most of them were eventually delivered by the end of 1987. Due to many revisions of some of the papers, the process of proofreading did not end before March 1988.

ACKNOWLEDGEMENTS

I am most grateful to those who contributed to this volume. It is a great honor for me to have gained and kept their confidence in the process of editing this volume.

Also, I would like to thank the originally invited contributors for their response and especially those among them who were planning to or actually began to write a paper, but had to resign because they eventually found themselves unable to meet the deadline(s) set for this volume.

I would like to express my gratitude to Andrew Hodges, Janos Makowsky, and Oswald Wiener, who had a decisive influence on this project; Andrew Hodges through his biography of Alan Turing[3] and discussions on Turing and his work; Janos Makowsky through his continuous intellectual and moral support – he also made several important suggestions for the list of invited contributors; Oswald Wiener through stimulating discussions during the last ten years on the topics of this volume, and much more.

Furthermore, I would like to thank

Martin Davis for allowing his essay "Mathematical Logic and the Origin of Modern Computers" to be included in this volume. It is reprinted here with permission by the Mathematical Association of America, which is gratefully acknowledged. Also, he pointed out the work of Koppel to me;

Robert Schrader for his constant encouragement and for his support of my work, including the editing of this volume;

Rudolph Lawrence Stuller for numerous discussions during his years at the "Institut";

Robert Fittler for lectures and conversations on mathematical logic;

John Archibald Wheeler for memorable conversations and the interest he showed in this volume;

James B. Hartle for discussions and his hospitality at the ITP, Santa Barbara, in the summer of 1985;

Renate Rothe for mail and telephone support;

Silvia Hanko, my helpful assistant at "mental images", for typing many of the letters and manuscripts on the computer;

and

Danny Lewis who did a magnificent job in copy editing and in typesetting the whole volume in TeX.

---

3 Andrew Hodges, *Alan Turing, The Enigma.* London: Burnett Books and Hutchinson, and New York: Simon and Schuster (1983)

Finally, I would like to thank Theo Kammerer and Alexander Unverzagt, of Verlag Kammerer & Unverzagt, as well as Oxford University Press for making this book possible.

Rolf Herken   Berlin, March 1988

Note to the First Reprint (1995). In this first reprint various errors have been corrected which were pointed out to the editor by the authors and readers of the book. The editor is indebted to Rudolf Siegle that the book is now published by Springer-Verlag, Wien, after being out of print for a period of two years. *R.H.*

Part I

Alan Turing and the Turing Machine

Andrew Hodges

By marking its fiftieth anniversary, this volume recognizes the long-lasting influence of the Turing machine concept. By collecting together new contributions from so many fields, it signals the exceptionally wide scope of that influence. This brief essay, intended to recall and honor the person of Alan M. Turing himself, will likewise mark not only the historical moment of Turing's publication, but will seek to convey the breadth of thought that underlay his introduction and development of the machine concept.

Alan Turing was born on 23 June 1912, and after submitting to a traditional British upper-middle-class schooling, won a scholarship to King's College, Cambridge University. He graduated with distinction in mathematics in 1934. Within less than a year he was elected to a Fellowship of King's College, his dissertation being a proof of the Central Limit Theorem, and it might have appeared that his interests would focus on the theory of probability.

In fact, however, Turing's research was by no means confined to one field, and in particular he had taken an interest since at least 1933 in the foundations of mathematics. In early 1935 he attended a course on mathematical logic given by the distinguished Cambridge topologist M.H.A. Newman, which ended with a full treatment of Gödel's proof that no system of axioms for arithmetic could be both consistent and complete, in the sense Hilbert had defined. But Hilbert's program for the foundations of mathematics had also called for a discussion of whether an axiomatic system could be *decidable,* i.e., whether there existed some definite method which in principle could be applied to decide the truth of any mathematical assertion; and this, the *Entscheidungsproblem,* still remained unresolved. Turing was attracted by Hilbert's challenge, worked on it alone, and astonished Newman by presenting him in April 1936 with a paper entitled "On Computable Numbers, with an Application to the Entscheidungsproblem", which introduced the abstract machine construction that we now celebrate.

The central difficulty in setting the question of decidability lay in defining satisfactorily what was to be meant by a "definite method" for solving mathematical problems. It was in resolving this issue that Turing showed his greatest originality. Turing seized upon the idea that the very nature of a definite method was that it had to be applied *mechanically,* and accordingly modeled the work of performing such a method in terms of the operation of a *machine.* He conceived the picture of a machine able to read and write symbols written on a tape, and refined this idea into a standard form. The concept of the Turing machine is now addressed in textbooks of logic and computer science, and a much closer exposition of its historical place in the development of mathematical logic is offered by other papers in this volume; the following comments are intended only to introduce the scope of Turing's thought. Even its original terminology hinted at this scope: machine "scanning" of paper tape; machines with finitely many "configurations"; machines with fixed "tables of behavior" suggested the fields of technology, quantum physics, and psychology.

Turing's experience and ingenuity as a mathematician showed itself in his ability to refine these general ideas into the precise formulation of the Turing machine concept, one which required only a very limited range of possible "behaviors". But Turing was not content with the purely mathematical tour de force involved in showing how to build up complex logical operations from a very small set. In arguing that machines possessing so limited a repertoire warranted identification with the most general possible class of "definite methods", Turing went outside mathematics proper. In several pages of nonmathematical discussion, Turing offered two general models for "definite methods" and argued that each could be modeled by the Turing machine. One was that of a person following a procedure, definite in the sense that at every stage a completely explicit "note of instructions" could be written down explaining what was to be done in such a way that another person could take up the work. But Turing clearly favored a far bolder appeal to the idea of a finite number of possible "states of mind", and the idea that any step of the process being followed must be describable in terms of observation of symbols, changing of symbols, and change of state of mind. According to this argument no explicit description of the process being followed need be given; it sufficed that the process was in some way embodied in the internal structure of the mind.

With this definition of computability it was possible to demonstrate the existence of uncomputable numbers – and to lead from this to resolve the Entscheidungsproblem in the negative. There cannot exist any general method for deciding mathematical questions such as Hilbert had hoped for. But we should note from the title that this was only *an application.* Turing focussed his paper on the concept of computability – indeed on "computable numbers", always suggesting a close connection with the human work of computation. Turing's discovery of this absolute standard of computability, with absolute limitations, was in itself a major discovery in mathematics. And so – although only time would show its full implications – was Turing's novel definition of a *universal* machine – a Turing machine that could perform the work of any other Turing machine, provided a description of the machine to be imitated were placed upon the tape for it to read.

Within a few days, however, Turing learned of one or both of Church *1936, 1936a,* which Turing ruefully described as "doing the same things in a different way" inasmuch as the mathematical work was parallel and arrived at the same conclusion regarding the Entscheidungsproblem. Turing submitted his paper for publication on 28 May 1936 but added an appendix dated 28 August demonstrating that his definition of computability was identical with Church's "effective calculability". The revised paper (Turing *1936-7*) was published at the turn of 1936 in the *Proceedings of the London Mathematical Society.* Church's review of it in the Journal of Symbolic Logic put the words "Turing machine" into print for the first time.

Such is the formal history – the history within mathematics. But it would be misleading if we concentrated solely on the relationship of papers within the field of mathematical logic. The essence of Turing's achievement was the discovery of a concept *with an application to* logic, rooted in ideas which lay outside mathematics. Church had proposed an assumption – "Church's Thesis", that the "effectively calculable" should be identified with the general recursive functions. There were strong mathematical arguments leading to this view. But Turing found justifications for this thesis from thinking much more generally about the nature of practical computation and from the nature of mental processes. Turing's definition was modeled on what human beings could actually do.

Why was Turing, a complete outsider to the field, able to supply and apply so radical an approach to its outstanding question? Can we guess an informal prehistory to the Turing machine, explaining something of its origin? Unfortunately Turing wrote nothing to explain the development of his thought in the 1930's. But we are fortunate enough to have one piece of evidence which casts a strong light on Turing's inner concerns in his undergraduate period. This is an essay entitled "Nature of Spirit", which he wrote probably in 1932 (given in full in Hodges *1983*). A few sentences give the flavor:

> It used to be believed in Science that if everything was known about the Universe at any particular moment then we can predict what it will be through all the future ... More modern science however has come to the conclusion that when we are dealing with atoms and electrons we are quite unable to know the exact state of them...

Some of what Turing wrote was influenced by A.S. Eddington, whose 1928 book *The Nature of the Physical World* Turing had read at school. Eddington was a strong protagonist for the view that the advent of quantum mechanics and the demise of classical determinism meant that Mind could again be accorded autonomous power without conflict with physical law. Turing expanded on his physical picture of mental power.

... We have a will which is able to determine the action of the atoms probably in a small portion of the brain, or possibly all over it. The rest of the body acts to amplify this ... when the body dies the "mechanism" of the body holding the spirit is gone and the spirit finds a new body sooner or later perhaps immediately ...

To understand the significance of Turing's words it must be explained for whom this essay was written: it was a private communication to Mrs. Morcom, mother of Turing's school-friend Christopher Morcom whose early death in 1930 had robbed Turing of a person he greatly loved and admired. Turing wrote of how he believed her son to be still alive in spirit and helping him. These are sentiments which seem very foreign to the trenchant materialist and atheist Turing who emerged after 1936, but beneath the surprising contrast, we should recognize the common thread – a great seriousness about the sheer mystery of mental phenomena, and an equally serious conviction that they must be reconciled with a scientific world view.

The problem of mind is the key to 'Computable Numbers'. Somehow, it would appear, Turing sensed in the questions about definite, mechanical methods an opportunity to abstract and refine the notion of *being determined,* and apply this newly refined concept to the old question of mind. Somehow he perceived a link between what to anyone else would have appeared the quite unrelated questions of the foundations of mathematics, and the physical description of mind. The link was a scientific, rather than philosophical view; what he arrived at was a new materialism, a new level of description based on the idea of discrete states, and an argument that this level (rather than that of atoms and electrons, or indeed that of the physiology of brain tissue) was the correct one in which to couch the description of mental phenomena. It was to promoting and exploring this idea that he gave much of his subsequent life.

By the time the paper was published, Turing was himself at Princeton and so in a position to communicate his ideas to the most powerful figures in modern mathematics. He was disappointed by the response, for example noting how Weyl failed to remark on his work. But besides the factor of his personal diffidence, perhaps it could be said that putting his ideas in the form expected by the professional mathematical world was not really his highest priority. It is true that his principal work in 1937-8 was mathematical in form. Turing's work on "ordinal logics" (Turing *1939*) was an investigation of whether it was possible to escape the force of Gödel's incompleteness theorem by exploiting infinite systems of axioms. Turing also did work on the theory of the Riemann zeta-function – a subject which could hardly be more clearly placed at the classical center of mathematics (Turing *1943*). Yet the logical work was strongly motivated by an interest in understanding the idea of "intuition", which Turing considered identifying with the noncomputable steps involved in generating infinite axiom systems; and the Riemann zeta function problem motivated him to build from gear wheels a special-purpose calculator (Turing

1939a). While at Princeton he also constructed a binary multiplier with electric relays (Hodges *1983*). Back at Cambridge in 1939 he also joined the small group attending Wittgenstein's discussions of the foundations of mathematics (Wittgenstein *1976*). The mechanical and the psychological continued to hold a great attraction for him. As it happened, events conspired to develop these interests, in a way Turing could hardly have imagined in 1936.

From a conventional academic point of view, the Second World War interrupted Turing's career; but Turing was not a conventional academic and his career in 1939 was by no means settled; he had declined the possibility of pursuing an American post, but Cambridge held out no immediate prospect of a lectureship. In the event, Turing became the chief scientific figure in the British cryptological effort during the Second World War, with particular responsibility for deciphering the communications of German naval forces – a feat in itself of immense historical significance and intellectual credit.

Turing found himself devising and mechanizing actual algorithms of a quite novel sophistication, giving full play to his logical ingenuity and his innovations in the theory of probability. Unfortunately government secrecy still prevents revelation of all but a smattering of these advances in nonnumerical computation (see Hodges *1983*, Good *1979*). We do know, however, of much active discussion between Turing and others involved in the cryptanalytic work, concerning the capacity of mechanical methods to perform tasks such as chess-playing. And in general terms, we know that mechanical methods (whether embodied in actual machines, or in people trained to perform routine tasks) overtook a great deal that had been hitherto regarded as the province of human judgment. It is hard to imagine an experience better calculated to stimulate Turing's fascination with the prospect of mechanizing mental tasks.

Turing also gained exposure to the most advanced technology of the early 1940's, including the then novel use of electronic components in digital operations. Indeed he spent much of the last year of the war in building with his own hands an electronic speech scrambler of his own, highly advanced, design (Hodges *1983*). He was very quick to observe that electronic components could provide the speed required to turn the idea of the universal Turing machine into a practical form. By 1945 he was speaking – typically – of "building a brain". His chance came rapidly with an appointment to the National Physical Laboratory, in which he was commissioned to design – in effect – an electronic computer. The detailed proposal that he wrote there at the turn of 1945 (Turing *1946*) cannot be accorded priority as the world's first; it referred to the EDVAC proposal of June 1945. But Turing borrowed from the American report only in details; his design was independent and flowed from his highly individual conception of a machine with a small number of simple operations from which all others could be built up not by adding to the hardware but by writing (as we would say now) appropriate software.

In his discussion of how to exploit the capacity of an electronic computer, Turing's report was well ahead of the parallel American effort; he was able to combine his extensive wartime experience of sophisticated routines with his grasp of the power of a universal machine to effect symbolic manipulations of any kind. Turing had used symbolic abbreviations for Turing machine specifications in 'Computable Numbers', and for him it was only a short step to observe that a universal machine could itself do the work of expanding such abbreviations, thus formulating the idea of computer languages and programming in the modern sense. (Indeed, it is hard in reading the 1936 paper, in which Turing had to employ the mentality of the modern programmer so as to think himself through the operations of a "blind" machine, to remember that computer programming did not then exist.)

The "notes of instructions" described in 1936 now became the lines of programs, practical programs that he began to write immediately. But just as immediate was his concern to explore the prospect of what he called "intelligent machinery" – what would normally now be described as "artificial intelligence". That is, he wished from the beginning to promote and exploit the thesis that all mental processes – not just the processes which could be explicitly described by "notes of instructions" – could be faithfully emulated by logical machinery. He was particularly interested in the capacity of the computer to modify its original instructions in the light of experience, arguing that this process could be considered as on a par with human learning. On these subjects Turing's own lucid writing (Turing *1946, 1947, 1948*) demands attention. (These important documents were unpublished during Turing's lifetime but are now, at last, all accessible.) It is quite astonishing to see how, in just ten years after 'Computable Numbers', he had translated its ideas into a powerful, prophetic overview of the potential for computer technology. For a short while, Turing's position at the National Physical Laboratory seemed to give him a leading role in creating this new technology, but the frustration he experienced led him to resign in 1948. He accepted instead a position at Manchester University which gave him a free hand to use, though not to design, the electronic computer that was being developed there.

Turing thereafter did little – in fact surprisingly little – to build up the modern science of computation. Although Turing *1946* had spelt out the idea of programming languages, and although indeed his work on "abbreviated code instructions" at the NPL had developed these ideas, at Manchester he left this entirely to others and worked himself in machine code expressed in a hideous base-32 notation. His short paper (Turing *1949*) on "Checking a large routine", given at the inauguration of the Cambridge EDSAC computer, is now seen as anticipating ideas of program proof not developed until the 1960's. It illustrates the mathematical power that he could easily have donated to the developing discipline, but it is the only such illustration.

Although retiring from his leading role in automatic computation, Turing made no retreat from his idea that mental processes are correctly described in the logical model independently of the particular physical embodiment, and so can be embodied in a physical form other than the brain. Turing *1950* – a paper written for the philosophical rather than the mathematical reader – offered a particularly clear specification of the logical model, and brought the development of Turing's thought to its fullest form. Although Church's thesis is sometimes described as the Church-Turing thesis, so as to recognize Turing's parallel contribution, we might more accurately enunciate a distinct Turing thesis with a somewhat different content: the Turing thesis is that the discrete-state-machine model is the relevant description of one aspect of the material world – namely the operation of brains.

Turing made a robust, indeed provocative, defense of this view and its implications. Pushing his thesis as far as he could, he opened up new issues and arguments. His continuing discussion of "thinking" and "intelligence" tended always to enlarge the scope of what was to be considered relevant. In 1936, his argument had centered on the carrying out of algorithms, in the work of 1946–1948 chess-playing (much discussed in wartime work) became his paradigm of intelligence, a principal point being that a successful chess-playing machine would have to evolve algorithms never explicitly supplied to it. In Turing *1950,* the arguments turned on the eventual success of the "intelligent machinery" in the much more ambitious task of sustaining general conversation.

Turing led up to this picture through a discussion which in effect tackled the problem of drawing a line around what was to be called "intelligence". The famously irreverent opening of his paper describes an "imitation game" in which an "interrogator" questions two unseen people, a man and a woman, and decides which is the woman on the basis of written replies – in fact by teleprinter communication– in which both man and woman are claiming to be the woman. Turing then went on to imagine a similar game in which human and machine compete to assert their human status under the same conditions. In fact, the apparent analogy is curiously inexact and diverts attention from the idea Turing actually wishes to put across. In the first game, a successful imitation proves nothing at all; we know that physical gender is *not* determined by teleprinter responses. In the second game, however, the very point of Turing's argument is that the successful imitation of intelligence in teleprinter messages *does* prove something, for it *is* intelligence. Intelligence (as opposed to gender, physical strength, or other qualities) is effectively defined as that which can be manifested by the communication of discrete symbols. As Turing put it, he wished to draw a line between mental faculties and others such as "shining in beauty competitions" or "racing against an aeroplane". This was achieved by the conditions of the game, as "drawing a fairly sharp line between the physical and intellectual capacities of a man."

Turing *1950* then suggests the means by which a machine might be led to acquire such intelligence, taking the view that it would be necessary to supply a system of rules by explicit programming, rather than expect learning to evolve entirely by allowing random mutations of a Turing machine program to be "rewarded" or "punished" by a "teacher". But behind these suggestions for how to proceed within the discrete-state-machine model of Mind, there lies the question which Turing raised by that phrase "fairly sharp line", affecting the very validity of that model. Can that line between the intellectual and physical faculties in fact be sharply drawn? The imaginary conversations between human and machine, as wittily concocted by Turing in this paper, employ ordinary language which relates to the external world. But human brains would seem to learn and apply such language through a complex interface of sensation and action relating to that world. Could a machine acquire language without a corresponding interface? Could the external world be reduced without loss to the symbols read by a Turing machine? Turing suggested this question was unimportant (referring to Helen Keller) yet in conclusion suggested that it might after all be best "to provide the machine with the best sense organs that money could buy." The problem of experience, of whether it is necessary for an entity with human "intelligence" to have also the human connection between mental and physical worlds, seems to me a key question about the validity of the Turing thesis. It is a question raised but not answered by Turing's rich, vivid, iconoclastic exploration of ideas, never confined to a safe compartment of thought.

Another question on which Turing left relevant but oddly insubstantial remarks, was that of whether the absolute limitations of computability have anything to do with the problems of mind and matter. He was curiously uninterested in his own discovery of absolutely undecidable questions within the discrete-state-machine model, being much more concerned to stress the power of the universal machine. There are no later references, for instance, to his 1937-8 thoughts on whether leaps of mental intuition can be said to correspond to the noncomputable steps involved in generating infinite axiom systems.

But Turing did describe other questions concerning the embodiment of the logical machine model in the physical world. Even in 'Computable Numbers' Turing gave a footnote relating the finite number of allowed symbols to the continuous infinity of physically possible symbols. The topology of "symbol space" appeared again in a curious digression in Turing *1947,* describing the physical consequences of recirculating the pulses in the acoustic delay lines at that time being developed for storage; a related argument came into his discussion of the physical principles of electrostatic storage at the Harvard conference of January 1947 (Harvard *1948*). In both cases he discussed the physical conditions required for discrete structure to be sustained. Turing *1948* offered a discussion of how thermodynamic considerations affected the translation of the universal machine concept into a physical form.

Turing *1950* introduced the discrete-state-machine model thus: "For instance in considering the switches for a lighting system it is a convenient fiction that each switch must be definitely on or definitely off. There must be intermediate positions, but for most purposes we can forget about them." This was a careful statement about the validity of modeling a physical system by the discrete state machine. Correspondingly, later in this paper Turing set up the actual physical continuity of the nervous system as a serious objection to his thesis. He countered it with the argument that the discrete machine could imitate the effects of continuity by means of random elements. This was a somewhat thin argument, for one thing because it failed to mention the question of quantum-mechanical uncertainty, for another because even within classical physics the effects of small initial variations are arbitrarily large, and it is not clear that the discrete state machine can correctly emulate such amplifying effects. Indeed, in another part of the same paper Turing actually drew attention to the contrast between the predictability of the discrete state machine, and the impossibility of obtaining predictions in classical physics because of how "the displacement of a single electron by a billionth of a centimeter at one moment might make the difference between a man being killed by an avalanche a year later, or escaping". Thus, we cannot feel that Turing had arrived at a complete theory of what he meant by modeling the mental functions of the brain by a logical machine structure. But we should give proper credit to Turing for raising such questions at all.

In 1949 Turing also made a new and important contribution to mathematics by showing the unsolvability of the word problem for semi-groups with cancellation (Turing *1950a*). Such work offered a path back to the heart of classical mathematics, had he wanted it. But it was a path that Turing did not follow, even though he watched its progress with interest. (His last published writing (Turing *1954*) was a popular article which explained the significance of such problems and surveyed new developments.) Nor did he turn back to number theory, although in a pioneering use of electronic computation he applied the Manchester machine to the Riemann zeta-function calculation (Turing *1953*). As before the war, his central interests were not primarily mathematical but scientific. He turned instead to a subject apparently unrelated to his earlier work – the mathematical modeling of morphogenesis. Here again he had the experience of jumping in as a complete outsider and making an important discovery. His chemical model led to nonlinear differential equations with the property of generating discrete spatial patterns from a spatially symmetric but unstable initial condition (Turing *1952*). Turing certainly saw this work as an attack on "the argument from design", by its accounting for phenomena prone to be advertised as beyond the bounds of material explanation. As such we can see it as a continuation by other means of his early concern to relate the most mysterious things in life to scientifically explicable structures. A more specific link with his earlier work is shown by his reference (a letter to J.Z. Young quoted in Hodges *1983*) to the question of how neural connections come to be physically formed in the brain. But perhaps we might also see another connection: symmetry-breaking provides the means for discrete structures to arise out of the physical continuum. In any case, this was a founding paper not only in mathematical biology but in the modern study of nonlinear classical dynamics – the subject hinted at in Turing's earlier reference to the "avalanche". Again, Turing pioneered the creative use of the computer by using it for numerical simulation of the nonlinear systems he postulated.

In 1954, Turing was examining the foundations of quantum mechanical physics – the subject in which Eddington had earlier engaged his attention so much, but which he had somewhat oddly omitted from his discussion of predictability in the logical machine model. In particular he was attracted to study the process which so mysteriously "reduces" a continuum of states to a discrete spectrum of observable values. Turing noticed the fact that subjecting a quantum-mechanical state to continuous observation and reduction has the effect of preventing dynamical evolution, pointing out that standard accounts of quantum mechanics gave no indication of how frequently such observations were supposed to take place. He expressed the idea that nonlinearity should enter into quantum mechanics (Gandy *1954*). Did he hope (as others since have hoped) to model reduction by a nonlinear process analogous to his chemical symmetry-breaking? Would he now try to relate the fundamental questions of classical and quantum physics to the discrete determinism of the Turing machine model? We can only guess, because Turing's death on 7 June 1954 ended his investigations.

Turing's suicide cannot be simply ascribed to his trial and enforced treatment by hormonal injections for homosexual conduct (then absolutely illegal under British law); these events had taken place two years earlier in 1952, and they had neither shamed nor cowed him. But Turing was in a unique position: his work had been central to extremely secret British and American operations in World War II, and indeed he had renewed such work in 1948. (After his arrest it was stopped.) Turing referred cryptically to an equally disturbing "crisis" in 1953 involving police surveillance. In my view a key question is whether he came to feel he was in an impossible moral position, unable to reconcile his demand for personal freedom with loyalty to the State.

Alan Turing offered an answer, a thesis: the identification of the world of Mind with that of the discrete state machine. It is a thesis with the immensely important feature that it can be explored in a practical manner, as research in Artificial Intelligence is currently doing, and for this alone we would be right to mark the anniversary of that founding paper. But he also left many open questions, difficult questions that cross boundaries between logic, physics, and the study of the human environment. Alan Turing's great work of 1936 shows us how fresh thought, motivated by taking profound questions completely seriously, can transcend and change our cultural compartments of thought. The passage of fifty years has not dimmed the brightness of that example. But his life, as well as his work, illustrates the extreme difficulty we face in delineating the features of human intelligence.

References

Turing, A.M.

1936-7 On computable numbers, with an application to the Entscheidungsproblem. *P. Lond. Math. Soc. (2)* **42** (1936-7) 230–265; received May 25, 1936, Appendix added August 28; read November 12, 1936; A correction, ibid., **43** (1937) 544–546.

1939 Systems of logic based on ordinals. *P. Lond. Math. Soc. (2)* **45** (1939) 161–228.

1939a Blue-print (in collaboration with D.C. MacPhail) in the Turing archive, King's College, Cambridge.

1943 A method for the calculation of the zeta-function. *P. Lond. Math. Soc. (2)* **48** 180–197 (received 1939).

1946 *Proposed Electronic Calculator,* Report to the Executive Committee of the National Physical Laboratory, 19 February 1946, reprinted and annotated in Turing *1986.*

1947 Lecture to the London Mathematical Society, 20 February 1947, reprinted in Turing *1986.*

1948 *Intelligent Machinery,* report to the Executive Committee of the National Physical Laboratory, 28 September 1948, reprinted with some errors of transcription in *Machine Intelligence* **5**, eds. B. Meltzer and D. Michie (Edinburgh University Press, 1969), pp. 3–23.

1949 In: *Report of a Conference on High Speed Automatic Calculating Machines,* Univ. Math. Lab., Cambridge; reprinted and annotated by F.L. Morris and C.B. Jones, *Ann. Hist. C.* **6**, 139–143.

1950 Computing machinery and intelligence. *Mind* **59** (1950) 433–460.

1950a The word problem in semi-groups with cancellation. *Ann. Math. (Princeton)* **52** (1950) 491–505.

1952 The chemical basis of morphogenesis. *Phi. T. Roy. B* **237** (1952) 37–72.

1953 Some calculations of the Riemann zeta-function. *P. Lond. Math. Soc. (3)* **3** (1953) 99–117.

1954 Solvable and unsolvable problems. *Peng. Sci.* **31** (1954) 7–23.

1986 *A.M. Turing's ACE Report of 1946 and Other Papers,* ed. R.E. Carpenter and R.W. Doran, vol. 10 in the Charles Babbage Institute Reprint Series for the History of Computing. Cambridge, MA: The MIT Press (1986).

Church, A.

1936 An unsolvable problem of elementary number theory. *Am. J. Math.* **58** (1936) 345–363.

1936a A note on the Entscheidungsproblem. *J. Symb. Log.* **1** (1936) 40–41.

Hodges, A.

1983 *Alan Turing: The Enigma.* London: Burnett Books and Hutchinson, and New York: Simon and Schuster (1983).

Gandy, R.O.

   1954 Letter to M.H.A. Newman, now in the Turing archive at King's college, Cambridge; quoted in Hodges *1983*.

Good, I.J.

   1979 Studies in the history of probability and statistics. XXXVII. A.M. Turing's statistical work in World War II. *Biometrika* **66** (1979) 393–396.

Harvard

   1948 *Proceedings of a Symposium on Large-Scale Digital Computing Machinery*, Annals of the Computation Laboratory, Harvard University, vol. XVI.

Wittgenstein, L.

   1976 *Wittgenstein's Lectures on the Foundations of Mathematics, Cambridge 1938*, ed. C. Diamond. Hassocks, Sussex: The Harvester Press (1976).

Turing's Analysis of Computability, and Major Applications of It

Stephen C. Kleene

The aim of this article is to present to readers, not assumed to have any previous acquaintance with Turing's work, both (1) the basic ideas in Turing's analysis of computability, and (2) what seems to me the most straightforward way of developing from it some significant features of the structure of mathematics. The application I use of his computability, and the path I follow in the developments from it, are somewhat different than in Turing's own writing. Other investigators contributed. I shall annotate this as we proceed.

1. *Algorithms (Decidability and Computability)*

In mathematics, we are often interested in having a general *method* or *procedure* for answering any one of a certain class of questions; for example, the class of questions, "Is n a prime number?". It is an infinite class of questions, one question for each positive integer $1, 2, 3, \ldots$ as the value of n. A question is selected from the class by picking such a value of n. And as a general method to test whether the n picked is a prime number, the mathematician answers "No" if $n = 1$; "Yes" if $n = 2$ or $n = 3$; and if $n \geq 3$, tries to divide n evenly by successive positive integers ≥ 2 up (if necessary) to the last one whose square is $\geq n$, answering "No" if he succeeds with one of them, otherwise answering "Yes".

In this example, the questions are yes-or-no questions. We may have the like with what-questions. For example, "What is the nth prime number?". The obvious method involves determining the n-th prime for $n = 1, \ldots, m$ before seeking it for $n = m+1$. It is a theorem of Euclid that, to any number a, there is a prime between $a + 1$ and $a! + 1$ inclusive. So, if we have already found the m-th prime, call it p, we can find the $m + 1$-st prime by testing for primality the numbers $p + 1, p + 2, \ldots, p! + 1$ in order till we find one that is prime. Actually, the upper bound $p! + 1$ for the numbers we may need to test is unnecessary. It is used in Euclid's proof of his theorem that there are infinitely many primes, which I presupposed in posing the class of questions, "What is the n-th prime number?" with the tacit understanding that there is one for each positive integer n. And when we are determining the primes in order, we only need to use previous primes as the divisors in testing for primality.

Of course, we have been taking it for granted that we already have a method for answering all questions of the infinite class, "Does b divide a?" for positive integers a and b. This method uses the method we have for finding the quotient and remainder on dividing a by b. This indeed is the long-division process that we learned in our elementary schooling. If the remainder is 0, the answer to "Does b divide a?" is "Yes"; if not, "No".

Actually, there is a more efficient arrangement of the work of constructing a list of the primes among the positive integers up to a given one. This is the sieve of Eratosthenes (c. 225 B.C.). We start with a list of those positive integers, omitting 1. Then we cancel the multiples of 2 after 2 itself, then the multiples of the next uncancelled number (which is 3) after itself, and so on. The numbers that do not fall through the sieve, i.e., are not cancelled, are the primes.

What do we mean by a method for answering any one of a given infinite class of questions? I think we can agree that such a method is given by a set of rules or instructions, describing a procedure that works as follows. *After* the procedure has been described, if we select *any* question from the class, the procedure will then tell us how to perform successive steps, so that after a finite number of them we will have the answer to the question selected. In particular, immediately after selecting the question from the class, the rules or instructions will tell us what step to perform first, unless the answer to the question selected is immediate. After our performing any step to which the procedure has led us, the rules or instructions will *either* enable us to recognize that now we have the answer before us and to read it off, *or else* that we do not yet have the answer before us, in which case they will tell us what step to perform next. In performing the steps, we simply follow the instructions like robots; no ingenuity or mathematical invention is required of us.

Such methods as I have just described have had a prominent role in mathematics. They have been called "algorithms". The term "algorithm" is a corruption of the last part of the name of the ninth-century Arabian mathematician Abu Abdullah abu Jafar Muhammad ibn Musa al-Khowarizmi from the Khowarizm oasis in central Asia. Also the term "algebra" comes from the words "al jabr" occurring in the Arabic title of one of his writings.

Algorithms and algebra did not originate with al-Khowarizmi, though he made respectable contributions to both subjects. The recognition of algorithms goes back at least to Euclid (c. 330 B.C.). There is a method, called "Euclid's greatest common divisor algorithm", for finding the greatest common divisor of two positive integers a and b; and this is used in an algorithm for answering the questions, "Does the equation $ax + by + c = 0$, where a, b, c are positive integers, have a solution in integers for x and y?". Equations whose solutions are sought in integers are called "Diophantine", after Diophantus (c. 250 A.D.).

When we have an algorithm for an infinite class of yes-or-no-questions, the property $R(a)$ or relation $R(a, b)$ or $R(a_1, \ldots, a_n)$ whose holding is in question is called *decidable* and the algorithm is also called a *decision procedure*. When we have an algorithm for an infinite class of what-questions, the function $\phi(a_1, \ldots, a_n)$ whose value is in question is called *computable* and the algorithm is also called a *computation procedure*.

An algorithm must be described fully *before* we start picking questions from the class; so its description must be finite. Thus an algorithm is a *finitely* described procedure, sufficient to guide us to the answer to any one of *infinitely* many questions, by *finitely* many steps in the case of each question.

With a finite class of questions, the situation is trivial: in principle, an algorithm always exists, consisting simply of a list or table of the answers to the questions.

We don't have to go far to find examples of infinite classes of questions for which the existence of algorithms is problematical.

From now on, I shall take my variables like $a, b, c, \ldots, x, y, z, \ldots$ to range over the *natural numbers* (or *nonnegative integers*) $0, 1, 2, \ldots$, except as may be indicated otherwise. I shall use "(Ex)" to abbreviate "there exists an x such that" or "for some x".

Suppose $R(a, x)$ is some decidable relation between two natural numbers a and x; i.e., an algorithm exists to decide, for each choice of values of a and x, whether or not $R(a, x)$ holds. What about $(Ex)R(a, x)$? This is a property of one natural number a. Is this property decidable? An algorithm for it is not given simply by the meaning of the symbols. All that the symbols suggest directly is that, to try to find out whether or not $(Ex)R(a, x)$ holds for a given a, we ask successively for $x = 0, 1, 2 \ldots$ (with that value of a) whether or not $R(a, x)$ holds. We can answer each of these questions, so far as we pursue them, since (as we assumed) $R(a, x)$ is decidable. If eventually we obtain "Yes" for some x, we answer "$(Ex)R(a, x)$?" with "Yes". But suppose we keep on getting "No" till doomsday. Then we fail to answer "$(Ex)R(a, x)$?". For we won't know whether doomsday simply came too soon, or if there is indeed no natural-number value of x making $R(a, x)$ true (with the given value of a).

Although the meaning of $(Ex)R(a, x)$ does not provide an algorithm directly, it may be possible, for a particular choice of the relation $R(a, x)$, to prove some theorems which will lead to an equivalent formulation of when $(Ex)R(a, x)$ holds from which an algorithm can be recognized. For example, if (with a given decidable $R(a, x)$) it can be proved that, whenever (for a given a) $R(a, x)$ holds for some x, there is such an $x \leq \phi(a)$ where $\phi(a)$ is a given computable function of a, then we would have an algorithm for $(Ex)R(a, x)$ by searching just (if need be) till we reach $\phi(a)$.

2. *The Totality of Algorithms*

Various problems whether algorithms exist for given infinite classes of questions have engaged the attention of mathematicians from early mathematical history. For indeed, mathematicians can claim the greatest success in treating a mathematical subject represented by an infinite class of questions when they have devised a method for answering all the questions of the class, that is an algorithm. Why not the best? But mathematicians haven't always succeeded in their quests for algorithms.

So it is surprising that only in the present century did people begin to think intensively about exactly what an algorithm is, or about what decidability or computability really mean. The idea of an algorithm, or of a decision or computation procedure, is sufficiently real so that in two thousand years of examples mathematicians have had no trouble in agreeing in each particular case of an infinite class of questions with a procedure they had in hand that the procedure is an algorithm for the class of questions or is not. But this history – these examples – do not provide a concrete picture of what the totality of all possible algorithms looks like. Without such a picture there is no possibility of showing that for some infinite class of questions an algorithm does not exist.

Such a picture was first provided in precise terms by several investigators in the 1930's, actually in three equivalent ways (and several further equivalents surfaced in the 1940's and 1950's). Here I shall paint the picture along the lines used by Alan M. Turing, a British mathematician who lived from 1912 to 1954.

Let me first sharpen our project a bit. As remarked, the case of a class of only finitely many questions doesn't interest us. For a class of infinitely many questions, I shall suppose the questions can be listed or "enumerated" using the natural numbers as the indices, thus: "Q_0?", "Q_1?", "Q_2?", ..., "Q_a?", ... (Infinite classes of questions which can't be so listed – which are "uncountable" – arise in mathematics; but the consideration of them is beyond the scope of this article.) Say, first, the questions are what-questions. The answers to the questions must (for this article) come from a given class of objects that can be likewise listed, thus: $c_0, c_1, c_2, \ldots, c_b, \ldots$. So it suffices to consider classes of questions of the form "$\phi(a) = $?" with answers "$b$", where a is the subscript of "Q_a?" locating the question in the first list, and b is the subscript for the object c_b in the second list used in answering it. Actually, it is convenient to consider more generally functions $\phi(a_1, \ldots, a_n)$ with n natural-number variables (for any positive integer n) taking natural numbers as values (which we call *number-theoretic functions*). For, such functions are frequently what we want algorithms for primarily, and not just by going over from our original questions to their subscripts a in a list "Q_0?", "Q_1?", "Q_2?", ..., "Q_a?", ..., and from our original answers "c_b" to their subscripts b.

For yes-or-no-questions, by letting 0 be a code for the answer "Yes" and 1 for the answer "No", we correlate what-questions (with only two possible answers, "0" and "1").

So our project can be narrowed to: *What number-theoretic functions $\phi(a_1, \ldots, a_n)$ are computable?*

3. Turing Computability of Number-theoretic Functions

I answer this question using Turing's ideas. Actually he worked instead primarily with: *What real numbers have computable binary decimal expansions?* He used his machines to print the successive digits ad infinitum on alternate squares of a 1-way infinite tape, while the intervening squares were reserved for temporary notes serving as scratchwork in the continuing computation.

The application, as we give it now, of Turing machines to characterize the computable number-theoretic functions $\phi(a_1, \ldots, a_n)$ was worked out in detail by Kleene for his seminar at Wisconsin in the spring term of 1940–41, and published in Chapter XIII of his *1952* book.[1] It is closer in some respects to Post *1936* than to Turing *1936-7*.

Let us recall my description in § 1 of a method or procedure for answering any one of a given (countably) infinite class of questions. I talked about "steps". What are we manipulating in these "steps"? We might say mental objects. This is so in the case of computations so simple that we perform them mentally, maybe with some motions of the lips. But except in very simple calculations, we need to work with symbols on paper (or some equivalent).

How many symbols do we need? Or indeed, better, how many can we have? Surely we can learn to recognize unambiguously only a finite number of symbols. In his pioneering paper *1936-7*, Turing wrote, "If we were to allow an infinity of symbols, then there would be symbols differing to an arbitrarily small extent." In contrast to *analog* computing, as for example by a slide rule, where there is a whole continuum of positions on each of the sliding scales, the computations we are considering are of number-theoretic functions $\phi(a_1, \ldots, a_n)$, where the possible values of the variables $a_1, \ldots, a_n$, and the possible answers b, are discrete (they do not merge with one another). So using symbols differing to an arbitrarily small extent would be inappropriate to our project. We are considering *digital* computation rather than *analog* computation.

The symbols (that is, occurrences of symbols from a given finite list) are what we observe before, and again after, any step in a computation. In actual practice,

---

1 An italicized date in conjunction with the name of an author constitutes a reference to the References at the end of this article.

as for example in performing a long division, we may have an array of symbols, and shift our attention from one place on the paper to another. If the array is large, we may not be able to take it all in at once. Let us consider the paper, or whatever we use instead, as a space to accommodate symbols, divided into parts which I call *cells*, each of which is what we observe at a given moment. We can construe what is written on a cell as a single symbol, even if it is a (finite) complex of our primary symbols.

The cells must have some kind of a regular geometrical arrangement, or our procedure in computing would become chaotic, frustrating its progress toward a determinate result. If one thinks hard about this, and entertains various alternative arrangements of the cells which come to mind, one can see that whatever can be done with any reasonable arrangement can be done with a linear arrangement, open-ended say to the right.[2] Let us call the cells then *squares* on a *tape,* beginning with a first (*leftmost*) square and with successive squares ad infinitum to the right.

A person in computing is not constrained to working from just what he sees on the square he is momentarily observing. He can remember information he previously read from other squares. This memory consists in a state of mind, his mind being in a different state at a given moment of time depending on what he remembers from before. Turing wrote, "the number of states of mind which need to be taken into account is finite. ... If we admitted an infinity of states of mind, some of them will be 'arbitrarily close' and will be confused."

From such reflections on how a human computer works, Turing formulated his concept of computing machines (now called *Turing machines*), to which I proceed without further ado.

A Turing machine is an agency or apparatus or black box with the following features.

In describing its operation, I distinguish discrete *moments* of time, numbered $0, 1, 2, \ldots$.

The machine is supplied with a linear *tape,* ruled in *squares,* infinite to the right. The tape is threaded through the machine so that at each moment just one square is observed by the machine or *scanned.*

The machine at each moment of time is in one of a fixed list of $k + 1 \, (\geq 2)$ states $q_0, \ldots, q_k$. (These correspond to the states of mind of a human computer.) I call q_0 the *passive* state, and $q_1, \ldots, q_k$ *active* states.

Each square of the tape is capable of having printed on it any one of a fixed list of $l \, (\geq 1)$ symbols $s_1, \ldots, s_l$, or of being *blank,* which I write s_0; so there are $l + 1$ possible *square conditions* $s_0, \ldots, s_l$.

At each moment of time, the total (*tape vs. machine*) *situation* consists of, **first,** a particular printing on the tape (i.e., which squares are printed, and each with which

---

2 This is elaborated in Kleene *1952* under (c) on pp. 379–381.

of $s_1, \ldots, s_l$), **second,** a particular position of the tape in the machine (i.e., which square is scanned), and **third,** a particular machine state (i.e., which of $q_0, \ldots, q_k$ is the state of the machine). I call the situation *active,* if in it the state is active; *passive,* if in it the state is passive.

As we shall use the machines, only a finite number of the squares will be printed at any moment; so the situations will constitute finite objects.

If the situation at a given moment t is active, the machine performs an *act* between the moment t and the next moment $t + 1$, whereby it is determined what the situation will be at Moment $t + 1$. The act consists of three parts, each of which possible changes one element of the situation (these being the only changes between Moment t and Moment $t + 1$); **first,** the choice of one of $s_0, \ldots, s_l$ to be the condition $s_{i'}$ at Moment $t + 1$ of the square which was scanned at Moment t (say that square was in condition s_i at Moment t); **second,** a possible shifting of the tape in the machine, so that at Moment $t + 1$ the scanned square is one square left of (L), the same as $(C$ for "center"), or one square right of (R), the square scanned at Moment t; and **third,** the choice of one of $q_0, \ldots, q_k$ to be the machine state $q_{j'}$ at Moment $t + 1$ (say the machine state was q_j at Moment t). A machine act can thus be described by a triple

$$s_{i'} \quad \begin{matrix} L \\ (C) \\ R \end{matrix} \quad q_{j'} \quad \text{or briefly} \quad i' \quad \begin{matrix} L \\ (C) \\ R \end{matrix} \quad j' .$$

If the situation at Moment t is passive, no act is performed, so the situation at Moment $t + 1$ is the same as at Moment t.

What act does a given machine perform from a given active situation? The act performed is determined normally (i.e., outside of one exceptional kind of situation) by the scanned square condition s_i and the machine state q_j at that moment. This pair $s_i q_j$, or briefly ij, I call the *configuration* (*active,* if q_j is active). The exceptional kind of situation is when the configuration $s_i q_j$ calls for a move left (L), as part of the act normally performed, but the scanned square is already the leftmost square of the tape. Then the machine instead "jams", i.e., it goes to the passive state q_0 with nothing else changed.

In fact, as we shall ordinarily use the machines, it will never happen that the configuration calls for a move left from the leftmost square.[3]

With $l + 1$ scanned square conditions and k active states, there are $(l+1)k$ active configurations. A particular Turing machine is specified (in its behavior, which is

---

3 Without relying on this, we could easily fix matters so that the configuration determines the act in all situations without exception. Let the leftmost square bear a special mark $*$, doubling the list of square conditions to $s_0, \ldots, s_l, *s_0, \ldots, *s_l$. From a configuration $*s_i q_j$, the act is then $*s_i C q_0$ if $s_i q_j$ calls for a motion left, and otherwise the same as from $s_i q_j$ except with a $*$ on the first member of the triple unless $s_i q_j$ calls for a motion right.

all we care about) by a *machine table,* showing for each active configuration what act shall be performed (if possible).

I illustrate this by giving (in Figure 1) at the top a table for a machine, call it "$\mathfrak{G}$", with eleven active states (I write them simply "1", ..., "11" instead of "q_1", ..., "q_{11}") and one symbol s_1, which is a tally mark "|". To follow the action of this machine, we must assume some situation as obtaining at Moment 0. I take it to be as shown first at the bottom left, with the machine in State 1, scanning the rightmost of two squares each printed with a tally, the tape being otherwise blank. This situation is active; and we find the act Machine $\mathfrak{G}$ will perform from Moment 0 by entering the table in the first row (since its state is 1) and the second column (since the scanned square bears "|"). There we find "$1R2$"; so the machine leaves that tally unchanged (1), moves right on the tape (R), and assumes its second active state (2). Thus we get the tape vs. machine situation at Moment 1 shown next. Similarly we arrive at the next two situations.

To save space in showing situations, let me change my format to omit the picture of the tape, simply write 0's and 1's for blank and tally-printed squares, and indicate the state by a superscript.

At the same time, I change to indicate the conditions of the squares only as far out as the rightmost one thus far scanned in the operation of the machine (all the rest we know are blank). As we remarked, our tape vs. machine situations are finite. In this format, I show in Figure 1 at the right the situations all the way down to Moment 23. Since the state at Moment 23 is the passive one 0, the situation will be the same at all subsequent moments.

Our announced objective was to apply the Turing machine concept to characterizing the computation of a number-theoretic function $\phi(a_1, \ldots, a_n)$. First, I take $n = 1$. The idea is that an algorithm or computation procedure for $\phi(a)$ can be embodied in a Turing machine which, when we feed it a tape on which a particular value of the variable a is represented, will perform acts terminating in a situation in which the corresponding value $b = \phi(a)$ of the function is also represented.

First, we must agree on how to represent the natural numbers $0, 1, 2, \ldots$ on the tape. I take the most primitive representation, using only the tally symbol "|". However, because I don't want 0 to be represented by no tallies (which would be invisible on the tape), I use one tally to represent 0, two to represent 1, three to represent 2, etc., thus:

$$0 \quad 1 \quad\ 2$$
$$,\quad |, \quad |\ |, \quad \ldots .$$

For a given machine to *compute* a given function $\phi(a)$, it must behave as follows. After picking a value of a, we start the machine off at Moment 0 by putting a tape in it with a represented by $a + 1$ tallies on consecutive squares beginning with the next-to-the-leftmost square, with the tape otherwise blank, and with the machine

Machine state	Scanned square condition	
	0	1
1	$0C0$	$1R2$
2	$0R3$	$1R9$
3	$1L4$	$1R3$
4	$0L5$	$1L4$
5	$0L5$	$1L6$
6	$0R2$	$1R7$
7	$0R8$	$0R7$
8	$0R8$	$1R3$
9	$1R9$	$1L10$
10	$0C0$	$0R11$
11	$1C0$	$1R11$

Tape vs. machine situation (left)

Moment	
0	situation 1
1	situation 2
2	situation 3
3	situation 4

Tape vs. machine situation (right)

Moment							
0	0	1	1^1				
1	0	1	1	0^2			
2	0	1	1	0	0^3		
3	0	1	1	0^4	1		
4	0	1	1^5	0	1		
5	0	1^6	1	0	1		
6	0	1	1^7	0	1		
7	0	1	0	0^7	1		
8	0	1	0	0	1^8		
9	0	1	0	0	1	0^3	
10	0	1	0	0	1^4	1	
11	0	1	0	0^4	1	1	
12	0	1	0^5	0	1	1	
13	0	1^5	0	0	1	1	
14	0^6	1	0	0	1	1	
15	0	1^2	0	0	1	1	
16	0	1	0^9	0	1	1	
17	0	1	1	0^9	1	1	
18	0	1	1	1	1^9	1	
19	0	1	1	1^{10}	1	1	
20	0	1	1	0	1^{11}	1	
21	0	1	1	0	1	1^{11}	
22	0	1	1	0	1	1	0^{11}
23	0	1	1	0	1	1	1^0

Figure 1. Table for Machine ${\mathfrak{G}}$ and example of its action.

in its first active state 1 scanning the rightmost of those $a + 1$ tallies. If it is a machine that computes our function, it must eventually stop (that is, assume the passive state 0) with the corresponding function value $\phi(a)$ represented by $\phi(a) + 1$ tallies on squares occurring consecutively to the right after leaving one square blank following the representation of a, with the tape otherwise blank, and with the scanned square the one bearing the rightmost of the $\phi(a) + 1$ tallies. This must happen for each choice of the natural number a.

I illustrate this taking $\phi(a)$ to be the successor function $a + 1$ (simple, but important!). First, take $a = 1$. So the machine, started off thus

should eventually stop thus

$(a = 1, \phi(a) = 2)$. Indeed, the machine $\mathfrak{S}$, whose table I gave in Figure 1, does just this, giving us the answer "2" to our question "$1 + 1 = $?" at Moment 23. If Machine $\mathfrak{S}$ does the like for every value of a, then it computes the function $\phi(a) = a + 1$.

In fact it does! To see this, let us break down its action into groups of steps fitting a pattern, as I shall first illustrate for the case $a = 1$ which I displayed in Figure 1.

Pat. 1 (used only initially). Machine $\mathfrak{S}$, if it is in State 1 scanning a rightmost printed square (at Moment 0), thereupon goes right two squares (it keeps count by changing successively to States 2 and 3), prints there, and returns left one square, assuming State 4 (at Moment 3).

Pat. 2. What does Machine $\mathfrak{S}$ do from a situation in which it has just become in State 4 (as at Moments 3 and 10)? It goes left till it first encounters a printed square after one or more blank squares (it changes to State 5 just after being as at Moments 3 and 11 on a blank square). Then it probes the square next left of that printed square (going to State 6 there) to see whether it is not blank or is blank, going respectively to State 7 or 2 on the aforesaid printed square (as at Moment 6 or 15).

Pat. 3. Starting in State 7 on a printed square with a blank square next right and some printed squares further right (Moment 6), it erases that printed square, goes right till it comes to a blank square after one or more printed squares, prints on it, and returns left one square assuming State 4 (Moment 10).

Pat. 4 (used only terminally). Starting in State 2 on a printed square with a blank square next right and some printed squares further right (Moment 15), it goes right printing on all but the last of the consecutive blank squares that were just right of it, and continues over the next group of printed squares to print on the first blank square following, where it goes to State 0 (Moment 23).

Let us apply this analysis of our machine's operation to a little more complicated example, say with $a = 3$, not showing all the consecutive situations but just those at the beginnings and ends of the parts which follow a pattern (Figure 2).

	0	1	1	1	1^1							
Pat. 1												
	0	1	1	1	1	0^4	1					
Pat. 2												
	0	1	1	1	1^7	0	1					
Pat. 3												
	0	1	1	1	0	0	1^4	1				
Pat. 2												
	0	1	1	1^7	0	0	1	1				
Pat. 3												
	0	1	1	0	0	0	1	1^4	1			
Pat. 2												
	0	1	1^7	0	0	0	1	1	1			
Pat. 3												
	0	1	0	0	0	0	1	1	1^4	1		
Pat. 2												
	0	1^2	0	0	0	0	1	1	1	1		
Pat. 4												
	0	1	1	1	1	0	1	1	1	1	1^0	

Figure 2. Action of Machine $\mathfrak{S}$ in computing $a + 1$ for $a = 3$.

In summary our machine $\mathfrak{S}$, after printing a 1 to start the second group of 1's or tallies, copies the 1's of the first group into the second group. It keeps track of the 1's copied by erasing each as it copies it, except the last (leftmost) 1, upon encountering which it restores those erased before copying that.

To compute a function $\phi(a_1, \ldots, a_n)$ with any $n \geq 1$ variables, we start the machine at Moment 0 with the n-tuple $a_1, \ldots, a_n$ represented on the tape by groups of $a_1+1, \ldots, a_n+1$ consecutive tallies, preceded and separated by single blank squares, with all other squares blank, and the square bearing the rightmost of these tallies scanned in State 1. To compute ϕ, the machine must, whatever the choice of the values of $a_1, \ldots, a_n$, eventually stop with the $n + 1$-tuple $a_1, \ldots, a_n, \phi(a_1, \ldots, a_n)$ similarly represented.

Although we are using only one symbol, the tally "|", in questioning a Turing machine with an n-tuple of natural numbers and in receiving the answer, our definition of Turing computability permits the machine a list of $l \geq 1$ symbols $s_1, \ldots, s_l$. We take s_1 to be the tally "|". If $l > 1$, the other symbols $s_2, \ldots, s_l$ are allowed to be printed on the tape at intermediate moments of a computation. In fact, it can be proved that every function $\phi(a_1, \ldots, a_n)$ thus computable can be computed by a Turing machine with only the one symbol "|", as in the above illustration and those I shall give presently.[4]

4. *Turing's Thesis*

What I am claiming after Turing – call it "Turing's thesis" – is that every number-theoretic function $\phi(a_1, \ldots, a_n)$ for which there is an algorithm – which is intuitively computable (one sometimes says "effectively calculable") – is *Turing computable*; that is, there is a Turing machine which computes it in the manner I have described.

It may seem anomalous, then, that it should be so much of an exercise to see that such a simple function as $a + 1$ is Turing computable. However, we undertook to treat $a + 1$ right at the beginning of our study of Turing machines. And we used only one symbol "|". Using more symbols would tend to make the Turing machine action better resemble computation as we are used to doing it.

Our illustration may unduly suggest that the computer is restricted to taking an ant's eye view of its work, squinting at the symbol on one square at a time. However, when more than the one symbol "|" is used, we can interpret what we mean by a symbol liberally. Thus the Turing-machine squares can correspond to whole sheets of paper. If we employ sheets ruled into 20 columns and 30 lines, and authorize 99 primary symbols, there are $100^{600} = 10^{1200}$ possible square conditions, and we are at the opposite extreme. The schoolboy doing arithmetic on $8\frac{1}{2}$ by $12''$ sheets of ruled paper would never need, and could never utilize, all this variety.

Another representation of a Turing machine tape is as a stack of IBM cards, each card regarded as a single square for the machine.

There is, however, an intrinsic difficulty which the theory of computing must face. We are defining a function $\phi(a)$ to be Turing computable if there is a Turing machine which will compute its value for *all* choices of the value of a, and similarly with $\phi(a_1, \ldots, a_n)$. The schoolboy computing on sheets of paper is not normally asked to work with numbers too big to go on one of his sheets. If he had to do

---

4 See Kleene *1952*, the bottom of p. 361.

that, he too would have to work in a groping manner as our machine $\mathfrak{G}$ does in computing $a + 1$, though using more than the one symbol "|" he would not need to do so for as small values of a.

Turing wrote,

> ... we cannot regard the process of recognition as a simple process. This is a fundamental point and should be illustrated. In most mathematical papers the equations and theorems are numbered. Normally the numbers do not go beyond (say) 1000. It is, therefore, possible to recognize a theorem at a glance by its number. But if the paper was very long, we might reach Theorem 157767733443477; then, further on in the paper, we might find "... hence (applying Theorem 157767733443477) we have ... ." In order to make sure which was the relevant theorem we should have to compare the two numbers figure by figure, possibly ticking the figures off in pencil to make sure of their not being counted twice.

Incidentally, you surely recognize that Turing machines can be used to perform computations on all sorts of finite sequences of symbols, not just strings of tallies representing natural numbers. This is an elegant feature of his machine concept.

Mathematicians deal with idealized systems of objects, obtained by extrapolating for the purposes of their thought from what people encounter in the real world. They imagine the infinite sequence $0, 1, 2, \ldots$ of the natural numbers, and thus they get a beautiful theory with an elegant logical structure, though in the actual counting of discrete objects we can never use more than finitely many of them.

Computation, theoretically considered (to be performable for all possible values of the independent variables), is idealized. Turing's analysis takes this idealized aspect of it into account. A Turing machine is like an actual digital computing machine, except that (1) it is error free (i.e., it always does what its table says it should do), and (2) by its access to an unlimited tape it is unhampered by any bound on the quantity of its storage of information or "memory".

5. Some Turing Computable Operations

As we have illustrated by our 12-state machine $\mathfrak{G}$ to compute $a + 1$, it can be a by no means trivial exercise to construct a Turing machine to compute even a quite simple function for which we have an algorithm. Scholars have verified that Turing machines exist embodying all sorts of algorithms.

One must begin by handling various simple operations.

For example, the table for a machine to copy a number a represented leftmost on the tape (otherwise blank) and scan the copy in *standard position* (i.e., on the rightmost tally) is obtained from the table for our machine $\mathfrak{G}$ (which copies a number adding 1 to it) simply by changing the entry in line 11 and column 0 from $1C0$ to $0L0$.

We can similarly construct a machine to copy not the leftmost of a sequence of numbers, but say the *m*-th from the left. Or the copy may be put after a one-space gap, meaning that two blank squares separate the copy from the preceding number (which was rightmost before the copying). We may later erase all the numbers from a given space back to such a one-space gap and close up the resulting many-space gap.

Such simple operations constitute useful building blocks in rendering by Turing machines algorithms as they are commonly encountered. I shall illustrate.

One common form of definition is a *primitive recursion,* such as

$$\begin{cases} \phi(0, b) = \psi(b), \\ \phi(a+1, b) = \chi(a, \phi(a, b), b). \end{cases}$$

Specifically, this is a (primitive) recursion on *a* with one parameter *b*. (In general, there may be *n* parameters $b_1, \ldots, b_n$ for any $n \geq 0$.) If $\psi(b)$ and $\chi(a, c, b)$ are previously defined functions, this defines the function $\phi(a, b)$. If we already have algorithms for $\psi(b)$ and $\chi(a, c, b)$, an algorithm for computing $\phi(a, b)$ consists in computing successively (using the algorithms for ψ and χ)

$$\begin{array}{lll} \phi(0, b) & \text{as } \psi(b) & (\text{briefly } f_0 = \psi(b)), \\ \phi(1, b) & \text{as } \chi(0, \phi(0, b), b) & (\text{briefly } f_1 = \chi(0, f_0, b)), \\ \phi(2, b) & \text{as } \chi(1, \phi(1, b), b) & (\text{briefly } f_2 = \chi(1, f_1, b)), \\ \cdots \end{array}$$

until we get $\phi(a, b)$ (briefly f_a) for the given value of the recursion variable *a*, this all being done with a fixed given value of the parameter *b*.

Dedekind in *1888* and Peano in *1889* used primitive recursions to define successively the sum $a + b$, the product $a \cdot b$ and the exponentiation function a^b of two natural numbers *a* and *b*, thus (where *b* is the recursion variable and *a* the parameter),[5]

$$\begin{cases} a + 0 = a, \\ a + (b + 1) = (a + b) + 1, \end{cases}$$
$$\begin{cases} a \cdot 0 = 0, \\ a \cdot (b + 1) = (a \cdot b) + a, \end{cases}$$
$$\begin{cases} a^0 = 1, \\ a^{b+1} = a^b \cdot a. \end{cases}$$

---

5 The term "recursion" was employed by Skolem in *1923* and Hilbert in *1926;* and "primitive recursion" was introduced by Péter in *1934* to distinguish it from other kinds of recursion. Dedekind, Peano and Skolem worked with the positive integers.

Let us consider how a Turing machine $\mathfrak{M}_\phi$ can be constructed to compute the function $\phi(a, b)$ defined by the primitive recursion displayed above, assuming we already have machines $\mathfrak{M}_\psi$ and $\mathfrak{M}_\chi$ to compute $\psi(b)$ and $\chi(a, c, b)$.

The machine $\mathfrak{M}_\phi$ I have in mind, started out in the standard way with any given choice of values of a and b represented on the tape, will print more and more numbers to the right after leaving a one-space gap, until finally it prints the desired value f_a of $\phi(a, b)$. Then all the numbers between that and the gap will be erased, and the resulting many-space gap will be closed up.

In Figure 3, I show the numbers in the order in which they will be printed on the tape. Each letter like "c" stands for $c + 1$ consecutive tally-printed squares, and each comma "," stands for one blank square. After a number has just been printed, its rightmost tally will be scanned.

$, a, b, , a, b, f_0$ $\qquad\qquad$ (where $f_0 = \psi(b)$)

$, a, \begin{cases} f_0 & \text{if } a = 0. \\ 0, f_0, b, f_1 & \text{(where } f_1 = \chi(0, f_0, b)) \quad \text{if } a \neq 0 \end{cases}$

$, a - 1, \begin{cases} f_1 & \text{if } a - 1 = 0. \\ 1, f_1, b, f_2 & \text{(where } f_2 = \chi(1, f_1, b)) \quad \text{if } a - 1 \neq 0 \end{cases}$

$, a - 2, \begin{cases} f_2 & \text{if } a - 2 = 0. \\ 2, f_2, b, f_3 & \text{(where } f_3 = \chi(2, f_2, b)) \quad \text{if } a - 2 \neq 0 \end{cases}$

$, a - 3, \begin{cases} f_3 & \text{if } a - 3 = 0. \\ 3, f_3, b, f_4 & \text{(where } f_4 = \chi(3, f_3, b)) \quad \text{if } a - 3 \neq 0 \end{cases}$

$\cdots$

Figure 3. Action of Machine $\mathfrak{M}_\phi$ in computing $\phi(a, b)$.

Notice the one-space gap represented by the double comma in the first line. The machine $\mathfrak{M}_\psi$ which computes $\psi(b)$, started on a tape bearing just ",b" leftmost, will come to ", b, f_0" where $f_0 = \psi(b)$, and in doing so it doesn't look left of the blank square preceding "b". So, started instead in the situation ", a, b, , a, b", the ", a, b, , a" to the left of the ", b" won't distract it in its computation of f_0. Hence, after arranging for our desired machine $\mathfrak{M}_\phi$ to begin by copying "a, b" to obtain ", a, b, , a, b" we can hitch that part of it into $\mathfrak{M}_\psi$. To do so, we annex the table for $\mathfrak{M}_\psi$ with its active states renumbered to run up not from 1 but from the next number after those used to program the copying of "a, b". The first of these new state numbers is substituted for the passive state 0 of a machine which would stop after that copying. So we get to f_0 (end of the first line), and it constitutes no problem to add further lines to the table we are building for $\mathfrak{M}_\phi$ so that $\mathfrak{M}_\phi$ will then copy a (the 3rd number back), and test whether a is 0 or not. The result of this test is expressed by one of two different states being assumed next.

The state to be assumed in the case that $a = 0$ will lead to the copying of the second number back (which in the situation we have just reached is f_0), and thereafter (not shown in Figure 3) to the erasing of all printing to the left of that copy back to the one-space gap, and to the closing up of the resulting many-space gap with the machine assuming the passive state scanning the rightmost tally. Thus, as desired, we get ", 0, b, f_0" when $a = 0$.

If our machine $\mathfrak{M}_\phi$ has gone to the state representing $a \neq 0$, it is to print 0, copy the 3rd number back (f_0 in our situation) and b (the 5th number back), feed into the machine $\mathfrak{M}_\chi$ for computing $\chi(a, c, b)$ with result f_1 in our situation. The part of our machine $\mathfrak{M}_\phi$ thus far set up, which has gotten us to f_1 in our present case that $a \neq 0$, will for this case be hitched into a machine which copies the 4th number back (a in our situation) decreased by 1 (which it can do since we are in the case $a \neq 0$), tests whether the result ($a - 1$) is 0 or not, if it is copies the 2nd number back (f_1), after which it proceeds as our machine did with f_0 in the upper case of the second level of Figure 3; but if it is not 0, copies the 5th number back (0) increased by 1, the third number back (f_1) and the fifth number back (b), feeds into $\mathfrak{M}_\chi$ getting f_2 and finally feeds into itself by assuming the same state as we had when we first got f_1 after finding that $a \neq 0$. So it will copy the 4th number back (which is now $a - 1$) decreased by 1 (using $a - 1 \neq 0$) with result $a - 2$, test whether that is 0 or not, etc., performing for the 4th level of Figure 3 the operations corresponding to the 3rd level; and so on at every level, until finally the process is terminated after the upper case ($= 0$) obtains.[6]

Note that, if the like is true of $\mathfrak{M}_\psi$ and $\mathfrak{M}_\chi$, then $\mathfrak{M}_\phi$ in its action in computing $\phi(a)$ never tries to go left from the leftmost square, and uses no symbol other than the tally "|".

Scholars in the manner illustrated have convinced themselves that all possible algorithms for computing number-theoretic functions can be embodied in Turing machines (Turing's thesis). The ingredients that are basically necessary were all provided by Turing: a fixed finite number of symbols, a fixed (perhaps very large) number of states, actions determined by the condition of one scanned square and the state in accordance with the constitution of the particular machine (i.e., its table), unlimited space (on the tape) for receiving the questions and reporting the answers and temporarily storing scratchwork, and of time (the moments) for completing the computations.

---

6 The construction of a machine $\mathfrak{M}_\phi$ is spelled out fully in Kleene *1952*, p. 373, with preceding material (with "$y, x_2, \ldots, x_n$" for any $n \geq 1$ instead of "a, b").

6. The Existence of Unsolvable Decision and Computation Problems

We know of problems on which mathematicians have worked unsuccessfully, sometimes even for thousands of years, which later were shown to be unsolvable. One of these is to find a ruler and compass construction à la Euclid for trisecting any angle. Only after about 2,250 years, in the 1930's, was it proved that such a construction does not exist. Surely, it is a major advance in mathematics when mathematicians get a grip on an enterprise on which they have been engrossed that tells them that what they were trying so hard to do can't be done, or, in other cases, that something can't be done with a certain body of methods they were using, so that others will have to be brought to bear if they are to have any hope of success.

A result of the introduction of an accepted exact characterization (as by Turing computability) of the totality of all algorithms is that it then became possible to prove that there are infinite classes of questions for which no algorithms exist: i.e., that there are *unsolvable decision problems* and *unsolvable computation problems*. What was being sought as a solution does not exist, though of course that fact itself is a solution in another sense. That is, the problem "Find an algorithm for this class of questions" is unsolvable, but the problem "Tell me whether an algorithm exists for this class of questions and if so find one" is solved, with the answer "No".

Let us interrupt the development to sketch the history. As remarked in § 2, Turing *computability* was one of three equivalent ways of characterizing exactly the functions for which algorithms exist which appeared in the 1930's. The concepts used in the other two were Church-Kleene λ-*definability* (developed at Princeton University in 1932 and 1933) and Herbrand-Gödel *general recursiveness* (presented by Gödel in his *1934* lectures at the Institute for Advanced Study in Princeton). Their equivalence was established by Kleene in *1936a* (with his *1936*). Church in *1936* proposed that the number-theoretic functions for which there are algorithms (the "effectively calculable" functions) are exactly the λ-definable functions, or equivalently the general recursive functions, which has come to be known as "Church's thesis".[7] Post in *1936*, who knew of the work at Princeton but not of Turing's work, gave much the same analysis as Turing *1936-7* (without detailed development). Church in *1936* gave examples of unsolvable decision problems for an infinite class of questions arising in the theory of λ-definability (the existence of which is "Church's theorem") and Kleene in *1936* (knowing of Church's result) gave ones in the theory of general recursiveness, and Turing in *1936-7* in his theory of computability. From these results it was a straightforward step to the unsolvability of some decision problems arising in logic. Specifically, both Church in *1936a* and Turing in *1936-7* showed the unsolvability of the classical decision problem for provability in the pure predicate calculus.

---

7 All this is elaborated in Kleene *1981*.

Turing's part in this was done independently of (and slightly later than) the work at Princeton. Indeed, Turing learned of that work (including the results of Church *1936* and *1936a*) only in May 1936 or slightly earlier just as he was ready to send off the manuscript of his *1936-7* for publication, when an offprint or offprints arrived at Cambridge University of Church *1936* or of *1936* and *1936a* both. Thereupon he added references to it in his manuscript (received May 28, 1936), and an appendix (dated August 28, 1936) sketching a proof of the equivalence of his computability to λ-definability, which he expanded in *1937*.

Church wrote me on May 19, 1936 (thus shortly after he had enunciated his thesis and obtained his first unsolvability result), "What I would really like done would be my results or yours used to prove the unsolvability of some mathematical problems of this order not on their face specially related to logic." This hope was fulfilled beginning in *1947* when Post (USA) and Markov (USSR), independently of each other, showed "the word problem for semi-groups" to be unsolvable. The "word problem for groups", a celebrated problem for algebraists, who had failed in intensive efforts to solve it, was shown to be unsolvable in a 143 page paper by Novikov (USSR) in *1955,* and the "homeomorphism problem for four-dimensional manifolds" in topology by Markov in *1958.* Thus the undecidability results of the 1930's began a development encompassing a wide range of applications in the mainstream of algebra and topology.

I proceed now to extract consequences from Turing's thesis as I have formulated it for number-theoretic functions, beginning with the existence of an unsolvable computation problem in Theorem 2 and of an unsolvable decision problem in Theorem 3.

We have seen that a particular Turing machine, so far as we are interested in it, is fully described by a table, giving the act it is to perform from each of the $(l+1)k$ active configurations for it.

Now I assign a positive integer, called its "index", to any Turing machine with a given table. For any natural numbers $b_0, \ldots, b_n$, let $< b_0, \ldots, b_n > = p_0^{b_0} \cdot \ldots \cdot p_n^{b_n}$ where $p_0, p_1, \ldots, p_i, \ldots = 2, 3, \ldots, p_i, \ldots$ are the prime numbers in order of magnitude. The *index* of a machine $\mathfrak{M}$ shall be $< c_1, \ldots, c_k >$ where k is the number of $\mathfrak{M}$'s active states, each $c_m = < d_{m0}, \ldots, d_{ml} >$ where l is the number of $\mathfrak{M}$'s symbols, and each

$$d_{mn} = < i, \quad \begin{matrix} L \\ (C), \\ R \end{matrix} \quad j > \qquad \text{for} \qquad \text{``}i, \quad \begin{matrix} L \\ (C), \\ R \end{matrix} \quad j\text{''}$$

the respective entry in its table with L,C,R rendered now as $0, 1, 2$. Thus the index of the machine $\mathfrak{G}$ with the table given in Figure 1 is

$$<<< 0, 1, 0 >, < 1, 2, 2 >>, << 0, 2, 3 >, < 1, 2, 9 >>, \ldots ,$$
$$<< 1, 1, 0 >, < 1, 2, 11 >>> .$$

Likewise I assign an *index* to a computation carried through any moment *m*. It is $< u_0, \ldots, u_m >$ where each u_t represents the situation at Moment *t*, being $< v_t, r_t, q_t >$ where v_t represents the printing on the tape (that is, what we show at the right in the example in Figure 1 without the superscript, written now as a single number), r_t gives the position of the scanned square (the one bearing the superscript in Figure 1) counting the leftmost square as the 0th, and q_t indicates the machine state (the superscript in Figure 1). For example in Figure 1, $u_6 = << 0, 1, 1, 0, 1 > , 2, 7 >$.

We have been applying Gödel's method of numbering to our complexes of symbols. Turing did the like with another method of numbering.

Now I let "$T(i, a_1, \ldots, a_n, x)$" stand for the following $n + 2$-ary relation:

i is the index of a Turing machine (call it "Machine $\mathfrak{M}_i$") and *x* is the index of a computation by $\mathfrak{M}_i$ for $a_1, \ldots, a_n$ as the arguments down through a moment at which it has just completed to give a value (call that value "$\phi_i(a_1, \ldots, a_n)$").

Much of the time I shall be employing these notations for $n = 1$, writing "$a_1, \ldots, a_n$" then as "a".

The quantity $\phi_i(a_1, \ldots, a_n)$ here is not defined for every $n + 1$-tuple of values of $i, a_1, \ldots, a_n$. So it is what I call a *partial* number-theoretic function of $i, a_1, \ldots, a_n$. Indeed, for each $i, a_1, \ldots, a_n$, it is defined exactly when $T(i, a_1, \ldots, a_n, x)$ holds for some *x*. Thus, recalling from § 1 that "(Ex)" is short for "there exists an *x* such that ",

$$\{\phi_i(a_1, \ldots, a_n) \quad \text{is defined}\} \leftrightarrow (Ex)T(i, a_1, \ldots, a_n, x).$$

With a fixed value of *i*, $\phi_i(a_1, \ldots, a_n)$ as a function of $a_1, \ldots, a_n$ may not (depending on *i*) be defined for every *n*-tuple of values of $a_1, \ldots, a_n$. So ϕ_i is a partial function of *n* variables – the partial function which $\mathfrak{M}_i$ computes, if *i* is the index of a Turing machine. I use the term "partial function" to include the ordinary or "total" number-theoretic functions. A given partial number-theoretic function $\phi(a_1, \ldots, a_n)$ is computed by a Turing machine $\mathfrak{M}$ exactly if, whenever $\mathfrak{M}$ is applied (in our standard way) to an *n*-tuple $a_1, \ldots, a_n$, then $\mathfrak{M}$ will eventually stop with $a_1, \ldots, a_n, \phi(a_1, \ldots, a_n)$ represented on the tape (in our standard way) in case $\phi(a_1, \ldots, a_n)$ is defined, and $\mathfrak{M}$ will fail to stop with the *n*-tuple $a_1, \ldots, a_n, b$ for any *b* represented on the tape in case $\phi(a_n, \ldots, a_n)$ is undefined. In the latter case, $\mathfrak{M}$ either runs forever or stops in some other kind of situation. If *i* is not the index of a Turing machine, ϕ_i is the totally undefined function of *n* variables.

Theorem 1: *The $n + 2$-ary relation $T(i, a_1, \ldots, a_n, x)$ is decidable.*

Proof. I begin by taking "decidable" in the intuitive sense. I must show that there is a decision procedure (intuitively considered) by which, given any $n + 2$-tuple $i, a_1, \ldots, a_n, x$ of natural numbers, we can decide whether or not $T(i, a_1, \ldots, a_n, x)$ holds for it.

To decide, first we test whether i has the right structure to be the index of a Turing machine $\mathfrak{M}_i$. If it doesn't, we declare that $T(i, a_1, \ldots, a_n, x)$ is false. If it does, we can read from it the table for $\mathfrak{M}_i$, using which we can apply $\mathfrak{M}_i$ to compute for $a_1, \ldots, a_n$ as the arguments, to see that we obtain successively the situations represented by successive positive exponents in x (which, for $T(i, a_1, \ldots, a_n, x)$ to be true, must be of the form $p_0^{u_0} \cdot \ldots \cdot p_m^{u_m}$ with $u_0, \ldots, u_m$ all > 0), up to the last one (represented by u_m), at which we look to see if the printing on the tape is of the form "$a_1, \ldots, a_n, b$" for some b (all of this being leftmost) with the tape otherwise blank and the rightmost of the tallies scanned in State 0 (assumed at Moment m for the first time). If so, $T(i, a_1, \ldots, a_n, x)$ is true (with $\phi_i(a_1, \ldots, a_n) = b$); otherwise $T(i, a_1, \ldots, a_n, x)$ is false (with $\phi_i(a_1, \ldots, a_n)$ undefined). So $T(i, a_1, \ldots, a_n, x)$ is decidable intuitively.

So now, by Turing's thesis, Theorem 1 holds with "decidable" meaning precisely Turing decidable. It is a straightforward exercise (using techniques such as were illustrated in § 5) to construct a Turing machine which calculates the (total) function $\tau(i, a_1, \ldots, a_n, x)$ taking 0 or 1 as value according as $T(i, a_1, \ldots, a_n, x)$ is true or false.

Theorem 2: *The function $\psi(a)$ defined by*

$$\psi(a) = \begin{cases} \phi_a(a) + 1 \text{ if } (Ex)T(a, a, x) \text{ (\textbf{Case 1}),} \\ 0 \text{ otherwise (\textbf{Case 2})} \end{cases}$$

is uncomputable.

Remark. The displayed equation does define $\psi(a)$ as an ordinary ("total") number-theoretic function, using in Case 1 the implication

$$(Ex)T(a, a, x) \rightarrow \{\phi_a(a) \text{ is defined }\}$$

which is included in the equivalence displayed preceding Theorem 1 (for $n = 1$ and $i = a$).

Proof of Theorem 2. The proof will necessarily rest on Turing's thesis.

Assume, for reductio ad absurdum, that ψ is computable. Then by the thesis, there is a machine $\mathfrak{M}_p$ which computes it. So ψ is the function ϕ_p computed by $\mathfrak{M}_p$, i.e., for all a, $\psi(a) = \phi_p(a)$, and in particular, $(*) \psi(p) = \phi_p(p)$. Since $\phi_p(p)$ is defined, using the other implication

$$\{\phi_a(a) \text{ is defined }\} \rightarrow (Ex)T(a, a, x)$$

we get $(Ex)T(p, p, x)$. So Case 1 of the definition of ψ applies for p as the value of a, and thus $\binom{*}{*}\psi(p) = \phi_p(p) + 1$. The equations $(*)$ and $\binom{*}{*}$ contradict each other. So, by reductio ad absurdum, ψ is not computable.

Theorem 3: *The property $(Ex)T(a, a, x)$ is undecidable.*

Proof. Why doesn't the definition of $\psi(a)$ in Theorem 2 give us an algorithm for computing it? Only because of our want of an algorithm for deciding, for any a, which case applies, i.e., whether $(Ex)T(a, a, x)$ holds (Case 1) or not (Case 2).

To elaborate, suppose there were such an algorithm. Take any a. If that algorithm gives "Yes" for "$(Ex)T(a, a, x)$?" (Case 1), then we can run the situations for Machine $\mathfrak{M}_a$, started with a on the tape at Moment 0, out to the moment at which $T(a, a, x)$ holds for x the index of the computation, read off from the tape the value $\phi_a(a)$ thus computed, and add 1 to get $\psi(a)$. If we get "No" for "$(Ex)T(a, a, x)$?" (Case 2), then immediately $\psi(a) = 0$.

So by Theorem 2, there is no algorithm for deciding $(Ex)T(a, a, x)$.▽

Theorem 3 is a counterpart of Church's theorem, based on his thesis, from *1936a*. His unsolvable problems are easily seen to be likewise of the form $(Ex)R(a, x)$ with $R(a, x)$ decidable. (That $T(a, a, x)$ is decidable is immediate from Theorem 1.) Many of the decision problems on which mathematicians had been working can be put in the same form. So further developments such as I sketched earlier in this section could be expected.

7. Gödel's Theorems

The part of the mathematical community attuned to foundational issues was much shaken, several years before the appearance of Church's thesis and theorem in *1936* and the like in Turing *1936-7*, by Gödel's *1931* results. I shall indicate now how the work of Church or Turing provided the means for putting Gödel's results in a general setting.

I begin by observing that an "algorithmic theory" is one kind of a theory the Greeks had, for certain quite restricted mathematical subjects or classes of questions. Another is an "axiomatic-deductive theory", with which students are familiar from the example of Euclid's geometry. In such a theory, one starts with some propositions called *axioms* which are accepted as true immediately from their meanings, and *deduces* others called *theorems* (whose truth we might not have been ready to accept immediately) by using *logical inferences* which we accept as propagating truth forward. Thus an axiomatic-deductive theory provides a means for recognizing the truth of various propositions.

Of course algorithmic theories may receive vindication from axiomatic-deductive theories which establish the correctness of the answers they provide.

Later we shall touch on the role of axiomatic-deductive theories as methods for finding answers to questions of a class.

In modern times (beginning in the 1880's and 1890's) axiomatic-deductive theories, as typified by Euclid's geometry (c. 300 B.C.), have been sharpened up.

Some of Euclid's proofs were criticized as defective in that the theorems did not follow from only the mathematical assumptions formulated in the axioms. Unstated assumptions had been sneaked in through Euclid's use of figures. (Thus some anti-Euclidean jokesters proved "false" theorems like "Every triangle is isosceles" by using "fudged" figures, which nothing stated in Euclid's axioms prohibited.) These gaps in Euclid's treatment were mended, mainly by Pasch in *1882* and Hilbert in *1899.*

If one is going to police a theory so that all the mathematical assumptions are explicitly stated, there is no reason for not being equally careful in stating the logical principles used in its deductions or proofs. One can achieve rigor in both these components of the theory only after being equally careful about the syntax of the language that is used in stating the propositions. In doing all this, we wind up by employing a symbolic language specially designed for the theory in hand. You are familiar with the use of formulas in mathematics, which you are used to seeing run in with ordinary English words. But now we shall put everything into formulas. We call the result a *formal system.*

In a formal system, our purpose would not be accomplished if we had to rely on the meanings, rather than just the forms, of the symbols and the expressions built from them. For the question would arise whether unstated assumptions are being sneaked in through the meanings. The symbols, and formulas suitably built from them, will have meanings, providing an *interpretation* of the formal system, which makes the system interesting to us and by which we recognize it to be a *formalization* of some portion of the preexisting "informal" mathematics. But we will have rules specifying from only their forms as linguistic objects which strings of (occurrences of) the symbols are *formulas,* and which strings of formulas are *proofs* (of the last formula of the string). Formulas of which proofs exist are *provable* (or are *theorems*) in the formal system.

Since these rules must provide us with complete fully-usable definitions of the classes of the formulas and of the proofs, they must afford us algorithms applying to the symbolic objects for the classes of questions, "Is this string of symbols a formula?" and "Is this string of formulas a proof?". The provision, in the process of establishing a formal system, of such algorithms is a feature of formal systems essential to our purposes in formalizing. This was perhaps only recognized explicitly since the late 1930's after the general concept of algorithm came into currency.

Formal systems formalizing portions of mathematics as it had been developed informally were given by Frege (*1883, 1903*), by Peano (*1894–1908*), and by White-head and Russell in their monumental *Principia Mathematica, 1910–13*.

As the most elementary nontrivial domain for applying these ideas, I take elementary number theory. This subject, in which systems of objects like the positive integers, or the natural numbers, or all the integers, are investigated, has had a long history going back to Euclid and before. In this century, especially since the two volumes of Hilbert and Bernays, *Grundlagen der Mathematik (Foundations of Mathematics), 1934, 1939,* we have become used to having the elementary theory of the natural numbers represented in a formal system. I shall call this system, in any one of several familiar versions, N.[8] I will speak more generally of a formal system S embracing elementary number theory, which can be just N or may include a formalization of other parts of mathematics, as for example real-number analysis.

Consider any such formal system S embracing elementary number theory – one which, as we may say, is "adequate" for expressing the propositions, and carrying out the proofs, of elementary number theory.

For each particular natural number $0, 1, 2, \ldots$ as the value of a, $(Ex)T(a, a, x)$ is a proposition of elementary number theory. For each of $a = 0, 1, 2, \ldots$, we can find a respective formula C_a of S which expresses the proposition $(Ex)T(a, a, x)$ under the interpretation of S. Indeed, there should be an algorithm for finding C_a from a (it could be embodied in a Turing machine operating on the natural number a to produce the corresponding formula C_a).

For a given a, if $(Ex)T(a, a, x)$ is true, there is an informal proof of it which consists in verifying for the right x that $T(a, a, x)$ holds. Assuming the adequacy of S for natural number theory, that proof can be "formalized" as a proof in S of the formula C_a expressing $(Ex)T(a, a, x)$. Let me use the convenient abbreviation "$\vdash_S C_a$" to say "C_a is provable in S". So we have, for every a,

(a) $$(Ex)T(a, a, x) \rightarrow \vdash_S C_a.$$

This much, I argue, is true for every formal system adequate for elementary number theory. In a detailed study, it can be established for particular such systems S by elementary reasoning.[9]

We would like S to have the feature that only true formulas are provable in it ("correctness"). Clearly a system not having this feature would be "subversive".

---

8 A treatment is in Kleene *1952* Part II.

9 E.g., for the N of Kleene *1952*, on pp. 243–244, using a device of Gödel *1931* (his *1934* β-function) and the "primitive recursiveness" of $T(i, a_1, \ldots, a_n, x)$ (hence of $T(a, a, x)$) mentioned below (late in § 8).

This includes the following, for all a, where "$\neg$" is the symbol in S for "not", and "$\overline{}$" expresses "not" in our informal language:

(b) $$\vdash_S C_a \rightarrow (Ex)T(a, a, x),$$

(c) $$\vdash_S \neg C_a \rightarrow \overline{(Ex)T(a, a, x)}.$$

Theorem 4[10]: *Let S be a formal system adequate for elementary number theory, as for example N, so that it has the foregoing features up through our statement that (a) holds for all a. A number p can be found such that:*

(i)   *If (c) holds for all a, then $\overline{(Ex)T(p, p, x)}$.*
(ii)   *If (c) holds for all a, then not $\vdash_S \neg C_p$.*
(iii)   *If (b) and (c) hold for all a, then not $\vdash_S C_p$.*

Proof. Since for a particular formal system, there are algorithms for recognizing when a sequence of its symbols is a formula, and when a sequence of its formulas is a proof, we can put all the sequences of formulas into an infinite list, and go through this list checking which are proofs. Thus all the proofs are brought into a list (obtained by suppressing from the aforesaid list the members which are not proofs). Also we have an algorithm for finding C_a, and hence $\neg C_a$, from a. Hence there is an algorithm which, applied for a given value of a, searches through the proofs in S in their order in the last-mentioned list, looking for a proof of $\neg C_a$, and if it finds one writes 0, but otherwise never stops searching. Now I apply Turing's thesis in the form relating to the computation of a partial number-theoretic function (of one variable, in this application) to claim that there is a Turing machine, say $\mathfrak{M}_p$, which carries out this algorithm. So $\mathfrak{M}_p$, applied to a, eventually stops giving a value (0 in fact) exactly if there is a proof of $\neg C_a$. Thus, for all a,

(d) $$(Ex)T(p, a, x) \leftrightarrow \vdash_S \neg C_a.$$

Now I prove (i)–(iii) successively.

(i)   Assume (c), and for reduction ad absurdum, $(Ex)T(p, p, x)$. Then by (d) $\vdash_S \neg C_p$, whence by (c) $\overline{(Ex)T(p, p, x)}$, contradicting our assumption.
(ii)   Assume (c). By (i) $\overline{(Ex)T(p, p, x)}$, when by (d) not $\vdash_S \neg C_p$.
(iii)   Assume (b) and (c). By (i), $\overline{(Ex)T(p, p, x)}$, whence by (b), not $\vdash_S C_p.\triangledown$

Since C_p expresses $(Ex)T(p, p, x)$ under the interpretation of the symbolism of S, (i)–(iii) of the theorem can be combined in the following statement: *If (b) and (c) hold for all a, then a formula C_p of S can be found such that* (by (iii) and (ii)) *neither C_p nor its negation $\neg C_p$ is provable in S (thus C_p is "formally undecidable" in S), though* (by (i)) *$\neg C_p$ is true.*

---

10 By incorporating a device of Rosser *1936* using a bit more complicated decidable relation than $T(a, a, x)$, the hypotheses in (i)–(iii) can be simplified to "S is simply consistent" (see before Theorem 5 below).

What does the true but unprovable formula $\neg C_p$ express? $\neg C_p$ expresses $\overline{(Ex)}T(p,p,x)$ or equivalently $(x)\overline{T}(p,p,x)$; and by substituting p for a in (d) and negating both sides,

$$\overline{(Ex)}T(p, p, x) \leftrightarrow \{ \text{ not } \vdash_S \neg C_p\}.$$

So $\neg C_p$ expresses its own unprovability.

The title of Gödel's *1931* paper, translated into English, is "On formally undecidable sentences of *Principia Mathematica* and related systems I."

In setting up a formal system S, we hope to have encompassed all the methods – the axioms and principles of deduction – for the mathematical theory we are formalizing. Will the body of these methods, as formalized in S, suffice for deciding as to the truth or falsity of each proposition of the theory, by there being a proof in S either of the formula C expressing the proposition or of the formula $\neg$C expressing its negation? In the case of elementary number theory, "No"! Neither C_p nor $\neg C_p$ is provable in S. Of course, having proved Theorem 4, we then know that $\neg C_p$ is true, so we can get a stronger system than S by adding $\neg C_p$ as a new axiom! This of course is assuming (b) and (c), which are included in our supposition that S respects truth (or is correct).

In brief, S can't be both correct and complete. Assuming its correctness, S is *(simply) incomplete,* i.e., there is a pair of formulas C and $\neg$C of S without free variables neither of which is provable in S.

Our Theorem 4 is a version of Gödel's "(first) incompleteness theorem" of *1931*. In this version, I have generalized his theorem to apply to all formal systems S adequate for elementary number theory, without regard to the details of the formalization. (In describing a particular formal system, there is an immense amount of detail.) We have assumed only the very general features of S outlined before I stated Theorem 4. These features are closely connected with the purposes for which formal systems were devised, in this case ones formalizing (at least) number theory. In *1931,* one could be uncertain whether Gödel's incompleteness theorem applies to systems quite remote in their details from *Principia Mathematica.* Another respect in which the theorem is generalized is that we have a fixed preassigned number-theoretic property $(Ex)T(a, a, x)$ such that the undecidable sentences for all systems S (under our very general conditions) express one or another instance of this property. Thus the theory of just this one property (as I remarked with another example in a talk in 1935) provides inexhaustible scope for the exercise of mathematical ingenuity.

Gödel's (first) *1931* incompleteness theorem presented a very sobering set-back to the program Hilbert had been promoting of embodying all mathematics up to a certain level in a formal system, and proving the system free of contradictions by "safe" methods. Contradictions had turned up in the higher reaches of mathematics around the beginning of this century, so mathematics did have a problem of purging itself of perjury.[11]

A formal system is (*simply*) *consistent* if in it no pair of contradictory formulas C and ¬C are both provable. As a detailed study shows, in any formal system S adequate for elementary number theory, it is possible to find a formula, call it "Consis", expressing the simple consistency of S.

Theorem 5: *If the system S of Theorem 4 is simply consistent, then* not $\vdash_S$ Consis, *i.e., the formula expressing that fact is not provable in S.*

Proof. I begin by establishing

(1) $\{S$ is simply consistent $\} \rightarrow \{(c)$ holds for all $a\}$.

So I assume S is simply consistent, $\vdash_S \neg C_a$ and $(Ex)T(a, a, x)$, and endeavor to deduce a contradiction. But from $(Ex)T(a, a, x)$ by (a), $\vdash_S C_a$. This with $\vdash_S \neg C_a$ contradicts that S is simply consistent.

Combining (1) with (i) in Theorem 4,

(2) $\{S$ is simply consistent $\} \rightarrow \overline{(Ex)}T(p, p, x)$.

The informal reasoning by which we have established (2) is entirely elementary. So if S is, as we are assuming, adequate for elementary number theory, it should be possible to formalize the proof of this implication in S.[12] Using "⊃" as the formal symbol in S for implication "→" and remembering that Consis expresses "S is simply consistent" and that C_p expresses "$(Ex)T(p, p, x)$", formalizing the proof of (2) in S gives

(3) $\vdash_S$ Consis $\supset \neg C_p$.

To complete the proof of Theorem 5, assume that S is simply consistent, and for reductio ad absurdum that $\vdash_S$ Consis. By the latter with (3), $\vdash_S \neg C_p$. By the former with (1), (c) holds for all a. Thence by (ii), not $\vdash_S \neg C_p$, contradicting $\vdash_S \neg C_p$.▽

Theorem 5 is Gödel's second incompleteness theorem of *1931*, again generalized in respect to the system S. By this theorem, the "safe" methods that Hilbert had been proposing to use in proving a formal system consistent cannot, for a system embracing at least elementary theory, be methods all of which are formalized in the system. Hilbert's program for proving consistency must find among its "safe" methods ones outside of those formalized in the system.

---

11 Cf. e.g., Kleene *1952*, § 11.

12 This was done in detail for their system N by Hilbert and Bernays *1939*, pp. 283ff., especially pp. 306–324.

8. The Arithmetical Hierarchy

A property of n natural numbers was called *arithmetical* by Gödel *1931* if it can be expressed in terms of constant and variable natural numbers, the addition and multiplication functions $+$ and $\cdot$, equality $=$, and the logical symbolism of the first-order predicate calculus. From the results of Gödel and Kleene with the Church-Turing thesis, the class of the arithmetical properties of one variable a coincides with those expressible in the first display in Theorem 7 (and similarly with n variables $a_1, \ldots, a_n$).[13]

Theorem 6: *For each decidable relation $R(a_1, \ldots, a_n, x)$, there is a number p such that, for all $a_1, \ldots, a_n$,*

$$(Ex)R(a_1, \ldots, a_n, x) \leftrightarrow (Ex)T(p, a_1, \ldots, a_n, x).$$

Proof. Let $\mu x R(a_1, \ldots, a_n, x)$ be the least x such that $R(a_1, \ldots, a_n, x)$ if $(Ex)R(a_1, \ldots, a_n, x)$, and be undefined otherwise. This is a computable partial number-theoretic function. So by Turing's thesis, it is computed by a Turing machine, say $\mathfrak{M}_p$; so $\mu x R(a_1, \ldots, a_n, x)$ is $\phi_p(a_1, \ldots, a_n)$. Using the condition for $\phi_i(a_1, \ldots, a_n)$ to be defined (displayed preceding Theorem 1),

$(Ex)R(a_1, \ldots, a_n, x)$

$$\leftrightarrow \{\mu x R(a_1, \ldots, a_n, x) \text{ is defined}\}$$

$$\leftrightarrow (Ex)T(p, a_1, \ldots, a_n, x). \triangledown$$

We can describe Theorem 6 as an "enumeration theorem" for the n-ary relations of the form $(Ex)R(a_1, \ldots, a_n, x)$ with $R(a_1, \ldots, a_n, x)$ decidable. Thus $(Ex)T(i, a_1, \ldots, a_n, x)$ is an $n + 1$-ary relation of like form such that, taking $i = 0, 1, 2, \ldots$ in it, we get a list or "enumeration" of all those n-ary relations. Thus $(Ex)T(i, a_1, \ldots, a_n, x)$ is an "enumerating relation" for them.

In stating my last theorem, I use "(x)" to express "for all x". The prefixes (Ex), (x), (Ey), (y), etc. are called "quantifiers". In proving the theorem, I use the following laws of logic:

$$\overline{(Ex)}A(x) \leftrightarrow (x)\overline{A}(x), \quad \overline{(x)}A(x) \leftrightarrow (Ex)\overline{A}(x), \quad \overline{\overline{A}} \leftrightarrow A.$$

Theorem 7: *Consider the following forms of properties of a natural number a, where each R is a decidable property or relation of its variables:*

$$R(a) \quad \begin{matrix} (Ex)R(a, x) & (x)(Ey)R(a, x, y) & (Ex)(y)(Ez)R(a, x, y, z) & \cdots \\ (x)R(a, x) & (Ex)(y)R(a, x, y) & (x)(Ey)(z)R(a, x, y, z) & \cdots \end{matrix}$$

---

13 Details are in Kleene *1952* §§ 48, 49, 57.

To each form after the first, there is a property of a of that form which is neither expressible in the "dual" form (i.e., the form which is paired with it) nor in any of the forms with fewer quantifiers. Indeed,

$$(Ex)T(a, a, x) \quad (x)(Ey)T(a, a, x, y) \quad (Ex)(y)(Ez)T(a, a, x, y, z) \quad \ldots$$
$$(x)\overline{T}(a, a, x) \quad (Ex)(y)\overline{T}(a, a, x, y) \quad (x)(Ey)(z)\overline{T}(a, a, x, y, z) \quad \ldots$$

are such properties of the respective forms after the first.

Proof. Take for example the form $(x)(Ey)R(a, x, y)$.

I shall show that the property $(x)(Ey)T(a, a, x, y)$ (of this form) is not expressible in the dual form $(Ex)(y)R(a, x, y)$. Suppose it were, i.e., that, for all a,

(A) $(x)(Ey)T(a, a, x, y) \leftrightarrow (Ex)(y)R(a, x, y)$

for some decidable R. By Theorem 6, for this R there is a p such that, for all a,

(B) $(Ey)\overline{R}(a, x, y) \leftrightarrow (Ey)T(p, a, x, y)$.

From (A) and (B) I deduce contradictory expressions for the proposition $(Ex)(y)R(p, x, y)$, thus:

(C) $(Ex)(y)R(p, x, y) \leftrightarrow (x)(Ey)T(p, p, x, y),$

(D) $(Ex)(y)R(p, x, y) \leftrightarrow \overline{\overline{(Ex)}}(y)R(p, x, y)$

$$\leftrightarrow \overline{(x)}(Ey)\overline{R}(p, x, y) \leftrightarrow \overline{(x)}(Ey)T(p, p, x, y).$$

$(x)(Ey)T(a, a, x, y)$ is not expressible in a form with fewer quantifiers, for example as $(x)R(a, x)$ with a decidable $R(a, x)$. For if $(x)(Ey)T(a, a, x, y) \leftrightarrow (x)R(a, x)$, then $(x)(Ey)T(a, a, x, y) \leftrightarrow (Ey)(x)(R(a, x) \ \& \ y = y)$, and $R(a, x) \ \& \ y = y$ is decidable.}▽

This theorem gives a hierarchy of forms for properties of a natural number a such that increasing collections of properties are expressible using the more complicated forms (and different collections in the two forms of equal complexity).

Church's theorem and Gödel's theorem can be associated with two of the forms.

As an undecidable property for Church's theorem (our Theorem 3), we can take any of the properties shown at the bottom in Theorem 7, in particular the one $(Ex)T(a, a, x)$ we used in Theorem 3.

By arguments given above under the generalized Gödel theorem (our Theorem 4), for any number-theoretic property $P(a)$ reasonably expressible in S by formulas D_a, having all and only the formulas D_a provable which are true amounts to saying that, for all a,

$$P(a) \leftrightarrow \vdash_S D_a \leftrightarrow (EX)[X \text{ is a proof in } S \text{ of } D_a]$$
$$\leftrightarrow (Ex)[x \text{ is the index of a proof in } S \text{ of } D_a]$$
$$\leftrightarrow (Ex)R(a, x) \text{ (for a certain decidable } R\text{).}$$

This presupposes a choice of some system of indexing proofs in S by natural numbers, on the basis of proofs being strings of strings of symbols from a given list.[14] We get the last line because a formal system must have an algorithm for recognizing proofs, and for reasonable choices of a property $P(a)$ (such as $(x)\overline{T}(a, a, x)$ at the bottom in Theorem 7) an algorithm for finding D_a from a.

In summary, *having a formal system S in which all and only the true instances of a property P(a) are "provable" (i.e., the formulas D_a expressing them in S are provable in S) entails P(a) being expressible in the form $(Ex)R(a, x)$ with a decidable R.*

By Theorem 7, the property $(x)\overline{T}(a, a, x)$ $(\leftrightarrow \overline{(Ex)T}(a, a, x))$ is not so expressible, and neither are any of the properties shown at the bottom there with two or more quantifiers.

In particular, using $(x)\overline{T}(a, a, x)$ or equivalently $\overline{(Ex)T}(a, a, x)$ as the $P(a)$ and $\neg C_a$ as the D_a, the a's for which $\neg C_a$ is provable do not coincide with those for which $\overline{(Ex)T}(a, a, x)$ is true: there is a p such that *either* $\overline{(Ex)T}(p, p, x)$ and not $\vdash_S \neg C_p$ *or else* not $\overline{(Ex)T}(p, p, x)$ and $\vdash_S \neg C_p$. But (c) eliminates the latter, so we have (i) and (ii) of Theorem 4.

I also included (iii) there (which follows from (i) by the correctness), in keeping with Gödel's theme of "formal undecidability".

For $(Ex)T(a, a, x)$ as the $P(a)$ and C_a as the D_a, by (a) and (b) all and only the true instances are "provable".[15]

Thus Church's theorem and the generalized Gödel theorem are two instances of when, if we desire to have a necessary and sufficient condition of a certain kind for a property $P(a)$, it may be impossible to get such a condition if we are asking for it to come from too low in the hierarchy. There are levels of mathematical concepts, concepts at a higher level having no equivalents at lower levels.

Let me continue with a bit more of the history. We could have continued from our proof of Theorem 1 to show easily that the relation $T(i, a_1, \ldots, a_n, x)$ is "primitive recursive", i.e., its "representing function" $\tau(i, a_1, \ldots, a_n, x)$ is "primitive recursive" (Kleene *1952*, p. 219). Furthermore, there is a primitive recursive function $v(x)$ such that, whenever $T(i, a_1, \ldots, a_n, x)$ holds, $\phi_i(a_1, \ldots, a_n) = v(x)$. This gives

---

14 Using the technique of Gödel numbering which we employed in formulating $T_i(a_1, \ldots, a_n, x)$ in § 6.

15 A proof of Theorem 4 based on Theorem 3 is afforded by adding to (a)–(c) the supposition for reductio ad absurdum that, for all a,

(ā) $\overline{(Ex)T}(a, a, x) \to \vdash \neg C_a$,

so we would have all and only the true instances "provable" both of $\overline{(Ex)T}(a, a, x)$ (expressed by $\neg C_a$) and of $(Ex)T(a, a, x)$ (expressed by C_a). For then (using the argument of Turing *1936-7*, the fourth paragraph of § 11) we would get an algorithm for $(Ex)T(a, a, x)$, contradicting Theorem 3, thus. After picking a, search through the proofs in S for a proof either of C_a (finding which we answer "Yes") or of $\neg C_a$ (finding which we answer "No").

the substance, in our development of Turing's computability, of the "normal form theorem" which Kleene gave in *1936* using general recursiveness (and using partial recursiveness in *1952,* p. 330). Indeed, in our present treatment (using "μx" as in the proof of Theorem 6), we have for each n and i, $v(\mu x T(i, a_1, \ldots, a_n, x))$ as the normal form of the partial function $\phi_i(a_1, \ldots, a_n)$ of n variables. Altogether, this material affords a counterpart, in our development of Turing's ideas, of the "universal machine" which he gave in *1936-7.* Our universal machine is, for each n, the one that computes $\phi_i(a_1, \ldots, a_n)$ as a partial function of $n + 1$ variables.

Versions of our Theorems 3 and 4 were given in Kleene *1943* (and less simply in *1936*) in terms of the theory of general recursiveness.

The arithmetical hierarchy (our Theorem 7) was given likewise by Kleene in *1943* (1940 abstract in *Bulletin of the American Mathematical Society*), where as here he placed Church's theorem and the generalized Gödel theorem in relation to it, and independently by Mostowski in *1947.* Since Addison *1958* and Mostowski *1959,* Σ_k^j has become standard notation for the properties expressible in the k-quantifier form with an existential quantifier first, and Π_k^j with a universal quantifier first, with $j = 0$ for the arithmetical hierarchy; and with $j > 0$ for the hierarchies (going much further up) with higher types j of variables quantified which Kleene subsequently introduced and studied, adapting the idea of an "oracle" introduced by Turing in *1939.*[16]

Gödel at first did not accept Church's thesis (cf. Davis *1982*). Turing's arguments eventually persuaded him. Thus, in his Postscriptum to the Davis *1965* printing of his *1934* lectures, he says on p. 71, "... due to A.M. Turing's work, a precise and unquestionably adequate definition of the general concept of formal system can now be given, [and] the existence of undecidable arithmetical propositions and the non-demonstrability of the consistency of a system in the same system can now be proved rigorously for *every* consistent formal system containing a certain amount of finitary number theory."

As reported in Wang *1974,* p. 84,

> Gödel points out that the precise notion of mechanical procedures[17] is brought out clearly by Turing machines producing partial rather than general recursive functions. In other words, the intuitive notion does not require that a mechanical procedure should always terminate or succeed. A sometimes unsuccessful procedure, if sharply defined, still is a procedure, i.e., a well determined manner of proceeding. Hence we have an excellent example here of a concept which did not appear sharp to us but has become so as a result of a careful reflection. The resulting definition of the concept of mechanical by the sharp concept of "performable by a Turing machine" is both correct and unique. Unlike the more complex concept of always-terminating mechanical procedures, the unqualified

---

16 Some indications of all this are on pp. 63–64 of Kleene *1981.*

17 "alias 'algorithm[s]' or 'computation procedure[s]' or 'finite combinatorial procedure[s]' " (Gödel, in Davis *1965,* p. 72).

concept, seen clearly now, has the same meaning for the intuitionists as for the classicists. Moreover it is absolutely impossible that anybody who understands the question and knows Turing's definition should decide for a different concept.

Partial recursive functions were introduced in Kleene *1938* and Turing machines were applied to compute them in his *1952* Chapter XIII. Under what Turing in *1936-7* § 10 called possibly the simplest way of defining a computable function $\phi(n)$ of an integral variable n, $\phi(n)$ is *computable* if there is a computable sequence in which 0 appears infinitely often, with $\phi(n) =$ the number of figures 1 between the n-th and the $(n + 1)$-th figure 0 [for $n = 0$, before the first]. That method is inapplicable to partial (not necessarily total) functions.

Allowing for the differences in Turing's context (he wants his machines to run ad infinitum printing on alternate squares an infinite sequence of 0's and 1's), in effect his "universal machine" computes a non-total partial function (the values being dual expansions of real numbers). The arguments for this function are the standard descriptions (S.D's) of machines $\mathfrak{M}$ (being the machine tables codified). When the universal machine is supplied a tape at the beginning of which the S.D of a machine $\mathfrak{M}$ is printed, the universal machine prints the sequence of 0's and 1's which is computed by $\mathfrak{M}$ if $\mathfrak{M}$ does compute such a sequence; and otherwise does not print such a sequence. Just as our undecidable property (Theorem 4) was that of a number a being one for which the partial function $\phi_a(a)$ is defined, he has that of the S.D of a machine $\mathfrak{M}$ being one for which this partial function is defined.

Later in Wang *1974* (p. 325) Gödel is reported as objecting that "Turing 'completely disregards' that

(G)   'Mind, in its use, is not static, but constantly developing.'

. . . Gödel granted that

(F)   The human computer is capable of only *finitely* many internal (mental) states.

holds '*at each stage* of the mind's development', but says that

(G)′   '. . . there is no reason why this number [of mental states] should not converge to infinity in the course of its development.' "

I have condensed this as in Webb *1980*, p. 222. Another statement of it is in Remark 3 of Gödel's posthumous publication *1972a*. The recent literature has extensive and very meticulous discussions of this, delving into the nature of mental and mechanical procedures and computer theory, as in Webb *1980* and *1987* (which is an introductory note to the publication of Gödel *1972a*), Gandy *1980*, Dahlhaus and Makowsky *1985*, and in the present volume in the article by Penrose. Here I shall confine myself to a brief response, much as in Kleene *1987*.

If one chooses to believe (G)′, it could imply that there would be no end to the possibilities for the human mind to invent stronger and stronger formal systems that would, in the face of Gödel's ever-renewing incompleteness theorem, prove more and more number-theoretic propositions of the form $(x)\overline{T}(a, a, x)$.

But, as I have said, our idea of an algorithm has been such that, in over two thousand years of examples, it has separated cases when mathematicians have agreed that a given procedure constitutes an algorithm from cases in which it does not. Thus algorithms have been procedures that mathematicians can describe completely to one another *in advance* of their application for various choices of the arguments. How could someone describe completely to me *in a finite interview* a process for finding the values of a number-theoretic function, the execution of which process for various arguments would be keyed to more than the *finite* subset of our mental states that would have developed by the end of the interview, though the total number of our mental states might converge to infinity if we were immortal. Thus Gödel's remarks do not shake my belief in the Church-Turing thesis in the **Public Processes Version** of Hofstadter *1980*, p. 562 (stated there for deciding whether a property holds rather than for calculating the value of a function).

Gödel continued from (G)', "Now there may be exact systematic methods of accelerating, specializing, and uniquely determining this development, e.g. by asking the right questions on the basis of a mechanical procedure", though he admitted, "the precise definition of a procedure of this kind would require a substantial deepening of our understanding of the basic operations of the mind." I think it is pie in the sky! And I would add, to do us any good, such a procedure would in turn need to be a process finitely describable in advance of its application (an algorithm), and then the combination of the two algorithms would constitute one algorithm which by my lights would need to be finitely describable in advance, on the basis of a finite budget of presently-known mental states, thus under the Church-Turing thesis.

References

Addison, John W.

> 1958 Separation principles in the hierarchies of classical and effective descriptive set theory. *Fund. Math.* **46** (1958) 123–135.

Church, Alonzo

> 1936 An unsolvable problem of elementary number theory. *Am. J. Math.* **58** (1936) 345–363.

> 1936a A note on the Entscheidungsproblem. *J. Symb. Log.* **1** (1936) 40–41. Correction, ibid., 101–102.

Dahlhaus, E., and J.A. Makowsky

> 1986 Computable directory queries. In: *Proceedings of CAAP '86*, 11th Colloquium on Trees and Algebra in Programming, Nice, March 24–26, 1986, ed. P. Franchi-Zannettacci, Lecture Notes on Computer Science 214, pp. 254–265. Berlin: Springer-Verlag (1986).

Davis, Martin (ed.)

1965 *The Undecidable: Basic Papers on Undecidable Propositions, Unsolvable Problems and Computable Functions.* Hewlett, NY: Raven Press (1965).

1982 Why Gödel didn't have Church's thesis. *Inf. Contr.* **54** (1982) 3–24.

Dedekind, Richard

1888 *Was sind und was sollen die Zahlen?* Braunschweig: Vieweg & Sohn (1888).

Frege, Gottlob

1893 *Grundgesetze der Arithmetik, begriffsschriftlich abgeleitet,* vol. 1. Jena: H. Pohle (1893).

1903 Ibid., vol. 2 (1903).

Gandy, Robin

1980 Church's thesis and principles for mechanisms. In: *The Kleene Symposium,* pp. 123–148. Amsterdam/New York/Oxford: North-Holland Publ. Co. (1980).

Gödel, Kurt

1931 Über formal unentscheidbare Sätze der Principia Mathematica und verwandter Systeme I. *Monats. Math. Ph.* **38** (1931) 173–198.

1934 On undecidable propositions of formal mathematical systems. Mimeographed lecture notes, taken by Kleene and Rosser. Printed with revisions in Davis *1965,* pp. 39–74.

1972a Some remarks on the undecidability results. Forthcoming in *Kurt Gödel, Collected Works,* vol. II. New York: Oxford University Press, and Oxford: Clarendon Press.

Hilbert, David

1899 *Grundlagen der Geometrie,* 7th ed. Leipzig/Berlin: Teubner-Verlag (1930).

1926 Über das Unendliche. *Math. Annal.* **95** (1926) 161–190.

Hilbert, David, and Paul Bernays

1934 *Grundlagen der Mathematik,* vol. I. Berlin: Springer-Verlag (1934).

1939 Ibid., vol. II (1939).

Hofstadter, Douglas R.

1980 *Gödel, Escher, Bach: An Eternal Golden Braid.* Vintage Books Edition. New York: Random House (1980).

Kleene, Stephen C.

1936 General recursive functions of natural numbers. *Math. Annal.* **112** (1936) 727–742. For an erratum, a simplification and an addendum, see Davis *1965,* p. 253 (read "$(Ey)T_1(\theta(q), \theta(q), y)$" in place of "$(y)T_1(\theta(b), \theta(b), y)$").

1936a λ-definability and recursiveness. *Duke Math. J.* **2** (1936) 340–353.

1938 On notation for ordinal numbers. *J. Symb. Log.* **3** (1938) 150–155.

1943 Recursive predicates and quantifiers. *T. Am. Math. S.* **53** (1943) 41–73. For a correction and an addendum, see Davis *1965,* pp. 254 and 287.

1952 *Introduction to Metamathematics,* 9th reprint. Amsterdam: North-Holland Publ. Co. (1988).

1981 Origins of recursive function theory. *Ann. Hist. C.* **3** (1981) 52–67. Corrections in Davis *1982,* footnotes 10 and 12.

1987 Reflections on Church's thesis. *Notre Dame J. Form. Log.* **28** (1987) 490–498.

Markov, A.A.

1947 On the impossibility of certain algorithms in the theory of associative systems. *Comptes rendus (Doklady) de l'Académie des Sciences de l'URSS, n.s.* **55** (1947) 583–586 (tr. from the Russian).

1958 Nerazrešimost´ problemy gomeomorfii (Unsolvability of the problem of homeomorphy). In: *Proceedings of the International Congress of Mathematicians,* Edinburgh, 1958, pp. 300–306.

Mostowski, Andrzej

1947 On definable sets of positive integers. *Fund. Math.* **34** (1947) 81–112.

1959 On varying degrees of constructivism. In: *Constructivity in Mathematics, Proceedings of Colloquium, Amsterdam, 1957,* pp. 178–194. Amsterdam: North-Holland Publ. Co. (1959).

Novikov, P.S.

1955 On the algorithmic unsolvability of the word problem in group theory. In: *AMS Transl. (2)* **9** (1958) 1–222. Original in Russian, 1955.

Pasch, Moritz

1882 *Vorlesungen über neuere Geometrie.* Leipzig: Teubner-Verlag (1882).

Peano, Guiseppe

1889 *Arithmetices Principia, Nova Methodo Exposita.* Turin: Bocca (1889).

1894-1908 *Formulaire de Mathématiques,* Introduction and five volumes, edited by Peano and written by him and seven collaborators. Turin (1889).

Péter, Rózsa

1934 Über den Zusammenhang der verschiedenen Begriffe der rekursiven Funktion. *Math. Annal.* **110** (1934) 612–632.

Post, Emil L.

1936 Finite combinatory processes – formulation I. *J. Symb. Log.* **1** (1936) 103–105.

1947 Recursive unsolvability of a problem of Thue. *J. Symb. Log.* **12** (1947) 1–11.

Rosser, J. Barkley

1936 Extensions of some theorems of Gödel and Church. *J. Symb. Log.* **1** (1936) 87–91.

Skolem, Thoralf

1923 Begründung der elementaren Arithmetik durch die rekurrierende Denkweise ohne Anwendung scheinbarer Veränderlichen mit unendlichem Ausdehnungsbereich. *Skrifter utgit av Videnskapsselskapet i Kristiana, I. Matematisk-naturvidenskabelig klasse 1923,* no. 6.

Turing, Alan Mathison

1936-7 On computable numbers, with an application to the Entscheidungsproblem. *P. Lond. Math. Soc. (2)* **42** (1936-7) 230–265. A correction, ibid. **43** (1937) 544–546. A useful critique is in the appendix to *Post 1947.*

1937 Computability and λ-definability. *J. Symb. Log.* **2** (1937) 153–163.

1939 Systems of logic based on ordinals. *P. Lond. Math. Soc. (2)* **45** (1939) 161–228.

Wang, Hao

1974 *From Mathematics to Philosophy.* London: Routledge and Kegan Paul, and New York: Humanities Press (1974).

Webb, Judson Chambers

1980 *Mechanism, Mentalism and Metamathematics. An Essay on Finitism.* Dordrecht/Boston/London: D. Reidel (1980).

1987 Introductory note to Remark 3 of Gödel *1972a.* Forthcoming in *Kurt Gödel, Collected Works,* vol. II (see Gödel *1972a*).

Whitehead, Alfred North, and Bertrand Russell

1910-13 *Principia Mathematica.* vol. 1 1910, vol. 2 1912, vol. 3 1913. Cambridge, England: Cambridge University Press (1910-13).

The Confluence of Ideas in 1936

Robin Gandy

Abstract. This paper seeks to explain the almost simultaneous appearance in 1936 of several independent characterizations of the notion of effective calculability. It traces the history of that notion from Babbage's work in the 1830's to the planning of electronic digital computers in the 1940's. Particular attention is paid to the history of decision problems and to the significance of Turing's work.

1. Introduction

In 1936 papers by Church, Kleene, Turing, and Post were published. Each of the authors proposed a precise definition for the seemingly vague notion of effective calculability; the definitions can be shown to be equivalent. The first three authors exhibit problems which are effectively undecidable; Church *1936a* and Turing both showed that the Entscheidungsproblem is such a problem. Details of the exact chronology of publication is given in part I of the list of references.

It is not uncommon in mathematics – and in the other sciences – for concepts, methods, and theorems to be discovered independently and almost simultaneously; the development of the calculus by Newton and Leibniz is a classic example. There is, so to speak, something in the air which different people catch. In this paper I try to answer the following questions:

QUESTION 1.1. What was in the air in the early 1930's?

QUESTION 1.2. What connections are there between the abstract work in the papers cited and actual mathematical calculation (by hand or by machine)?

QUESTION 1.3. The problems dealt with in the 1936 papers largely arose from the work of Hilbert and his school at Göttingen; why did that school contribute so little to their solution?

QUESTION 1.4. Eventually Turing's work has proved to be of greater mathematical and philosophical significance than that of the other authors; why?

Here are three motives for trying to give a precise definition to a vague, intuitively perceived notion:

1.5.(1) The notion may be clearly defined in some contexts; one wishes, for greater generality, to extend the definition to a wider range of contexts. Example: the notions of area and volume.

1.5.(2) One may hope that greater precision and/or a wider range of application will increase one's power to obtain positive results. Example: extension of 'integer' and 'prime' from the rationals to other algebraic number fields.

1.5.(3) If one wishes to obtain negative results, to show that something is not true of the notion, then one must give a precise definition of the notion so as to delimit its extent. Example: 'constructible by ruler and compasses'.

I shall argue that (3) was the most important – though not the only – motive behind the discoveries of 1936. But I shall first consider the work of Babbage and his followers, for which (1) and especially (2) were the motives; 'increased power' here meant literally 'increased mechanical power'.

Remark. I shall be concerned with the climate of opinion at various times. My knowledge of this comes from (mostly well-known) published sources. Mathematicians are usually guarded in their public expressions of opinion; so it is possible that the study of manuscript material might show that some of my assertions are misleading or false. The reader will find a summary of my conclusions in Section 14.

2. *Babbage and His Followers*

Babbage conceived the idea of an 'Analytic Engine' around 1834. His idea was that it could be programmed (by punched cards) to perform any of the kinds of computations which were then done, with much drudgery, by humans. He therefore gave considerable thought to what operations were needed. If we consider the Analytic Engine as a register machine (Babbage called the registers 'variables') and suppose for simplicity that the registers contain natural numbers rather than signed decimal approximations to reals, then the operations considered by Babbage can be (inductively) characterized as follows:

2.1.(1) The arithmetic functions $+$, $-$, $\times$ are operations (where $x - y = 0$ if $y \geq x$).

2.1.(2) Any sequence of operations is an operation.

2.1.(3) *Iteration.* The *n*-fold iteration of an operation *P* (where *n* is the number in a specified register whose contents is not affected by *P*) is an operation.

2.1.(4) *Conditional iteration.* If *P* is an operation and *T* is a test on the numbers in certain registers, then the result of iterating *P* until *T* succeeds is an operation.

2.1.(5) *Conditional transfer.* If *P* and *Q* are operations then the result of doing *P* if a test *T* succeeds, *Q* if it fails, is an operation.

Notes. (A) If one considers instead of Babbage's engine a machine whose registers may contain arbitrarily large numbers, then the operations which can be programmed using (1)–(3) are just the primitive recursive functions. Since (4) allows one to program the least number operation, and since (3) is covered by (4), it follows, by Kleene's normal form theorem, that the functions which can be calculated by (1), (2), and (4) are precisely those which are Turing computable.

(B) Further one can replace (1) either by the successor and zero functions, using equality as the only test; or by successor and predecessor with test for zero as the only test.

(C) A classic example of conditional iteration, certainly familiar to Babbage, is the Euclidean algorithm[1]. The Greeks knew that in its geometric form (Euclid X.3) it will fail to terminate if the two given lines are incommensurable: the original halting problem!

(D) The above principles are all explicitly stated and illustrated by examples in the accounts of the Analytic Engine in Babbage *1864*, Menabrea *1842*, and Lady Lovelace *1843*. I have used the texts (and page numbers) as printed in Morrison and Morrison *1961*.

Although Babbage mentions conditional transfer (67–68), he, with a natural respect for well-structured programming, uses only conditional iteration, which he calls 'backing', in his illustration (65–67) of how the engine can exhibit judgement. Menabrea does exponentiation as iterated multiplication (240–241); actually he uses conditional iteration, counting down on the exponent. He states conditional

---

1 I do not know when 'algorithm' acquired its current meaning. Originally it applied only to calculation with Indian (Arabic) numerals as treated of in the works of Al-Khwarizmi; as 'augrim' it appears, for instance, in Chaucer. This is the only meaning (with the etymologically more correct spelling 'algorism') recognized by the O.E.D. and its 1933 supplement. Dirichlet used 'Euclidean algorithm' in his lectures on number theory, first published by Dedekind in 1863. Venn used 'algorithmic' analogically, in his *Symbolic Logic* (1881). Littré in his *Dictionnaire de la Langue Français* (1885) gives: "En termes d'algèbre, procédé de calcul."

transfer explicitly (240), allowing that a 'go to' instruction may have to be executed by ringing a bell to summon the attendant; he gives an example of its use (241). Lady Lovelace introduces rather inappropriate notations for iteration (277, 280). She gives a program (286–295) for calculating the (first three nonvanishing) Bernouilli numbers. This raises a point which is interesting both theoretically and biographically. Lady Lovelace uses a course-of-values recursion. She claims (293), quite falsely, that the program will only need a fixed number of registers. (Falsely because the formula expresses the next nonvanishing Bernouilli number in terms of *all* the previous ones). Dorothy Stein, in her riveting biography of Lady Lovelace (*1985*), quotes from a letter (to Babbage) which evidently refers to her claim:

> I have (I think very judiciously and warily) touched on the only departures from *perfect* identity which *could* exist during the repetitions ... & yet I have not *committed* myself by saying if these departures would require to be met by the introduction of one or more new cards, or not, but have simply indicated that as the variations follow a regular rule, they would be easily provided for. I think I have done it admirably and diplomatically. (*Here* comes in the *intrigante* and the *politician!*) (p. 111)

Of course Gödel's (or Cantor's?) device of coding a sequence of numbers by a single number was not available to her, and in any case is of no use when there is a bound on the number which can be stored in a single register. Babbage contemplated (69) a machine with a thousand registers, each containing fifty decimal digits. This would have allowed the calculation of the first few hundred Bernouilli numbers with reasonable accuracy

2.2. *Babbage's Thesis.* Babbage writes:

> These two memoirs [viz. the account by Menabrea and Lady Lovelace's notes to her translation of it] furnish, to those who are capable of understanding the reasoning, a complete demonstration – *That the whole of the development and operations of analysis are now capable of being executed by machinery.* (op. cit., p. 69; Babbage's italics)

Babbage, in his work on general algebra and functional equations, had shown his ability to think in abstract terms. If, then, one had led him to speculate (not difficult!) on what could be done with an abstract machine, free from limitations on its storage, he would surely have assented to a version (based on Sections 2.1.(1)–(5)) of Church's thesis.

The 'complete demonstration' proceeds – apart from some more or less airy remarks by Lady Lovelace (e.g., "We may say most aptly, that the Analytic Engine *weaves algebraical patterns* just as the Jacquard-loom weaves flowers and leaves" – p. 252) wholly by the analysis of examples. I discuss the unsatisfactory nature of such heuristic arguments in Section 9.7.

2.3. *Subsequent developments.* Babbage was concerned both with the general principles of his engine and with its actual construction. But the gap between theory and practice remained unbridged (and probably unbridgeable) until the 1940's.[2] Thus the construction of a universal machine which could, in principle, carry out any process of calculation, was, as the committee appointed by the British Association to report "on the advisability of constructing ... Mr. Babbage's Analytical Machine ..." said, in 1879, "not more than a theoretical possibility" (Randell *1982*, pp. 55–65). So Babbage's motive for characterizing what can be calculated was, so to speak, not operative, and there would have been little reason for pure mathematicians to consider the matter seriously. (Cayley and W.K. Clifford both served on the mentioned committee.)

P.E. Ludgate [1909] and L. Torres y Quevedo [1914] both made proposals for universal calculating machines; they took over from Babbage the operations described in Section 2.1, but their work was not widely known. Other authors, concerned with more practical machines, referred to Babbage's work. Examples from Randell *1982* are: M. d'Ocagne [1922], L. Couffignal [1933], V. Bush *1936*, H.H. Aiken [1964] (which is an unpublished memorandum of 1937). But the emphasis is on programming a fixed iterable sequence of arithmetical operations. The fundamental importance of conditional iteration and conditional transfer for a general theory of calculating machines is not recognized, though the principles may be used in very particular contexts; for example conditional iteration of subtraction is often used in the operation of division.

Conclusions. Babbage asserted what was, in effect a version of Church's thesis. His work was never entirely forgotten, but its theoretical importance – its importance, so to speak, as software – was little recognized, and was often obscured by the concern with hardware. In Babbage's own writings software and hardware are often discussed together: see, for example, Babbage *1837*. Menabrea's account and Lady Lovelace's notes on it are distinguished by the fact that they deal only with software. Interestingly enough it appears that their work was not known even to some of those who were familiar with Babbage *1864*: see p. 191 of Randell *1982*.[3] Thus there is no reason to suppose that the mathematicians involved in the discoveries of 1936 (including Hilbert, Bernays, Gödel, and Herbrand) were familiar with Babbage's theoretical ideas (as embodied in the principles 2.1.); certainly none of them refer to his work. It was only after the development of electronic digital computers that

---

2 For details of attempts to put the theory into practice the reader is referred to Randell *1982*, which is a superb collection of texts and of commentaries on them and which includes a detailed bibliography. Names with dates in square brackets refer to his bibliography.

3 And although Bush, in his survey of computing machinery (*1936*), praises Babbage as a pioneer, he does not discuss the principles underlying the design of the Analytic Engine.

Babbage's theoretical contributions became widely known; for example, Hartree *1949* devotes a whole chapter to the Analytic Engine.

3. *Decision Problems and Their Solvability*

I now turn to the possibility of negative results as a motive for characterizing calculability. What is required is a clearly stated (type of) problem which can in some special cases be settled by calculation, but for which a uniform general computational method of solution seems unlikely.

3.1. The classic example is that posed by Diophantine equations. Here is Hilbert's statement of it[4]:

> 10. Entscheidung der Lösbarkeit einer diophantischen Gleichung
> Eine diophantische Gleichung mit irgendwelchen Unbekannten und mit ganzen rationalen Zahlkoeffizienten sei vorgelegt: *man soll ein Verfahren angeben, nach welchem sich mittels einer endlichen Anzahl von Operationen entscheiden läßt, ob die Gleichung in ganzen rationalen Zahlen lösbar ist.*

The crucial word here is 'Verfahren': 'procedure' is the natural translation, although in legal contexts 'process' would be more correct.

3.2. Dehn in *1911* states the word problem for groups (and some related problems) in similar terms to Hilbert, but uses 'Methode' in place of 'Verfahren'.

3.3. Thue in *1914*, stating what we should now call the word problem for semigroups, also uses 'Methode'. But he further requires that the (finite) number of operations used in the decision process should be calculable ('berechenbar') in advance. I discuss the significance of this requirement in Section 4.2.

3.4. *The Entscheidungsproblem.* This was first considered as a problem in the algebra of logic. In § 3 of his *1915* Löwenheim, in effect, solves the decision problem for first- and second-order monadic calculus.[5] In § 4 he proves that any statement of second-order logic is equivalent to one in which only symbols for binary relations occur. Of this reduction he writes:

> We can gauge the significance of our theorem by reflecting upon the fact that every theorem of mathematics, or of any calculus that can be invented, can be written as a relative equation; the mathematical theorem then stands or falls according to whether the equation is satisfied or not. This transformation of arbitrary mathematical theorems

---

4 From the lecture on Mathematical Problems he gave to the International Congress at Paris in 1900; published in *1901*.
5 In the preface in van Heijenoort *1967* only the first-order result is mentioned. But Theorem 5 states the second-order result.

into relative equations can be carried out, I believe, by any one who knows the work of Whitehead and Russell[6]. Since, now, according to our theorem the whole relative calculus can be reduced to the binary relative calculus, it follows that we can decide whether an arbitrary mathematical equation is true provided we can decide whether a binary relative equation is identically satisfied or not. (From the translation printed in van Heijenoort *1967*, p. 246)

H. Behmann (who had written his doctoral thesis under Hilbert) argues in his *1922* that Schröder modeled his calculus of relations too closely in the style of algebra and analysis; Behmann rewrites some of the problems and results in terms of first- and second-order predicate calculus. He states the most general form of the Entscheidungsproblem as follows:

Es soll eine ganz bestimmte allgemeine Vorschrift angegeben werden, die über die Richtigkeit oder Falschheit einer beliebig vorgelegten, mit rein logischen Mitteln darstellbaren Behauptung nach einer endlichen Anzahl von Schritten zu entscheiden gestattet, oder zum mindesten dieses Ziel innerhalb derjenigen – genau festzulegenden – Grenzen verwirklicht werden, innerhalb deren seine Verwirklichung tatsächlich möglich ist. (p. 166)

A nice exercise for translation into English, perhaps:

A quite definite generally applicable prescription is required which will allow one to decide in a finite number of steps the truth or falsity of a given purely logical assertion; or at least precise limits should be given within which an effective prescription of this kind can be found.

Behmann remarks that, according to the thesis of logicism, the general problem is equivalent to the problem of deciding which mathematical propositions are true. He goes on to consider the Entscheidungsproblem for first and second order logic and, using a mixture of quantification theory and the calculus of relations, solves both problems for the monadic case. It is odd that he refers neither to Löwenheim *1915*, nor to Skolem *1919* (which includes elegant solutions for both problems); perhaps this omission reflects his distaste for the algebra of logic.

Hilbert in his *1930* (based on an address given to the International Mathematical Congress at Bologna in 1928) mentions the Entscheidungsproblem for first-order logic and refers to the work of Löwenheim and Behmann. The classic statement of the Entscheidungsproblem (for first-order predicate calculus) is in Hilbert and Ackermann *1928*:

---

6 In contemporary terms this transformation can be stated as follows: let ϕ and ψ be formula in the second-order language of set theory, and let α be the least ordinal such that ϕ holds in V_α; then one can construct a formula χ of second-order predicate calculus such that χ is satisfiable if and only if ψ holds in V_α.

Das Entscheidungsproblem ist gelöst, wenn man ein Verfahren kennt, das bei einem vorgelegten logischen Ausdruck durch endlich viele Operationen die Entscheidung über die Allgemeingültigkeit bzw. Erfüllbarkeit erlaubt. (p. 73)

(The Entscheidungsproblem is solved if one knows a procedure which will permit one to decide, using a finite number of operations, on the validity, respectively the satisfiability of a given [first-order] logical expression.)

A natural reaction to Hilbert's tenth problem would be to doubt that it could be settled either positively or negatively. One might well believe that as mathematics progressed more and more decidable cases would be discovered, each with its own particular method for decision, and even that for any particular equation a time would come when it would fall under a known decidable case; but one might also believe that at no time would the list of decidable cases be exhaustive, and hence that there could be no uniform decision procedure.

Further, new methods of solution have involved new ideas (continued fractions, Diophantine approximation, elliptic functions, and so on); and although the application of these methods to a particular equation have involved, so far, only known processes of calculation such as could be performed, in principle, by the Analytic Engine, one might speculate that some as yet undiscovered conceptual framework for a decision might require some as yet undiscovered process of calculation.

The excessive use of conditionals and subjunctives in the previous paragraphs suggests, and was meant to suggest, that many mathematicians would keep an entirely open mind about the solvability of the Diophantine (and other) problems. I now consider more dogmatic positions.

4. Optimism

Hilbert, impressed by the increasing power, range, and rigor of mathematics, raised his 'banner with a strange device' – not 'Excelsior' but 'In der Mathematik ist kein Ignorabimus'. I think he first waved it aloft at the Paris Conference of 1900. He was still flying it at his last published address (*1930a*) – and the interested listener can hear it on a record included in Reidemeister *1971*. The last words of *1930a* (given at Königsberg on September 5, 1930) are:

... ein unlösbares Problem überhaupt nicht gibt. Statt des törichten Ignorabimus heiße ich im Gegenteil unsere Losung:

Wir müssen wissen,
Wir werden wissen.

(and the last two lines were engraved on his tombstone).

Of course this belief in the inexorable march of mathematics does not entail a belief that there will be a uniform method of solution for every problem. In *1901*

Hilbert discusses negative solutions (absence of a procedure) for such problems as the duplication of the cube by ruler and compasses. But his wording of the tenth problem does suggest that he expected a positive solution. And although in his published work he was wary of committing himself, there is good evidence that he believed, at least until 1931, that the Entscheidungsproblem would be shown to be decidable.[7] Hilbert's optimism was shared by his group at Göttingen. After the statement of his results for the monadic case, Behmann in his *1922* writes:

> ...Als ein weiterer wäre etwa ins Auge zu fassen, auch beliebige Beziehungen zwischen Individuen zu berücksichtigen – wofür mir ein gangbarer Lösungsweg bis so weit noch nicht bekannt ist. (p. 187)

> (A further extension to look at would be got by taking into consideration arbitrary relations between individuals. I do not *as yet* know of any possible approach to a solution in this case.) [My italics]

In his *1930* Bernays exhibits the optimism of the Göttingen school very clearly:

> Von der Zahlentheorie, wie sie durch die Peanoschen Axiome, mit Hinzunahme der rekursiven Definition, abgegrenzt wird, glauben wir, daß sie in diesem Sinne deduktiv abgeschlossen ist; die Aufgabe eines wirklichen Nachweises hierfür ist aber noch völlig ungelöst. (Bernays *1976*, p. 59)

> (We believe that [first-order] number theory as given by Peano's axioms supplemented by recursive definitions is deductively closed [i.e., is complete] but the problem of proving this is by no means solved.)

Remarks. (1) In the paper Bernays avoids technical details, but it is plain from the context that he is referring to the first-order theory.

(2) We now know (but Bernays did not) that completeness, even for Σ_1^0 sentences, implies the solvability of the Entscheidungsproblem.

(3) Bernays adds the following footnote to his definition of "deduktiv abgeschlossen":

> Man beachte, daß diese Forderung der deduktiven Abgeschlossenheit noch nicht so weit geht wie die Forderung der Entscheidbarkeit einer jeden Frage der Theorie...

> (The requirement of completeness does not go so far as the requirement of decidability.)

To our ears this sounds just thoughtless, but I think Bernays has two sensible ideas in mind. First, he is being careful *not* to assume that the Entscheidungsproblem is decidable; for, if it were, the decidability of (first-order) arithmetic truth would follow immediately from the completeness of Peano arithmetic. Second, he has in mind that a decision procedure should be effective in the strong sense that one should be able to give, in advance, a bound on the number of steps in an application of the

---

7 See, in particular, the discussion by Feferman on pp. 208–212 of Gödel *1986*, which is concerned with the closely connected problem of the completeness of first-order arithmetic.

procedure. The mere *fact* of completeness would give no clue as to how long the search through proofs might take. Since this ambiguity in the meaning of 'effective' turns up, like a bad penny, over and over again I will discuss it now.

4.2. *Effectiveness.* Thue (cf. Section 3.3.), and possibly Bernays, may have had in mind that a bound on the number of steps should be calculated from the specification of a particular instance of the decision problem by the use of familiar functions; and of course in *practice* this is absolutely necessary. But in general what is being required is that "the process terminates after a finite number of steps" should be interpreted constructively. In his *1918* Hilbert discussed, in connection with decision problems, the gap between 'pure' existence proofs and the calculation of bounds. Since then many authors have sought to restrict the word 'computable' to situations where bounds can be given, though they are often unclear about what they mean by 'given'. But in a *general* theory I do not think it is sensible to mix together constructive and nonconstructive notions of existence. If one accepts, on whatever grounds, that a process terminates after a finite number of steps, then one should also accept, on the same grounds, that the number of steps can, in principle, be computed – one only needs a clock![8] This point was (first?) made forcibly and concisely by Church in footnote 10 of his *1936.* Kreisel in his ground-breaking *1951* and *1952* showed how it was sometimes possible to pass from the evidence of existence (a nonconstructive proof) to a computation (using transfinite recursion) of the relevant bound. And, since then, the classification of computable functions according to the strength of the formal system necessary for the proof of their being total has become an industry. (For further information see Smorynski *1979, 1982.*)

4.3. *Conclusion.* Although Hilbert and his colleagues did not assume that the Entscheidungsproblem is solvable, nevertheless, before 1931 at least, they hoped so, and were optimistic about the power of 'Beweistheorie' to settle this and related problems. Because of this, and because of the extensions of the notion of recursiveness – to transfinite types and transfinite ordinals – sketched in Hilbert *1926* they did not, I believe, think that the time was ripe to seek an exhaustive classification of effective methods, nor to set limits on the power of those methods; see also the remarks made at the end of Section 8.

---

8 A constructive mathematician may have good grounds for believing that the process doesn't fail to terminate, without believing that it does terminate. Markov used his principle to jump over this gap.

5. *Skepticism*

In his *1908* Brouwer writes:

> It follows that the question of the validity of the principium tertii exclusi is equivalent to
> the question *whether unsolvable mathematical problems can exist.* There is not a shred
> of proof for the conviction which has sometimes been put forward that there exist no
> unsolvable mathematical problems. (Brouwer *1975*, p. 109)

And he gives as an example of a problem the solvability of which is doubtful:
"Do there occur in the decimal expansion of π infinitely many pairs of consecutive equal digits?" (p. 110). Later (e.g., in *1923*) he takes it for granted that such
questions cannot be answered. Hilbert might perhaps have retorted that "wir müssen
wissen" does not apply to facts about the decimal expansion of π; also Brouwer's
examples are single questions rather than decision problems. However, this is easily
rectified. Example: to decide whether or not, for each k, there is in the first k^k digits
of the decimal expansion of π a stretch of k consecutive digits in which there are
fewer than $[k/9]$ occurrences of the digit 7.

5.1. According to his *1965*, Post had, in the early 1920's, convinced himself that
there are generated (i.e., recursively enumerable) sets whose finiteness problem is
undecidable (i.e., which are not recursive). See Section 13.3 below.

5.2. A strong candidate for an undecidable decision problem is the Entscheidungsproblem; we have already seen that Löwenheim and Behmann argued that
a decision procedure for second-order logic would give a decision procedure for
mathematical truth. In his *1927* von Neumann uses a similar argument to show the
total implausibility of a decision procedure for first-order predicate calculus. The
system he is considering uses axioms which are got by substitution from axiom
schemata; *modus ponens* is the only rule. Quantification is covered by including the
axiom schemata for Hilbert's ϵ-symbol. Of these systems von Neumann writes:

> Es scheint also, daß es keinen Weg gibt, um das allgemeine Entscheidungskriterium dafür,
> ob eine Normalformel *a* beweisbar ist, aufzufinden. (Nachweisen können wir freilich
> gegenwärtig nichts. Es ist auch gar kein Anhaltspunkt dafür vorhanden, wie ein solcher
> Unentscheidbarkeitsbeweis zu führen wäre.) Diese Ungewißheit hindert uns aber nicht
> daran, festzustellen: Heute ist es nicht allgemein zu entscheiden, ob irgendeine gegebene
> Normalformel *a* (bei der im folgenden zu beschreibenden Axiomregel) beweisbar ist
> oder nicht. Und die Unentscheidbarkeit ist sogar die Conditio sine qua non dafür, daß
> es überhaupt einen Sinn habe, mit den heutigen heuristischen Methoden Mathematik zu
> treiben. An dem Tage, an dem die Unentscheidbarkeit aufhörte, würde auch die Mathe-
> matik im heutigen Sinne aufhören zu existieren; an ihre Stelle würde eine absolut mecha-
> nische Vorschrift treten, mit deren Hilfe jedermann von jeder gegebene Aussage entschei-
> den könnte, ob diese bewiesen werden kann oder nicht. (von Neumann *1961*, pp. 265–266)

(So it appears that there is no way of finding a general criterion for deciding whether or not a well-formed formula is a theorem. (We cannot at the moment prove this. We have no clue as to how such a proof of undecidability would go.) But this ignorance does not prevent us from asserting: As of today we cannot in general decide whether an arbitrary well-formed formula can or cannot be proved from the axiom schemata given below. And the contemporary practice of mathematics, using as it does heuristic methods, only makes sense because of this undecidability. When the undecidability fails then mathematics, as we now understand it, will cease to exist; in its place there will be a mechanical prescription for deciding whether a given sentence is provable or not.)

Remarks: (1) Von Neumann allows for a great richness of notation in the systems he considers. For example the system of *Principia Mathematica* and von Neumann's own axiomatization of set theory would fall under his definition.

(2) Unlike Hilbert and Ackermann, von Neumann is talking about provability, not validity.

(3) Actually, mechanizability is not enough to stop the standard practices of mathematics. Despite Tarski's decision procedure (which, in general, requires a number of steps which is an iterated exponential of the parameters), ordinary mathematical methods go on being used to deal with problems which could in principle be settled by Tarski's procedure.

5.3. A similar robust skepticism was expressed by Hardy in his Rouse Ball lecture given at Cambridge in 1928 and published as *1929*. He says:

> Suppose, for example, that we could find a finite system of rules which enabled us to say whether any given formula was demonstrable or not. This system would embody a theorem of metamathematics. There is of course no such theorem and this is very fortunate, since if there were we should have a mechanical set of rules for the solution of all mathematical problems, and our activities as mathematicians would come to an end. (p. 16)

Hardy might have read von Neumann's paper, or have heard of it from F.P. Ramsey; in his lecture he discusses Hilbert's ideas at length, but he does not refer to von Neumann. Because he wrote the text for a lecture he gives no precise references.

6. The Triumph of Skepticism

It is an agreeably ironic fact that Gödel's first public announcement of his incompleteness results was made at Königsberg the day before Hilbert gave the address whose conclusion was quoted above in Section 4.[9] In his *1931* Gödel states:

> Theorem IX. *In any of the formal systems mentioned in Theorem VI, there are undecidable problems of the restricted functional calculus* (that is formulas of the restricted functional calculus for which neither validity nor the existence of a counterexample is provable). (Gödel *1986*, p. 187)

Since the systems considered are ω-consistent extensions of a formulation P of the simple theory of types, they are powerful enough to define and prove simple facts about the satisfaction relation for the predicate calculus. For any such system Σ Gödel constructs a formula ϕ_Σ which is satisfiable, but for which this fact cannot be proved in Σ. As a consequence, given any proposed algorithm α for the Entscheidungsproblem and any system Σ, either it cannot be proved in Σ that α always gives an answer, or it cannot be proved in Σ that its answer is always correct.

Thus Gödel's result meant that it was almost inconceivable that the Entscheidungsproblem should be decidable: a solution could, so to speak, only work by magic.

6.1. The new evidence was taken account of by various people. Thus Herbrand writes in his *1931a*:

> Either the first problem [the Entscheidungsproblem] is solvable; or else there are unsolvable problems of arithmetic ... Recent results lead us to think that it is the second alternative that holds; this would involve the falsity of an extremely widespread idea. (Herbrand *1971*, p. 276)

And Schütte (who had just obtained his doctorate under Hilbert) writes in *1933* that a general solution of the Entscheidungsproblem is probably not possible. Hilbert and Bernays in *1934* however keep, so to speak, a stiff upper lip: "So we are far away from a general solution of the Entscheidungsproblem" (p. 132).

6.2. A natural question to ask is why neither Gödel nor von Neumann *proved* the undecidability of the Entscheidungsproblem. Von Neumann was certainly excited by Gödel's lecture at Königsberg and was the first person fully to understand Gödel's results (see Wang *1981*). But by then he was mainly interested in other branches of mathematics – after 1931 he did not, for many years, publish any papers about logic. More importantly, I think, his mind moved too fast for him to be able to tackle what

---

9 See the dates given in 'Rundschau' in Erkenntnis, Vol. 1, pp. 80–81, Gödel *1986*, pp. 196–205, and Dawson *1984*.

is, in effect, a philosophical problem: one needed to reflect long and hard on the idea of calculability.

Gödel was, presumably, aware of the following fact: if C is an enumerated class of total functions (such as the Herbrand-Gödel functions of Gödel *1934*) which are uniformly formally reckonable in a system of the kind considered in *1931*, then a decision algorithm for the predicate calculus would allow one to diagonalize out of C; hence such an algorithm cannot belong to C. But to turn this into a proof of undecidability one has to know that C includes all effectively calculable functions – all functions which can be calculated by finite means. I discuss in Section 9.7 the reservations which Gödel may have had about proposed candidates (including his own) for C. But Gödel was primarily concerned – in opposition to the climate of the time – with the analysis of nonfinitist concepts and methods. This is very clearly set out in a letter of 1967 to Hao Wang (Wang *1974*, pp. 8–9). But a concern with nonfinitary reasoning is not what is needed for an analysis of calculations. Gödel admired and accepted Turing's analysis, but it is not surprising that he did not anticipate it.[10] Indeed, to the end of his life he believed that we might be able to use nonfinitary reasoning in (nonmechanical) calculations; see Section 10.5(2)(iv).

7. Work at Princeton 1931–1935

Kleene *1981* and Davis *1982* include detailed accounts of the development of the lambda-calculus and of the theory of Herbrand-Gödel recursive functions; my text is, to a certain extent, a commentary on theirs.

7.1. *Church's Postulates.* In the first half of this century an up-and-coming mathematical logician was expected to win his spurs by producing his own individual system of logic; this should be – at least apparently – consistent and should provide a foundation for as much of standard mathematics (more particularly, analysis) as the author believed to be sound. Church's *1932* and *1933* describe such a system. Here are some remarks about it.

(1) Like the system of Frege's Grundgesetze and unlike most axiomatized set theories, existence of mathematical objects is guaranteed by liberal rules for term formation, not by postulates of existence.[11]

---

10 Feferman, in his *1984*, suggests that Gödel, always cautious about the work he published, feared that even if he had made such an analysis it could not "be made convincing to the mathematical public of his day."

11 *Principia Mathematica* lies somewhere between. Since most terms are incomplete symbols, statements about them imply a postulate of existence.

(2) Both for logical and mathematical purposes the fundamental notion is that of function. From any term M one can form the functional abstract $\lambda x M$, provided that x occurs in M; and one can apply any term M (considered as a function) to any term N. Thus there is no distinction between functions and objects. Primitive symbols are introduced corresponding to logical operations: there is no distinction between 'term' and 'formula'. In such a system it is natural to construe the natural numbers as iterators ($\lambda f x \cdot f x$, $\lambda f x \cdot f(f x)$, and so on) rather than as cardinals or ordinals. If Church's program had been successful he would have been able to develop arithmetic without postulating an axiom of infinity; the program failed for the same reason as did Frege's.

(3) The system is entirely type free, but Church had hoped to avoid contradiction by allowing that some terms (in particular those which asserted properties of viciously circular terms) would be meaningless. The system does have a model: one in which all terms denote the same object (the 'True'). But in their *1935* Kleene and Rosser showed that *every* term is provable. Martin Davis remarks: "Not exactly what one dreams of having one's graduate students accomplish for one."

7.2. *The Lambda Calculus.* In 1933 Kleene gradually discovered the astonishing power of definition of the untyped λ-calculus. He has given (in a taped discussion printed in Crossley *1975*, and in his *1981*) graphic accounts of his discovery that the predecessor function is λ-definable. Church had doubted that this could be so. Indeed in the typed λ-calculus it cannot be defined without using a descriptions operator (see Church *1940* and Schwichtenberg *1976*).

Kleene also exploited the combinators which had been introduced in Schönfinkel *1924* and developed in Curry *1929, 1930,* and Rosser *1935*. At the end of his doctoral dissertation (accepted in September 1933, rewritten in 1934, and published as *1935*) Kleene showed that one can λ-define the sequence of *theorems* (not merely of their Gödel numbers) of any formal system which extends the λ-calculus by adding a finite number of primitive constants and a finite number of axioms and (finitary) rules of procedure. This suggested to Kleene and Rosser (towards the end of 1933) that any system which contained the λ-calculus and a certain minimum of quantification theory would be inconsistent. They proved this for Church's system early in 1934 and they published the proof in *1935*. Curry *1941* gives a readable account of the ideas on which their proof rests.

7.3. *Subsequent History.* In December 1933 Church gave a lecture on Richard's paradox, published as his *1934*. In it he describes the method (diagonalizing over the provably total number-theoretic functions) which, together with Kleene's results, forms the basis for the proof of inconsistency.[12]

Church also, in this lecture, sketches a system in which possible inconsistencies might be avoided by using a hierarchy of formal implication operators. This was outlined in his *1935* and developed in detail in his *1936b*.

Thus in 1934 the interest of the group shifted from systems of logic to the λ-calculus and certain mild extensions of it: the λ-κ and λ-δ calculi. A further and important new direction for their work was provided by the lectures (his *1934*) which Gödel gave from February to May at Princeton; I discuss the impact of these below in Section 9.

Finally in his lecture Church uses the term 'intuitively definable function' but does not discuss exactly how this phrase should be understood, although it seems (see the last paragraph on p. 357) that he would rule out functions whose values could only be calculated by solving unsolved number-theoretic problems. And soon after using the above phrase he writes:

> Hence it appears possible that there should be a system of symbolic logic containing a formula to stand for every definable function of positive integers, and I fully believe that such a system exists. (p. 358)

It is certainly conceivable that he had in mind a rather narrow sense of '(intuitively) defined'[13] and believed that the λ-calculus, or some mild extension of it, would provide 'such a system'.

7.4. *Church's Proposal.* According to Davis (*1982*, p. 8 – based on correspondance with Kleene) Church first proposed his thesis, identifying 'effectively calculable' with 'λ-definable' sometime in February 1934. I discuss the thesis below in Section 9; here I will consider some motives which may have led Church to make it.

(1) It was obvious that λ-definable functions were calculable, and it was known that some nonprimitive recursive functions were λ-definable.[14] Thus it was natural to look for a satisfactory (not merely an ad hoc) generalization. When one first works with the λ-calculus it seems artificial and its methods devious; but later one appreciates its simplicity and power.

(2) In footnote 3 of *1934* Gödel writes:

> The converse [to the fact that the value of a primitive recursive function can be computed by a finite procedure from the argument(s)] seems to be true if, besides recursions ac-

---

12 The only additional things which Kleene and Rosser use in their *1935* are some simple properties of conjunction and of type-free formal implication; these are available in Church's system of postulates. In particular $\lambda yz \cdot \forall x (yx \supset zx)$ is represented by a primitive symbol.

13 Turing has written in the margin of his copy: "Does 'defined intuitively' mean computable or might it include the class non comp functions?" [sic].

14 Kleene *1935* shows that Ackermann's function is λ-definable; see footnote on p. 224.

cording to the scheme (2) recursions of other forms (e.g., with respect to two variables simultaneously) are admitted. This cannot be proved, since the notion of finite computation is not defined, but it serves as a heuristic principle.

Martin Davis has suggested (in conversation) that if Gödel made this remark in the lectures, it may have suggested to Church that an (all-embracing) characterization of 'computable function' might be attainable, and he would then see λ-definability as a strong candidate. But the quotation I gave at the end of Section 7.3 suggests that Church already had in mind the desirability and the possibility of such a characterization (via a system of logic).[15]

(3) It would obviously be a comfort to Church if something of foundational importance should emerge from his work; that a phoenix should arise from the ashes of 'A Set of Postulates'.

(4) In his lecture on the Richard paradox he shows that he is aware that there are connections between a characterization of definable (and/or provably total) functions and unsolvable problems. The possibility of proving the Entscheidungsproblem undecidable must surely have glimmered, however faintly.

8. Recursive Functions

The use of recursion in computation is at least as old as Euclid, though the word is recent. When I was at school I was taught how to use *recurrence relations*. Hilbert in *1905* uses *rekurrente;* in *1923* (for the first time?) he uses *Rekursion.* The first use of *recursive* in English that I know of is in Ramsey *1928*. Dedekind realized that (primitive) recursion constituted a scheme for defining functions which needed justification; this he gave in *1888*, calling it *definition by induction.* Skolem in *1923* showed that many number-theoretic functions are primitive recursive; he uses *rekurrierend.*

In *1926* Hilbert greatly extended the use of the term by using transfinite types and – in effect – transfinite recursion. Indeed he claims that it is consistent to suppose that all number-theoretic functions are recursive in this extended sense. Ackermann *1928* considers functions which can be defined using primitive recursion at all finite types.[16] He gives a definition of a particular function (which Hilbert had described in *1926*) by primitive recursion at type 2, gives an alternative definition using double, nested recursion and shows that it is not primitive recursive. Gödel *1931* gives the

---

15 See also Church's letter quoted in Section 9.7.

16 It may be remarked that these fall under Dedekind's scheme of definition, since he does not restrict the values of the function being defined to any particular domain.

definitive definition and development of the primitive recursive functions, which he calls just *rekursiv*.

Rósza Péter (or Politzer) showed (*1934*, abstracted in *1932*) that various more powerful-seeming schemes for recursion do not in fact lead outside the class of primitive recursive functions. She then investigated (*1935, 1936,* and later papers), in laborious detail, various more general types of recursion. I think that she invented the term *primitive Rekursion* which occurs in her *1934*. It is also used by Hilbert and Bernay *1934* in a particular section (pp. 326–343) in which reference is made to Péter's work. Kleene, in his *1936*, introduced the term *primitive recursive function.*

8.1. *Herbrand.* The various suggestions which Herbrand made for more general kinds of recursion are discussed in admirable detail in van Heijenoort *1982*, from which, for the first suggestion, I quote:

> La première fois, au debut de 1931, il écrit que, selon l' 'intuitionisme' (et dans ce passage il entendait par ce mot le finitisme de Hilbert, qu'il avait adopté) 'toutes les fonctions introduites devront être effectivement calculables pour toutes les valeurs de leurs arguments, par des opérations décrites entièrement d'avance' (*Herbrand 1931a,* page 187, ou *1968,* page 210 [or *1971,* p. 273]). (p. 72)

This is the first occurrence I know of the phrase 'effectively calculable' (but see Section 4.2 above). I think Herbrand expected that the development of proof theory would introduce effectively calculable functions which (like Ackermann's function) were not primitive recursive; so he was interested in possible forms of definition for such functions.

In a letter to Gödel in 1931 he suggested that such a function f might be defined by a set of equations between terms involving f and symbols for previously defined functions, provided that the equations determined a unique total function. Unless the proviso is interpreted constructively the definition will not yield a method for calculating f. At the end of the Princeton lectures (May 1934) Gödel quoted Herbrand's suggestion, and strengthened it by requiring that each value of f be obtained from the equations by the substitution of numerals for variables and for variable-free terms to which they have already been proved equal. This defines the class of 'Herbrand-Gödel' recursive functions.

In *1931b* (dated July 14 and received by the editors on the day on which Herbrand died in a climbing accident) Herbrand proposed strengthening first-order number theory by adding symbols for calculable functions (cf. the quotation from Bernays in Section 4 above). He writes:

> On pourra aussi introduire un nombre quelconque de fonctions $f_i x_1 x_2 \ldots x_{n_i}$ avec des hypothèses telles que:
> (a) *Elles ne contiennent pas de variables apparentes;*

(b) *Considérées intuitionnistiquement, elles permettent de faire effectivement le calcul de $f_i x_1 x_2 \ldots x_{n_i}$ pour tout système particulier de nombres; et l'on puisse démontrer intuitionnistiquement que l'on obtient un résultat bien déterminé.* (*1931b*, p. 5; *1968*, pp. 226–227; *1971*, pp. 291–292)

Since Herbrand goes on to suggest that allowing such functions will yield a system for which Gödel's second incompleteness theorem does not hold, it is plain that the condition (b) is not to be confined to any particular formal system; van Heijenoort discusses this at some length. I think one can paraphrase the passage as follow:

One may introduce symbols for functions subject to quantifier-free conditions, provided these conditions allow one to calculate (finitistically) the values of each function and provided one has an intuitively acceptable proof that the conditions imply that the function is total.

The development of proof theory has shown the correctness of Herbrand's insight. To extract the maximum information from a formal proof in, for example, Peano arithmetic, one uses functions defined by complicated recursions (as in Ackermann *1940* and Kreisel *1952*); the 'finitist' calculations proceed by transfinite recursion, and the 'intuitively acceptable' proof of totality by transfinite induction (up to ϵ_0). Further, if one has accepted that the functions computable in Peano arithmetic are total, then it may seem intuitively all right to accept that the function got by diagonalizing over them is also total. And this fact allows one to prove the consistency of Peano arithmetic (cf. Paris-Harrington *1977*, and Ketonen and Solovay *1981*). One might, with only a little exaggeration, call this idea of producing a hierarchy of recursive functions, and linking the 'intuitively acceptable' arguments for justifying their introduction with 'finitist' proofs of consistency of formal systems of increasing strength, *Herbrand's program* (but he had other programs as well). It sprung from, and is closely connected with, the work of Hilbert and his group.

But the program does not give a definitive characterization of 'effectively calculable', nor does it lead to a negative solution of the Entscheidungsproblem. This is because the hierarchy of functions considered is essentially an open one; Herbrand (*1971*, p. 296) explicitly states that it is impossible to describe outright all (procedures for defining) such functions. It may be possible, witness Gödel's process of successive substitutions, to lay down once and for all a finitist procedure for computing specific values, but an outright determination of 'is total' must depend on the arithmetic *truth* of Π_2^0 sentences; however, the whole point of Hilbert's and Herbrand's work was to avoid any reference to the truth of quantified sentences – which indeed, in their more dogmatic moments, they regarded as without meaning. This is a further reason, in addition to those given above in Section 4.3, for the fact that Hilbert and his followers did not contribute to the discoveries of 1936.[17]

9.   *Church's Thesis*

9.1. *The Connections with the Herbrand-Gödel Definition.* It did not take long for the group at Princeton – Church, Kleene, and Rosser – to realize that 'λ-definable' and 'Herbrand–Gödel recursive' are equivalent. They worked both independently and collaboratively to prove this; the exact timing is a little confused; it is discussed in Davis *1982*. The proofs of equivalence were published in Kleene *1936a*, which uses material from Kleene *1936* – including Kleene's normal form; *1936a* was received by the editors in July 1935. Outlines of proofs of equivalence are given in Church *1936*.

I think the group must have seen that equivalence proofs were likely to be forthcoming already in 1934. For both systems, the computation of a value proceeds by steps from an initial set up (which includes numerals for the arguments) to a final set up, which can be recognized as final and which includes the numeral for the value. Gödel *1931* provides the apparatus which makes it obvious that the steps (interpreted in terms of Gödel numbers) are primitive recursive. Kleene had shown how all the consequences derived from an initial set up by such steps can be effectively enumerated.[18] Hence all that is required is a search through this enumeration; it is not too difficult to define a 'μ-operator' by equations. This line of argument is very close to the one Kleene gives in *1981* in his account of how he discovered his normal form (p. 60). Davis in *1982* suggests that in April 1935, when Church talked about his thesis to the American Mathematical Society, he was not yet convinced that every recursive function is λ-definable; for in the abstract of the talk, which Davis prints, he uses the Herbrand-Gödel definition. This suggestion does not in fact conflict with my opinion, for Church was always cautious and a stickler for detail,[19] and the details, which are wearisome, may not at that stage have been filled in.

9.2. *Church's Statement of the Thesis.* Church makes his proposal in the form of a definition:

---

17 But it should be remembered that Bernays, who was at Princeton in 1935–36, read, criticized, and suggested improvements to the papers of Church, Kleene, and Turing, all of whom acknowledge his help.

18 See Theorem 17II of his *1935*. The enumeration is by a λ-term; but the method of construction also yields a primitive recursive function.

19 Church's meticulous attention to detail was plainly inherited by Kleene and Rosser. But some of it, less expectedly, rubbed off on Turing, who once got quite agitated when I wrote '$x_1 \ldots x_n$' in place of '$x_1, x_2, \ldots, x_n$'.

We now define the notion, already discussed, of an effectively calculable function of positive integers by identifying it with the notion of a recursive function of positive integers[18] (or of a λ-definable function of positive integers). This definition is thought to be justified by the considerations which follow, so far as positive justification can ever be obtained for the selection of a formal definition to correspond to an intuitive notion.

[18] The question of the relationship between effective calculability and recursiveness (which it is here proposed to answer by identifying the two notions) was raised by Gödel in conversation with the author. The corresponding question of the relationship between effective calculability and λ-definability had previously been proposed by the author independently. (Church *1936*, p. 356, or Davis *1965*, p. 100)

The interest and significance of the definition lie in its justification. Post rightly remarks:

Actually the work already done by Church and others carries this identification considerably beyond the working hypothesis stage. But to mask this identification under a definition hides the fact that a fundamental discovery in the limitations of the mathematicizing power of Homo Sapiens has been made and blinds us to the need of its continual verification. (Post *1936*, footnote 8)[20]

In his *1943* Kleene restated Church's definition as 'Thesis I' (p. 60), and in his *1952* refers to it as Church's Thesis (p. 317), by which name it – and some connected hypotheses – has been known ever since.

Kleene in *1952* martials the arguments for the thesis; we discuss them here, paying particular attention to Church's paper.

9.3. *The Argument by Example.* Church does not explicitly make use of this argument, though I think he assumes that the interested reader would be familiar with some of the results of Kleene *1935*, and with familiar mathematical examples of definition by recursion.

Remarks. (1) From the results of Kleene *1935* it is readily seen that the class of λ-definable functions are closed under Babbage's principles as set out above in Section 2.1.

(2) The logicians had produced examples of calculable functions (such as Ackermann's) which would not have arisen (in 1936!) in standard mathematics; but these functions *were* defined by recursion.

(3) Kleene's dissertation shows, as we might now put it, the power of the λ-calculus as a high-level programming language. A λ-term can take other λ-terms (subroutines), instead of numbers, as inputs; and, if taken in the right order, successive conversions do correspond to a natural way of implementing the program which the

---

20 Church replied to this criticism in his review of Post's paper; see Section 10.4 below.

λ-term represents. Of course Kleene did not think in these terms, but this way of putting it helps one to appreciate the intuitions which led Church and Kleene to believe in the truth of the thesis.

9.4. *The Step-by-step Argument.* This is Church's chief argument (pp. 358–360). He considers the evaluation of a value fm of a function either by the application of an algorithm, or by the derivation of $fm = n$ from axioms about f in some formal system. In each case the evaluation proceeds in a series of steps. If each step is recursive then f will be recursive. As evidence for the premise he points out that Gödel had shown that the steps in a proof in Gödel's system P are primitive recursive.

9.5. *The Argument by Confluence.* In footnote 3 (p. 346) Church writes:

> The fact, however, that two such widely different and (in the opinion of the author) equally natural definitions of effective calculability turn out to be equivalent adds to the strength of the reasons adduced below for believing that they constitute as general a characterization of this notion as is consistent with the usual intuitive understanding of it.

(Kleene was so impressed by this argument that he worked very hard to give different characterizations of his concept 'recursive function of finite type' and to prove them equivalent.)

9.6. *The Criterion of the Failure of Diagonal Arguments.* Kleene (in *1981*) writes:

> When Church proposed his thesis, I sat down to disprove it by diagonalizing out of the class of the λ-definable functions. But, quickly realizing that the diagonalization cannot be done effectively, I became overnight a supporter of the thesis. (p. 59)

Presumably Church had also realized that, at least, there was no *obvious* way of deciding which λ-terms represent total number-theoretic functions. And once he had proposed the thesis, his analysis of the Richard paradox (see Section 7.3) must have convinced him that there was no *effective* decision procedure. I discuss his development of this idea in Section 11 below.

The most forceful argument for Church's thesis is Turing's analysis, which I discuss in Section 10.

9.7. *Objections to the Arguments.* In a letter to Kleene in November 1935 Church wrote:

> In regard to Gödel and the notions of recursive and effective calculability, the history is the following. In discussion [sic] with him the notion of lambda-definability, it developed that there was no good definition of effective calculability. My proposal that lambda-definability be taken as a definition of it he regarded as thoroughly unsatisfactory ... Evidently it occurred to him later that Herbrand's definition of recursiveness, which has no regard to effective calculability, could be modified in the direction of effective calculability and he made the proposal in his lectures. At that time he did specifically raise the

connection between recursiveness in this new sense and effective calculability, but said he did not think that the two ideas could be satisfactorily identified 'except heuristically'. (Davis *1982*, p. 9)

Post also (see Section 9.2) regarded Church's thesis merely as a working hypothesis. Here are some reasons for not regarding the arguments (9.3–9.6) as conclusive.

(1) The argument by example does justify the heuristic[21] value of the thesis, and was entirely suitable for Babbage's purpose. But it cannot settle the philosophical (or foundational) question. It might happen that one day some genius established an entirely new sort of calculation. An analogy will perhaps help to show the force of this objection.

During the early part of this century it was realized that intuitive notions of 'limit' and 'continuous' could be applied in contexts quite different from those in which they had first been given precise definitions (viz. the theory of functions of real and complex variables). It seemed, and in standard texts such as Bourbaki's 'Éléments de Mathématique' still seems, that these notions could be *definitively* characterized by introducing axioms for a 'topological space' and then defining the notion in terms of the primitive terms which occur in the axioms. However, cases have arisen in which there is a clear intuitive notion of continuity which cannot be so defined. For example, the notion of continuity which Kreisel *1959* uses in introducing his 'continuous functionals of finite type', and certain refinements of it, are not topological. But there is a generalization, 'filter space', in which Kreisel's notion of continuity can be simply and directly expressed.[22] Despite much evidence from examples, and from the confluence of definitions – in terms of open sets, neighborhoods, and operations of closure – 'topological space' is not *the* definitive concept.

(2) A similar doubt arises in connection with the step-by-step argument. An entirely new kind of algorithm, or an entirely new kind of rule of proof, might proceed by (irreducible) steps which were not recursive. How can one make unassailable predictions about the future development of mathematics? (Turing showed how.)

(3) The argument by confluence makes it likely that an interesting and important class of functions has been singled out. But, like the other arguments, it cannot reach out beyond contemporary mathematical experience.

---

21 The OED gives 'heuristic: serving to find out or discover'.
22 Also certain kinds of limit space will serve. For a full discussion see Hyland *1979*.

10.   *Turing's Work*[23]

Turing graduated in the summer of 1934. While still an undergraduate he had worked out for himself (in ignorance of Lindeberg's work) a rigorous proof of the central limit theorem; this work secured his election, in the spring of 1935, to a fellowship at King's College, Cambridge. At this time Turing was interested in branches of both pure mathematics and theoretical physics. He went to lectures on mathematical logic by M.H.A. Newman and learned from them of Gödel's work and the Entscheidungsproblem. Like Hardy (Section 5.3) Newman used the word 'mechanical'. In his obituary of Turing *1955* Newman writes:

> To the question 'what is a "mechanical" process?' Turing returned the characteristic answer 'Something that can be done by a machine' and he embarked on the highly congenial task of analysing the general notion of a computing machine. (p. 256)

It *is* characteristic of Turing to pass from the abstract use of 'mechanical' (the OED gives "performed without the exercise of thought or volition") to the concrete use. Indeed in *1939* Turing says:

> We may take this statement literally, understanding by a purely mechanical process one which could be carried out by a machine. (p. 166)

Nevertheless Newman's remark gives a *misleading account of Turing's line of thought;* see Section 10.3.(3).

10.1. *Turings's Analysis.* Turing considers 'computable (real) numbers'; in fact, what he is considering is total, computable functions of a positive integral argument with values 0 or 1. He starts off his detailed analysis (pp. 249–258) by saying:

> The real question at issue is "What are the possible processes which can be carried out in computing a [real] number?"

This is significantly different from the question ("What is a computable function?") which other authors asked. Turing, so to speak, has pointed himself in the true direction.

(1) He then considers the actions of an abstract human being who is making a calculation; he pictures him as working on squared papers as in "a child's arithmetic book". He argues – much too briefly – that nothing will be lost by supposing that the calculation is carried out on a potentially infinite tape divided into cells in each of which a single symbol (or none) may be written.

---

23 A description, more detailed in some respects, will be found in Kleene's contribution to this volume. For facts about Turing's life, aside from my own memories, I am indebted to Andrew Hodges' *1983:* a detailed and penetrating – if sometimes a little fanciful – biography.

(2) By considering the limitations of our sensory and mental apparatus, Turing arrives at the following restrictions on the actions of the computor.[24]

  (i) There is a fixed upper bound to the number of distinct symbols which can be written in a cell.

 (ii) There is a fixed upper bound to the number of contiguous cells whose contents the computor can take in ('at a glance', one might say) when he is deciding, at a given stage, what to do. Turing shows by an example that for a normal human being – the reader – this bound, for a linear arrangement, is less than 15.

(iii) At each step the computor may alter the contents of only one cell, and there is a fixed upper bound to the distance the computor can move to get to this cell from the scanned cells; so we may as well suppose it is one of them.

 (iv) There is a fixed upper bound to the distance through which the scan can be moved between steps. (Turing's argument is rather indirect). Moving the scan is part of the action.

  (v) There is a fixed upper bound to the number of 'states of mind' of the computor; his state of mind, together with the contents of the scanned cells, determine the action he takes and his next state of mind. In place of a 'state of mind' Turing admits that the computor might leave an instruction as to how to continue (p. 253). Thus the computor must follow a fixed, finite set of instructions.

(3) All this may be summed up as follows: The computation proceeds by discrete steps and produces a record consisting of a finite (but unbounded) number of cells, each of which is blank or contains a symbol from a finite alphabet. At each step the action is local and is locally determined, according to a finite table of instructions.

(4) Turing easily shows that the behavior of the computor can be exactly simulated by a Turing machine. He uses the convention that the binary digits of the number being computed are printed on alternate squares, the others being used for rough working. He abandons this convention whenever it suits him to do so.

10.2. *Turing's Other Arguments.* He gives a wide range of examples of functions, numbers (e.g., real algebraic numbers and the real zeros of Bessel functions – Babbage would have applauded) and operations which are computable.[25] He only gives one example of the argument by confluence: he shows that if a function is for-

---

24 I use 'computor' for a human being, 'computer' for a machine.

25 Turing planned a sequel to his paper in which he would develop computable analysis. He told me (in the 1950's) that a particular reason for his not pursuing this topic was the fact, pointed out by Bernays and discussed briefly by Turing in his Correction to *1936-7*, that not all real numbers have unique representations as binary decimals. He also told me that after he had returned to Cambridge in 1938 he was asked to examine a doctoral dissertation by R.L. Goodstein in which recursive analysis

mally calculable in the predicate calculus from given axioms[26] then it is computable (p. 252). After he had submitted his paper, Newman received an offprint of Church *1936*;[27] Turing then sketched, as an appendix, a proof of equivalence between 'computable' and 'λ-definable'.

10.3. *Discussion.* (1) I suppose, but do not know, that Turing, right from the start of his work, had as his goal a proof of the undecidability of the Entscheidungsproblem. He told me that the 'main idea' of the paper came to him when he was lying in Grantchester meadows in the summer of 1935. The 'main idea' might have either been his analysis of computation, or his realization that there was a universal machine, and so a diagonal argument to prove unsolvability.

(2) Turing's analysis does much more than provide an argument for Church's thesis; *it proves a theorem.* Here is a first version:

> *Any function which can be calculated by a human being can be computed by a Turing machine.*

The proof is quite as rigorous as many accepted mathematical proofs – it is the subject matter, not the process of proof, which is unfamiliar. However – as often with published mathematical proofs – there are gaps which need to be filled in. (A) It is not totally obvious that calculations carried out in two (or three) dimensions can be put on a one-dimensional tape and yet preserve the 'local' properties. But, as I have stated them in Section 10.1(2) the undeniable limitations on human perception and action apply just as forcibly to two- or three-dimensional (a book is, after all, three-dimensional) records. And the limitations show that the change in the record which is made in a single step can be specified by a *totally finite* (and hence certainly recursive) function. So, without further ado, one can use Church's step-by-step argument. (I owe this observation to John Shepherdson; see his contribution to this volume.) (B) It is not necessary to the argument, as Turing already indicated, to suppose that mathematicians always work in a deterministic manner with a finite number of states of mind. What is required is to realize that when they are *calculating* (not when they are speculating or conjecturing) they are following a given routine – a fixed set of instructions. If someone just guesses, or appeals to a hitherto unmentioned rule 42,[28] we do not use the words 'computing' or 'calculating'. To avoid ambiguities

---

(including measure theory) was developed, but which was largely in error because the ambiguity of representation had not been taken account of; Goodstein, in conversation with me, confirmed this.

26 This notion was introduced in Gödel *1931* (for relations) as 'Entscheidungsdefinit'. There have been many variations on it, and generalizations of it were used by Mostowski and Kreisel as a characterization of the (generalized) concept of 'recursive'. See Kreisel *1965* and references cited therein. Also see Gödel's remarks about its absolute character in his *1936*.

27 For a touching account of this event see Hodges *1983*, pp. 112–113.

28 "All persons more than a mile high to leave the court."

in the meaning of 'mechanical', I shall use the word 'routine'. Finally, from now on I shall use the word 'computable' to refer to any of the equivalent notions 'Turing computable', 'λ-definable', 'Herbrand-Gödel' recursive. So, Turing proved:

10.3.1. Theorem: *Any function which can be calculated by a human being following a fixed routine is computable.*

Turing certainly believed the converse of this theorem; but some people have denied it.[29] We are not concerned here with the – very important – practical limitations on time and space, but with what can be done regardless of such limitations. Objections to the converse of the theorem are then just the bad penny of Section 4.2 turning up again. So I will cut short any further argument and restate Church's Thesis as:

10.3.2. Turing's Theorem: *Any function which is effectively calculable by an abstract human being following a fixed routine is effectively calculable by a Turing machine – or equivalently, effectively calculable in the sense defined by Church – and conversely.*

(3) *Turing's analysis makes no reference whatsoever to calculating machines.* Turing machines appear as a result, as a codification, of his analysis of calculation by humans.[30] It is true that Turing was fascinated by machines and mechanical devices; this may have influenced his thoughts before he arrived at his analysis, and it certainly meant that he was pleased to be able to present the *results* of his analysis as giving the literal meaning to 'mechanical' – see the quotation at the beginning of Section 10. But he evidently did not think that ideas derived from actual calculating machines were sufficiently relevant to be worth mentioning in his paper; and indeed machines in existence in 1936 were not very relevant. I have argued, in *1980,* that it is by no means obvious that the limitations described in Section 10.2 apply to mechanical devices; Turing does not claim this.

It has been suggested (e.g., by Hyman on page 255 of his *1984*) that Turing was influenced by Babbage's *ideas.* I think this can be ruled out: when Turing learned about them in the 1940's he talked of them with enthusiasm (Hodges *1983,* p. 297). Had he known about them in 1936 he would certainly have referred to Babbage in his paper. He may have known that Babbage planned an Analytic Engine, either from the exhibit in the Science Museum in Kensington, or from the article 'Calculating

---

29 I remember R.L. Goodstein maintaining in conversation – he was more guarded in print – that only primitive recursive functions are calculable. According to my account, what he should have said was, for example, that Ackermann's equations do not define a total function.

30 Post considers instead a 'worker' who moves up and down the tape.

Machines' in the 11th edition of the Encyclopaedia Britannica.[31] But neither of these sources gives information about the theoretical principles – in particular, the use of conditional instructions – on which the design of the engine was based.

(4) At the beginning of *1936-7* Turing writes:

> The "computable" numbers may be described briefly as the real numbers whose expression as a decimal are calculable by finite means. (p. 230)

This, by itself, is not a sufficient characterization. A computor who used unanalyzed intuition, or used an oracle in an impenetrable black box, would not be using nonfinite means. What is the minimal supplement one can make to 'finite means' so as to capture exactly the limitations of computors and computers? Various answers have been proposed, but not, I think, a definitive one.

(5) Besides providing the definitive meaning of 'computable function', Turing's paper also makes precise the notion of 'formal system'. The development of symbolic logic had focused attention on systems of formulae and proofs in which the rules of formation and inference depend only on the physical form of the expressions, not at all on their meaning. The simplest way of applying Turing's ideas is to say that a formal system is one in which the process of checking that expressions are correctly formed and that proofs are correct can be carried out by a Turing machine. Turing himself briefly considers 'choice machines' (p. 232) – nowadays referred to as 'nondeterministic' – and shows that, by using a systematic search through all allowable choices, a choice machine can be simulated by a Turing machine (footnote ‡ on p. 252). He claims to have written the instruction table for a machine which will enumerate the theorems (containing only symbols from a fixed finite list) of the predicate calculus (footnote § on p. 252).[32]

(6) It is almost true to say that Turing succeeded in his analysis because he was not familiar with the work of others. In the version submitted in May 1936 the only logical works referred to are Hilbert and Ackermann *1928* (which Turing erroneously assigns to 1931) and Gödel *1931*. The approach is novel, the style refreshing in its directness and simplicity. The bare-hands, do-it-yourself approach does lead to clumsiness and error. But the way in which he uses concrete objects such as exercise books and printer's ink to illustrate and control the argument is typical of his insight and originality. Let us praise the uncluttered mind.

10.4. *The Reception of Turing's Ideas.* Turing himself was modest about his achievement:

---

31 In later life he was fond of consulting this work, which he had inherited from his father. I, in my turn, have inherited this copy and find it still an invaluable work of reference.

32 And in his *1937* he gives, in full detail, the program for a machine which will enumerate the formulae which can be obtained by λ-reduction from a given term of the λ-calculus.

The identification of 'effectively calculable' functions with computable functions is possibly more convincing than an identification with the λ-definable or general recursive functions. (Turing *1937*, p. 153)

Church, in his review of Turing's paper, writes:

[Turing's notion] has the advantage of making the identification with effectiveness in the ordinary (not explicitly defined) sense evident immediately – i.e., without the necessity of proving preliminary theorems. (Church *1937*)

But Church is slightly less dogmatic in his review of Post's paper, preferring, apparently, definition to further argument, and defending this move against Post's criticism (see Section 9.2).

[Post] takes this identification as a "working hypothesis" in need of continual verification. To this the reviewer would object that effectiveness in the ordinary (not explicitly defined) sense has not been given an exact definition, and hence the working hypothesis in question has not an exact meaning. To define effectiveness as computability by an arbitrary machine subject to the restrictions of finiteness would seem to be an adequate representation of the ordinary notion, and if this is done the need for a working hypothesis disappears. (ibid.)

It is not quite clear whether Post did accept Turing's analysis; the matter is discussed in Section 13.4.

Hilbert and Bernays in their *1939* do not mention Turing, nor do they discuss Church's thesis. They take 'regelrecht auswertbar' (formally evaluable) as the definition of 'berechenbar' (calculable) (p. 393); their conditions on formal systems amount to requiring that all syntactic relations and operations should be primitive recursive (p. 255).

In his *1952* Kleene gave a definitive account of Church's and Turing's arguments, although he did not assert, as I have done, that Turing's work constitutes the proof of a theorem. Since then there has been no serious dispute among mathematical philosophers about the correctness of suitably guarded statements (such as Theorem 10.3.2) of Church's thesis. In particular, Gödel accepted unreservedly Turing's characterization of 'calculable by a mechanical process' and of 'formal system'.[33]

10.5. *Philosophical Significance of Turing's Analysis.* (1) Turing's work is a paradigm of philosophical analysis: it shows that what appears to be a vague intuitive notion has in fact a unique meaning which can be stated with complete precision. After Turing one can no longer, on this topic, use the ploy (made so familiar by Professor Joad whenever he appeared on the BBC's "Brains Trust") of 'it all depends on what you mean'. Of this Gödel said:

---

33 See Gödel *1946* (part of which is quoted in Section 10.5), the note added to the translation of *1931* in van Heijenoort *1967* and the postscriptum added to his *1934* in Davis *1965*; these last two are reproduced in Gödel *1986*.

Tarski has stressed in his lecture (and I think justly) the great importance of the concept of general recursiveness (or Turing's computability). It seems to me that this importance is largely due to the fact that with this concept one has for the first time succeeded in giving an absolute definition of an interesting epistemological notion, i.e., one not depending on the formalism chosen. In all other cases treated previously, such as demonstrability or definability, one has been able to define them only relative to a given language, and for each individual language it is clear that the one thus obtained is not the one looked for. For the concept of computability, however, although it is merely a special kind of demonstrability or decidability [sic] the situation is different. By a kind of miracle it is not necessary to distinguish orders, and the diagonal procedure does not lead outside the defined notion. This, I think, should encourage one to expect the same thing to be possible also in other cases (such as demonstrability or definability). (Gödel *1946*, p. 84)

(2) Turing's analysis can be seen as part of the general movement, already in evidence in the seventeenth century, towards a mechanistic – or physicalistic – account of human thought and behavior. This is a large subject, but I think there are a few points, relevant to the questions posed in Section 1 which are worth making.

  (i) Hodges records that Turing's imagination was stimulated by reading, when about ten years old, a book called *Natural Wonders Every Child Should Know* (by E. T. Brewster) in which the body (including the brain) is described as a machine (Hodges *1983*, pp. 11–13).

 (ii) Turing's account of the limitations of our sensory and mental apparatus is concerned with perceptions and thoughts, not with neural mechanisms. And there is no suggestion that our brains act like Turing machines.

(iii) In their seminal *1943* McCulloch and Pitts asserted, justifiably, that the control mechanism of the universal Turing machine could be simulated by a finite assembly of idealized neurons, connected to each other by synapses incorporating delays. Their idealized neurons can be thought of as logical gates, and were used in von Neumann *1945* and in Turing *1945* as the building blocks for the construction of (the executive parts of) universal digital computers. McCulloch and Pitts also described how a strictly bounded amount of memory could be provided by cycles of neurons. The actual neural mechanisms for memory are still mysterious.

(iv) Gödel and Post always believed that a true account – an acceptable theory – of human mathematical intelligence must be nonmechanical. In particular, Gödel has argued (see Wang *1974*, pp. 325–326) that in our ability to handle abstract concepts we are not subject to the restrictions described by Turing. These only apply when we are dealing with (potentially) concrete objects such as strings of symbols.

 (v) Turing certainly believed that, ultimately, our thought processes are consequences of (or, at least are tied to) physical processes. But this belief does not require one to reject theories of intelligence which use nonmechanical notions. Turing was not a dogmatist; what he asserted, in the discussions in the early 1950's (see his *1950*) was that, by and large, human intelligence could be simulated by machines, where the 'machines' use probabilistic methods and are so constructed that they are capable of making mistakes.

(3) Another contribution of Turing's analysis to philosophy has been, I think, too little noticed. Wittgenstein discussed (ad nauseam?) how it is that we can assert that a rule has been followed *correctly* when different applications of it are made to different cases. In one sense of 'case' each application is different from all others – it happens at a different time; and in that sense the problem is about how we can say that different occurrences of a word mean 'the same' or actions on different occasions are 'the same'. But Wittgenstein (and Kripke)[34] often use as examples rules, such as addition, where, on the common sense account, the cases *are* different. Turing's analysis shows how following such a rule consists in the iteration of one of a fixed number of behavior patterns: to follow a rule is to repeatedly carry out a recipe. This does not solve the problem in the sense described above, but it does show that the use of rules with potentially infinitely many cases of application is just a rhetorical device, which does not really raise a new sort of problem. Turing attended Wittgenstein's lectures in 1939 and frequently contributed to the discussion (Wittgenstein *1976*). Wittgenstein talked about following a rule; I find it rather odd that Turing did not make – in some form or other – the point I have just made.

11. Diagonalization and Undecidability

The arguments given by Church, Kleene, and Turing all have the same structure and can be summarized as follows:

(1) One shows that there is a uniform process U which, provided with a (putative) specification for a calculable function F and a numerical argument m, will calculate the value fm if this is defined.

(a) For the λ-calculus U is a λ-term such that Ufm conv n if and only if Fm conv n where f is the Gödel number of the term F and x is the numeral (iterator) corresponding to the number x.[35]

---

34 Kripke *1982* is a very readable account of the problem; Baker and Hacker *1984* and *1985* argue that Kripke misrepresents Wittgenstein in a number of ways. Wittgenstein was, as usual, more concerned to reveal the inadequacy or error of proposed solutions than to give a final solution of his own.

35 Church actually combines steps (1) and (2). However, his construction of **e** (p. 361) is easily modified to obtain a suitable term U.

(b) For the equational calculus the process U enumerates the (Gödel numbers of the) consequences of the equations which define F until an equation of the form $Fm = n$ is found.

(c) For Turing machines U is just the universal machine.

(2) One shows that if one could decide the (Π_2^0) problem: does a given putative specification actually specify a total function (or, in Turing's terminology a circle-free machine), then, using U, the diagonal argument would give a contradiction.

(3) If every Σ_1^0 problem were decidable, then so would be every arithmetic, and in particular every Π_2^0, problem. Each author does in fact give a specific undecidable Σ_1^0 problem:

(a) Does a given formula of the λ-calculus have a normal form?

(b) Does the set of equations (specifying F) with Gödel number f determine a value for $F(f)$?

(c) Does a specified machine ever print 0?

(4) Church and Turing give first-order axioms for their respective processes of computation, and thus prove the undecidability of the Entscheidungsproblem. This seems to us a straightforward (first-year graduate) exercise in formalization. But they both made mistakes, which were pointed out by Bernays.

11.1. *Remark.* All the work under discussion so far is concerned with total functions.[36] Once it had been realized that one had to deal with computations which might not terminate, it would be natural to consider partial functions. In fact these did not get into print until Kleene *1938,* though the work had already been done in 1936 – see Kleene *1981,* p. 60. Since then *general recursive* has been used to refer to total functions, in contrast not – as originally intended – with primitive recursive, but with partial recursive. Allowing partial functions was essential for the further development of the subject.[37]

---

36 Church introduces 'potentially recursive functions' but, as he uses them, they are just total functions over the domain of Gödel numbers of well-formed formulae.

37 In particular partial functions are necessary for the inductive definition of (the graph of) a function, and so for the proof of fixed point theorems, in particular the 'first recursion theorem' of Kleene *1952.* Dedekind had already used them in his justification of 'definition by induction'; it is surprising that (as far as I have been able to discover) this method did not resurface until Kleene's *1952;* although in his *1936a* Kleene does give a λ-definition of the fixed-point operator (see (19) on p. 346). The second recursion theorem (fixed points in terms of Gödel numbers) was used and proved by Kleene in his *1938.* The first recursion theorem is the keystone of Platek's 1966 theory of recursion on abstract structures. Inductively defined sets (of which Post's canonical systems are an example) and relations provide the foundation for various generalizations of recursion theory. Both these ideas could have been developed for number-theoretic functions before 1936. Perhaps everyone concerned had forgotten, or did not know of Dedekind's work; his axioms have been often attributed to Peano.

12. From Turing Machines To Electronic Computers

There is no doubt that the work of 1936, more particularly Turing's paper, influenced the design and development of high speed digital computers.

Certainly, the title of Turing's paper and the fact that its first section is headed 'computing machines' encouraged people concerned with the design of computers to read it, or at least to look at it; but, the ideas it contains would have been equally important if Turing, like Post, had avoided the use of the word 'machine'.

I think these ideas, in order of importance, are as follows:

(1) The elementary steps are extremely simple, and have specifications of a fixed length.

(2) The universal machine is a stored-program machine; that is, unlike Babbage's all-purpose machine, the mechanisms used in reading a program are of the same kind as those used in executing it.

(3) Conditional instructions are no different from unconditional ones.

(4) The operation is easily adapted to binary storage and working.[38]

It is (1) and (4) that made it possible for Pitts and McCulloch to show that the control mechanism of a Turing machine can be simulated by a finite network of 'neurons' (gates with delays). The λ-calculus and the equational calculus both use 'substitute a given term for a given variable in a given expression' as an elementary step; this cannot have a total specification of fixed length. On the other hand the λ-calculus is a stored program device, since there is no difference between the program (a λ-term) and the successive stages in its computation. Although the earlier designs for computers (in particular the EDVAC) only allow rather restricted use of conditional instructions, the use of gates with two or more inputs does in fact reflect the conditional nature of the elementary steps of a Turing machine.

12.2. There is some controversy about the exact extent of the influence of Turing's ideas on the design of electronic computers in the U.S.A. I record a few facts:

(1) By the late 1930's von Neumann had become familiar with Turing's ideas and was enthusiastic about them (Randell *1972*, p. 10; Hodges *1983*, p. 145; Davis *1988*.[39])

---

38 Presumably Turing realized, when he wrote the paper, that, like Post, he could have used 'mark' and 'blank' as the only symbols; but his table for the universal machine would have been totally unreadable if he had done so. Kleene used binary Turing machines in his *1952*.

39 There is no evidence that there was ever a meeting at which they exchanged ideas about the possibilities of designing 'universal' electronic computers. They did meet early in 1947, but by that time they had written their respective reports. In a perceptive footnote (*1983*, fn. 5.26 on pp. 555–

(2) In 1945 von Neumann wrote his 'First draft of a report on the EDVAC'. In this he makes considerable use of the idealized neuronal networks of McCulloch and Pitts *1943*. He does not explicitly refer to Turing *1936-7*;[40] but (as with Turing's universal machine) the program is stored in (a special part of) the memory. This report circulated widely and was influential. Turing read it in the summer of 1945.

(3) Most of the essential ideas for the design of electronic computers – binary working, use of logical circuits and stored programs – were developed independently by various people, from 1936 onwards; see Randell *1982*, chapters VII and VIII, and Burks *1980*. The most important and influential place for the construction of electronic computers in the 1940's was the Moore School of Engineering; the first working large scale (18,000 valves) electronic computer was the ENIAC. The original proposal (in April 1943) was by Mauchly and Eckert. The machine was finished at the end of 1945. In the meantime plans were being made for 'EDVAC-type' machines; von Neumann became a consultant there in September 1944. Much of the work on ENIAC and EDVAC was classified.

It has become, it seems, impossible to discern a linear flow of ideas; probably there was no such thing. Much acrimony (see Eckert *1980*, and Mauchly *1980*) and a protracted legal battle developed from the question of who told what to whom.

(4) According to Randell *1980*, in 1942-43 Turing and Newman and a group of mathematicians and engineers at Bletchley Park discussed, out of office hours, Turing's universal machine, Babbage's plans for the Analytic Engine, and the possibilities of artificial intelligence. Turing played some part in the design of the first machines built under Newman's direction. The final machines ('Colossi') can claim to be the first medium-sized (2,000–3,000 valves) fairly flexible, programmable electronic computers. (The Mark II Colossus came into service in June 1944.)

(5) Turing worked at the National Physical Laboratory from 1945 to 1948 designing a computer. His report on the ACE (*1945*) was submitted in March 1946. Turing was influenced by von Neumann *1945*; but he describes, in fair detail, the design of a quite specific machine, and, ahead of the times, proposed that many of the operations of the machine should be effected by writing subroutines, rather than by building special, single-purpose units.[41] In his lecture *1947* to the London Mathematical Society he traces the connections between the design of the ACE and his 1936-7 paper.

---

556) Hodges argues that such a meeting would not, in any case, have been likely to have been of great importance; each would develop his own ideas.

40 But in his Hixon lecture in 1948 (*1951*) he gives Turing his full due.

41 Hodges discusses at length the ideas of the ACE report (*1982*, pp. 317–333). A discussion of technical points, and of the history of the report, by the editors, is in Turing *1986*.

(6) At the same time, in Manchester, Newman was responsible for, and contributed to, the design of what became the Manchester Mark I machine; see his *1948*; Turing jointed[42] him in the autumn of 1948. His chief interest was in using (and helping others to use) the machine. Although he wrote the programmer's handbook he did not contribute to the design of the machine or of programming languages for it. Hodges (*1983*, p. 401) lists some of the things which Turing could have worked on but did not.

12.3. Though it may be difficult to trace the precise influence of Turing *1936-7* on the design and development of high-speed digital computers, the fundamental importance of that paper for the *theory* of computation is clear. Turing machines (and modifications of them) still provide the standard setting for the definition of the complexity of computation in terms of bounds on time and space; together with the neural nets of McCulloch and Pitts they provided the foundations of the theory of automata; together with the generated sets of Post *1943* they provided the foundation for the theory of formal grammars.

13. *Post's Contribution*

13.1. In the early 1920's Post developed a correct general theory of formal systems. A first formulation of his notion of a formal system, derived from his presentation of many-valued propositional calculi, appears in his *1921*. He submitted an account of his work for publication in 1941; this eventually appeared as *1965*. His *1943* is, in effect, an extract from the larger work. Davis, who had studied under Post, gives, in his *1982*, a clear account of the technical development of the work, so I shall concentrate here more on its philosophical side.

Post became interested in marking out the boundary between what can be done in mathematics by purely formal means from the more important part which depends on understanding and meaning.[43]

Post writes:

> But perhaps the greatest service the present account could render would stem from its stressing of its final conclusion that *mathematical thinking is, and must be essentially creative*. It is to the writer's continuing amazement that ten years after Gödel's remarkable achievement current views on the nature of mathematics are thereby affected only to the point of seeing the need of many formal systems, instead of a universal one. Rather has it seemed to us inevitable that these developments will result in a reversal of the entire

---

42 A gem from the first proof.

43 There is an analogy here with Wittgenstein's investigation of (the limits of) language in his *Tractatus*. He wrote in 1917: "The poem by Uhland is really magnificent. And this is how it is: if only you do not try to utter what is unutterable then nothing gets lost." (Engelman *1967*, p. 7)

axiomatic trend of the late nineteenth and early twentieth centuries, with a return to meaning and truth. Postulational thinking will then remain as but one phase of mathematical thinking. (*1965*, p. 345)

13.2. *Canonical and Normal Systems.* Post's idea of a formal system is that it will generate (or 'produce') a particular set of words: for example, well-formed formulae ('enunciations') or theorems ('assertions'). I will call words which contain free variables as well as letters from a given finite alphabet *open* words, while words on the given alphabet are *closed* words. A *production scheme* has the form

$$W_1 \text{ and } W_2 \text{ and } \ldots W_r \text{ produce } W_0,$$

where the W_i are all open words, and it is required that each variable of W_0 appears in one of the premises $W_1, \ldots, W_r$. A *canonical system* consists of a finite number of production schemes and a finite set of initial closed words. If L is a canonical system I will denote by L^* the set of words obtained from the initial words by successive application of (closed substitutional instances of) the production schemes; Post calls these the assertions of L. In a *normal system* the production schemes have the particular form '$W_1 x$ produces $x W_2$' where W_1 and W_2 are closed words. Post proves that for any canonical system L with alphabet A there is a normal system N with a larger alphabet B such that L^* consists of just those words of N^* which contain only letters from A. He later states what Davis calls

Post's Thesis:

> Every generated set of sequences on a given set of letters $a_1, a_2, \ldots, a_\mu$ is a subset of the set of assertions of a system in normal form with primitive letters $a_1, a_2, \ldots, a_\mu$, $a'_1, a'_2, \ldots, a'_{\mu'}$, i.e., the subset consisting of these assertions of the normal system involving only the letters $a_1, a_2, \ldots, a_\mu$. (p. 405)

'Generated set' here is the intuitive concept; one might replace it by 'theorems of a formal system (or of a system of symbolic logic)' – understood in an intuitive sense.

13.3. *Incompleteness and Undecidability Results.* Post realized that it is possible to construct a universal normal system K, though he does not give details. By a diagonal argument he shows that there cannot be a finite process for deciding whether or not a given word belongs to K^* (pp. 406–407). Later he identifies the decidability of L^* with the existence of a normal system which will generate the complement of L^* (p. 412). A *normal deductive system* is a normal system L which includes K as a subsystem, and which is such that if S is any normal system which is specified by a word $\hat{S}$ of L, and if any word W on the alphabet of S is specified by a word $\hat{W}$ of L, then $a\hat{S}a\hat{W}$ belongs to L^* if and only if W belongs to S^*, while if $b\hat{S}b\hat{W}$ belongs to L^*, then W belongs to the complement of S^* (p. 415 – I have used a somewhat different notation from Post's). Now the diagonal argument yields two incompleteness theorems.[44]

Theorem 1: *Any normal deductive system L is incomplete in the sense that there will by a system S and a word W belonging to the complement of S^* such that $b\hat{S}b\hat{W}$ does not belong to L^*.*

Theorem 2: *Given any normal deductive system L_1 one can find a stronger system L_2 which produces more true statements (of the form $b\hat{S}b\hat{W} \in L_2^*$) than does L_1.*

By appealing to his thesis Post makes these theorems more vivid by stating them as slogans:

(A) "A complete system of symbolic logic is impossible." (p. 416)

(B) "The logical process is essentially creative." (p. 417)

13.3. *Discussion.* (1) Post believed that the whole system of *Principia Mathematica* could be presented as a normal system. (This would be a very arduous task: it would involve giving precise rules for substitution and for the use of typical ambiguity). He had also seen how the elementary theory of strings (or words) could be handled within *PM* (see footnote 8; in the language of Part V of *PM* strings are 'pseudo-series'). Post planned to use these facts to deduce that *Principia Mathematica* is incomplete (see p. 418), and it seems (Davis *1982*, p. 22) that he gave lectures on this at Columbia University.

(2) However, to prove that a normal system L is normal deductive one must also show that $a\hat{S}a\hat{W} \in L^*$ if and only if $W \in S^*$. If we regard the production of $a\hat{S}a\hat{W}$ by L as a proof in L of '$W \in S^*$', then this condition requires that we can prove '$W \in S^*$' in L if and only if '$W \in S^*$' is true. But $W \in S^*$ is an existential sentence. In other words we have to prove that L is ω-consistent.[45] Post does not mention consistency; I suppose that he took it for granted that the theorems of *Principia Mathematica* are true.

(3) From (2) it can be seen, either by arithmetization or by working in the elementary theory of concatenation, that Theorem 1 is extremely close to Gödel's first incompleteness theorem; the fundamental difference is that Post is using a more liberal notion of system than Gödel did in *1931*.

(4) In discussing slogan (B) Post writes:

> It makes of the mathematician much more than a kind of clever being who can do quickly what a <u>machine</u> could do ultimately. We see that a <u>machine</u> would never give a complete

---

44 Post indicates how these theorems could be proved, but does not claim to have given proofs.
45 More precisely, as with Gödel's proof, what is required is that the system be '1-consistent'.

logic; for once the machine is made <u>we</u> could prove a theorem that it does not prove. (p. 417)[46]

Although a poor anticipation of the future power of computers, this is a very concise anticipation of an argument used by Lucas in his *1961* (and by others) in the discussion of minds and machines. The force of the argument, so it seems to me, lies not so much in the fact that given a system which we *know* to be consistent we can construct (in a quite mechanical way) a more powerful system, but in the fact that we can recognize some systems to be consistent long before we know how to prove that they are. And Post himself, in footnote 100 (written, presumably, in 1941 rather than in the 1920's), after dismissing the 'mere' addition of a new unprovable sentence, writes:

> For the entire development should lead away from the purely formal as the ideal of a mathematical science, with a consequent return to postulates that are to be self evident properties of the new meaningful mathematical science under consideration. (p. 416)

In the remarks (from his workbooks) which Post prints as an appendix, he is groping towards a better understanding of the creative process. In particular he considers the introduction of notation for ever larger ordinals (p. 423). This suggests to us a comparison with Turing *1939*; Turing quite explicitly thinks of the recognition that a given ordering is a well-ordering as a creative act, and the construction of a logic from the well-ordering as a mechanical process. Post in his 1941 footnotes does not refer to Turing's paper, but he does refer to it in his *1944*.

13.4. *Post's Arguments For His Thesis.* In *1921* and in the first two sections of *1965* the enunciations and assertions are terms (built up from symbols for functions and individuals). However, in section 4 he shows how systems which generate terms can be reduced to systems which generate strings, and so it is not necessary to distinguish between the two kinds of system.

(1) Post uses the argument by example; he has shown that the theorems of m-valued propositional calculi and the quantification theory of *Principia Mathematica* can be produced by canonical and hence by normal systems. And in footnote 18 to *1943* he says, "It appeared obvious to the writer that all of *Principia Mathematica* could likewise be reduced to a system in canonical form C."

---

46 I don't think too much should be read into Post's use of the word 'machine' in this passage; there is an obvious sense in which the production of words in a normal system is a mechanical process. In the appendix there are a few remarks about machines. He considers a machine with finitely many parts and finitely many relations between them. In footnote 112 to this passage Post writes: "Apart from [the fact that he is describing a nongrowing machine], this fails to be an anticipation of the Turing machine ... in its very attempt to allow for the structure of the machine." Perhaps he had in mind something like the 'mechanical devices' of Gandy *1980*.

(2) He uses the argument by confluence; a system of one of the sorts considered can be simulated by any of the others. He also realized (footnote 79) that any canonical system could be axiomatized in the first-order predicate calculus, though he seems not to have realized (footnote 90) that this would make the undecidability of the Entscheidungsproblem a corollary of Theorem 1.

(3) Finally, he considers a very powerful-seeming generalization of the production rules (p. 404): words $W_1, \ldots, W_r$ produce W_0 provided $W_0, W_1, \ldots, W_r$ satisfy logical conditions $f_0(W_0), \ldots, f_r(W_r)$ which may contain free (but not bound) occurrences of the variables which occur in the W's. Here 'logical' means expressible by a formula of *PM*. If no further conditions were imposed this would mean for example, that the set of (numerals for) numbers satisfying any formula of *PM* would be a generated set. So he requires that wherever a substitution instance of a generalized scheme is applied to produce a new (closed) word, all the sentences $f_0(W_0), \ldots, f_r(W_r)$ can be proved in *PM*. He can then argue that a generalized canonical system can be reduced to a canonical one. I believe that he thought, at the time, that this argument provided powerful evidence for his thesis (or, as he would put it, his working hypothesis); the statement of it quoted above in Section 13.2 occurs immediately after the above argument.

(4) However, in the Appendix he writes:

> Establishing [the thesis] is not a matter of mathematical proof but of psychological analysis of the mental processes involved in combinatory mathematical processes. Because these seemed to be sufficiently simple to be exhaustively described, the writer gave up the direct use of <u>Principia</u> <u>Mathematica</u> as a partial verification of the characterization in question. (p. 415)

It must seem to us that 'psychological analysis' of the steps by which theorems are proved in any system of symbolic logic would lead, by an argument similar to – perhaps simpler than – Turing's, to a complete justification of Post's thesis. From the 1941 footnotes it is plain that Post did not think so, although it appears that he did think that Church and Turing had correctly characterized the notion of decision procedure. For example:

> The writer cannot overemphasize the fundamental importance to mathematics of the existence of absolutely unsolvable combinatory problems. ... The fundamental new thing is that for combinatory problems the given set of instruments is in effect the only humanly possible set. (footnote 1)

On the other hand, he writes:

> However should Turing's finite number of mental states bear up under adverse criticism, and an equally persuasive analysis be found for all humanly possible modes of symbolization, then the writer's position, while still tenable in an absolute sense, would become largely academic. (footnote 9)

I think the explanation of this apparently ambiguous attitude is that Post is making a sharp distinction between decision procedures and procedures for generating correct (not merely formally provable) statements. Thus a generated set would be something which would work creatively. In footnote 6 he writes:

> While the Turing simplifications ... may make the detailed development envisioned in the appendix unnecessary for the analysis of process ... it is doubtful if Turing considerations can replace such a development in the analysis of proof.

And in footnote 1 he writes:

> A fundamental problem is the question of the existence of absolutely undecidable propositions.

This was the problem he hoped one day to solve. It is a problem which most people today have stopped (alas?) thinking about.

13.5. *Post's 1936 Paper.* Post's formulation of finite combinatory processes differs only in unimportant ways from Turing's machines.

 (i) The work is done by a worker, not a machine.

 (ii) Only one symbol (besides blank) is used.

(iii) A particular case of a decision problem, or a particular argument for a function is entered (encoded) at a standard position on the tape.

(iv) The worker decides the case, or gives out the value, by halting with the encoded answer at the standard position.

 (v) Post also considers continuous working, with successive members of a generated set appearing successively on the tape.

Post does not analyze nor justify his formulation, nor does he indicate any chain of ideas leading to it. In footnote 4 of his *1965* he writes:

> The more detailed discussion of mandates of lowest type all but anticipated the writer's published note of 1936.

The work on mandates was done in 1924. He was led to introduce them when considering quantifiers; but I do not know what they were. Evidently in 1936 he knew of Church's work (to which he refers) and this led him to write up and publish his work. The importance he attached to psychological analysis and psychological fidelity leads one to suppose that he must have thought, however tentatively, along the same lines as Turing. More cannot be said.

14. Conclusions

14.1. Since the Babylonians, mathematicians have always been much concerned with calculation, and have sought for, and found, ways of making calculation easier and of reducing the solution of problems to a matter of computation. Leibniz dreamt not only of the invention of all-purpose calculating machines, but also of the reduction of all reasoning to calculation.

Babbage was the first to show how Leibniz's dream could be realized – how the analogical meaning of 'mechanical' could be translated into its concrete meaning. What he realized was that all calculations depend on the combination and ITERATION of simple operations, and that the iteration must be controlled by testing whether or not certain CONDITIONS are satisfied. Babbage's thesis (Section 2.2) asserts that any calculation which a mathematician may wish to make can be made, in principle, by a machine. Taken in its context it is heuristic rather than philosophical.

14.2. The philosophical importance of iteration as the source of the natural numbers, and of calculations with them, was realized by Dedekind. Iteration for Frege and Russel is, so to speak, hidden in the definition of the ancestral. Frege does not even define multiplication in the *Grundgesetze,* and both there and in *Principia Mathematica* addition is defined for (all) cardinals, not just for numbers.

14.3. During the nineteenth century the algebrization of many parts of mathematics and of logic became an industry, and led to much attention being given to algorithmic processes. Kronecker, who sought to algebrize analysis, believed in his more dogmatic moments that the only objects in mathematics were those which could be described in finite terms (without reference to infinite collections) and that only decidable predicates of them were fully meaningful. The algorithmic approach to logic reached a summit with the invention of decision procedures for first- and second-order monadic predicate calculus (Section 3.4).

14.4. Also in the nineteenth century the importance of axiomatic systems became clear. In a given branch of mathematics things are true not because of the essential nature of the objects studied, but because of conditions (axioms) which we, perhaps arbitrarily, impose on them. As Bertrand Russell once remarked: "Thus mathematics may be defined as the subject in which we never know what we are talking about nor what we are saying is true."[47] If not truth, by what other criteria can an axiomatic system be judged? For Hilbert, at least, there was only one answer: by its consistency. For his axioms for (various) geometries he had proved consistency by interpreting the axioms in the theory of the reals. But for the reals (and the natural numbers) this method could not be applied. So, starting with *1905*, he developed his

---

47 In "Mathematics and the Metaphysicians", reprinted in *Mysticism and Logic.*

'Beweistheorie'. The logical apparatus as well as the mathematical subject matter had to be formalized; formal proofs are then finite objects. Metamathematics is the study of these objects, and here Hilbert accepted Kronecker's restrictions to 'finitistic' methods. Operations on formal proofs (such as reduction to a normal form) are to be (in an intuitive sense of the word) recursive; relations between (parts of) proofs are to be decidable; the prime method of metamathematical proof is induction on the construction of formal proofs. Although (prior to 1931) Hilbert and his group did not code proofs by numbers, the connection with numerical calculation was fairly clear. By careful arrangement of the data, algebraic and symbolic operations can be reduced (as Babbage had realized) to numerical ones. A general theory of proofs would require the investigation and classification of effectively calculable processes.

14.5. In his eagerness to promote his Beweistheorie, Hilbert sought to use it to prove the consistency of the continuum hypothesis; in doing so he introduced a kind of hierarchy of (new sorts of) recursive functions. Some of the functions thus introduced were plainly calculable; but the extent of the hierarchy, and the extent of that part of it which includes only calculable functions, was not (and is not!) clear. He, and the group around him, probably believed – they certainly hoped – that the Entscheidungsproblem was decidable, and that its characteristic function would appear in the calculable part of the hierarchy. These opinions, I think, answer Question 1.3.

14.6. But the growing realization that a solution to the Entscheidungsproblem would yield a procedure for solving many (or even all) mathematical problems meant that some, in particular von Neumann, believed there could be no such solution; a belief which became almost a certainty when Gödel published his *1931*. To *demonstrate* this certainty it was essential to set precise limits to 'effectively calculable'. This (combined with a natural interest in mechanical calculation) is what was 'in the air' (Question 1.1) for Turing.

14.7. In the early 1920's Post, with *Principia Mathematica* as a model, developed a theory of formal systems, and realized that some of them would have unsolvable decision problems. But he did not publish his work until much later. Church's announcement (in 1935) stimulated him to publish his *1936*.

14.8. Church's type-free system, in which 'function' was the key concept, was designed to provide a foundation for logic and mathematics. Unlike its rivals it made iteration the starting point for arithmetic. Church did not, to begin with, realize that the logic free part of the system – the λ-calculus – was a thing of depth and power; it was Kleene's researches which revealed this. With hindsight one can say that it provides a true foundation for Babbage's key idea – conditional iteration – and because it is type-free it provides for a wider range of calculable functions than did Dedekind's 'definition by induction'. Church knew of the work on primitive and more general recursions, and of the importance of such functions in proof theory, and of the need to set limits on 'effective calculability'. This, and an appreciation of the power of the λ-calculus, was what was 'in the air' for him. With insight and boldness he proposed his thesis.

14.9. Soon after, Gödel, in his lectures at Princeton, working out a suggestion of Herbrand's, proposed the characterization of (what Kleene was to call) general recursive functions. The key idea here is that of computability in a formal system; it goes back to Hilbert's *1905,* and had been developed and exploited by Gödel. What is 'in the air' here is Hilbert's requirement (a consequence of consistency) that the numerical results of calculations in the 'meaningless', infinitary, formal systems (such as analysis) should also be true in the meaningful finitist sense. What was lacking was a thoroughly cogent general definition of 'formal system'.

14.10. All the work described in Sections 14.3–14.9 was based on the mathematical and logical (and *not* on the computational) experience of the time. What Turing did, by his analysis of the processes and limitations of calculation by human beings, was to clear away, with a single stroke of his broom, this dependence on contemporary experience, and produce a characterization, which – within clearly perceived limits – will stand for all time. That is the answer to Question 1.4.

14.11. What Turing also did was to show that calculation can be broken down into the iteration (controlled by a 'program') of extremely simple concrete operations; so concrete that they can be easily described in terms of (physical) mechanisms. (The operations of the λ-calculus are much more abstract). This is part of the answer to Question 1.2.

14.12. It is interesting to speculate on what would have happened if history had been different; if powerful all-purpose computers had been developed, while work on foundations and on logic had remained in a primitive state and had not soared into the blue sky. In time, without doubt, a general theory of computation – including undecidability theorems – would have developed. But almost a century separates the practical use of steam engines from Carnot's work and the beginning of thermodynamics. A not much shorter time was required from the first long distance telegraphic communication to the development of communication theory, permitting communication to be made more reliable by methods other than the brute force (and expensive) one of increasing the power. And even in the theory of computing there is a surprisingly long interval between the *use* of random access storage and the development (by Shepherdson and Sturgis in *1963*) of a smooth general *theory.*

14.13. In the other direction, from pure research to application, the traffic is often more rapid: only *half* a century from Maxwell's discovery of the displacement current to the beginning of wireless telegraphy. Because of the huge resources made available in the Second World War for the exploitation of scientific ideas, the interval from 1936 to the construction of stored-program digital computers was abnormally short. It is hard to gauge the extent of the influence of Turing's work, but influence there was. And as the design of computers and of programing languages became more sophisticated, so the influence of the work of Church, Kleene, and Turing grew: another part of the answer to Question 1.2.

14.14. Turing's use of evident human limitations, the discovery by McCulloch and Pitts that neuronal nets can act (as we should now say) as finite automata, and the rapid increase in the power of computers, have led to a widespread acceptance of computers as *the* model for the workings of the central nervous system and for intelligent behavior. But the gap between human and artificial intelligence is still immense. Gödel and Post believed that a satisfactory theory of mathematical intelligence must take account of nonfinitary and creative reasoning. Should we hope for another Turing to produce such a theory?

References

I. The papers of 1936:

Church, Alonzo
> 1936 An unsolvable problem of elementary number theory. *Am. J. Math.* **58** (1938) 345–363. Presented to the American Mathematical Society, April 19, 1935; abstract in *B. Am. Math. S.* **41** (May 1935).

> 1936a A note on the Entscheidungsproblem. *J. Symb. Log.* **1** (1936) 40–41; received April 15, 1936. Correction, ibid. (1936) 101–102; received August 13, 1936.

Kleene, Stephen C.
> 1936 General recursive functions of natural numbers. *Math. Annal.* **112** (1936) 727–742. Presented by title to the American Mathematical Society September 1935; abstract in *B. Am. Math. S.,* July 1935.

Post, Emil L.
> 1936 Finite combinatory processes – Formulation 1. *J. Symb. Log.* **1** (1936) 103–105; received October 7, 1936; presented to the American Mathematical Society January 1937; abstract in *B. Am. Math. S.,* November 1936.

Turing, Alan M.
> 1936-7 On computable numbers, with an application to the Entscheidungsproblem. *P. Lond. Math. Soc.* (2) **42** (1936-7) 230–265; received May 25, 1936, Appendix added August 28; read November 12, 1936; A correction, ibid., **43** (1937) 544–546. Turing's paper appeared in Part 2 of vol. 42 which was issued in December 1936.

Note: All of these are photographically reproduced in Davis *1965*.

II. Other works consulted or cited:

Ackermann, W.

    1928 Zum Hilbertschen Aufbau der reellen Zahlen. *Math. Annal.* **99** (1928) 118–133. Translation in van Heijenoort *1967.*

    1940 Zur Widerspruchsfreiheit der Zahlentheorie. *Math. Annal.* **117** (1940) 162–194.

Adams, R.G.

    1983 *A History of the Theory of Recursive Functions and Computability with Special Reference to the Developments Initiated by Gödel's Incompleteness Theorems.* Ph.D. thesis submitted to the Council for National Academic Awards (U.K.) from the Hatfield Polytechnic. (I have found this a useful and accurate work of reference.)

Anderson, A.R. (ed.)

    1964 *Minds and Machines.* Englewood Cliffs, NJ: Prentice-Hall Inc. (1964).

Babbage, C.

    1837 On the mathematical powers of the calculating engine. Unpublished manuscript, printed in Randell *1982.*

    1864 *Passages From the Life of a Philosopher.* London (1864). Selections reprinted in Morrison and Morrison *1961.*

Baker, G.P., and P.M.S Hacker

    1984 *Skepticism, Rules and Language.* Oxford: Basil Blackwell (1984).

    1985 *Wittgenstein: Rules, Grammar and Necessity,* vol. 2 of An Analytical Commentary on the Philosophical Investigations. Oxford: Basil Blackwell (1985).

Behmann, H.

    1922 Beiträge zur Algebra der Logic, insbesondere zum Entscheidungsproblem. *Math. Annal.* **86** (1922) 163–229.

Bernays, P.

    1930 Die Philosophie der Mathematik und die Hilbertsche Beweistheorie. *Blätter für Deutsche Philosophie* **4** (1930–31), 326–367. Reprinted in Bernays *1976.*

    1976 *Abhandlungen zur Philosophie der Mathematik.* Darmstadt: Wissenschaftliche Buchgesellschaft (1976).

Brouwer, L.E.J.

    1908 De onbetrouwbaarheid der logische principes. *Tijd. Wijsb.* **2** (1908) 152–158. Translated in Brouwer *1975.* pp. 107–111.

    1923 Über die Bedeutung des Satzes von ausgeschlossenen Dritten in der Mathematik, insbesondere in der Funktionstheorie. *J. rein. Math.* **154** (1923) 1–7. Reprinted in Brouwer *1975;* translated in van Heijenoort *1967,* pp. 335–342.

    1975 *Collected Works,* vol. 1, ed. A. Heyting. Amsterdam: North-Holland Publ. Co. (1975).

Burks, A.W.

  1980 From ENIAC to the stored-program computer: Two revolutions in computers. In: Metropolis et al. *1980,* pp. 311–344.

Bush, V.

  1936 Instrumental Analysis. *B. Am. Math. S.* **42** (1936) 649–669.

Church, A.

  1932 A set of postulates for the foundation of logic. *Ann. Math. (2)* **33** (1932) 346–366.

  1933 A set of postulates for the foundation of logic (second paper). *Ann. Math. (2)* **34** (1933) 839–864.

  1934 The Richard paradox. *Am. Math. Mo.* **41** (1934) 356–361.

  1935 A proof of freedom from contradiction. *P. Nat. Ac. Sci.* **21** (1935) 275–281.

  1936b Mathematical Logic. Mimeographed notes of lectures given at Princeton October 1935 – January 1936.

  1937 Reviews of Turing *1936-7* and Post *1936. J. Symb. Log.* **2** (1937) 42-43.

  1940 A formulation of the simple theory of types. *J. Symb. Log.* **5** (1940) 56–68.

Crossley, J.N. (ed.)

  1975 *Algebra and Logic. Papers from the 1974 Summer Research Institute of the Australian Mathematical Society, Monash University.* Lecture Notes in Mathematics 450. Berlin: Springer-Verlag (1975).

Curry, H.B.

  1929 An analysis of logical substitution. *Am. J. Math.* **51** (1929) 363–384.

  1930 Grundlagen der kombinatorischen Logic. *Am. J. Math.* **52** (1930) 789–836.

  1941 The paradox of Kleene and Rosser. *Trans. AMS* **50** (1941) 454–516.

Davis, M.

  1965 *The Undecidable: Basic Papers on Undecidable Propositions, Unsolvable Problems and Computable Functions,* ed. M. Davis. New York: Raven Press (1965).

  1982 Why Gödel didn't have Church's thesis. *Inf. Contr.* **54** (1982) 3–24.

  1988 Mathematical logic and the origin of modern computers. This volume; also in: *Studies in the History of Mathematics.* Washington, DC: Mathematical Association of America (1987).

Dawson, J.W.

  1984 Discussion on the foundation of mathematics. *Hist. Phil. Log.* **5** (1984) 111–129.

Dedekind, R.

  1888 *Was sind und was sollen die Zahlen?* Braunschweig (1888).

Dehn, M.

  1911 Über unendliche diskontinuierliche Gruppen. *Math. Annal.* **71** (1911) 116–144.

Eckert, J.P.

    1980 The ENIAC. In: Metropolis et al. *1980,* pp. 525–539.

Engelmann, P.

    1967 *Letters from Ludwig Wittgenstein with a Memoir.* Oxford: Basil Blackwell (1967).

Feferman, S.

    1984 Kurt Gödel: conviction and caution. *Phil. Nat.* **21** (1984) 546–562.

Gandy, R.O.

    1980 Church's thesis and principles for mechanisms. In: *The Kleene Symposium,* eds. J. Barwise, J.J. Keisler, and K. Kunen, pp. 123–145. Amsterdam: North-Holland Publ. Co. (1980).

Gödel, K.

    1931 Über formal unentscheidbare Sätze der *Principia Mathematica* und verwandter Systeme I. *Monats. Math. Ph.* **38** (1931) 173–198. Translated in van Heijenoort *1967,* pp. 596–616; reprinted, with same translation in Gödel *1986.*

    1934 On undecidable propositions of formal mathematical systems. Mimeogr. notes by S.C. Kleene and J.B. Rosser of lectures given at Princeton. First published with a Postscriptum in Davis *1965;* reprinted in Gödel *1986.*

    1936 Über die Länge von Beweisen. *Ergebnisse eines mathematische Kolloquiums* **7** (1936) 6. Translated in *Davis 1965;* reprinted and translated in Gödel *1986.*

    1946 Remarks before the Princeton bicentennial conference on problems in mathematics, 1946. First published in Davis *1965.*

    1986 *Collected Works,* vol. I, ed. S. Feferman et al. Oxford: Oxford University Press (1986).

Hardy, G.H.

    1929 Mathematical Proof. *Mind* **38** (1929) 1–25. Reprinted in G.H. Hardy's *Collected Works,* vol. VII.

Hartree, D.R.

    1949 *Calculating Instruments and Machines.* Urbana, IL: University of Illinois Press (1949).

Herbrand, J.

    1931a Unsigned note on Herbrand's thesis written by himself. Published in *Ann. U. Paris* **6** (1931) 186–189. Reprinted in Herbrand *1968;* translated in Herbrand *1971.*

    1931b Sur la non-contradiction de l'arithmétique. *J. rein. Math.* **166** (1931) 1–8. Reprinted in Herbrand *1968;* translated in van Heijenoort *1967* and Herbrand *1971.*

    1968 *Écrits Logiques,* ed. J. van Heijenoort. Paris: Presses Universitaires de France (1968).

    1971 *Logical Writings,* ed. W.D. Goldfarb. Dordrecht: Reidel Publ. Co. (1971).

Hilbert, D.

    1901 Mathematische Probleme. *Arch. Math. Physik (3)* **1** (1901) 44–63 and 213–237. Reprinted in *Hilbert 1935.*

    1905 Über die Grundlagen der Logik und der Arithmetik. In: *Verhandlungen des dritten internationalen Mathematiker-Kongresses,* Heidelberg vom 8. bis 13. August 1904, pp. 174–185. Leipzig: Teubner (1905). Reprinted in Hilbert *1930b;* translated in van Heijenoort *1967.*

    1918 Axiomatisches Denken. *Math. Annal.* **78** (1918) 405–415. Reprinted in Hilbert *1935.*

    1923 Die logischen Grundlagen der Mathematik. *Math. Annal.* **88** (1923) 151–165. Reprinted in Hilbert *1935.*

    1926 Über das Unendliche. *Math. Annal.* **95** (1926) 161–190. Translated in van Heijenoort *1967.*

    1930 Probleme der Grundlagen der Mathematik. *Math. Annal.* **102** (1930) 1–9. Reprinted in Hilbert *1930b.*

    1930a Naturerkennen und Logik. *Naturwissen.* **18** (1930) 959–963. Reprinted in *Hilbert 1935.*

    1930b *Grundlagen der Geometrie,* 7th edition. Leipzig: Teubner (1930).

    1935 *Gesammelte Abhandlungen,* vol. III. Berlin: Springer-Verlag (1935).

Hilbert, D., and W. Ackermann

    1928 *Grundzüge der theoretischen Logik.* Berlin: Springer-Verlag (1928).

Hilbert, D., and P. Bernays

    1934 *Grundlagen der Mathematik,* vol. I. Berlin: Springer-Verlag (1934).

    1939 *Grundlagen der Mathematic,* vol. II. Berlin: Springer-Verlag (1939).

Hodges, A.

    1983 *Alan Turing: The Enigma.* London: Burnett Books (1983).

Hyland, J.M.E.

    1979 Filter spaces and continuous functions. *Ann. Math. Log.* **16** (1979) 101–143.

Hyman, A.

    1984 *Charles Babbage: Pioneer of the Computer.* Oxford: Oxford University Press (1984).

Ketonen, J., and R.M. Solovay

    1981 Rapidly growing Ramsey functions. *Ann. Math. (2)* **113** (1981) 267–314.

Kleene, S.C.

    1935 A theory of positive integers in formal logic. *Am. J. Math.* **57** (1935) 153–173 and 219–244.

    1936a λ-definability and recursiveness. *Duke Math. J.* **2** (1936) 340–353. Abstract in *B. Am. Math. S.* **41,** July 1935.

1938 On notations for ordinal numbers. *J. Symb. Log.* **3** (1938) 150–155.

1943 Recursive predicates and quantifiers. *Trans. AMS* **53** (1943) 41–73. Reprinted in Davis *1965.*

1952 *Introduction to Metamathematics.* Amsterdam: North Holland Publ. Co. (1952).

1981 Origins of recursive function theory. *Ann. Hist. C.* **3** (1981) 52–67.

Kleene, S.C., and J.B. Rosser

1935 The inconsistency of certain formal logics. *Ann. Math.* *(2)* **36** (1935) 630–636. Abstract in *B. Am. Math. S.,* January 1935.

Kreisel, G.

1951 On the interpretation of non-finitist proofs I. *J. Symb. Log.* **16** (1951) 241–267.

1952 On the interpretation of non-finitist proofs II. Interpretation of number theory. Applications. *J. Symb. Log.* **17** (1952) 43–58.

1959 Interpretation of analysis by means of functionals of finite type. In: *Constructivity in Mathematics,* ed. A. Heyting, pp. 101–128. Amsterdam: North-Holland Publ. Co. (1959).

1965 Model theoretic invariants: applications to recursive and hyperarithmetic operations. In: *The Theory of Models,* eds. J.W. Addison, L. Henkin, and A. Tarski, pp. 190–205. Amsterdam: North-Holland Publ. Co. (1965).

Kripke, S.A.

1982 *Wittgenstein on Rules and Private Languages.* Oxford: Basil Blackwell (1982).

Countess of Lovelace, Augusta Ada

1843 Sketch of the Analytical Engine Invented by Charles Babbage, by L.F. Menabrea of Turin, Officer of the Military Engineers, with notes upon the Memoir by the Translator. *Taylor's Scientific Memoirs* **3** (1843). Reprinted in Morrison and Morrison *1961.*

Löwenheim, L.

1915 Über Möglichkeiten im Relativkalkül. *Math. Annal.* **76** (1915) 447–470. Translated in van Heijenoort *1967.*

Lucas, J.R.

1961 Minds, machines and Gödel. *Philosophy* **36** (1961) 112–127. Reprinted in *Anderson 1964.*

Mauchly, J.W.

1980 The ENIAC. In: Metropolis et al. *1980,* pp. 541–550.

McCulloch, W.S., and Pitts, W.

1943 A logical calculus of the ideas immanent in nervous activity. *B. Math. Biophy.* **5** (1943) 115–133.

Menabrea, L.F.

1842 Notions sur la machine analytique de M. Charles Babbage. *Bibliothèque Universelle de Genève* **82** (1842) 325–376. Translated in Lovelace *1843.*

Metropolis, N., J. Howlett, and G.-C. Rota (eds.)

    1980 *A History of Computing in the Twentieth Century.* New York: Academic Press (1980).

Morrison, P., and E. Morrison (eds.)

    1961 *Charles Babbage and his Calculating Engines: Selected Writings by Charles Babbage and Others.* New York: Dover Publ. Inc. (1961).

Newman, M.H.A.

    1948 General principles of the design of all-purpose computing machines. *P. Roy. Soc. (A)* **195** (1948) 271–274.

    1955 Alan Mathison Turing 1912–1954. *Biographical Memoirs of Fellows of the Royal Society* **1** (1955) 252–263.

Paris, J.B., and Harrington, L.

    1977 A mathematical incompleteness in Peano arithmetic. In: *Handbook of Mathematical Logic,* ed. J. Barwise, pp. 1133–1142. Amsterdam: North-Holland Publ. Co. (1977).

Péter, R. (or R. Politzer)

    1932 Rekursive Funktionen. In: *Verhandlung des internationalen Mathematiker-Kongresses Zürich 1932,* 336–337.

    1934 Über den Zusammenhang der verschiedenen Begriffe der rekursiven Funktionen. *Math. Annal.* **110** (1934) 612–632.

    1935 Konstruktion nichtrekursiver Funktionen. *Math. Annal.* **111** (1935) 42–60.

    1936 Über die mehrfache Rekursion. *Math. Annal.* **113** (1936) 489–527.

Post, E.L.

    1921 Introduction to a general theory of elementary propositions. *Am. J. Math.* **43** (1921) 163–185. Reprinted in van Heijenoort *1967.*

    1943 Formal reductions of the general combinatorial decision problem. *Am. J. Math.* **65** (1943) 197–215.

    1944 Recursively enumerable sets of positive integers and their decision problems. *B. Am. Math. S.* **50** (1944) 248–316. Reprinted in Davis *1965.*

    1965 Absolutely unsolvable problems and relatively undecidable propositions: Account of an anticipation. (Submitted for publication in 1941). Printed in Davis *1965,* pp. 340–433.

Ramsey, F.P.

    1928 On a problem of formal logic. *P. Lond. Math. Soc. (2)* **30** (1928) 338–384.

Randell, B.

    1972 On Alan Turing and the origins of digital computers. In: *Mach. Intell. 7,* eds. B. Melzer and D. Michie, pp. 3–20. Edinburgh: Edinburgh University Press (1972).

    1980 The COLOSSUS. In: Metropolis et al. *1980,* pp. 47–92.

1982 *The Origins of Digital Computers,* Selected Papers, ed. B. Randell, 3rd ed. Berlin: Springer-Verlag (1982).

Reidemeister, K. (ed.)

1971 *Hilbert: Gedenkband.* Berlin: Springer-Verlag (1971).

Rosser, J.B.

1935 A mathematical logic without free variables. *Ann. Math. (2)* **36** (1935) 127–150; *Duke Math. J.* **1** (1935) 328–355.

Schönfinkel, M.

1924 Über die Bausteine der mathematischen Logik. *Math. Annal.* **92** (1924) 305–316. Translated in van Heijenoort *1967.*

Schütte, K.

1933 Untersuchungen zum Entscheidungsproblem der mathematischen Logik. *Math. Annal.* **109** (1933) 572-603.

Schwichtenberg, H.

1976 Definierbare Funktionen im λ-Kalkül mit Typen. *Arch. Math. Log. Gr.* **17** (1976) 113-114.

Shepherdson, J.C.

1988 Mechanisms for computing over arbitrary structures. This volume.

Shepherdson, J.C., and H.E. Sturgis

1963 Computability of recursive functions. *J. As. Comp. Mach.* **10** (1963) 217–255.

Skolem, T.A.

1919 Untersuchungen über die Axiome des Klassenkalküls and über Produktions- und Summationsprobleme, welche gewisse Klassen von Aussagen betreffen. *Videnskaps Akademi (or–selkapets) i Kristiana (or Oslo) (Mat.-Naturv. Klasse) Skrifter No. 3* (1919) 37pp.

1923 Begründung der elementare Arithmetik durch die rekurrierende Denkweise ohne Anwendung scheinbarer Verändlichen mit unendlichem Ausdehnungsbereich. Ibid. No. 6 (1923) 38pp.

Smorynski, C.A.

1979 Some rapidly growing functions. *Math. Intell.* **2** (1979) 149–154.

1982 The varieties of arboreal experience. *Math. Intell.* **4** (1982) 182–189.

Stein, D.

1985 *Ada: A Life and a Legacy.* Cambridge, MA: The MIT Press (1985).

Thue, A.

1914 Probleme über Veränderungen von Zeichenreihen nach gegebenen Regeln. *Videnskaps-Akademi i Kristiana (Mat.-Naturv. Klasse) Skrifter No. 10* (1914) 34pp.

Turing, A.M.
  1937 Computability and λ-definability. *J. Symb. Log.* **2** (1937) 153–163.
  1939 Systems of logic based on ordinals. *P. Lond. Math. Soc. (2)* **45** (1939) 161–228.
  1945 Proposal for development in the mathematics division of an Automatic Computing Engine (ACE). Report E882, Executive Committee NPL. Re-issued in 1972 with a forward by D.W. Davies as NPL report, *Com. Sci.* **57**. Printed in *Turing 1986.*
  1947 The automatic computing engine. Lecture given to the London Mathematical Society on February 20, 1947. Printed in Turing *1986.*
  1950 Computing machinery and intelligence. *Mind* **59** (1950) 433–460. Reprinted in Anderson *1964.*
  1986 *A.M. Turing's ACE Report of 1946 and Other Papers,* ed. R.E. Carpenter and R.W. Doran. Cambridge, MA: The MIT Press; and Los Angeles: Tomash Publishers (1986).

van Heijenoort, J.
  1967 *From Frege to Gödel: A Source Book in Mathematical Logic 1879–1931.* Cambridge, MA: Harvard University Press (1967).
  1982 L'Oeuvre logique de Jacques Herbrand et son contexte historique. In: *Proceedings of the Herbrand Symposium: Logic Colloquium '81,* ed. J. Stern, pp. 57–85. Amsterdam: North-Holland Publ. Co. (1982).

von Neumann, J.
  1927 Zur Hilbertschen Beweistheorie. *Math. Z.* **26** (1927) 1-46. Reprinted in von Neumann *1961,* vol. I.
  1945 First Draft of a report on the EDVAC. Contract No. W-670-ORD-492, Moore School of Electrical Engineering, University of Pennsylvania. Excerpts printed in Randell *1982.*
  1951 The general and logical theory of automata. In: *Cerebral Mechanisms in Behaviour: the Hixon Symposium September 1948, Pasadena,* ed. L.A. Jeffress, pp. 1–31. New York: Wiley & Sons (1951). Reprinted in von Neumann *1961,* vol. V.
  1961 *Collected Works,* vols. I–VI., ed. A.H. Taub. Oxford: Pergamon Press (1961).

Wang, Hao
  1974 *From Mathematics to Philosophy.* London: Routledge & Kegan Paul (1974).
  1981 Some facts about Kurt Gödel. *J. Symb. Log.* **46** (1981) 653–659.

Wittgenstein, L.
  1976 *Wittgenstein's Lectures on the Foundations of Mathematics, Cambridge 1939,* ed. C. Diamond. Hassocks, Sussex: The Harvester Press (1976).

Turing in the Land of O(z)

Solomon Feferman[*]

After having published his fundamental and justly famous paper "On Computable Numbers, with an Application to the Entscheidungsproblem" (*1936-37*), followed in 1937 by a detailed proof of the equivalence between his notion of computability and Church's notion of λ-definability, Alan Turing produced a third quite distinctive and substantial paper in mathematical logic, entitled "Systems of Logic Based on Ordinals" (*1939*). There Turing took up quite systematically the natural idea of trying to overcome the Gödelian incompleteness of formal systems, by means of transfinite iteration of principles which serve to overcome incompleteness locally. In a brief introduction to his 1939 paper, Turing explains his aims very succinctly:

> The well-known theorem of Gödel *1931* shows that every system of logic is in a certain sense incomplete, but at the same time it indicates means whereby from a system L of logic a more complete system L' may be obtained. By repeating the process we get a sequence $L, L_1 = L', L_2 = L_1', \ldots$ each more complete than the preceding. A logic L_ω may then be constructed in which the provable theorems are the totality of theorems provable with the help of the logics $L, L_1, L_2, \ldots \ldots$ Proceeding in this way we can associate a system of logic with any constructive ordinal. It may be asked whether such a sequence of logics of this kind is complete in the sense that to any problem A there corresponds an ordinal α such that A is solvable by means of the logic L_α. I propose to investigate this question in a rather more general case, and to give some other examples of ways in which systems of logic may be associated with constructive ordinals.

The basic technical notion engaged to this end was that of "ordinal logic", and Turing's investigation of it was characteristically original and penetrating. Though he achieved some partial completeness results, these appealed to the use of certain external information about ordinal representations which subverted the basic aim.

---

[*] Guggenheim Fellow 1986–87. This paper was written during the first quarter of 1987 while the author was a visitor at the Forschungsinstitut für Mathematik of the E.T.H. in Zürich. The support of the Guggenheim Foundation and the generous assistance provided by the Forschungsinstitut are greatly appreciated.

Moreover, Turing proved that one could not have both completeness and invariance, another apparent desideratum for ordinal logics. Thus, despite the naturalness of the questions raised, and the great progress that Turing made on dealing with them in his single-handed effort, the end results were disappointing. It would be some twenty years before his work would be taken up by others as a point of departure for further advances.

The circumstances under which Turing wrote his ordinal logics paper are of historical interest and help explain the delay in full attention which it received. The basic idea for the work came from Alonzo Church, who proposed it as a topic for a Ph.D. thesis at Princeton University, where Turing spent the years 1936–1938 in his sole extended period away from England.

While it does not appear that Church contributed beyond this in any substantive way to Turing's thesis, it was probably at his behest that the entire development was carried out in the framework of Church's λ-calculus. From a technical point of view this may be considered retrogressive, as the Herbrand-Gödel-Kleene approach to notions of effectiveness through the theory of general and partial recursive functions was then already taking over as the preferred path of investigation and exposition. Thus Turing's work on ordinal logics in the symbolism of the λ-calculus looked more difficult to understand and use than it ought to have, if it had been couched in different terms.[1] Though Turing was to do further work in logic, intermittently with his war work in cryptanalysis and his post-war work on the design and construction of digital computers, he never returned to the subject of ordinal logics. There are, however, some suggestive connections between his thoughts in *1939* and his later visionary thoughts on machine intelligence.

My purpose in the following is primarily to tell, for a general audience, how Turing came to write his paper on ordinal logics and something about the nature of his achievements there. In a Technical Appendix, I review his work and subsequent developments in this subject in terms suitable for readers with a background in recursive function theory.

1.

The story of how Turing came to write his paper on ordinal logics is contained in Andrew Hodges' excellent biography, *Alan Turing, The Enigma* (*1983*, pp. 90–146). It is retold in the following in condensed form, drawing extensively on Hodges for the relevant biographical details, as well as Kleene *1981* and Feferman *1986* for the development of logic and recursion theory in this period.

---

1 The introduction to the reprinting of Turing *1939* in Davis *1965*, pp. 155–222 still calls it a "profound and difficult paper".

The story begins with Turing's major achievement, his work on computability, carried out in 1935–36 soon after he had become a fellow of King's College Cambridge at the age of 23. Earlier in 1935 Turing had attended a course on the Foundations of Mathematics given by the topologist M.H.A. Newman. Among other things, Newman explained Hilbert's problems concerning consistency, completeness, and decidability of various formal axiomatic systems, as well as Gödel's incompleteness results for sufficiently strong such systems. Turing had already been interested in mathematical logic but had been working primarily in other areas of mathematics, especially group theory. Newman's course served to focus his interests in logic; in particular, Turing became intrigued by the *Entscheidungsproblem* (decision problem) for the first order predicate (or functional) calculus, and this came to dominate his thought from the summer of 1935 on (Hodges *1983,* pp. 94–96). In grappling with this problem he was led to conclude that the solution must be negative; but in order to demonstrate that, he would have to give an exact mathematical analysis of the informal concept of *computability by a strictly mechanical process.* This Turing achieved by mid-April 1936, when he delivered a draft of his paper, "On Computable Numbers, with an Application to the Entscheidungsproblem", to Newman. At first Newman was skeptical of Turing's analysis, thinking that nothing so straightforward in its basic conception as the Turing machines could be used to answer this outstanding problem. However, he finally satisfied himself that Turing's notion did indeed provide the most general explanation of finite mechanical process, and he encouraged the publication of the 'Computable Numbers' paper (ibid., pp. 109–112).

Neither Newman nor Turing were then aware that the question of analyzing the notion of *effective calculability* had occupied the attention of Gödel, Herbrand, and especially Church since the early 1930's. This side of the story is well told by Kleene in his 1981 paper on the origins of recursive function theory.[2] Kleene was a Ph.D. student of Church at Princeton from 1931 to 1933 (along with Rosser). Church was promoting a universal system for logic and mathematics in the framework of the lambda (λ)-symbolism for defining functions. Intuitively, $\lambda x t[x]$ is supposed to denote a function f whose value for each y is the result $t[y]$ of substituting y for x in the term $t[x]$. Terms are built up from variables and constants by *application s(t)* without restriction; thus "self-application" $x(x)$ is permitted. Church set for Kleene the problem of developing the theory of positive integers in his formalism, using an identification of the integers with certain λ-terms. The initial steps were rather difficult (even the predecessor function posed a problem), but once the first hurdles were cleared, Kleene was able to show more and more number-theoretic functions definable by the conversion processes of λ-terms. But Church's original system was shown before long (by Kleene and Rosser in 1934) to be inconsistent, and attention

---

2 Cf. also Kleene's introductory notes to Gödel *1934* in Gödel's *Collected Works, Volume I,* pp. 338–345, as well as Davis *1982.*

was then narrowed to a demonstrably consistent subsystem, which came to be called the λ-*calculus.*[3] The consistency of this subsystem was established by Church and Rosser, with some inputs by Kleene. As it happened, Kleene had already achieved all of his previous definability results in this restricted calculus. It was clear from the Church-Rosser consistency property that every λ-definable function (in the sense given by convertibility in the λ-calculus) is effectively calculable; moreover, Kleene was able to show each example of an effectively calculable function which came to mind to be λ-definable. In view of Kleene's success in meeting all such challenges, Church was led to make the proposal, which has come to be known as *Church's Thesis,* that the λ-definable functions comprise all the effectively definable functions. In the words of Kleene: "When Church proposed this thesis, I sat down to disprove it by diagonalizing out of the class of the λ-definable functions. But quickly realizing that the diagonalization cannot be done effectively, I became overnight a supporter of the thesis" (Kleene *1981,* p. 59).

Gödel (coming from the University of Vienna) visited the Institute for Advanced Study in Princeton during the year 1933–34, and in the spring of 1934 he gave lectures on his incompleteness results. Notes for these lectures were taken by Kleene and Rosser, and, after corrections by Gödel, were circulated at the time. They were subsequently reproduced in Davis' collection (*1965*) of basic papers on the "undecidable" and computable functions, and more recently in Volume I of Gödel's *Collected Works.*[4] During the course of these lectures, Gödel presented a definition, based on a suggestion of Herbrand, of *general recursiveness* as an analysis of the most general concept of computability, using systems of equations. However, at that time Gödel regarded this identification only as a "heuristic principle".[5] During this same period, Church had conversations with Gödel in which he advanced his own proposal, but Gödel regarded this as "thoroughly unsatisfactory"; still, Church replied that "if [Gödel] would propose any definition of effective calculability which seemed even partially satisfactory [Church] would undertake to prove that it was included in lambda-definability" (Kleene *1981,* p. 59). It was only as a result of Turing's later work on mechanical computability that Gödel would accept this identification, albeit indirectly, through a chain of equivalences.

The first step in this chain was the demonstration by Church and Kleene that λ-definability for functions of positive integers is equivalent to general recursiveness. Thus when Church finally announced his Thesis in published form, in Church *1936,* he formulated the Thesis in terms of general recursiveness but bolstered it by the fact of this equivalence (see Davis *1982,* pp. 10–11). Church applied his Thesis in

---

3 There are many variants of the λ-calculus, such as the λ-*K* calculus, λ-*I* calculus, etc.; these will not be distinguished here. The classic presentation of this subject is of course in Church *1941*; an up-to-date and comprehensive exposition is provided by Barendregt *1984.*

4 These notes are designated as Gödel *1934* here and in Gödel *1986.*

5 Gödel *1934,* footnote 3 (p. 348 in Gödel *1986*).

the *1936* paper to demonstrate the effective unsolvability of various mathematical and logical problems, including the decision problem for sufficiently strong formal systems. A year later (15 April 1936, just when Turing was presenting the draft of *his* paper to Newman), Church submitted his *1936a* with its simple proof of the unsolvability of the decision problem for the first-order functional calculus of logic.

The dismaying news of Church's work reached Cambridge in May 1936. At first this seemed to pre-empt Turing's analysis of computability and his result on the *Entscheidungsproblem;* but Turing's definition of computability was sufficiently different from that of Church as to warrant separate publication. Thus Turing's paper was submitted after all, on 28 May 1936;[6] in August 1936 he tacked on an appendix sketching a proof of the equivalence between *his* notion of computability with that of λ-definability, thus establishing the second link in the chain of equivalences.

Independently of Turing, but with knowledge of Church's Thesis, Post also proposed in 1936 a definition of mechanical computability which was very close to that of Turing. From the conceptual point of view these are now generally regarded as providing the most convincing analyses of the informal notion of effective or mechanical computability. Since 1936 there have been further defined notions of effectiveness, all shown to be equivalent, but there is hardly any other which carries the same immediate conviction that the sought-for general definition is at hand. Indeed, Gödel later championed Turing's definition in his postscripts to his paper on incompleteness and the 1934 lectures (on the occasion of their reprinting). For example, in the latter he says that "... due to A.M. Turing's work, a precise and unquestionably adequate definition of the general concept of formal system can now be given. ... Turing's work gives an analysis of the concept of 'mechanical procedure' (alias 'algorithm' or 'computation procedure' or 'finite combinatorial procedure'). This concept is shown to be equivalent with that of a 'Turing machine.' "[7] On the other hand Gödel never remarked in print concerning his negative views about Church's Thesis as initially formulated.

Church himself agreed early on that Turing's definition was conceptually superior. In his 1937 review[8] of Turing's paper, after referring to the equivalence proof in the appendix, Church says: "As a matter of fact, there is involved here the equivalence of three different notions: computability by a Turing machine, general recursiveness in the sense of Herbrand-Gödel-Kleene, and λ-definability in the sense of Kleene and the present reviewer. Of these, the first has the advantage of making the identification with effectiveness in the ordinary (not explicitly defined) sense evident immediately – i.e., without the necessity of proving preliminary theorems.

---

6 And, according to Hodges *1983*, p. 113, Church himself ended up being the referee.

7 Davis *1965*, pp. 71-72. Gödel first stated such views publicly in his 1946 Princeton Bicentennial lecture reproduced in Davis *1965*, pp. 84–88 (also to appear in Gödel's *Collected Works, Volume II*).

8 *J. Symb. Log.* (2) (1937), pp. 42-43.

The second and third have the advantage of suitability for embodiment in a system of symbolic logic."

For the actual development of the (abstract) theory of computation, where one must build up a stock of particular functions and establish various closure conditions, both Church's and Turing's definition are equally awkward and unwieldy. In this respect, general recursiveness is superior, and Kleene's particular schemata for that (using primitive recursion and the minimum operator μ) are the most tractable. In Kleene's words: "I myself, perhaps unduly influenced by rather chilly receptions from audiences around 1933–35 to disquisitions on λ-definability, chose, after general recursiveness had appeared, to put my work in that format. ... I cannot complain about my audiences after 1935 ... ." Still, he defends λ-definability as having "... the remarkable feature that it is all contained in a very simple and almost inevitable formulation, arising in a natural connection with no prethought of the result" (Kleene *1981*, p. 62).

2.

Back to Turing in Cambridge, May 1936: he had been understandably let down by Church's scoop, but gratified that he had independently arrived at the same result concerning the *Entscheidungsproblem,* and that his own work on computability was recognized (if by no one else than Newman and Church) to have independent merit. What next? Newman recommended that he go study with Church in Princeton, and wrote Church to see if a visit and any sort of grant could be arranged. Turing also applied for a Procter fellowship, offered by Princeton. There were three of these altogether, one designated for Oxford, one for Cambridge, and one for the Collège de France. Turing failed to win the one for Cambridge, and no other money was forthcoming from Princeton. Still, he thought he could just manage on his fellowship funds from Kings College ($300 p.a.) and decided to spend the year 1936–37 in Princeton; he arrived there at the end of September 1936 (Hodges *1983*, pp. 112–117).

Turing was well aware of the growing importance of the Princeton Mathematics Department as a world center for mathematics. It had already been a leader on the American scene when the department was greatly enriched, in the early 1930's, by the establishment of the Institute for Advanced Study (IAS). The two shared the same facilities (Fine Hall) until 1940, so that the lines between them were blurred and there was a great deal of interaction. The IAS had been built up by Oswald Veblen, with Einstein as one of its first members. It benefited particularly as a result of the exodus of mathematicians and physicists from Germany (and especially from the great Mathematical Institute in Göttingen) following the advent of Nazism. Among the mathematical luminaries that Turing found on his arrival in

Princeton were Einstein, von Neumann, Weyl, Courant, Hardy, and Lefschetz. (Of these, the first three were members of the IAS, Courant and Hardy were visitors, and Lefschetz was in the Princeton department.) Turing wrote home saying that "the mathematics department here comes fully up to expectations". But in logic, he had hoped to find – besides Church – Gödel, Bernays, Kleene and Rosser ". . . who were here last year [but] have left" (ibid., p. 117). Turing was particularly disappointed to have missed Gödel. As already mentioned, Gödel had visited the IAS in 1933–34; he commenced a second visit in the Fall of 1935 but left after a brief period due to illness. He was to visit the IAS once more, in 1939, before becoming a member in 1940; apparently Gödel and Turing never did meet.[9] Bernays, who *had* visited in 1935–36, was noted as Hilbert's collaborator on *Beweistheorie,* the proposed means for carrying out Hilbert's foundational program, while Kleene and Rosser (who were, previously, Ph.D. students of Church) had left to take up positions elsewhere. Others whom Turing might have looked to were von Neumann and Weyl; both had taken an active interest in logic in earlier years (of which, see below), but were totally engaged in mainstream mathematics at the time of Turing's arrival on the scene.

Thus Turing was left to draw on Church for direction in the further pursuit of his studies. He dutifully attended Church's lectures "which were rather on the ponderous and laborious side" and took notes on Church's theory of types. Turing also met with him in person from time to time, but Church was ". . . a retiring man himself, not given to a great deal of discussion." Turing was himself on the shy side and, indeed, he was described as a "confirmed solitary". The brief characterizations of Church's style and personality, from Hodges (p. 119), are fair enough, as far as they go. But it should be added that Church was (and is) noted for the great care and precision of his writing and lecturing, and these virtues probably benefited Turing – whose own writing was rough-and-ready and prone to minor errors.

At Church's suggestion, Turing gave a lecture to the Mathematics Club on his work on computability, but the turnout was disappointing. Then in January 1937, when Turing's paper on computability appeared in print, he received only two requests for reprints, one from England and one from Germany. So, for the time being, his work did not attract much attention.

In the period January-April 1937, Turing worked on three relatively brief papers. One (eventually to become *1937*) was to give more detail about his proof of the equivalence of his notion of computability with that of λ-definability. The other two papers were on group theory; and one of those solved negatively a problem concerning possible approximations of continuous groups, which proved to be of interest to von Neumann. During this period, Dean Eisenhart of the Princeton Mathematics De-

---

9 See Feferman *1986*, pp. 8 and 11. Turing was also never to meet Kleene (according to Kleene *1981*) nor, apparently, either Bernays or Rosser.

partment urged Turing to stay on for a second year and to apply again for a Procter fellowship (worth $2000 p.a.). He was supported in this by von Neumann, whose letter of recommendation praised Turing for his work on almost periodic functions and continuous groups but made no mention whatever of his work on computability. This time, Turing succeeded in obtaining the fellowship and, having failed to get a position in Cambridge for which he had also applied, decided to stay in Princeton for an extra year and do Ph.D. work under Church. For the dissertation, it was proposed that Turing take up an idea that had been broached in Church's course, relating to Gödel's incompleteness theorems; presumably this was the kernel of Turing's work on ordinal logics.

3.

Turing returned to England for the summer of 1937 and devoted himself to three projects during that period. The first was to complete his paper *1937* on computability and λ-definability. The second was to pursue Church's idea for his Ph.D. thesis, looking to ordinal logics as a way to "escape" Gödel's incompleteness theorems. His third project was in a part of mathematics that seemed to have nothing to do with logic, namely the field of analytic number theory. Here the problem that Turing took up concerned the famous Riemann Hypothesis (R.H.) about the zeros of the Riemann zeta-function, and certain of its consequences. As his point of departure, Turing took the work of the Cambridge mathematician Skewes, who had shown in 1933 that if the R.H. is true then $n/\log n < \pi(n)$ (the number of primes $\leq n$) for some $n \leq 10^{10^{10^{34}}}$, though massive numerical work had previously suggested that $n/\log n$ overestimates $\pi(n)$ for all n. Turing hoped either to lower Skewes' enormous bound or eliminate R.H. as a hypothesis. Though in the end he thought he had achieved both, and wrote a draft of his work, Turing never published the resulting paper.[10] Turing's interest in the R.H. was to surface repeatedly in the following year. As we shall see, it was to provide an example of a mathematical statement of simple logical form in his paper-to-be on ordinal logics. Moreover, during his year (1937–38) back in Princeton, Turing had ideas for the design of an "analogue" machine for calculating the zeros of the Riemann zeta-function, similar to one used in Liverpool for calculating the tides. And much later (1953) he was to publish a paper entitled "Some Calculations of the Riemann Zeta-function", which

---

10 According to a communication from Robin Gandy, the paper circulated after Turing's death and a number of errors were found. However, these errors were eventually all corrected by Cohen and Mayhew (cf. ftn. 11).

provided a concrete computational method to check that the R.H. holds up to a certain point.[11]

Turing did make good progress on his thesis topic in 1937 and even hoped to finish it by Christmas of that year. In this work he developed Church's suggestion of ordinal logics in a systematic way. The aim was to overcome the incompleteness phenomena discovered by Gödel: with each sufficiently correct effectively generated formal axiomatic system (or "logic") L is associated a true but unprovable statement A_L of the (Π_1^0) form $\forall x R(x)$, expressing that a certain (primitive) recursive property R holds for all integers x. If we start with an initially given system L_1 whose theorems concerning integers are all correct, and adjoin A_{L_1} (call it A_1) to form L_2 ($= L_1 \cup \{A_1\}$), then Gödel's incompleteness result applies again to L_2, so that associated with L_2 is the true but unprovable A_{L_2} (call it A_2). We can iterate this construction arbitrarily often, to form $L_n = L_1 \cup \{A_1, \ldots, A_{n-1}\}$. But $L_\omega = L_1 \cup \{A_1, \ldots, A_n, \ldots\}$ is still effectively generated (and correct), so we must proceed further into the transfinite in order to overcome incompleteness. In order to maintain effective generation, one will pass to a transfinite limit ordinal α and system L_α only when α is the limit of an effectively presented sequence $\alpha_1, \alpha_2, \ldots, \alpha_n, \ldots$ and the systems L_{α_n} are already obtained. In order to make this precise one has to deal with a system of effective representations of ordinals in the integers. Such a system O of *constructive notations a for ordinals* α had been developed by Church and Kleene in 1935 (published as their *1937*) in the framework of the λ-calculus. The idea for Turing's thesis was to investigate the construction and degree of completeness of sequences $\Lambda = < L_a \mid a \in O >$ of logics associated with constructive notations a for ordinals, which would increase in strength as the ordinal of a increased, thus overcoming incompleteness – at least locally. The main question was whether one could thereby overcome incompleteness globally. A secondary question was whether such a logic would be *invariant,* i.e., whether the extent of $\Lambda(a)$ (i.e., its set of theorems) could depend only on the classical ordinal α associated with a; this question must be considered since there are in general many notations a for the same α. Turing considered several natural ways in which ordinal logics could be constructed: (i) Λ_P, obtained by successively adjoining statements directly overcoming Gödel incompleteness at each stage a; (ii) Λ_H, a form of transfinite type theory; and (iii) Λ_G (after Gentzen), obtained by adjoining principles of transfinite induction and recursion up to a at each level $\Lambda_G(a)$. Turing's main results were that: (1) Λ_P is complete for true Π_1^0 statements, and (2) under quite general conditions, an

---

11 See Hodges *1983,* pp. 133–142, and note 3.40, p. 548, for Turing's work on the Skewes number; according to the latter, Turing's method was later used by Cohen and Mayhew in 1968 to reduce the Skewes number to $10^{10^{529.7}}$, but the number had already been reduced in 1966 to 1.65×10^{1165} by Lehman, using another method. For Turing's later work on calculating the zeros of the Riemann zeta function see Hodges *1983,* note 7.19, p. 562.

ordinal logic Λ can't be both invariant and complete (even for Π_1^0 statements). Thus, e.g., Λ_P, being Π_1^0 complete, is not invariant, while Λ_H is necessarily incomplete since it is invariant. Turing had hoped to strengthen (1) to a completeness result for true Π_2^0 sentences, i.e, statements of the form $\forall x \exists y R(x, y)$ with R primitive recursive. This class formally includes various statements of mathematical interest, such as the Riemann Hypothesis (by Turing's analysis of the problem). However, he was unable to achieve such an improved completeness result.[12] Still, his partial completeness result (1) could have been regarded as meeting the initial aim of "overcoming" the incompleteness phenomena discovered by Gödel, since these only concerned true but unprovable Π_1^0 statements. Even so, Turing was rightly dissatisfied with this partial completeness result, since it shifted the problem of settling the truth of Π_1^0 statements by axiomatic means to that of recognizing whether what appears to be a notation a for a constructive ordinal actually is one – and that problem is at least as complicated as determining which Π_1^0 statements are true. (For, if a is formally given as a limit of a sequence $a(n)$, $n = 1, 2, 3, \ldots$, we have $a \in O$ if and only if $\forall x$ [$a(x) \in O$ and $a(x)$ represents an increasing sequence], with 'x' ranging over integers. Even the question whether a represents ω is already at least as complicated as the most general Π_1^0 problem. And, as would be shown later by Kleene *1955*, the problem, given any a whether $a \in O$, is more complicated than any arithmetical problem, even using an unlimited number of numerical quantifiers.) Turing was also rightly disappointed with his general incompleteness result (2) for invariant logics. However, he had succeeded in a relatively brief time in making remarkable progress on the topic proposed by Church and had laid the ground for all future investigations in this direction.

The entire thesis was couched in the formalism of the λ-calculus, very likely at Church's behest, but this was no obstacle for Turing, who had already proved himself to be quite adept at working with the system. Unfortunately, understanding of the mechanics of the λ-calculus was limited to a rather small audience, and this would in turn affect the reception of Turing's thesis work (as was the case with Kleene earlier, see above). Kleene had already shifted the development of the theory of constructive ordinal notations to the more perspicuous framework of general recursiveness in his paper *1938* first reported in 1936. But Turing did not evidence familiarity with this alternative approach, and even if he were familiar with it, Church may not have allowed him any choice in the matter.[13]

---

12 In fact, in my subsequent work (*1962*) to be described briefly in the Technical Appendix below, I showed that Λ_P is incomplete for Π_2^0 sentences.

13 Concerning this, Robin Gandy has commented in a personal communication that "even without Church's influence I think Turing might well have preferred [the system of notations based on] the λ-calculus to Kleene's O. He enjoyed devising the appropriate λ-formulae ... [though] incidentally, some of the λ-formulae in his 'Ordinal Logics' paper are wildly wrong."

Turing's hopes to finish the thesis by Christmas of 1937 were not to be realized: "Church made a number of suggestions which resulted in the thesis being expanded to an appalling length"[14]. Of course, Church would not knowingly tolerate rough or imprecise formulations and proofs, let alone mistakes – and the published version (*1939*) of the thesis show that Turing went far to meet such demands while still putting his own characteristic stamp on the exposition. The thesis was finally submitted in May 1938 and an oral examination on his work was held at the end of that month. The committee consisted of Church, Lefschetz, and Bohnenblust, and Turing's performance was noted as being excellent. The Ph.D. degree itself was granted in June 1938. But subsequently "he made little use of the [doctoral] title, which had no application at Cambridge, and which elsewhere was liable to prompt people to retail their ailments" (Hodges *1983*, p. 145). Moreover, beyond publishing his thesis in 1939, Turing was to pursue the topic of ordinal logics no further. Perhaps he felt that it led up a blind alley, or he simply could not see what the next steps would be. In any case he had plenty of other problems and ideas to occupy his attention in the following year.

4.

The only ones in Princeton, besides Church, who were really in a position to appreciate Turing's accomplishments in logic were von Neumann and Weyl. Both were noted for their wide-ranging and deep contributions to many fields of mathematics, and as it happens both had previously done some work in logic. In Weyl's case, this had been (in a sense) a one-shot affair, with his monograph *Das Kontinuum* (*1918*), which gave predicative foundations for classical analysis. But he continued to maintain an interest in the philosophy of mathematics and relevant foundational results, and he wrote extensively on these subjects throughout his life.[15] On the other hand, in the 1920's von Neumann had made several contributions to proof theory à la Hilbert and also to the axiomatization of set theory (from which Bernays was to derive *his* theory of sets and classes). Moreover, von Neumann was one of the first to appreciate the significance of Gödel's incompleteness results. In fact, it is reported that he obtained the second incompleteness theorem – on the unprovability of the consistency statement of a consistent theory in that theory – independently of Gödel, once he had learned of Gödel's first incompleteness theorem. Von Neumann was also noted for his brilliance, phenomenal speed and memory, and his extensive knowledge of the mathematical literature. Yet he said that after 1931 he never read

---

14 Turing, quoted in Hodges *1983*, p. 145.
15 Weyl's contributions to logic and his philosophical views are discussed in my paper *1987*.

another paper in logic.[16] Even if that were the case, it is a bit surprising that while von Neumann had recognized Turing's mathematical talents outside of logic he had shown no knowledge or appreciation of Turing's work in logic when he wrote on Turing's behalf in 1937. Still, von Neumann thought sufficiently highly of Turing to offer him in May 1938 a position as his assistant at the Institute for Advanced Study, at $1500 a year. Both materially and scientifically this should have been very attractive; it could have been the start of an academic career for Turing in America. But he turned it down in favor of returning to King's College, where his fellowship was renewed. His period in America, particularly in Princeton, had not been particularly happy, and his decision to resume a position at King's meant that he could return home, which he did in July 1938.

In Hodges' engaging metaphor, Turing was Dorothy, von Neumann the Wizard, and Princeton the Emerald City in the *Wizard of Oz*. But: "His departure from the land of Oz was rather different from that in the fable. The Wizard was not a phoney, and had asked him to stay. While Dorothy had disposed of the Wicked Witch of the West, in his case it was the other way around. Though Princeton was fairly secluded from the orthodox, Teutonic side of America, it shared in a kind of conformity that made him ill at ease. ... In one way, however, he resembled Dorothy. For all the time, there was something that he could do, and which was just waiting for the opportunity to emerge."[17]

5.

Before turning, in conclusion, to a discussion of the significance of Turing's work on ordinal logics, I want to examine a number of side aspects of a semi-technical character. For, outside of its main line of development, his paper contains several interesting digressions and original observations; there are also some curious failures on Turing's part to observe the obvious. We shall review these aspects here, more or less following their order of appearance in the 1939 paper (to which I refer by section number); the more technical points have been freely recast in modern recursion-theoretic terms.

---

16 For the relation of von Neumann to Gödel see Feferman *1986*, pp. 6 and 12. For von Neumann's statement about his subsequent divorce from logic see Hodges *1983*, note 3.6, p. 546.

17 For the relation between Turing and von Neumann see Hodges *1983*, pp. 144–146; the quote itself is from pp. 145–146. Beginning in 1939, von Neumann was to recognize the fundamental importance of Turing's work on computability (op. cit., p. 145). For his further appreciation of it in connection with his own eventual contributions to the design and development of general purpose electronic digital computers see op. cit., pp. 304 and 388.

I have already mentioned that Turing shows (§ 3) the Riemann Hypothesis to be equivalent to a Π_2^0 statement $\forall x \, \exists y R(x, y)$ with R (primitive) recursive.[18] In also discussing types of mathematical propositions which can be put in this form, he further shows that statements of the form $\forall x[F(x) = 0]$ with F general recursive are Π_2^0. His argument uses Kleene's normal form theorem for general recursive functions, which represents each such F in the form $F(x) = U(\mu y T(e, x, y))$ where U is a primitive recursive function, T is a 3-placed primitive recursive relation, e is the Gödel number of a system of recursive equations E which defines F (or alternatively of a Turing machine M_e which computes F), and 'μy' abbreviates the 'least y such that'. Basically, Turing's argument is that $F(x) = 0 \Leftrightarrow \exists y[T(e, x, y) \wedge \forall z < y \neg T(e, x, z) \wedge U(y) = 0]$, thus putting $\forall x[F(x) = 0]$ in Π_2^0 form. But he might have observed that, as F is total, we also have $F(x) = 0 \Leftrightarrow \forall y[T(e, x, y) \wedge \forall z < y \neg T(e, x, z) \Rightarrow U(y) = 0]$, and this shows that $\forall x[F(x) = 0]$ is in Π_1^0 form.

What is more on the mark is Turing's next assertion that each statement: "machine M is circle-free", is Π_2^0, and conversely, that every Π_2^0 statement is equivalent to one of this form. By definition (in his *1936–37*), M is *circle-free* if it computes a total function from the positive integers into the set consisting of 0, 1 (thus representing a real number in binary form). But by composing any partial function F with the sign function $sg(x+1) = 1$, $sg(0) = 0$, we see that F is total if and only if $\lambda x.sg(F(x))$ is total. Hence the class of statements of the form M_e computes a total function (with no restriction on values), ranges through Π_2^0. Turing *could* have moved immediately to this conclusion from his use of Kleene's normal form theorem simply by observing that M_e computes a total function if and only if $\forall x \exists y \, T(e, x, y)$.

Next, the brief section § 4 contains a striking new idea put to a curious use. The aim here is to produce a problem which is not Π_2^0. This is trivial by a cardinality argument, but instead, Turing introduces a new notion (which is to change the face of recursion theory) namely, that of computability relative to an *oracle*. He begins by saying: "Let us suppose that we are supplied with some unspecified means of solving number-theoretic [Π_2^0] problems; a kind of oracle as it were. ... With the help of the oracle we could form a new kind of machine (call them *o*-machines), having as one of its fundamental processes that of solving a given number-theoretic problem." He then shows more specifically how to define computability by an *o*-machine and, by a direct extension of his argument in *1936–37*, that the problem of determining whether an *o*-machine is *o*-circle free is not solvable by an *o*-machine and hence not by the oracle *o* itself.

---

18 In the view of Robin Gandy, with which I certainly concur, this result of Turing's on the arithmetical classification of the R.H. was quite striking for the time (even if it may not have struck anyone *then*). Actually, the R.H. can be expressed in Π_1^0 form, as was subsequently shown by Kreisel; a simple explicit Π_1^0 representation is given in the paper Davis, Matijasevič, and Robinson *1976*, p. 335.

Turing did nothing further with the idea of o-machines, either in this paper or afterward. But Post *1944* took it as his basic notion for a theory of *degrees of unsolvability* and properly credited Turing with the result that for any problem (about integers) there is another of higher degree of unsolvability; and eventually, the idea of transforming computability from an *absolute* notion into a *relative* notion would serve to open up the entire subject of generalized recursion theory.[19]

Later (in § 9) after Turing defines ordinal logics $\Lambda =< L_a | a \in O >$, he takes as one of his main aims that of establishing completeness with respect to Π_2^0 propositions, i.e., of showing that if A is Π_2^0 and true then L_a proves A for some $a \in O$. Now, having already pointed out (§ 7) that O is not computable (to say the least, as we know), Turing asserts: "We might hope to obtain some intellectually satisfying system of logical inference [for deriving Π_2^0 statements] with some ordinal logic. Gödel's theorem shows that such a system cannot be wholly mechanical; but with a complete ordinal logic we should be able to confine the nonmechanical steps entirely to verifications that particular formulae [of the λ-calculus] are ordinal formulae." (Here Turing prefigures his later extended discussion of "the purpose of ordinal logics" which we take up below.) This statement directly follows a brief discussion as to what problems could be solved if we had an oracle for telling us, given a, whether or not $a \in O$. But he does not put these together to analyze the logical complexity of $\exists a\ [a \in O \land L_a \vdash A]$ relative to such an oracle o (simply that it is Σ_1^0 relative to O).

Immediately after the quoted expression of hope, Turing says: "We might also expect to obtain an interesting classification of number-theoretic [Π_2^0] theorems according to 'depth'. A theorem which required an ordinal α to prove it would be deeper than one which could be proved by the use of an ordinal β less than α." He goes on to say that "however, this presupposes more than is justified", and carries the idea no further. But here Turing anticipated, at least programmatically, the classification by ordinals of the provably (total) recursive functions of various formal systems, obtained later by proof-theoretical work. This has recently been carried over to the classification by depth (or logical strength) of Π_2^0 statements emerging from combinatorial mathematics.[20]

The core of Turing's paper, containing his incompleteness and partial completeness results is in § 9; this will be analyzed in the Technical Appendix below. Directly following, in § 10, there is another digression, this time concerning "constructive" analogues of Cantor's continuum hypothesis, the set-theoretical formulation of which

---

19 Two of the many directions in generalized recursion theory are treated in Barwise *1975* and Fenstad *1980*.

20 For an introduction to the ordinal classification of the provably recursive functions of various subsystems of analysis see Feferman *1977*. For the logical strength of certain Π_2^0 combinatorial statements see Paris and Harrington *1977*, and the survey paper Simpson *1986*.

is that there is a 1-1 correspondence between the set $P(\omega)$ of all subsets of ω (or equivalently of all sequences of 0's and 1's) and the set of all ordinals less than the least uncountable ordinal ω_1. Here, for the constructive analogue of $P(\omega)$, Turing takes the set of all computable sequences of 0's and 1's (or the description numbers of machines which compute these sequences), and for ω_1 he substitutes ω_1^{CK}, the least ordinal not constructibly countable in the sense of Church-Kleene, i.e., the least ordinal not represented by a notation in O. Then he asks whether it is possible to set up a computable one-one correspondence between these sets; more precisely, to find a (partial) recursive function F such that for each $a \in O$ and each n, $F(a, n)$ is 0 or 1 and such that $\forall n[F(a, n) = F(a', n)]$ if and only if $|a| = |a'|$ (i.e. a, a' represent the same ordinal). The answer, as Turing shows, is negative; the proof, which is not difficult, transfers a technique that he had applied in §9 to establish the incompleteness of invariant ordinal logics (see the Appendix below for more details).

As Turing points out, there is "... great ambiguity concerning what the constructive analogue of the continuum hypothesis should be", and he addresses only one possible formulation of it. He says that the suggestion for this came indirectly from F. Bernstein and that a related problem was suggested by Bernays. But Turing might also have been inspired by Hilbert's 1926 paper, "On the Infinite" (to which he refers in a different connection), where Hilbert attempted to establish an ordinal-recursive classification of integer functions in his abortive "solution" of the continuum problem.

In any case, here again Turing anticipates later work, on the classification of recursive functions by means of hierarchies. It is customary in that work to consider some simple relative computability relation $f \leq g$ between functions, such as that f is primitive recursive in g (or even weaker), and to seek assignments $f_a = \lambda n F(a, n)$ to each $a \in O$ with the property that f_a increases with the ordinal of a. Here the invariance required by Turing for his version of the continuum hypothesis is weakened to $|a| = |a'| \Rightarrow f_a \equiv f_{a'}$ (i.e, $f_a \leq f_{a'} \leq f_a$). It turns out that the general theory of such classifications, with both incompleteness and completeness results, runs entirely parallel to the theory of ordinal logics, as I showed in my paper *1962a*; more details about this are given in the Technical Appendix below.

6.

In conclusion, we turn to Turing's discussion of the purpose of ordinal logics (§ 11), which takes up little more than two pages. Here it is best to quote directly: "Mathematical reasoning may be regarded rather schematically as the exercise of a combination of two faculties, which we may call *intuition* and *ingenuity*." (There is a nice footnote to this sentence: "We are leaving out of account that most important faculty which distinguishes topics of interest from others ...") He goes on to explain that "intuition consists in making spontaneous judgements which are not the result of conscious trains of reasoning. These judgements are often but by no means invariably correct ... The exercise of ingenuity in mathematics consists in aiding the intuition through suitable arrangements of propositions. ... When these are really well arranged the validity of the intuitive steps which are required cannot seriously be doubted." In Turing's view, there is no sharp, objective line between these two "faculties" – the parts they play differ "from occasion to occasion, and from mathematician to mathematician". Formal logic helps remove this "arbitrariness"; the formal rules are supposed to be chosen so that inferences are always intuitively valid. Moreover, the exercise of ingenuity is then given more definite shape in the search for admissible chains of inference to make up a proof. "In pre-Gödel times it was thought by some that it would ... be possible to carry this programme to such a point that ... the necessity for intuition would then be entirely eliminated." But Gödel's incompleteness theorems have shown this is impossible, and one turns naturally instead to "nonconstructive" systems of logic in which "not all the steps in a proof are mechanical, some being intuitive". Ordinal logics provide examples of such: "When we have an ordinal logic, we are in a position to prove number-theoretic theorems by the intuitive steps of recognizing formulae [of the λ-calculus] as ordinal formulae [representing well-orderings]. ... We want it to show quite clearly when a step makes use of intuition and when it is purely formal. The strain put on the intuition should be a minimum. Most important of all, it must be beyond all reasonable doubt that the logic leads to correct results whenever the intuitive steps [i.e., recognition of notations for ordinals] are correct." Turing concludes this discussion by considering his ordinal logics Λ_P and Λ_H with respect to this last criterion, and after putting them in question, moves on at the end of his paper to consider a new type of ordinal logic Λ_G. His judgements about these particular ordinal logics do not concern us here; rather we shall examine his general conception about the whole enterprise.

In modern logical discussions, ingenuity in its normal sense is put outside the purview of the subject, just as Turing set the question of the interest of results aside, though for working mathematicians both of these are critical to what counts as "good" mathematics. Logic simply hopes to answer the question: "What counts as mathematics?", while disregarding all questions of value. In "pre-Gödel times", it thought to find the answer in that which can be formalized, or, more fully, that which can be formalized in an axiomatic system which is justified on the grounds of some basic mathematical conception. However, even in those times, the question of *which* formal systems were thus justified received no common answer. And even if it had turned out that one could find a complete formal system for mathematics, the question of justification would still remain – though no doubt it would have seemed less compelling. In any case, Gödel's incompleteness theorem brought to center stage the question of what leads one to accept a formal system for the (necessarily partial) representation of mathematics. It is here that "intuition" would have to play its role, both in judging the acceptability of any proposed systems and in searching for new such systems. However, there was one aspect of the incompleteness theorem that suggested how the day might be saved without having to over-exercise the intuition: Gödel's examples of formally undecidable propositions can be decided directly by informal considerations. If a statement ϕ which "says" of itself that it is underivable in L is shown to be underivable in L, then it is evidently true and ought to be added to L; it is just that somehow such ϕ were "overlooked", so that the expanded system ought to be deemed acceptable if L is. Thus, if the passage from L to L' is obtained simply by adjoining such evidently correct statements to L, the acceptability of L' follows directly from that L. Also, if each of a sequence $L_{a_1} \subset L_{a_2} \subset \ldots \subset L_{a_n} \subset \ldots$ of increasing systems is recognized to be acceptable then $\cup_n L_{a_n}$ is evidently acceptable too. Hence, if we start with a basic acceptable system L_1, it seems that *all* we have to do to overcome Gödel's incompleteness is to iterate the passage L to L' transfinitely, as in ordinal logics, to obtain the collection of systems $< L_a \mid a \in O >$. But then the whole question of which formal systems ought to be admitted under this process shifts to the question: for which representations a of ordinals is it justified to accept L_a? Here there is an implicit use of the principle of transfinite induction on the set $\{b \mid b \leq_O a\}$ of notations up to and including a, applied to the informal predicate "L_b is acceptable", i.e., "L_b is correct according to a basic mathematical conception".

Thus the demand on "intuition" in recognizing "which formulae are ordinal formulae" is somewhat greater than Turing suggests. But even in his own terms, there is a failure to test his analysis of purpose against reality. Is it a "spontaneous judgement" without *any* "conscious train of reasoning" that leads one to recognize a complicated though computable ordering as being a well-ordering? Can one truly say that of familiar orderings for ε_0 or even ω^ω, let alone the more complicated orderings that have naturally emerged in modern proof theory (Γ_0, $\phi_{\varepsilon_\Omega+1}(0), \ldots$)? Surely not. And once *some* form of reasoning has to be admitted, then the whole question of which notations for ordinals are to be accepted (if one is to work with ordinal logics at all) must be re-examined. This was subsequently done, at the suggestion of Kreisel *1958*, by restricting attention, successively, to those notations a for which one has a *proof* in L_b for some $b <_O a$ that $a \in O$ (i.e., that a represents a well-ordering). These have come to be called *autonomous ordinal notations,* and the notion of ordinal logic restricted in this way, *autonomous recursive progressions of axiomatic theories.* Kreisel originally applied the notion of autonomous ordinal logics in his analysis of finitism, and I took it up (see my *1964* and *1967,* a.o.) in a corresponding analysis of predicativity. The idea of an autonomous progression more nearly approximates the process of finding out what is *implicit* in accepting a basic system L_1, i.e., of what ought one to accept, on the same fundamental grounds, if one accepts L_1. This theme is developed further in Kreisel *1970* and my forthcoming paper *198?*. In the latter I propose a new notion, that of *reflective closure* of an axiomatic theory, which I believe ought finally to take the place of autonomous progressions as a more realistic means to explain this idea. The reader is referred to these papers for further discussion, which is not possible here.

It is easily shown that the $\cup L_a[a$ autonomous] is recursively axiomatizable and hence incomplete by Gödel's theorem. Thus, whatever the value of the notion of autonomous progression (or reflective closure) for describing all that we ought to accept once we have made a basic conceptual commitment, we cannot hope thereby to answer the general question as to which axiomatic theories ought to be accepted according to the explanation, in a logical framework, of what constitutes mathematics. Ordinal logics provided the first model for attacking this question in any systematic way, and perhaps their greatest value lay in demonstrating the possibility of carrying out such an investigation at all. Turing's disappointment in their usefulness, and our latter-day disappointment for more sophisticated reasons ought not to diminish our appreciation for what, after all, was this remarkable character of Turing's achievement.

Postscript

It is natural to try to relate Turing's work on ordinal logics with his post-war thoughts on machine intelligence and human mental activity (particularly the doing of mathematics). For that, we have Turing's 1948 NPL report (reproduced in Turing *1969*), and the further sources provided by Hodges *1983* (especially Chapter 6); I shall refer freely to the ideas therein. I shall also be quite brief here, both because I feel less confident about this aspect of Turing's thought and because I do not feel that there is a strong connection. Turing, as is well known, had a mechanistic conception of mind, and that conviction led him to have faith in the possibility of machines exhibiting intelligent behavior. His idea seemed to be: if intelligent human activity depends only on the structure of the brain as a network of cells that are either activated or unactivated (and not on its physical embodiment in tissue) then whatever humans can do in this respect can, in principle, be mimicked by machines. But even for Turing, the structure of the brain qua machine, as complicated as it may be, does not by itself suffice for intelligent behavior. For that, he says, one also needs *external instruction* and *internal initiative*. Similarly, machines must be taught to think, be rewarded for success and punished for failure, so that they may learn from experience. Here one has a division between what is provided by the mechanism per se and what must be brought from outside in the form of a *program* that instructs the machine what to do. This echoes the division in ordinal logics between what is provided by the formal application of axioms and rules in any logic L_a and what must come from outside in order to recognize (by "intuition" or an "oracle") a as an ordinal notation that unlocks the L_a machinery. Also, once inside L_a, the application of "ingenuity" in searching for a proof of a conjecture may be compared with the required "initiative" for intelligent behavior. However, I would not push this analogy too far, since it is already a bit strained.

In sum, one might regard Turing's work on ordinal logics as a temporary shunting off from his main track of thought, from 1936–37 on through his war work on mechanical cryptanalysis and then, after the war, with the design of computers and the analysis of intelligent behavior in mechanical terms. Turing never tried to develop an over-all philosophy of mathematics and in the end did not seem to be really troubled by the problems that Gödel's theorem raised for a mechanistic theory of mind.[21] Indeed, I suspect from the history that he did not really have his heart in the Ph.D. work under Church, though the idea of Church's suggestion certainly appealed to him and, once engaged in it, he gave it the fullest of his mind.

7. *Technical Appendix*

The first part of the following explains Turing's notions and results from his *1939* paper in terms of the theory of (partial) recursive functions rather than the λ-calculus. This helps to make it more understandable to a wider audience and allows a simpler comparison with the work on ordinal logics that followed Turing's, which is briefly reviewed in the second part of the following. The reader is assumed to be familiar with ordinary recursion theory and the elements of the theory of constructive ordinals.

TURING'S WORK (RECAST)

All section references in the following, indicated by '§', are to Turing *1939*. After preliminaries in §§ 1–2 about effective calculability (in the λ-calculus) Turing singles out in § 3 a class of arithmetical propositions for special attention, namely what is now designated as the class of Π_2^0 statements $\forall x \exists y f(x, y) = 0$ with f primitive recursive (or, equivalently f general recursive). He gives this class the unfortunate name "number-theoretic theorems", which is misleading since not every such proposition is a theorem in the usual sense and not every number-theoretical (or arithmetical) statement in the usual sense is included in this class. (An inessential difference is that for us, variables 'n', 'm',..., 'x', 'y', 'z' range over the set ω of natural numbers $\omega = \{0, 1, 2, \ldots\}$, while Turing restricts himself to positive integers.) Some of the reasons for giving special attention to the class of Π_2^0 statements have been discussed in the main text above.

---

21 Gödel argues against Turing's mechanistic view of mind, which he believes to be based on a "philosophical error", in an unpublished remark to appear in Volume II of Gödel's *Collected Works* (to be referred to there as *1972a.3*). Judson Webb has prepared an interesting introductory note to this remark, discussing the issues involved.

In the following we shall have to deal with effective enumerations of various statements and sets. For this purpose we use Kleene's 'T' predicate $T(e, x, y)$ which gives in particular for $e = 0, 1, 2, \ldots$ all Σ_1^0 (r.e.) sets as $\{x | \exists y T(e, x, y)\}$ and all Π_2^0 propositions as $\forall x \exists y T(e, x, y)$. In addition, the statements $\exists y T(e, 0, y)$ ($\forall y \neg T(e, 0, y)$) express all Σ_1^0 (Π_1^0) propositions for $e = 0, 1, 2, \ldots$. Besides this 'internal' numbering (via the index e) of statements ϕ, we shall also use an 'external' numbering provided by the Gödel number $\ulcorner \phi \urcorner$ of ϕ. It is assumed that the language of arithmetic contains symbols for all primitive recursive functions (though these can eventually be defined in terms of $+$ and $\cdot$, by one of Gödel's results).

In §6, Turing defines a *logic,* in effect, to be any recursively enumerable (r.e.) set L of (numbers of) Π_2^0 sentences ϕ such that

$$\ulcorner \phi \urcorner \in L \Rightarrow \phi \text{ is true.}$$

We shall write $L \vdash \phi$ for $\ulcorner \phi \urcorner \in L$. A logic L' is said to be *at least as complete as* L if $L \subseteq L'$, and *more complete* if this inclusion is proper. This notion of logic is in one sense extremely general, since it only requires the set of theorems to be recursively enumerable, but it is rather restrictive in another sense that only logics whose theorems can be interpreted as Π_2^0 propositions are considered and, moreover, only such among them whose theorems are true. Note that Turing does not require closure under logical equivalence, let alone equivalence in number theory. Thus, on his definition, L' might be more complete than L and still not express any more truths among its theorems.

Turing says that by Gödel's incompleteness theorem, no one logic L can be complete, i.e., contain all true Π_2^0 sentences. In fact, this already holds for the Π_1^0 statements, by ordinary recursion-theoretic diagonalization. The latter formulation includes Gödel's incompleteness theorem for logical systems L in the usual sense, when L is consistent and contains a sufficient amount of arithmetic.

For the construction of ordinal logics, Turing notes that if $F(n)$ is an effective enumeration of logics (i.e., F is recursive and $F(n)$ is an index of L_n as a Σ_1^0 set for each n) then $L = \cup_n L_n$ is also a logic; moreover, we have an effective method to pass from an index f for F to an index for L.

In §7, Turing deals with the effective representation of ordinals in two ways. The first way is, in effect, via the set W of numbers e for recursive well-orderings $\leq_e$. the second is via the class of Church-Kleene notations for ordinals, originally defined in the λ-calculus, twice modified to a recursion-theoretic definition of a system S_3 of notations in Kleene *1938*. Here we write O for the set of notations given by S_3. This is generated inductively along with the relation $<_O$ in such a way that each $a \in O$ represents an ordinal $|a|$, with $|1| = 0, |2^a| = |a| + 1$ and $|3 \cdot 5^e| = \sup_{n < \omega}(|\{e\}(n)|)$; furthermore, $b <_O a \Rightarrow |b| < |a|$ and in the limit case $(3 \cdot 5^e) \in O$, we have $\forall n[\{e\}(n) <_O \{e\}(n + 1)]$. Kleene *1944* later showed that $<_O$ is the restriction to O of an r.e. relation $\prec$. One advantage of the original

Church-Kleene representation of the constructive ordinals in the λ-calculus is that recursive definition on these is built into the ordinal (in a suitable sense, an ordinal term a of the λ-calculus represents iteration of an arbitrary function up to a). The corresponding means of handling recursive definition on O is as a special case of the recursion theorem. Thus, one can define partial recursive operations $+_O$, $\cdot_O$, etc. with $|a +_O b| = |a| + |b|$, $|a \cdot_O b| = |a| \cdot |b|$. etc. On the other hand, operations of ordered sum $d +_W e$, and ordered product, $d \cdot_W e$, are directly definable for $d, e \in W$ in such a way that $|d +_W e| = |d| + |e|$ and $|d \cdot_W e| = |d| \cdot |e|$.

Using the effective enumeration of $\{x | x \preceq a\}$, with each a is effectively associated $K(a)$ for which $a \in O \Leftrightarrow K(a) \in W$; moreover, when $a \in O$ then $|a| = |K(a)|$. On the other hand, though the ordinals represented by elements of W are the same as those represented by elements of O, there is no simple effective ordinal-preserving embedding of W into O. This does hold though for a subset W_1 of W, consisting of those $e \in W$ for which we have (i) a specified initial element in $\leq_e$, (ii) a recursive function which tells for each $x \in \text{dom}(\leq_e)$ whether or not x is at a successor position or a limit position, and (iii) a recursive function which gives the predecessor of each successor element. From this we can also obtain a recursive function L which assigns to each limit element x a fundamental sequence $y_n = L(x, n)$ with $\forall n [y_n <_e y_{n+1}]$ so that x is the l.u.b. of the y_n's in $\leq_e$. With the additional information (i)–(iii) represented by another code e' we have an effective function K' which associates with each $(e, e') \in W_1$ an element $K'(e, e')$ of O such that $|e| = |K'(e, e')|$. It is easy to see that every element $\leq_e$ can be imbedded in a $\leq_{e^*}$ in W_1 with $| \leq e^*| = \omega \cdot | \leq_e |$, simply by putting a copy of ω in each position of $\leq_e$. (this class W_1 is not discussed by Turing.) The least ordinal not represented by an element of O (W, or W_1) is denoted ω_1^{CK} (where 'CK' abbreviates 'Church-Kleene').

The notion of *ordinal logic* is finally defined in § 8; in effect, this is any partial recursive function Λ such that for each $c \in W$, $\Lambda(c)$ is (the index of) a logic L_c. Turing then shows how a wide class of ordinal logics can be constructed when we have a class $\mathscr{C}$ of logics L in the more usual sense for which (i) there is a method of passing from any L in $\mathscr{C}$ to a more comprehensive system L' in $\mathscr{C}$, (ii) the formulas of L in $\mathscr{C}$ include the Π_2^0 statements, (iii) the validity of provable Π_2^0 statements is preserved in passing from L to L', and (iv) $\mathscr{C}$ is closed under effective unions. Assume given L_1 in $\mathscr{C}$ with index m_1 and that we have a (partial) recursive G such that if m is an index of L in $\mathscr{C}$ then $G(m)$ is an index of L'. Then we have a (partial) recursive function $\{\ell\}$ such that for each $a \in O$, $\{\ell\}(a)$ is an index of a logic L_a in $\mathscr{C}$ with: (i) L_1 given, (ii) $L_{2^a} = L'_a$, by $\{\ell\}(2^a) = G(\{\ell\}(a))$, and (iii) $L_{3 \cdot 5^e} = \bigcup_{n<\omega} L_{\{e\}(n)}$ by $\{\ell\}(3 \cdot 5^e) = 3 \cdot 5^{C(\ell, e)}$ where $\{C(\ell, e)\}(n) \simeq \{\ell\}(\{e\}(n))$. (*Note*: The domain of $\{\ell\}$ is r.e. and includes O; in fact $\{\ell\}$ can be chosen total.)

In order to obtain an ordinal logic Λ from this in Turing's sense, one must associate a logic $\Lambda(c)$ with each $c \in W$. His method of doing this (p. 192) is a bit artificial, since we don't have an effective ordinal-preserving map from W into O.

But this does work for the set W_1 described above, so that for $e = (e_0, e_1) \in W_1$ we can define $\Lambda(e)$ on W_1 as $\Lambda(K'(e))$, using the construction just given of Λ on O. Since certain results are sensitive to the domain of an ordinal logic, we shall distinguish these according as the domain is O, W_1, or W.

The specific examples of ordinal logics that Turing gives are all based on logical systems in the usual sense of the word, specified by axioms and rules of inference. For these we shall now use the letter 'S' with or without subscripts, and write $\Lambda = < S_a | a \in O >$ for an ordinal logic on O consisting of such systems (and similarly for those on W and W_1), $\mathrm{dom}(\Lambda) = O$ (resp. W, W_1), and finally $\Lambda(a) = S_a$ (or an index $\{\ell\}(a)$ for S_a), when $a \in \mathrm{dom}(\Lambda)$. We put $S_a \equiv S_{a'}$ if S_a and $S_{a'}$ prove the same theorems, and $S_a \subseteq S_{a'}$ if every theorem of S_a is a theorem of $S_{a'}$. Finally we write $S_a \underset{\neq}{\subset} S_{a'}$ if there is ψ with $S_{a'} \vdash \psi$ but $S_a \nvdash \psi$. (For Turing, 'theorem' is always qualified to be a Π_2^0 statement.)

For the first example, Λ_P, Turing takes as initial system S_1 the variant P of *Principia Mathematica* used by Gödel in his *1931*, though much weaker systems, such as Peano arithmetic, *PA*, would serve his purpose here just as well. For the collection $\mathscr{C}$ of systems from which the S_a are to be drawn he takes those given by r.e. extensions of S_1. Such systems can be primitive recursively axiomatized, and with each index of such an S we associate a primitive recursive formula $\mathrm{Proof}_S(x, y)$ which defines (in S_1) the proof relation P_S for S, where $P_S(m, n)$ holds if m is the Gödel number of a proof in S of the formula with number n. Let $\mathrm{Prov}_S(y)$ be the formula $(\exists x)\mathrm{Proof}_S(x, y)$. For S', Turing takes S together with all statements of the form

$$L\text{-Ref}_S(\phi) \qquad \mathrm{Prov}_S(\ulcorner \phi \urcorner) \to \phi$$

where ϕ is restricted to be a Π_2^0 sentence $\forall z \exists w R(z, w)$. Then the statement $L\text{-Ref}_S(\phi)$ is equivalent to $\forall x [\mathrm{Proof}_S(x, \ulcorner \phi \urcorner) \to \forall z \exists w R(z, w)]$ and hence to $\forall x, z \exists w [\mathrm{Proof}_S(x, \ulcorner \phi \urcorner) \to R(z, w)]$, so is again a Π_2^0 statement. By $L\text{-Ref}_S$ we mean the collection of statements $L\text{-Ref}_S(\phi)$; nowadays this is called the *local re-flection principle for S*. Let $S' = S \cup (L\text{-Ref}_S)$ for any S in $\mathscr{C}$. It should be noted that $L\text{-Ref}_S(\ulcorner 0 = 1 \urcorner)$ is equivalent to $\mathrm{Prov}_S(\ulcorner 0 = 1 \urcorner) \to (0 = 1)$, hence to $\neg \mathrm{Prov}_S(\ulcorner 0 = 1 \urcorner)$, in other words to a sentence Con_S which expresses the consistency of S. Thus $S' = S \cup (L\text{-Ref}_S)$ overcomes the Gödelian incompleteness of S. Moreover, $L\text{-Ref}_S$ expresses that whatever is provable in S is correct, so ought to be accepted (on informal grounds) as soon as S is accepted. More precisely, if each Π_2^0 theorem of S is true then each instance of $L\text{-Ref}_S$ is also true – and since the basic logical rules of inference preserve truth, it follows that every Π_2^0 theorem of S' is true.

In the preceding argument, one appeals to the notion of truth for all arithmetical formulas. To avoid this, Turing notes that S is valid (correct) for closed Π_2^0 theorems if it is valid for closed Σ_1^0 theorems (equivalently, if it is ω-consistent for Σ_1^0 sentences), and that the passage $S \longmapsto S \cup L\text{-Ref}_S$ preserves validity for Σ_1^0 theorems. For suppose $(L\text{-Ref}_S(\phi_0)) \wedge \ldots \wedge (L\text{-Ref}_S(\phi_k)) \to \exists x R(x)$ (with R primitive recursive) is provable from S; we may assume that each ϕ_i is *not* provable from S, for otherwise $S \vdash L\text{-Ref}_S(\phi_i)$. Then $S \vdash [\text{Prov}_S(\ulcorner \phi_0 \urcorner) \wedge \neg \phi_0] \vee \ldots \vee [\text{Prov}_S(\ulcorner \phi_k \urcorner) \wedge \neg \phi_k] \vee \exists x R(x)$, so $S \vdash \text{Prov}_S(\ulcorner \phi_0 \urcorner) \vee \ldots \vee \text{Prov}_S(\ulcorner \phi_k \urcorner) \vee \exists x R(x)$. This is equivalent to a Σ_1^0 sentence, hence true. But since none of $\text{Prov}_S(\ulcorner \phi_i \urcorner)$ is true it follows that $\exists x R(x)$ is true, as required.[22]

Now the ordinal logic $\Lambda_P = \langle S_a | a \in O \rangle$ is obtained from the general construction by taking $S_1 = P$, $S_{2^a} = S_a \cup L\text{-Ref}_{S_a}$ and $S_{3 \cdot 5^e} = \cup_{n < \omega} S_{\{e\}(n)}$.

The second example that Turing gives of an ordinal logic in § 8 is a form of transfinite type theory, with cumulative types, denoted Λ_H.[23] This has domain W, and is obtained as follows. Given any recursive well-ordering $\le_d$ $(d \in W)$, form a new ordering $\le_e$ with $e \in W$ of type $|e| = \omega + |d|$. Then for each $n \in \text{dom}(\le_e)$ we have a sequence of variables $x^{(n)}, y^{(n)}, \ldots$ of type n (i.e., type level $|n|$) in the ordinary sense. The atomic formulas for S_d are $x^{(n)} \in x^{(m)}$ when $n <_e m$. Unfortunately, Turing says nothing about the axiomatization of S_d, and even though the syntax is indicated to be cumulative, we could not make full use of this in the comprehension axioms, for which the obvious choices are those of the form $\exists x^{(m)} \forall x^{(n)} [x^{(n)} \in x^{(m)} \leftrightarrow \phi(x^{(n)})]$ for each $n <_e m$. Turing's syntax could be enriched by variables $\overline{x}^{(m)}, \overline{y}^{(m)}, \ldots$ for each m, intended to range over all $x^{(n)}$ for $n <_e m$. However, Turing's further results for Λ_H simply depend on the fact that it is an invariant ordinal logic on W (see below), as would also hold for such variant formulations.

Remark. The reason Turing takes $|e| = \omega + |d|$ seems to be to make S_1 for Λ_H of the same strength as finite type theory P. If we form a system $\Lambda_{H'}$, like Λ_H, but with $\Lambda_{H'}(d)$ having variables $x^{(n)}$ for $n \in \text{dom}(d)$ for each $d \in W$, then $\Lambda_{H'}$ becomes equivalent to finite type theory at level ω instead of level 1.

With § 9 we come to the core of Turing's paper; all the principal results are contained therein. An ordinal logic Λ is defined to be *complete* if for each true $\Pi_2^0 \phi$ there is some $a \in \text{dom}(\Lambda)$ with $\Lambda(a) \vdash \phi$. It is said to be *invariant up to* α (*up through* α) if for each $a, a' \in \text{dom}(\Lambda)$ with $|a| = |a'| < \alpha$ $(\le \alpha)$ we have $\Lambda(a) \equiv \Lambda(a')$. Λ is said to be *invariant* if it is invariant up to ω_1^{CK}. When Λ is invariant up through α we may write S_α for any S_a with $|a| = \alpha$. Note that both Λ_H and $\Lambda_{H'}$, just defined, are invariant. Λ is said to be *strictly increasing* (*up through* α) if whenever $a <_O a'$ (and $|a'| \le \alpha$) we have $S_a \subsetneq S_{a'}$. Note, for example, that Λ_P

---

22 Note that this argument does not depend on restricting the class of ϕ for which one takes $L - \text{Ref}_S(\phi)$.

23 'H' is for 'Hilbert'; the source given by Turing is Hilbert *1926*, where a form of transfinite type theory is indicated, though not explicitly. For Turing's later work on the theory of types, see Gandy *1977*.

is strictly increasing, with $S_{a'} \vdash \mathrm{Cons}_{S_a}$ for each $a <_O a'$. There are three general results in §9 concerning these notions, and one specific one for Λ_P. Turing does not number these, so I shall refer to them here as 9.1–9.4.

9.1 Theorem: *If Λ is invariant on W_1 (or O) then it cannot be strictly increasing.*

Proof. The idea of Turing's proof of 9.1 is as follows (except that we replace his use of ω^ω by that of ω^2). Let $\forall x \exists y T(e, x, y)$, $e = 0, 1, 2, \ldots$ be a Π_2^0 enumeration of the Π_2^0 statements. Then a recursive function $F : \omega \to W_1$ is to be defined such that

(i) $\forall x \exists y T(e, x, y)$ true $\Rightarrow |F(e)| = \omega^2$

(ii) $\forall x \exists y T(e, x, y)$ false $\Rightarrow |F(e)| < \omega^2$.

This can be done as follows, with $r_e = F(e)$. The domain of $\leq_{r_e}$ consists of all (x, z) such that $\forall u \leq x \exists y \leq z T(e, u, y)$, and for (x_1, z_1), $(x_1, z_2) \in \mathrm{dom}(\leq_{r_e})$ we put $(x_1, z_1) <_{r_e} (x_2, z_2) \Leftrightarrow x_1 < x_2 \vee (x_1 = x_2 \wedge z_1 < z_2)$. Now suppose Λ is invariant and strictly increasing up through $\omega^2 + 1$. Then we can find a ψ such that $S_{\omega^2+1} \vdash \psi$ but $S_{\omega^2} \nvdash \psi$; moreover $S_\alpha \nvdash \psi$ for $\alpha \leq \omega^2$ since the S_a's are increasing. Let $G(e) = F(e) +_W 1_W$; it follows that

$$\forall x \exists y T(e, x, y) \text{ is true } \Leftrightarrow S_{G(e)} \vdash \psi.$$

But $\{e : S_{G(e)} \vdash \psi\}$ is r.e., while the set $\{e | \forall x \exists y T(e, x, y)\}$ is not, so we have a contradiction. (The proof carries over to Λ with domain O by using the ordinal preserving map K' of W_1 into O.)

Here, Turing's concentration on Π_2^0 led him to formulate a weaker result than he could have obtained by the same idea, namely:

9.1′ Theorem: Λ *cannot be both invariant on W_1 (or O) and strictly increasing up through $\omega + 1$.*

Proof. Let $\forall y R(e, y)$ be a Π_1^0 enumeration of Π_1^0 statements (e.g., as $\forall y \neg T(e, 0, y)$). We shall construct recursive $F : \omega \to W_1$ satisfying

(i) $\forall y R(e, y)$ true $\Rightarrow |F(e)| = \omega$

(ii) $\forall y R(e, y)$ false $\Rightarrow |F(e)| < \omega$,

as follows. With $r_e = F(e)$, put $x \in \mathrm{dom}(r_e)$ if $\forall y < x R(e, y)$. Then put $x_1 <_{r_e} x_2 \Leftrightarrow x_1, x_2 \in \mathrm{dom}(r_e) \wedge x_1 < x_2$. The rest of the argument is as for 9.1.

The next two theorems show that even if we drop the assumption that $\Lambda(a)$ is increasing with a, we can't have invariance. These results are labelled 'A' and 'B' in §9.

9.2 Theorem: *If Λ is a W-ordinal logic and is invariant up to α then for any $\beta < \alpha$, $S_\beta \subseteq \cup_{n<\omega}S_n$.*

9.3 Theorem: *If Λ is an O-ordinal logic and is invariant up to α then for any $\beta < \alpha$, $S_\beta \subseteq \cup_{\gamma<\omega^2}S_\gamma$.*

Proof of 9.2. Suppose $\omega \leq \beta < \alpha$. Here we define $F : \omega \to W$ with:

(i) $\forall yR(e, y)$ true $\Rightarrow |F(e)| = \beta$

(ii) $\forall yR(e, y)$ false $\Rightarrow |F(e)| < \omega$.

First let $\leq_{r_0}$ be a recursive well-ordering of order-type β; we can enumerate $\text{dom}(r_0)$ without repetitions by a recursive f. Then take $F(e) = r_e$ with $\text{dom}(\leq_{r_e}) = \{x|\forall y < xR(e, y)\}$, and $x_1 <_{r_e} x_2 \Leftrightarrow f(x_1) <_{r_0} f(x_2)$. Thus if $\forall yR(e, y)$ is true and $F(e) = r_e$ we have $\text{dom}(r_e) = \omega$ and the relation $\leq_{r_e}$ is $\cong$ to $\leq_{r_0}$, so $|F(e)| = \beta$. But if $\forall yR(e, y)$ is false then $\text{dom}(<_{r_e})$ is finite. Now suppose $S_\beta \vdash \psi$ where no $S_n \vdash \psi$. Then $\forall yR(e, y) \Leftrightarrow S_{F(e)} \vdash \psi$, giving a contradiction.

Proof of 9.3. Suppose $b \in O$, $|b| = \beta$ with $\omega^2 \leq \beta < \alpha$. Here we prove by induction on β that $S_\beta \subseteq \cup_{\gamma<\omega^2}S_\gamma$. We can find $3 \cdot 5^c \leq b$ with $b = 3 \cdot 5^c +_O m_O$. Then take

$$\{c_e\}(n) \simeq \begin{cases} \{c\}(n) & \text{if } \forall y \leq n\, R(e, y) \text{ holds} \\ \{c\}(k) +_O n_O & \text{if } \neg R(e, k), \text{ and } \forall y < k\, R(e, y). \end{cases}$$

Thus $|3 \cdot 5^{c_e}| = |3 \cdot 5^c|$ if $\forall yR(e, y)$ and $|3 \cdot 5^{c_e}| < |3 \cdot 5^c|$ otherwise. Now suppose $S_\beta \vdash \psi$ where no $S_\gamma \vdash \psi$ for $\gamma < \omega^2$. Then no $S_\gamma \vdash \psi$ for $\gamma < \beta$, by the induction hypothesis. Take $F(e) = 3 \cdot 5^{c_e} +_O m_O$, so $|F(e)| = |b|$ if $\forall yR(e, y)$ and $|F(e)| < |b|$ o.w. Thus $\forall yR(e, y) \Leftrightarrow S_{F(e)} \vdash \psi$.

As an application of 9.2 we have that $\Lambda_{H'}$ yields no more than finite type theory (P), and both Λ_H and $\Lambda_{H'}$ are incomplete (for Π_1^0 sentences).

Turing calls 9.2, 9.3 *incompleteness theorems* (and he might well have put 9.1 under this rubric, too). After the proofs he comments: "This theorem [9.2 and 9.3] can no doubt be improved in many ways. However, it is sufficiently general to show that, with almost any reasonable notation for ordinals, completeness is incompatible with invariance." Presumably, by possible improvements, Turing had in mind more restrictive ways of providing notations for ordinals. The most restrictive way of all would be to select a unique notation for each ordinal $\alpha < \omega_1^{CK}$; in that case invariance is trivial. But the question then is: what is a reasonable unique notation system for ordinals? One answer to this (in later work reported below, namely for Π_1^1 paths through O), shows that we still have incompleteness. Thus Turing's remark may be considered prescient, though it would have required considerable development; however, he did not expand at all on his ideas in this direction.

Perhaps the most interesting of Turing's theorems is the following 9.4; at any rate, its proof has the most interesting twist.

9.4 Theorem: *Let* $S_a = \Lambda_P(a)$ *for* $a \in O$. *Then for each true* Π_1^0 *sentence* ϕ *we can find* $a \in O$ *with* $|a| = \omega + 1$ *and* $S_a \vdash \phi$.

Proof. Let $\phi = \forall y R(y)$ with R primitive recursive. Find an index e by the recursion theorem provably satisfying

$$\{e\}(n) \simeq \begin{cases} n_O & \text{if } (\forall y \leq n) R(y) \\ 3 \cdot 5^e +_O 1_O & \text{if } (\exists y \leq n) \neg R(y). \end{cases}$$

Then $\{e\}$ is clearly total (by induction on n). Let $b = 3 \cdot 5^e$ and $a = b +_O 1_O$. Thus either

(i) $\forall y R(y)$, in which case $a \in O$ and $|a| = \omega + 1$, or

(ii) $\exists y \neg R(y)$, in which case $S_b \equiv S_a$ and $(S_b \vdash \mathrm{Cons}_{S_b})$.

Of course in case (ii), $a \notin O$, but S_a is defined in general by the recursion theorem (not just for $a \in O$), and when a provably has the form $b +_O 1_O$ (= 2^b), then S_a provably contains the local reflection principle for S_b, and in particular $S_a \vdash \mathrm{Cons}_{S_b}$; hence, if it happens, as in this case, that $S_a \equiv S_b$, we have $S_b \vdash \mathrm{Cons}_{S_b}$. But then by Gödel's incompleteness theorem, S_b is not consistent. Formalizing this argument in arithmetic gives

(iii) $S_1 \vdash \exists y \neg R(y) \rightarrow \neg \mathrm{Cons}_{S_b}$, or

(iv) $S_1 \vdash \mathrm{Cons}_{S_b} \rightarrow \forall y R(y)$.

But since we are assuming $\forall y R(y)$ is actually true, we have by (i) that $a \in O$, $a = b +_O 1_O$ and $|a| = \omega + 1$ so $S_a \vdash \mathrm{Cons}_{S_b}$ and finally $S_a \vdash \forall y R(y)$, as was to be proved.

Turing's comment about this, at the end of § 9 is that: "The completeness theorem as usual is of no value. Although it shows, for instance, that it is possible to prove Fermat's last theorem with Λ_P (if it is true) yet the truth of the theorem would really be assumed by taking a certain formula as an ordinal formula [notation for an ordinal]." Clearly, in the above argument, $a \in O$ just in case ϕ holds, and we are simply building the truth of ϕ into a in such a way that $S_a \vdash \phi$.

It is an immediate corollary of 9.4 that Λ_P is not invariant up to $\omega + 1$ (already obtained in a different way in 9.1').

Before embarking on the proof of 9.4 , Turing said: "It is to be expected that this ordinal logic is complete [for Π_2^0 sentences]." However, this expectation turned out to be wrong, as I showed in my *1962* paper (see below).

When we discussed Turing's analysis of the purpose of ordinal logics (in Section 6 of the main text above), we mentioned his concern about the application of the informal notion of correctness to the ordinal logics Λ_P and Λ_H, since both of these take a theory P of finite types (Gödel's 1931 adaptation of *Principia Mathematica*) as S_1. As I pointed out, he could just as well have taken Peano Arithmetic *PA* (or even a fragment thereof) as the basis, without affecting any of the results. The fact that Gödel's incompleteness theorems apply just as well to extensions of *PA* instead of *P* was certainly known by Kleene and Rosser from Gödel's 1934 lectures at the IAS, and should have been known to Church. Of course, Turing could have realized this just as well by himself. The point is that his qualms about using Λ_P and Λ_H for this reason are misdirected. Be that as it may, in the final section of the paper, § 12, Turing turns to the formulation of more constructive ordinal logics Λ_G suggested by the work of Gentzen *1936*. Actually, he defines three of these, Λ_{G^i} for $i = 1, 2, 3$, the first of which he quickly ignores. Unfortunately, his description of the syntax of Λ_{G^2} and Λ_{G^3} is very obscure, but basically the idea is to take a free variable symbolism which comprises primitive recursive arithmetic, and when we have a primitive recursive well-ordering $\leq_e$, to incorporate definition by transfinite recursion and (a rule for) proof by transfinite induction on $\leq_e$ into $\Lambda_{G^2}(e)$.[24] The system Λ_{G^3} differs only in taking its domain to consist of those primitive recursive ordering relations $\leq_e$ for which there are no *recursive* descending sequences $F(1) >_e F(2) >_e \ldots$. Turing points out that this class of e's is arithmetically definable, and that Λ_{G^3} "... appears to be adequate for most purposes." Clearly he is ignorant of the existence of primitive recursive $\leq_e$ which are well-ordered with respect to recursive descending sequences but not well-ordered in full. There is however a point to his remark: in Gentzen-type consistency proofs of formal systems by induction on suitable $\leq_e$, it is only the weaker property that is required. But we should only want to use $\leq_e$ with $e \in W$, and in practice it is no easier to establish the weaker property ("no recursive descending sequences") than the full property ("no descending sequences").

SUBSEQUENT WORK

Here I shall simply explain some of the main results with hardly any indication of proofs, but with the references to such.

In my paper *1962* I dealt with *transfinite recursive progressions of axiomatic theories* $< S_a | a \in O >$, my re-christening of ordinal logics with domain O (calling them *progressions* for short).[25] However, I dealt only with theories in the ordinary logical sense, and assumed that the language of S_1 includes the full language of

---

24 Robin Gandy has pointed out that Λ_{G^2} contains the first rather general scheme for definition of number-theoretic functions by transfinite recursion. (R. Péter had previously given various particular schemes for such.)

25 My work on this was first reported in abstracts in the *J. Symb. Log.* (1958), pp. 105–106 and *ibid.*, 24 (1959), pp. 312–313.

arithmetic, and that $PA \subseteq S_1$.[26] After thus recasting Turing's work on the general construction of ordinal logics in the framework of ordinary recursion theory, I re-proved his completeness result (9.4 above) for Π_1^0 sentences, for a progression based on the local reflection principle (i.e, where each $S_{2^a} = S_a \cup L\text{-Ref}_{S_a}$, though already $S_{2^a} \vdash \text{Cons}_{S_a}$ suffices). Then I proved that for this progression $\cup_{a \in O} S_a$ can be axiom-atized over S_1 (non-constructively) by the set of all true Π_1^0 sentences. It follows in this case that there are true Π_2^0 sentences not provable in $\cup_{a \in O} S_a$ (against Turing's expectation).

I then considered progressions based on the passage $S \mapsto S' = S \cup U\text{-Ref}_S$, where for any (primitive recursive presented) S, $U\text{-Ref}_S$ is the *uniform reflection principle for S*, which consists of all instances

$$(U\text{-Ref}_S(\phi)) \qquad \forall x \text{Prov}_S(\ulcorner \phi(0^{(x)}) \urcorner) \rightarrow \forall x \phi(x).$$

Thus $U\text{-Ref}_S$ may be considered as a kind of formalized ω-rule over S, and we should accept it if we accept S (as being correct under any basic mathematical con-ception including that of the natural number system). For this stronger progression $< S_a | a \in O >$, I proved completeness with respect to *all* true arithmetical sentences, as long as $S_1 \supseteq PA$.[27] In fact, for these one only needs the S_a with $|a| < \omega^{\omega^\omega}$. Finally, I also obtained completeness along suitable paths: there is a path P through O (lin-early ordered by $<_O$ and closed under predecessors) such that $\cup_{a \in O} S_a \equiv \cup_{a \in P} S_a$; moreover P can be chosen recursive in O. In fact, one can already find a hyperarith-metic path P' *within* O of length $\omega^{\omega^\omega + 1}$ such that every true arithmetical sentence is provable in $\cup_{a \in P'} S_a$; clearly, such P' cannot be extended further in O.

It was shown in Feferman and Spector *1962* that under quite general conditions on a progression $< S_a | a \in O >$ (basically that it is increasing and that the S_a's are consistent and uniformly r.e. in a) we have: (i) If P is any Π_1^1 path through O then $\cup_{a \in P} S_a$ is incomplete with respect to Π_1^0 sentences, and (ii) such paths exist. The paper introduced a set $O^\star \supseteq O$ of 'non-standard' notations consisting of those a for which $\{x | x \preceq a\}$ is O-like, but which need only be well-ordered under hyperarithmetic descending sequences. Then any $a \in O^\star - O$ determines a Π_1^1 path $P = \{x | x \in O \wedge x \prec a\}$ through O. Since each S_b with $b \in P$ is consistent, the set $C = \{b | b \prec a \wedge S_b \text{ consistent}\}$ must overshoot P, and hence for $b \in C - P$ we have Cons_{S_b} not provable in $\cup_{a \in P} S_a$. The set $O^\star$ defined for this purpose turned out later to have a number of further uses.

This incompleteness result was improved in Kreisel *1972* as follows. We know that for r.e. consistent $S \supseteq PA$, the sets A which are binumerable (or numeralwise

---

26 Actually a fragment of *PA* suffices as starting point when constructing progressions based on exten-sion principles at least as strong as the uniform reflection principle (below), since then the induction scheme can be derived from very weak instances.

27 I also showed (*1962*, p. 314) that when S_1 contains a certain part of second-order analysis then we have completeness in the associated progression for all true Π_1^1 sentences.

definable) in S are exactly the recursive sets. That is, every recursive A is of the form $\{n|S \vdash \phi(n)\}$ where $\forall n[S \vdash \phi(n)$ or $S \vdash \neg\phi(n)]$, and conversely. Now Kreisel showed that if A is binumerable in $\cup_{a \in P} S_a$, where P is a Π_1^1 path through O, then A is still recursive. Hence if $A \in \Pi_1^0 - \Sigma_1^0$, say $n \in A \Leftrightarrow \forall y R(n, y)$, there must exist an n such that $S \not\vdash \forall y R(n, y)$, though $\forall y R(n, y)$ is true. Kreisel discussed this result with respect to the question whether Church's thesis holds for "h[human]-effective definitions" and not just for mechanical computability. His main point here is that if $\cup_{a \in P} S_a$ is taken as a model of human reasoning, for P a suitable Π_1^1 path through O, then the answer is positive and if it is such a model for a suitable path P recursive in O then (by my *1962*) the answer is negative. But there is no good argument to accept either one of these, and so the discussion is inconclusive.[28]

The technical result in Kreisel *1972* is very easy to obtain from Feferman and Spector *1962*, as follows. Let $S_P = \cup_{a \in P} S_a$ with P a Π_1^1 path through O, and suppose $S_P \vdash \phi(n)$ or $S_P \vdash \neg\phi(n)$ for each $n \in \omega$. Then for $b \in C - P$ (with C as above), we have $S_b \vdash \phi(n)$ or $S_b \vdash \neg\phi(n)$ for each n, so $\{n|S_P \vdash \phi(n)\} = \{n|S_b \vdash \phi(n)\}$, and this set is recursive.[29]

Now, as mentioned in section 5 of the main text above there is a treatment of completeness and incompleteness for hierarchies of recursive functions parallel to that for progressions of theories. In my paper *1962a* I considered quite general classes of such hierarchies $< F_a | a \in O >$, increasing under certain relations $a \leq_O a' \Rightarrow F_a \leq F_{a'}$ (such as that of F_a being primitive recursive in $F_{a'}$, or F_a being majorized by $F_{a'}$). I proved there under very general conditions that: (i) for each general recursive F we can find $a \in O$ with $|a| = \omega^2$ and $F \leq F_a$, (ii) there are (arithmetical) paths P within O such that $|P| = \omega^3$ and for any recursive F we can find $a \in P$ with $F \leq F_a$, and (iii) if P is a Π_1^1 path through O then some recursive $F \not\leq F_a$ for any $a \in P$. By (i) we cannot have invariance (uniqueness) of such hierarchies up through ω^2, though natural unique hierarchies can be given up to ω^2. There were also useful applications of such hierarchies extended to the non-standard notations O^* in that paper as well as in Harrison *1965*.

The main further work on progressions of theories is for *autonomous progressions*, which were described in Section 6 of the main text above, the purpose for which is quite different from Turing's original purpose for ordinal logics. The non-autonomous part of the subject is stalled as long as we don't have a convincing answer to the question: *what is a natural path through O?* This should be closely related to the standing open question coming from proof-theory:[30] *what is a natural well-ordering?*

---

28 Kreisel's discussion *1972* (unusually convoluted, as a whole) also takes up the question of Church's Thesis for h-effective definitions with respect to certain results about intuitionistic systems.

29 Visser *1981* obtained some improvements to Kreisel's result, for paths through or within O.

30 See, for example, Feferman *1987a*.

References

Barendregt, H.P.

   1984 *The Lambda Calculus. Its Syntax and Semantics,* 2nd ed. Amsterdam: North-Holland Publ. Co. (1984).

Barwise, J.

   1975 *Admissible Sets and Structures.* Berlin: Springer-Verlag (1975).

Church, A.

   1936 An unsolvable problem of elementary number theory. *Am. J. Math.* **58** (1936) 345–363 (reprinted in Davis *1965*).

   1936a A note on the Entscheidungsproblem. *J. Symb. Log.* **1** (1936) 40–41; correction, ibid., 101–102 (reprinted in Davis *1965*).

   1941 The calculi of lambda-conversion. *Ann. Math. S. (Princeton)* **6** (1941), 2nd printing 1951.

Church, A., and S.C. Kleene

   1936 Formal definitions in the theory of ordinal numbers, *Fund. Math.* **28** (1936) 11–21.

Davis, M.

   1965 *The Undecidable. Basic Papers on Undecidable Propositions, Unsolvable Problems and Computable Functions.* Hewlett, NY: Raven Press (1965).

   1982 Why Gödel didn't have Church's Thesis. *Inf. Contr.* **54** (1982) 3–24.

Davis, M., Yu. Matijasevič, and J. Robinson

   1976 Hilbert's tenth problem. Diophantine equations: Positive aspects of a negative solution. In: *Mathematical Developments Arising From Hilbert Problems,* ed. F.E. Browder. Providence, RI: AMS (1976) 323–78.

Feferman, S.

   1962 Transfinite recursive progressions of axiomatic theories. *J. Symb. Log.* **27** (1962) 259–316.

   1962a Classifications of recursive functions by means of hierarchies. *T. Am. Math. Soc.* **104** (1962) 101–122.

   1964 Systems of predicative analysis. *J. Symb. Log.* **29** (1964) 1–30.

   1968 Autonomous transfinite progressions and the extent of predicative mathematics. In: *Logic, Methodology and Philosophy of Science III,* eds. B. van Rootselaar and J.F. Staal, pp. 121–135. Amsterdam: North-Holland Publ. Co. (1968).

   1977 Theories of finite type related to mathematical practice. In: *Handbook of Mathematical Logic,* ed. J. Barwise, pp. 913–971. Amsterdam: North-Holland Publ. Co. (1977).

   1986 The life and work of Kurt Gödel. In: Gödel *1986*, pp. 1–36.

   1987 Infinity in mathematics: is Cantor necessary? in: *L'Infinito nella Scienza (Infinity in Science).* Rome: Istituto dello Enciclopedia Italiana (1987) 151–209.

1987a Proof theory: a personal report. In: *Takeuti 1987,* pp. 447–485.

198? Reflecting on incompleteness (in preparation).

Feferman, S., and C. Spector

1962 Incompleteness along paths in progressions of theories. *J. Symb. Log.* **27** (1962) 383–390.

Fenstad, J.E.

1980 *General Recursion Theory: An Axiomatic Approach.* Berlin: Springer-Verlag (1980).

Gandy, R.O.

1977 The simple theory of types. In: *Logic Colloquium 76,* eds. R.O. Gandy and J.M.E. Hyland, pp. 173–181. Amsterdam: North-Holland Publ. Co. (1977).

Gentzen, G.

1936 Die Widerspruchsfreiheit der Zahlentheorie. *Math. Annal.* **112** (1936) 493–565.

Gödel, K.

1931 Über formal unentscheidbare Sätze der *Principia mathematica* und verwandter Systeme I. *Monats. Math. Ph.* **38** (1931) 173–198 (reprinted in Davis *1965* and Gödel *1986.*).

1934 On undecidable propositions of formal mathematical systems. Lecture notes taken by S.C. Kleene and J.B. Rosser, reprinted in Davis *1965* and Gödel *1986.*

1986 *Collected Works, Volume I. Publications 1929–1936,* eds. S. Feferman, J.W. Dawson, Jr., S.C. Kleene, G.H. Moore, R.M. Solovay, and J. van Heijenoort. New York: Oxford University Press (1986).

Harrison, J.

1968 Recursive pseudo-well-orderings. *T. Am. Math. Soc.* **131** (1968) 526–543.

Hilbert, D.

1926 Über das Unendliche. *Math. Annal.* **95** (1926) 161–190.

Hodges, A.

1983 *Alan Turing: The Enigma.* London: Burnett Books, and New York: Simon and Schuster (1983).

Kleene, S.C.

1938 On notation for ordinal numbers. *J. Symb. Log.* **3** (1938) 150–155.

1944 On the forms of predicates in the theory of constructive ordinals. *Am. J. Math.* **66** (1944) 41–58.

1955 On the forms of predicates in the theory of constructive ordinals (second paper). *Am. J. Math.* **71** (1955) 405–428.

1981 Origins of recursive function theory. *Ann. Hist. C.* **3** (1981) 52–67.

Kreisel, G.

1958 Ordinal logics and the characterization of informal concepts of proof. In: *Proceedings of the International Congress of Mathematics at Edinburgh* (1958) 289–299.

1970 Principles of proof and ordinals implicit in given concepts. In: *Intuitionism and Proof Theory,* eds. J. Myhill, A. Kino, and R.E. Vesley. Amsterdam: North-Holland Publ. Co. (1958).

1972 Which number theoretic problems can be solved in recursive progressions on Π_1^1-paths through O? *J. Symb. Log.* **37** (1972) 311–334.

Paris, J., and L. Harrington

1977 A mathematical incompleteness in Peano arithmetic. In: *Handbook of Mathematical Logic,* ed. J. Barwise. Amsterdam: North-Holland Publ. Co. (1977).

Post, E.L.

1944 Recursively enumerable sets of positive integers and their decision problems. *B. Am. Math. Soc.* **50** (1944) 284–316.

Simpson, S.G.

1986 Unprovable theorems and fast-growing functions. In: *Logic and Combinatorics,* ed. S.G. Simpson, pp. 359–394. Providence, RI: American Mathematical Society (1986).

Takeuti, G.

1987 *Proof Theory,* 2nd. ed. Amsterdam: North-Holland Publ. Co. (1987).

Turing, A.M.

1936-37 On computable numbers, with an application to the Entscheidungsproblem. *P. Lond. Math. Soc. (2)* **42** (1936-7) 230–265; A correction, ibid. **43** (1937) 544–546 (reprinted in *Davis 1965*).

1937 Computability and λ-definability. *J. Symb. Log.* **2** (1937) 153–163.

1939 Systems of logic based on ordinals. *P. Lond. Math. Soc. (2)* **45** (1939) 161–228 (reprinted in *Davis 1965*).

1969 Intelligent machinery. In: *Machine Intelligence 5,* eds. B. Meltzer and D. Michie. Edinburgh: Edinburgh University Press (1969).

Visser, A.

1981 An incompleteness result for paths through or within O. *P. Kon. Ned. A* **84(2)** (1981) 237–243; also pub. in *Ind. Math.* **43.**

Weyl, H.

1918 *Das Kontinuum. Kritische Untersuchungen über die Grundlagen der Analysis.* Leipzig: Veit (1918).

Mathematical Logic and the
Origin of Modern Computers*

Martin Davis

The very word *computer* immediately suggests one of the main uses of these re-markable devices: an instrument of calculation. But it is a matter of widespread experience that modern computers can be used for many purposes having no evident connection with numerical computation. The main thesis of this article is that the source of this generalized conception of the scope of computers is to be found in the vision of a computer as an engine of logic implicit in the abstract theory of computation developed by mathematical logicians.

The connection between logic and computing is apparent even from the every-day use of language: the English word "reckon" means both to calculate and to conclude. Without trying to understand this connection in any very profound man-ner, we can certainly see that computation is a (very restricted) form of reasoning. To see the connection in the opposite direction, imagine our seeking to demonstrate to a skeptic that some conclusion follows logically from certain assumptions. We present a "proof" that our claim is correct, only to be faced by the demand that we demonstrate that our proof is correct. If we then attempt a proof that our previous "proof" was correct, we clearly are faced with an infinite regress. The way out that has been found is to insist on a purely algorithmic criterion for logical correctness – a proof is correct if it proceeds according to rules whose correct application can be verified in a purely computational manner.

There are many examples of important concepts and methods first introduced by logicians which later proved to be important in computer science. Tracing the paths along which some of these ideas found their way from theory to practice is a fascinating (and often frustrating) task for the historian of ideas. The subject of this paper is Alan Turing's discovery of the universal (or all-purpose) digital computer as a mathematical abstraction. This concept was introduced by Turing as part of

* This paper was first published in *Studies in the History of Mathematics,* ©1987 by The Mathematical Association of America, and is reprinted here with their permission.

his solution to a problem that David Hilbert had called the "principal problem of mathematical logic" (Hilbert and Ackermann *1928*). We will try to show how this very abstract work helped to lead Turing and John von Neumann to the modern concept of the electronic computer.

But first, before discussing the work of Alan Turing, we will see how some of the underlying themes of computer science had already appeared in the seventeenth century in the work of G.W. Leibniz (1646–1716).

1. Leibniz' Dream

It is striking to note the many different ways in which Gottfried Leibniz anticipated what later came to be central concerns in computer science. He made an important invention, the so-called *Leibniz wheel,* which he used as early as the 1670's to build a mechanical calculating machine that could add, subtract, multiply, and divide. He showed keen awareness of the great advantages to be expected from the mechanization of computation. Thus, Leibniz said of his calculator:

> And now that we may give final praise to the machine we may say that it will be desirable to all who are engaged in computations ... managers of financial affairs, merchants, surveyors, geographers, navigators, astronomers ... But limiting ourselves to scientific uses, the old geometric and astronomic tables could be corrected and new ones constructed. ... it will pay to extend as far as possible the major Pythagorean tables; the table of squares, cubes, and other powers; and the tables of combination, variations, and progressions of all kinds, ... Also the astronomers surely will not have to continue to exercise the patience which is required for computation. ... For it is unworthy of excellent men to lose hours like slaves in the labor of computation. (Smith *1929*, pp. 180–181)

Leibniz was one of the first (Ceruzzi *1983*, p. 40, footnote 11) to work out the properties of the binary number system, which of course has turned out to be fundamental for computer science. He proposed the development of a *calculus of reason* or *calculus ratiocinator* and actually proceeded to develop what amounts to a fragment of Boolean algebra (Parkinson *1966*, pp. 132–133; Davis *1983*, pp. 2–3). Finally, there was Leibniz' amazing program calling for the development of a universal language – a *lingua characteristica* – which would not only incorporate the calculus ratiocinator, but would also be suitable for communication and would include scientific and mathematical knowledge. Leibniz hoped to mechanize much of thought, saying that the mind "will be freed from having to think directly of things themselves, and yet everything will come out correctly" (Parkinson *1966*, p. xvii). Leibniz imagined problems in human affairs being handled by a learned committee sitting around a table and saying (Kneale and Kneale *1962*, p. 328): *"Calculemus"* i.e., "Let us calculate!"

The importance with which Leibniz regarded these projects is clear from his assessment:

> For if praise is given to the men who have determined the number of regular solids – which is of no use, except insofar as it is pleasant to contemplate – and if it is thought to be an exercise worthy of a mathematical genius to have brought to light the more elegant properties of a conchoid of cissoid, or some other figure which rarely has any use, how much better will it be to bring under mathematical laws human reasoning, which is the most excellent and useful thing we have. (Parkinson *1966,* p. 105)

It is at times amusing to imagine some great person from a past age reacting to one of the marvels of the contemporary world. Confronted by a modern computer, Leibniz would surely have been awestruck by the wonders of twentieth century technology. But perhaps he would have been better equipped than any other seventeenth century person to comprehend the scope and potential of these amazing machines.

2. *Alan Turing's Analysis of the Concept of Computation*

A century and a half ago, Charles Babbage already had conceived of an all-purpose automatic calculating machine, his proposed but never constructed *analytical engine.* Babbage's device was intended to carry out numerical computations of the most varied kind that arise in algebra and mathematical analysis. To emphasize the power and scope of his engine, Babbage remarked facetiously that "it could do everything but compose country dances" (Huskey and Huskey *1980,* p. 300). A contemporary computer expert seeking a figure of speech to bring home to a popular audience the widespread applicability of computers would select a different example. For we know that today's computers can perfectly well be programmed to compose country dances (although presumably not of the finest quality). While for Babbage it was self-evident that calculating machines could not be expected to compose dances, it does not strike us today as being at all out of the question. Clearly, our very concept of what constitutes "computation" has been altered drastically. We shall see how the modern view of computation developed out of the work in mathematical logic of Alan Turing.

Babbage never succeeded in constructing his engine, in large part because of the limitations of nineteenth century technology. In fact, it was only with some of the electro-mechanical calculators that began to be built during the 1930's (for example, by Howard Aiken at Harvard University) that Babbage's vision was fully realized. But during the 1930's and 1940's no one involved with this work suggested the possibility of designing an automatic computer that not only could do everything that Babbage had envisioned, but also could be used for commercial purposes, or for that matter, to "compose country dances". Even as late as 1956, Howard Aiken, himself a pioneer of modern computing, could write:

> If it should turn out that the basic logics of a machine designed for the numerical solution of differential equations coincide with the logics of a machine intended to make bills for a department store, I would regard this as the most amazing coincidence that I have ever encountered. (Ceruzzi *1983,* p. 43)

If Aiken had grasped the significance of a paper by Alan Turing that had been published two decades earlier (Turing *1936-7*), he would never have found himself in the position of making such a statement only a few years before machines that performed quite well at both of the tasks he listed were readily available.

Alan Mathison Turing was born on June 23, 1912 in London. His father was a civil servant in India, and Turing spent most of his childhood away from his parents.[1] After five years at Sherborne, a traditional English public school, he was awarded a fellowship to King's College at Cambridge University. Turing arrived at Cambridge in 1931. This was just after the young logician Kurt Gödel had startled the mathematical world by demonstrating that for any formal system adequate for elementary number theory, arithmetic assertions could be found that were not decidable within that formal system. In fact Gödel had even shown that among these "undecidable propositions" was the very assertion that the given formal system itself is consistent. This last result was devastating to Hilbert's program in the foundations of mathematics, which called for proving the consistency of more and more powerful formal systems using only very restricted proof methods, methods that Hilbert called *finitistic.* John von Neumann was probably the most brilliant of the young people who had been striving to carry out Hilbert's program. In addition to his contributions to Hilbert's program, von Neumann's intense interest in logic and foundations is also evidenced by his early papers on axiomatic set theory (von Neumann *1961*). However, after Gödel's discovery, von Neumann stopped working in this field. In the spring of 1935, Turing attended a course of lectures by the topologist M.H.A. Newman on *Foundations of Mathematics* in which Hilbert's program and Gödel's work were among the topics discussed. In particular, Newman called the attention of his audience to Hilbert's *Entscheidungsproblem,* a problem which Hilbert had called the "principal problem of mathematical logic".

In 1928, a little textbook of logic by Hilbert and Wilhelm Ackermann, entitled *Grundzüge der theoretischen Logik,* had been published. The book emphasized first order logic, the logic of *and, or, not, if … then, for all,* and *there exists,* which the authors called the *engere Funktionenkalkül.* The authors showed how the various parts of mathematics could be formalized within first order logic, and a simple set of

---

1 In his authoritative biography, Andrew Hodges (*1983,* p. 132) quotes Turing as having, on at least one occasion, attributed his homosexuality to his childhood in boarding schools in England far from his parents in India. (Hodges himself makes it clear that he does not accept this explanation.)

rules of proof was given for making logical inferences. They noted that any inference that can be carried out according to their rules of proof is also *valid,* in the sense that in any mathematical structure in which all the premises are true, the conclusion is also true. Hilbert and Ackermann then raised the problem of *completeness:* if an inference is valid (in the sense just explained), would it always be possible, using their rules of proof, to obtain the conclusion from the premises? This question was answered affirmatively two years later by Gödel in his doctoral dissertation at the University of Vienna. Another problem raised in the *Grundzüge* by Hilbert and Ackermann was the Entscheidungsproblem, the problem of finding an algorithm to determine whether a given proposed inference is valid. By the completeness theorem from Gödel's dissertation, this problem is equivalent to seeking an algorithm for determining whether a particular conclusion may be derived from certain premises using the Hilbert-Ackermann rules of proof. The Entscheidungsproblem was called the "principal problem of mathematical logic", because an algorithm for the Entscheidungsproblem could, in principle, be used to answer any mathematical question: it would suffice to employ a formalization in first-order logic of the branch of mathematics relevant to the question under consideration. Alan Turing's attention was drawn to the Entscheidungsproblem by Newman's lectures, and he soon saw how to settle the problem negatively. That is, Turing showed that no algorithm exists for solving the Entscheidungsproblem. The tools that Turing developed for this purpose have turned out to be absolutely fundamental for computer science.

If a positive solution of the Entscheidungsproblem would lead to algorithms for settling all mathematical questions, then it must follow that if there is even one problem that has no algorithmic solution, then the Entscheidungsproblem itself must have no algorithmic solution. Now, the intuitive notion of *algorithm* serves perfectly well when what we only need to verify is that some proposed procedure does indeed constitute a positive solution to a given problem. However, remaining at this intuitive level, we could not hope to prove that some problem has no algorithmic solution. In order to be certain that no algorithm will work, it would appear necessary to somehow survey the class of all possible algorithms. This is the task that Turing set himself.

Turing began with a human being who would carry out the successive steps called for by some algorithm; that is, Turing proposed to consider the behavior of a "computer". Here the word *computer* refers to a person carrying out a computation; this was how Turing (and everyone else) used the word in 1935. Turing then proceeded (Turing *1936-7*), by a sequence of simplifications, each of which could be seen to make no essential difference, to obtain his characterization of computability.

Turing's first simplification was to assume that "the computation is carried out on one-dimensional paper, i.e., on a tape divided into squares" since "it will be agreed that the two-dimensional character of paper is no essential of computation". Turing continued:

> I shall also assume that the number of symbols ... used ... is finite. ... this restriction ... is not very serious. It is always possible to use sequences of symbols in the place of single symbols. ... The behavior of the computer at any moment is determined by the symbols which he is observing, and his 'state of mind' at that moment. We may suppose that there is a bound B to the number of ... squares which the computer can observe at one moment. We will also suppose that the number of states of mind which need be taken into account is finite.[2]

Turing next argues that the entire computation can be thought of as consisting of atomic "simple" steps, each of which consists of:

1. a change of one symbol in one of the B "observed" squares (changes of more than one symbol can always be reduced to successive changes of a single symbol);
2. changes from the squares currently being "observed" to other squares no more than a distance of L squares away, where L is some constant; and
3. a change in the "state of mind" of the "computer".

The final step in Turing's analysis is the crucial remark that the outcome will in no way be altered if the human computer is replaced by a machine capable of a finite number of distinct states, or "m-configurations" as Turing called them, corresponding to the different states of mind the human computer possesses in the course of the computation. Each m-configuration is then completely characterized by a table which specifies the changes of types 1, 2, and 3 above, corresponding to each possible set of observed squares and symbols appearing in those squares. Hence, any algorithmic process can be carried out by machine, and, moreover, by a machine that conceptually is quite simple. As an addendum to his analysis, Turing pointed out that his machines could be simplified even further, permitting only one square at a time to be observed or "scanned" and allowing only a change of the observed square to the square immediately to the left or immediately to the right. Unlike the previous simplifications, this one can be justified rigorously: one can *prove* that nothing is lost by the restriction to the simpler machines.

We have arrived at the famous notion of a *Turing machine*: a Turing machine can be specified by a finite set of quintuples, each having one of the three forms:

$$p\alpha\beta Rq \quad \text{or} \quad p\alpha\beta Lq \quad \text{or} \quad p\alpha\beta Nq$$

Such a quintuple signifies that if the machine is in m-configuration p and the symbol α appears in the scanned square, then the machine will replace α by β, enter m-configuration q, and move one square to the right, move one square to the left, or not move at all, depending on whether the third symbol in the quintuple is R, L, or N, respectively. The quintuples which constitute a particular Turing machine

---

2 For discussion of the significance of this "finite mental states" hypothesis, see Wang *1974*, p. 93, and Webb *1980*, pp. 8 and 221–222.

determine the course of a computation. Turing's machines were to be *deterministic* in the sense that no two of its quintuples are allowed to begin with the same pair p, α, so that, at any particular time, no more than one action could be called for. If a machine were to be in m-configuration p scanning a symbol α, such that *no* quintuple of the machine begins with the pair $p\alpha$, then the computation would be at an end – the machine would "halt". At any stage of a computation by a Turing machine, there will be only a finite number of nonblank squares on the tape. But it is crucial that there be no a priori bound on this number. Thus, the tape is infinite, in the sense that additional (possibly blank) squares are always available to the right of the scanned square.

If Turing's analysis is accepted, then it may be concluded that any algorithmic process whatever, can be carried out by one of these Turing machines. In particular, if the Entscheidungsproblem has an algorithmic solution, then it can be solved by a Turing machine. Moreover, for the very reason that made it appropriate to call the Entscheidungsproblem "the principal problem of mathematical logic", if any problem at all can be shown to be unsolvable by Turing machines, the unsolvability of the Entscheidungsproblem should readily follow.

In obtaining his undecidable arithmetic propositions, Gödel had made use of a diagonal method that was formally similar to Cantor's proof that the set of real numbers is not countable. Therefore, it was quite natural that Turing should think of applying diagonalization. He called a real number[3] in the interval $(0,1)$ *computable* if there exists a Turing machine that, beginning with a blank tape, successively prints the infinite sequences of 0's and 1's constituting the number's binary expansion. (Symbols other than 0 and 1 printed by the machine are just ignored for this purpose.) A machine that computes a real number in this sense was called *circle-free*; one that does not (because it never prints more than a finite number of 0's and 1's) was called *circular*. Since there are only countably many distinct finite sets of quintuples, the set of computable real numbers is likewise countable. Yet superficially, it appears that one could argue the reverse. As Turing expressed it, speaking in terms of the *sequences* if 0's and 1's corresponding to computable real numbers:

> If the computable sequences are enumerable, let α_n be the n-th computable sequence, and let $\phi_n(m)$ be the m-th figure in α_n. Let β be the sequence with $1 - \phi_n(n)$ as its n-th figure. Since β is computable, there exists a number K such that $1 - \phi_n(n) = \phi_K(n)$ all n. Putting $n = K$, we have $1 = 2\phi_K(K)$, i.e., 1 is even. This is impossible. The computable sequences are therefore not enumerable. (Turing *1936-7*, p. 246; reprinted in Davis *1965*, p. 132)

Turing continued (the words in brackets are substituted for Turing's for expository reasons):

---

3 To include numbers outside the interval $(0,1)$, Turing simply declared any number of the form $n + x$ computable if n is an integer and x is a computable number in the interval $(0,1)$.

The fallacy in this argument lies in the assumption that β is computable. It would be true if we could enumerate the computable sequences by finite means, but the problem of enumerating computable sequences is equivalent to the problem of finding out whether a given [finite set of quintuples determines] a circle-free machine, and we have no general process for doing this in a finite number of steps. In fact, by applying the diagonal process argument correctly, we can show that there cannot be any such general process.

In other words if there were such a "general process", it could be used to delete non-circle-free machines from an enumeration of all possible finite sets of quintuples, thereby producing an enumeration of "the computable sequences by finite means". But then β, as defined above, would be computable, and we would be led to a contradiction. The only way out is to conclude that "there cannot be any such general process".[4] The problem of determining whether the Turing machine defined by a given finite set of quintuples is circle-free has no algorithmic solution! Of course, in this form the result depends on accepting Turing's analysis of the computation process. However, it is possible to state the result in the form of a rigorously proved theorem about Turing machines. For this purpose, let us imagine the very quintuples constituting a Turing machine themselves placed on a Turing machine tape. We could then seek to construct a Turing machine M which, begun with a set of quintuples defining any particular Turing machine N on its tape, will eventually halt with an affirmative or a negative message on its tape, according as N is or is not circle-free. Turing's argument can then be used to prove that there cannot be such a Turing machine M. (To make this entirely precise, it is necessary to be explicit as to how the quintuples, as well as the affirmative and negative output messages, are to be coded on the tape in terms of some finite alphabet. But this causes no difficulty.)

Turing next showed that there is no algorithm for the:

Blank Tape Printing Problem: *To determine whether a given Turing machine, starting with a blank tape, will ever print some particular symbol, say |.*

Turing's proof of this result proceeds by showing that if there were such an algorithm, then there must also be an algorithm for determining whether a given Turing machine is circle-free. This argument is a bit complicated,[5] and we outline a simpler proof that uses another diagonalization. First we show that there is no algorithm for the:

---

4 Turing recognized that although this proof is "perfectly sound", it "may leave the reader with a feeling that 'there must be something wrong'", and he therefore supplied another proof that does not so closely approach paradox.

5 A remark for the knowledgeable: the problem of determining whether a given Turing machine is circle-free is complete of degree $\mathbf{0}''$, and therefore can not be reduced to either of the printing problems.

General Printing Problem: *To determine whether a given Turing machine, starting with a given string of symbols on its tape, will ever print |.*

Suppose there were an algorithm for this problem. Then, in particular, there would be an algorithm to determine whether a given Turing machine, *starting with its own set of quintuples on its tape,* will ever print |. So, it would follow from Turing's analysis that a Turing machine M could be constructed that would respond to a set of quintuples on its tape by printing | if and only if the machine defined by that set of quintuples *never* prints | when started with its own set of quintuples on its tape. Now, what happens when M is started with *its own set of quintuples* on its tape? It eventually prints | if and only if it never prints |! This contradiction shows that there can be no algorithm for the general printing problem. Finally, if there were an algorithm for the blank tape printing problem, it could also be used to solve the general printing problem. Namely, given the quintuples constituting a Turing machine M together with the string of symbols σ on its tape, we can construct a machine N that, beginning with a blank tape, first prints the string σ, and then behaves exactly like M. So N will eventually print | beginning with a blank tape if and only if M will eventually print | beginning with the string σ on its tape.

Turing used the fact that there is no algorithm for the blank tape printing problem to show that Hilbert's Entscheidungsproblem is likewise unsolvable. With each Turing machine M, he associated a formula $\alpha(M)$ of first-order logic which, roughly speaking, describes the behavior of M starting with a blank tape. He constructed a second formula β which has the interpretation that the symbol | eventually appears on the tape. It was then not difficult to see that β follows from $\alpha(M)$ by the Hilbert-Ackermann rules if and only if the Turing machine M eventually prints |. Thus, an algorithm for the Entscheidungsproblem would lead to an algorithm for the blank tape printing problem.

The notion of Turing machine was developed in order to solve Hilbert's Entscheidungsproblem. But it also enabled Turing to realize that it was possible to conceive of a single machine that was capable of performing all possible computations. As Turing expressed it: "It is possible to invent a single machine which can be used to compute any computable sequence." Turing called such a machine *universal*. A Turing machine U was to be called universal if, when started with a (suitably coded) finite set of quintuples defining a Turing machine M on its tape, U would proceed to compute the very same sequence of 0's and 1's that M would compute (beginning with an empty tape). Now, intuitively speaking, there clearly exists an algorithm that does what is required of the universal machine U; the algorithm just amounts to carrying out the instructions expressed by M's quintuples. Thus, the existence of a universal Turing machine is a consequence of Turing's analysis of the concept of computation. On the other hand, it is a rather implausible consequence. Why should we expect a single mechanism to be able to carry out algorithms for "the numerical solution of differential equations" as well as those needed to "make bills for a department store"? However, Turing did not simply depend on the validity of his analysis. He proceeded to produce in detail the actual quintuples needed to define a universal machine. Thus, in the light of the apparent implausibility of the existence of such a machine, Turing was entitled to regard his success in constructing one as a significant vindication of his analysis. The universal machine U actually given by Turing can be thought of as being specified by what is nowadays called an *interpretative* program. U operates by scanning the coded instructions (that is quintuples) on its tape and then proceeding to carry them out. Of course, many interpretative programs have been constructed in recent years to make it possible to run programs written in such languages as BASIC, LISP, SNOBOL, and PROLOG, but Turing's was the first.

Turing's analysis provided a new and profound insight into the ancient craft of computing. The notion of computation was seen as embracing far more than arithmetic and algebraic calculations. And at the same time, there emerged the vision of universal machines that "in principle" could compute everything that is computable. Turing's examples of specific machines were already instances of the art of programming; the universal machine in particular was the first example of an interpretative program. The universal machine also provided a model of a "stored program" computer in which the coded quintuples on the tape play the role of a stored program, and in which the machine makes no fundamental distinction between "program" and "data". Finally, the universal machine showed how "hardware" in the form of a set of quintuples thought of as a description of the functioning of a mechanism can be replaced by equivalent "software" in the form of those same quintuples in coded form "stored" on the tape of a universal machine.

While working out his proof that there is no algorithmic solution to the Entscheidungsproblem, Turing did not suspect that similar conclusions were being reached on the other side of the Atlantic. In fact, Newman had already received the first draft of Turing's paper, when an issue of the *American Journal of Mathematics* arrived in Cambridge containing an article by Alonzo Church of Princeton University, entitled "An Unsolvable Problem of Elementary Number Theory". In this paper, Church had already shown that there were algorithmically unsolvable problems. His paper did not mention machines, but it did point to two concepts, each of which had been proposed as explications of the intuitive notion of computability or, as Church put it, "effective calculability". The two concepts were λ-*definability,* developed by Church and his student Stephen Kleene, and *general recursiveness,* proposed by Gödel (in lectures at the Institute for Advanced Study in Princeton during the spring of 1934) as a modification of an idea of J. Herbrand. The two notions had been proved to be equivalent, and Church's unsolvable problem was in fact unsolvable with respect to either equivalent notion. Although in this paper Church had not drawn the conclusion that Hilbert's Entscheidungsproblem was itself unsolvable with respect to these notions, volume 1 (1936), number 1 of the new quarterly *Journal of Symbolic*

Logic contained a brief note by Church in which he did exactly that. A later issue of volume 1 of the same journal contained an article by Emil Post, taking cognizance of Church's work, but proposing a formulation of computability very much like Turing's. Turing quickly showed that his notion of computability was equivalent to λ-definability, and he decided to attempt to spend some time in Princeton.

Thus, much of what Turing had accomplished amounted to a rediscovery of what had already been done in the United States. But his analysis of the notion of computation and his discovery of the universal computing machine were entirely novel, going beyond anything that had been done in Princeton. In particular, although Gödel had remained unconvinced by the evidence available in Princeton, that Church's proposal to identify effective calculability with the two equivalent proposed notions was correct, Turing's analysis finally convinced him.[6]

Turing was at Princeton for two academic years beginning in the summer of 1936. Formally, he was a graduate student, and indeed he did complete the requirements for a doctorate with Alonzo Church as his thesis adviser. His doctoral dissertation was his deep and important paper Turing *1939*, in which he studied the effect on Gödel undecidability of transfinite sequences of formal systems of increasing strength. This paper also introduced the key notion of an *oracle*, which made it possible to classify unsolvable problems,[7] and which is playing a very important role in current research in theoretical computer science.[8] Some writers have been confused about the circumstances under which Turing was a graduate student at Princeton, and have assumed that Turing's earlier work on computability had been done at Princeton under Church's supervision. A circumstance that may have helped lead to this confusion is that the published account (Turing *1936-7*) of the work on computability concludes with an appendix (in which a proof is outlined of the equivalence of Turing's concept of computability with Church's λ-definability) dated August 28, 1936 at "The Graduate College, Princeton University, New Jersey, U.S.A".[9]

---

6 For a discussion of some of the historical issues involved in these developments, as well as references, see Davis *1982*.

7 Thus suppose that we could somehow come to possess an oracle or "black box" which can tell us for a given set of quintuples whether the Turing machine defined by that set of quintuples is circle-free. Then it is not difficult to show that it is possible to construct a Turing machine which can solve either of the two unsolvable printing problems if only the machine is permitted to ask the oracle questions and make use of the answers. However, this will not work in the other direction. This is expressed by saying that the circle-free problem is of a *higher degree of unsolvability* than the printing problems.

8 Many important open questions in computer science ask whether certain inclusions between classes of sets of strings are proper. (The famous $P = NP$ problem is of this character.) In many cases (including the $P = NP$ problem), although the original question remains unresolved, it has proved possible to obtain answers when the problem is modified to permit access to suitable oracles.

9 Actually even this appendix must have been completed before Turing left England. Turing's departure was on September 23.

Fine Hall[10] in 1936 housed not only the mathematics faculty of Princeton University, but also the mathematicians who were part of the recently established Institute for Advanced Study. The great influx to the United States of scientists fleeing the Nazi regime had begun. The concentration of mathematical talent at Princeton during the 1930's came to rival and then surpass that at Göttingen, where David Hilbert held sway. Among those to be seen in the corridors of Fine Hall were Solomon Lefschetz, Hermann Weyl, Albert Einstein, and ... John von Neumann.[11] During Turing's second year at Princeton, he held the prestigious Procter Fellowship. Among the letters of recommendation written in support of his application was the following:

June 1, 1937

Sir,

 Mr. A.M. Turing has informed me that he is applying for a Proctor [sic] Visiting Fellowship to Princeton University from Cambridge for the academic year 1937–1938. I should like to support his application and to inform you that I know Mr. Turing very well from previous years: during the last term of 1935, when I was a visiting professor in Cambridge, and during 1936–1937, which year Mr Turing has spent in Princeton, I had opportunity to observe his scientific work. *He has done good work in branches of mathematics in which I am interested, namely: theory of almost periodic functions, and theory of continuous groups.* [emphasis added]

 I think that he is a most deserving candidate for the Proctor [sic] Fellowship, and I should be very glad if you should find it possible to award one to him.

 I am, Respectfully, John von Neumann (Hodges *1983,* p. 131)

Thus, as late as June 1937, either von Neumann was unaware of Turing's work on computability, or he did not think it appropriate to mention it in a letter of recommendation. There have been tantalizing rumors of important discussions between the two mathematicians about computing machinery, during the Princeton years, or later, during the Second World War. But there does not appear to be any real evidence that such discussions ever took place.[12] However, von Neumann's friend and collaborator Stanislaw Ulam, in a letter to Andrew Hodges (Hodges *1983,* p. 145),

---

10 Fine Hall in 1936 (and indeed through the 1950's) was a low-level attractive red brick building. The building where Princeton's mathematics department is housed today is also called Fine Hall; it is visible as a concrete tower from Highway US 1, a mile away.

11 Kurt Gödel, who had lectured at the Institute for Advanced Study during the spring of 1934, was unfortunately not in Princeton during Turing's stay. Gödel left Princeton in the fall of 1935 and did not return until after the Second World War had begun.

12 In the doctoral dissertation Aspray *1980,* (pp. 147–148) there is a reference to discussions between Turing and von Neumann at this time, on the question of whether "computing machines could be built which would adequately model any mental feature of the human brain". Aspray based his account on an interview with J.B. Rosser. However, in a conversation with the present author, Aspray explained that Rosser had not claimed to have himself overheard such discussions, and that Rosser had been unable to remember his source. Aspray indicated that he no longer believes that such a conversation actually occurred. In a recent letter, Alonzo Church indicates that he neither recalled nor could find any record of such "consultations". See also Randell *1972.*

mentioned a game that von Neumann had proposed during the summer of 1938 when he and Ulam were traveling together in Europe; the game involved "writing down on a piece of paper as big a number as we could, defining it by a method which indeed has something to do with some schemata of Turing's".[13] Ulam's letter also stated that "von Neumann mentioned to me Turing's name several times in 1939 in conversations, concerning mechanical ways to develop formal mathematical systems." On the basis of Ulam's letter it seems safe to conclude that, by the outbreak of the Second World War in September 1939, von Neumann was well aware of Turing's work on computability and regarded it as important.

When did Turing begin to think about the possibility of constructing a physical device that would be, in some appropriate sense, an embodiment of his universal machine? According to Turing's teacher, M.H.A. Newman, this was in Turing's mind from the very first. In an obituary article in *The Times,* Newman wrote:

> The description that he then gave of a "universal" computing machine was entirely theoretical in purpose, but Turing's strong interest in all kinds of practical experiment made him even then interested in the possibility of actually constructing a machine on these lines. (quoted in Hodges *1983,* p. 545)

In Princeton, Turing's "practical" interests included a developing concern with cryptanalysis. Possibly in this connection, he designed an electro-mechanical binary multiplier, and gaining access to the Physics Department graduate student machine shop,[14] he constructed various parts of the device, building the necessary relays himself. Another of Turing's interests during this period – an interest which combined the theoretical with practical computation – was the famous Riemann Hypothesis concerning the distribution of the zeros of the Riemann ζ-function. Shortly after Turing returned to England in the summer of 1938, he applied for and was granted $40 to build a special purpose analogue computer for computing Riemann's ζ-function,[15] which Turing hoped to use to test the Riemann hypothesis numerically (Hodges *1983,* pp. 138–140, 155–158). But even as Turing was beginning serious work on this machine, the Second World War intervened and moved him in quite another direction. The ζ-function machine was never completed.

---

13 This sounds very much as though von Neumann had anticipated the important Chaitin-Kolmogoroff notion of descriptive complexity.

14 The Palmer Physics laboratory was located next door to Fine Hall – there was even a convenient passageway joining the two buildings.

15 The design of this computer was based on that of a machine in Liverpool which was used to predict the tides.

3. *To Build a Brain*

Turing spent the war years at Bletchley Park, a country mansion that housed Britain's brilliant group of cryptanalysts. The Germans had developed improved versions of a commercial encrypting machine, the *Enigma*. The task of breaking the enigma code fell to the group at Bletchley Park. They were given a head start in their task by having access to the work of a group of Polish mathematicians who had succeeded with an earlier and considerably simpler version of the Enigma. Building on this work, Turing and Gordon Welchman (an algebraic geometer from Cambridge) progressed to the point where machines, called Bombes (the name first used by the Poles for their much more primitive device) could be built to decode everyday German military communications. Naval communications were Turing's special province, and by the summer of 1941, the information derived from the Bombes enabled the British Admiralty to defeat the German submarine offensive against Atlantic shipping that had been threatening to strangle a beleaguered Britain (Welchman *1982*; Hodges *1983*, pp. 160–210).[16] But this great success was a precarious one. It was clear that if the Germans introduced more complexity into their procedures, the Bombes would be overwhelmed. And so, more ambitious machines (which indeed turned out to be necessary) were constructed: first the Heath Robinson series, and later the Colossus. The latter, constructed in 1943 under the direct supervision of M.H.A. Newman, used vacuum tube circuits to carry out complex Boolean computations very rapidly. The Colossus contained 1500 tubes and was built in the face of skepticism on the part of the engineers that so many vacuum tubes could work together without a failure, long enough to get useful work done.

Thus, when the war ended, Turing had a solid basic knowledge of electronics, and was aware that large scale computing machines could be constructed using electronic circuits. The significance for Turing of this practical knowledge can not be fully grasped without taking into account the new conceptual framework for thinking about computing to which his work on computability had led him. For Turing had been led to conclude that computation was simply carrying out the steps in some "rule of thumb" process (as Turing expressed it in an address to the London Mathematical Society (Turing *1947*, p. 107)). A "rule of thumb" process is to be understood as one which can be carried out simply by following a list

---

16 Another of Turing's contributions to this effort was the invention of new statistical methods for dealing with the vast quantities of data contained in the Enigma "traffic". These methods were later rediscovered independently by the American statistician A. Wald who gave them the name *sequential analysis*. Surely there cannot be many Britains whose contribution to the ultimate victory approached Turing's. A little over a decade after the turning of the tide in the "Battle of the Atlantic", being duly convicted of performing acts "of gross indecency", Turing was sentenced in a British court of law to a one year probation term, during which time he was required to submit to a course of hormone treatments that amounted to a temporary chemical castration. Such was the hero's reward!

of unambiguous instructions referring to finite discrete configurations of whatever kind. Turing's work had also shown that, without loss of generality, one could restrict oneself to instructions of an extremely simple kind. Finally, it was possible to construct a single mechanical device capable, in principle, of carrying out any computation whatever. "It can be shown that a single machine ... can be made to do the work of all" (Turing *1947*, p. 112). As exciting as this prospect must have appeared, it was only part of Turing's remarkable vision. Turing dared to imagine not only that computation encompassed far more than mere calculation, but that it actually included the human mental processes that we call "thought". He was interested in much more than a machine capable of very rapid computation; Alan Turing wanted to build a brain. This vision had been the subject of much discussion at Bletchley Park, where Turing focused on chess as an example of human "thought" that should be capable of mechanization. The full scope of Turing's thought was only exposed to the public later, in Turing *1947* and in his now classical essay Turing *1950*. In 1947 Turing was already speaking of circumstances in which "one is obliged to regard the machine as showing intelligence. As soon as one can provide a reasonably large memory capacity it should be possible to experiment along these lines" (Turing *1947*, p. 123). As we shall see, Turing was eager to help build such machines. But for a second time, Turing's professional life was profoundly affected by developments in the Western Hemisphere.

4. *Von Neumann and the Moore School*

As has already been noted, by the summer of 1938 von Neumann was very much aware of Turing's work on computability. There is also evidence that, early on, he perceived that Turing's work had implications for the practice of computation. A wartime colleague of von Neumann recalled that "in about 1943 or 44 von Neumann was well aware of the fundamental importance of Turing's paper of 1936 ... and at his urging I studied it ... he emphasized ... that the fundamental conception is owing to Turing ..." (Hodges *1983*, pp. 145, 304). Herman Goldstine (who was von Neumann's close collaborator) said, "There is no doubt that von Neumann was thoroughly aware of Turing's work ..." (Goldstine *1972*, p. 174).

As with Turing, von Neumann's wartime work involved large-scale computation. But, where the cryptoanalytic work at Bletchley Park emphasized the discrete combinatorial side of computation, so in tune with Turing's earlier work, it was old-fashioned, heavy, number-crunching that von Neumann needed. Although he had tried to inform himself about new developments in computational equipment, von Neumann learned of the ENIAC project quite fortuitously on meeting the young mathematician Herman Goldstine at a railway station during the summer of 1944.

Von Neumann quickly became a participant in discussions with the ENIAC group at the Moore School in Philadelphia.

The Colossus with its 1500 vacuum tubes was an engineering marvel. The ENIAC with 18,000 tubes was simply astonishing. The conventional wisdom of the time was that no such assemblage could do reliable work; it was held that the mean free path between vacuum tube failures would be a matter of seconds. It was the chief engineer on the ENIAC project, John Prosper Eckert, Jr., who was largely responsible for the project's success. Eckert insisted on extremely high standards of component reliability. Tubes were operated at extremely conservative power levels, and the failure rate was kept to three tubes per week. The ENIAC was an enormous machine, occupying a large room. It was a decimal machine and was programmed by connecting cables to a plugboard (Burks and Burks *1981*), rather like an old-fashioned telephone switchboard.

By the time that von Neumann began meeting with the Moore School group, it was clear that there were no important obstacles to the successful completion of the ENIAC, and attention was focused on the next computer to be built, tentatively called the EDVAC. Von Neumann immediately involved himself with the problems of the *logical* organization of the new machine. As Goldstine (*1972*, p. 186) recalls, "Eckert was delighted that von Neumann was so keenly interested in the logical problems surrounding the new idea, and these meetings were scenes of greatest intellectual activity." Goldstine comments:

> This work on the logical plan for the new machine was exactly to von Neumann's liking and precisely where his previous work on formal logics [sic] came to play a decisive role. Prior to his appearance on the scene, the group at the Moore School concentrated primarily on the *technological* problems, which were very great; after his arrival he took over leadership on the *logical* problems. (Goldstine *1972*, p. 188)

A key idea emphasized in the meetings was that any significant advance over the ENIAC would require a substantial capacity for the internal storage of information. This was because communication with the exterior would be at speeds far slower than the internal electronic speeds at which the computer could function, and therefore constituted a potential bottleneck. Once again, John Eckert played a crucial role. He had previously shown how to modify a device called a delay line (originally developed by the physicist W.B. Shockley, who later invented the transistor) so as to be a working component of radar systems. These delay lines (which stored information in the form of a vibrating tube of mercury) were just what was needed.

The communication bottleneck just mentioned would be evident in the case of any computation involving the manipulation of large quantities of data. But it was even more crucial for the *instructions* that the computer would carry out. Indeed, it would make little sense for a computer to produce the results of a calculation rapidly, only to wait idly for the next instruction. The solution was to store the instructions internally with the data: what has come to be called the "stored program concept".

The computers of the postwar period differed from previous calculating devices in having provision for internal storage of programs as well as data. But they were different in another more fundamental way. They were conceived, designed, and constructed, not as mere automatic calculators, but as *engines of logic,* incorporating the general notion of what it means to be computable and embodying a physical model of Turing's universal machine. Whereas there has been a great deal of discussion concerning the introduction of the "stored program concept", the significance of this other great, but rather subtle, advance has not been fully appreciated. In fact, the tendency has been to use the single term "stored-program concept" to include all of the innovations introduced with the EDVAC design. This terminological confusion may well be responsible, at least in part, for the fact that there has been so much acrimony about who deserves credit for the revolutionary advances in computing which took place at this time (see for example the report Asdpray *1982*).

The key document in which this new conception of computer first appeared was the draft report von Neumann *1945* which quickly became known as the EDVAC Report. This report never advanced beyond the draft stage and is quite evidently incomplete in a number of ways. Yet it was widely circulated almost at once and was very influential. In fact the conception of computing machine it embodies has come to be known as the "von Neumann architecture". One element of controversy, which will probably never be fully resolved, is the question of how much of the EDVAC report represented von Neumann's personal contribution. Although Eckert and his consultant J.W. Mauchly later denied that von Neumann had contributed very much, shortly after the report appeared they wrote as follows:

> During the latter part of 1944, and continuing to the present time, Dr. John von Neumann ... has fortunately been available for consultation. He has contributed to many discussions on the logical controls of the EDVAC, has prepared certain instruction codes, and has tested these proposed systems by writing out the coded instructions for specific problems. Dr. von Neumann has also written a preliminary report in which most of the results of earlier discussions are summarized. ... In his report, the physical structures and devices ... are replaced by idealized elements to avoid raising engineering problems which might distract attention from the logical considerations under discussion. (Goldstine *1972,* p. 191; Metropolis and Worlton *1980,* p. 55)

Goldstine (apparently unaware of Turing's claim to be mentioned in this connection) comments:

> Von Neumann was the first person, as far as I am concerned, who understood explicitly that a computer essentially performed logical functions, ... he also made a precise and detailed study of the functions and mutual interactions of the various parts of a computer. Today this sounds so trite as to be almost unworthy of mention. Yet in 1944 it was a major advance in thinking. (Goldstine *1972,* pp. 191–192)

One way in which the EDVAC report betrays its unfinished state is by the large number of spaces clearly intended for references, but not filled in. Almost every page contains the abbreviation "cf." followed by a space. All the more significant is the one reference that von Neumann did supply: the reference, supplied in full, was to the paper McCulloch and Pitts *1943* in which a mathematical theory of idealized neurons had been developed. Von Neumann suggested that basic vacuum tube circuits could be thought of as physical embodiments of these neurons. Here there are two connections with Turing's ideas. The first, more obvious one, is that, like Turing, von Neumann was thinking of a computer as being like a brain (or at least a nervous system). In Ulam's letter to Hodges quoted above, Ulam alluded to this confluence, writing in a postscript: "Another coincidence of ideas: both Turing and von Neumann wrote of 'organisms' beyond mere computing machines."[17] But a more explicit connection with Turing's work becomes evident on further study. McCulloch (see von Neumann *1963*, p. 319) later stated that the paper which von Neumann did reference had been directly inspired by Turing *1936-7*. In fact, the paper itself cites the fact that a universal Turing machine can be modeled in a suitable version of the neural net formalism as the principal reason for believing in the adequacy of the formalism.

There is other evidence that von Neumann was concerned with universality in Turing's sense. Thus he spoke (Randell *1982*, p. 384) of the "logical control" of a computer as being crucial for its being "as nearly as possible *all purpose*". In order to test the general applicability of the EDVAC, von Neumann wrote his first serious program, not for numerical computation of the kind for which the machine's order code was mainly developed, but rather to carry out a computational task of a logical-combinatorial nature, namely the efficient sorting of data.[18] The success of this program helped to convince von Neumann that "it is legitimate to conclude already on the basis of the now available evidence, that the EDVAC is very nearly an 'all purpose' machine, and that the present principles for the logical controls are sound" (Goldstine *1972*, p. 209). Articles written within a year of the EDVAC report confirm von Neumann's awareness of the basis in logic for the principles underlying the design of electronic computers. The introduction to one such article states:

In this article we attempt to discuss [large scale computing] machines from the viewpoint not only of the mathematician but also of the engineer and the logician, i.e. of the ... person or group of persons really fitted to plan scientific tools. (Goldstine and von Neumann *1946*)

---

17 I am grateful to Andrew Hodges for making a copy of this letter available to me.
18 The sorting algorithm that von Neumann implemented belongs to the family of so-called "merge" sorts. For a very interesting discussion of this program and of the proposed EDVAC order code, see Knuth *1970*.

Another article (Burks, Goldstine, and von Neumann *1946*) clearly alludes to Turing's work, even as it indicates that purely logical considerations are not enough:

> It is easy to see by formal-logical methods that there exist codes that are in abstracto adequate to control and cause the execution of any sequence of operations which are individually available in the machine and which are, in their entirety, conceivable by the problem planner. The really decisive considerations from the present point of view, in selecting a code, are of more practical nature: simplicity of the equipment demanded by the code, and the clarity of its application to the actually important problems together with the speed of its handling those problems. It would take us much too far afield to discuss these questions at all generally or from first principles.

There has been much acrimony over the question of just what von Neumann had contributed; indeed, this question even became the subject of extensive litigation. Much of the controversy concerns the relative significance of the contributions of von Neumann on the one hand, and of Eckert and Mauchly on the other. In particular, some recent studies challenge the belief that von Neumann's technical contributions were of much importance. (See the semi-popular history Shurkin *1984* and the meticulously researched Stern *1981*.) It is not difficult to understand why this should be. The Turing–von Neumann view of computers is conceptually so simple and has become so much a part of our intellectual climate that it is difficult to understand how radically new it was. It is far easier to appreciate the importance of a new invention, like the mercury delay line, than of a new and abstract idea.

5. *The ACE*

Meanwhile, what of Turing? His mother (quoted in Hodges *1983,* p. 294) reports him saying "round about 1944" that he had plans "for the construction of a universal computer". During the war, he had been telling colleagues that he wanted to build a "brain". He proposed to construct an electronic device that would be a physical realization of his universal machine. Early in 1945, while on a trip to the United States, J.R. Womersley, Superintendent of the Mathematics Division of the National Physical Laboratory of Great Britain, was introduced to the ENIAC and to the EDVAC report. As early as 1938, Womersley had considered the possibility of constructing a "Turing machine using automatic telephone equipment". His reaction to what he had learned in the United States was to hire Alan Turing (Hodges *1983,* pp. 306–307). By the end of 1945, Turing had produced his remarkable ACE report (Turing *1945*). The excellent article Carpenter and Doran *1977* contains an analysis of the ACE report, comparing it in some detail with von Neumann's EDVAC report. They note that, whereas the EDVAC report "is a draft and is unfinished ... more important ... is incomplete ...", the ACE report "is a complete description of a computer, right down to the logical circuit diagrams" and even including "a cost estimate of $11,200". Not surprisingly, Turing showed that he understood the scope of universality. He suggested that his ACE might be able to play chess and to solve jigsaw puzzles. The ACE report contains explicit mention of features such as an instruction address register and truly random access to memory locations, neither of which is dealt with in the EDVAC report, although both are already to be found in Burks, Goldstine, and von Neumann *1946*. Although it is known (Goldstine *1972*, p. 218) that the ACE report quickly made its way across the Atlantic, it seems impossible to determine whether the ACE report influenced the American developments.

It should also be mentioned that the ACE report showed an understanding of numerous issues in computer science well ahead of its time. Of these perhaps the most interesting are microprogramming, and the use of a stack for a hierarchy of subroutine calls. We have already mentioned Turing's address to the London Mathematical Society in February 1947, in which he unveiled the scope of his vision regarding the ACE and its successors. In this talk he explained that one of the central conclusions of his earlier work on computability was that "the idea of a 'rule of thumb' process and a 'machine process' were synonymous ... Machines such as the ACE may be regarded as practical versions of ... the type of machine I was considering ... There is at least a very close analogy ... digital computing machines such as the ACE ... are in fact practical versions of the universal machine." Turing went on to raise the question of "how far it is in principle possible for a computing machine to simulate human activities". This led him to propose the possibility of a computing machine programmed to learn and permitted to make mistakes. "There are several theorems which say almost exactly that ... if a machine is expected to be infallible, it cannot also be intelligent ... But these theorems say nothing about how much intelligence may be displayed if a machine makes no pretence at infallibility."

Turing was much better at communicating in this visionary manner than he was in dealing with the bureaucrats who actually allocate resources, and he had considerable difficulty in getting his ideas put into practice. However, a machine embodying much of the design in the ACE report was eventually constructed, the Pilot ACE, and a successful commercial machine, the DEUCE, followed.

6. *Logic and the Future*

Since 1945, we have witnessed a number of breath-taking technological developments that have completely altered the physical form and the computational power of computers. I sit writing this essay using a text-editing program running on my personal microcomputer with a memory capacity of 5 million bits. The "johniac" computers on which I learned to program in the 1950's had a total memory of 41,000 bits and were only affordable by institutions. But the connection between logic and computing continues to be a vital one, and the lesson of universality, of the possibility of replacing the construction of diverse pieces of hardware by the programming of a single all-purpose device continues to be relevant. In fact the very existence of personal microcomputers is the result of this lesson being learned anew in the case of integrated circuit technology. In 1971, faced with the requirement for ever more complex and diverse "chips" by his employer, the Intel Corporation, Marcian Hoff found the solution: a single all-purpose programmable chip, and the microprocessor was born. We can foresee that this will happen again and again as technology continues its march towards faster and smaller components.

Acknowledgement. The research for this article was done while I was a John Simon Guggenheim Fellow (1983–84). The article is part of a comprehensive investigation of the relation between logic and computation. I am very grateful to the Guggenheim Foundation for its support. I also wish to thank Professors Harold Edwards and Esther Phillips who read an earlier draft and made many helpful suggestions.

References

Aspray, W.F.

1980 *From Mathematical Constructivity to Computer Science: Alan Turing, John von Neumann and the Origins of Computer Science in Mathematical Logic.* Doctoral dissertation, University of Wisconsin.

1982 History of the stored-program concept. Report on a session held on this subject as part of Pioneer Day at the National Computer Conference, June 1982. *Ann. Hist. C.* **4** (1982) 358–361.

Burks, A.W., H.H. Goldstine, and J. von Neumann,

1946 Preliminary discussion of the logical design of an electronic computing instrument, Institute for Advanced Study; reprinted in von Neumann *1963,* pp. 34–79.

Burks, A.W., and A.R. Burks

1981 The ENIAC: First general-purpose electronic computer. *Ann. Hist. C.* **3** (1981) 310–399.

Carpenter, B.E. and R.W. Doran

1977 The other Turing machine. *Comp. J.* **20** (1977) 269–279.

Carpenter, B.E. and R.W. Doran (eds.)

1986 *A.M. Turing's ACE Report of 1946 and Other Papers.* Cambridge, MA: MIT Press (1986).

Ceruzzi, P.E.

1983 *Reckoners, the Prehistory of the Digital Computer from Relays to the Stored Program Concept, 1933–1945.* Westport, CN: Greenwood Press (1983).

Davis, M.

  1965 *The Undecidable.* New York: Raven Press (1965).

  1982 Why Gödel didn't have Church's thesis. *Inf. Contr.* **54** (1965) 3–24.

  1983 The prehistory and early history of automated deduction. In: *Automation of Reasoning,* vol. 1, eds. J. Siekmann and G. Wrightson. Berlin: Springer-Verlag (1983).

Goldstine, H.H.

  1972 *The Computer from Pascal to von Neumann.* Princeton, NJ: Princeton University Press (1972).

Hilbert, D. and W. Ackermann

  1928 *Grundzüge der Theoretischen Logik.* Berlin: Julius Springer (1928).

Hodges, A.

  1983 *Alan Turing: The Enigma.* New York: Simon and Schuster (1983).

Huskey, V.R. and H.D. Huskey

  1980 Lady Lovelace and Charles Babbage. *Ann. Hist. C.* **2** (1980) 299–329.

Kneale, W. and M. Kneale

  1962 *The Development of Logic.* Oxford: Oxford University Press (1962).

Knuth, D.E.

  1970 Von Neumann's first computer program. *Comp. Sur.* **2** (1970) 247–260.

McCulloch, W.S. and W. Pitts

  1943 A logical calculus of the ideas immanent in nervous activity. *B. Math. Biophy.* **5** (1943) 115–133; reprinted in: McCulloch, W.S., *Embodiments of Mind,* pp. 19–39. Cambridge, MA: MIT Press (1965).

Metropolis, N., J. Howlett, and G.-C. Rota (eds.)

  1980 *A History of Computing in the Twentieth Century.* New York: Academic Press (1980).

Metropolis, N. and J. Warlton

  1980 A trilogy of errors in the history of computing. *Ann. Hist. Comp.* **2** (1980) 49–59.

Parkinson, G.H.R.

  1966 *Leibniz – Logical Papers.* Oxford: Oxford University Press (1966).

Post, E.

  1936 Finite combinatory processes. Formulation I. *J. Symb. Log.* **1** (1936) 103–105; reprinted in Davis *1965.*

Randell. B.

  1972 On Alan Turing and the origins of digital computers. *Mach. Intell.* **7** (1972) 3–20.

Randell, Brian (ed.)

1977 Colossus: Godfather of the computer. *New Sci.* **73** (1977) 346–348; reprinted in Randell *1982*, pp. 349–354.

1982 *The Origins of Digital Computers, Selected Papers,* (third edition). Berlin: Springer-Verlag (1982).

Shurkin, J.

1984 *Engines of the Mind: A History of the Computer.* W.W. Norton & Company (1984).

Smith, D.E.

1929 *A Source Book in Mathematics.* McGraw-Hill (1929).

Stern, N.

1980 John von Neumann's influence on electronic digital computing, 1944–1946. *Ann. Hist. C.* **2** (1980) 349–362.

1981 *From Eniac to Univac: An Appraisal of the Eckert-Mauchly Machines.* Digital Press (1981).

Turing, A.M.

1936-7 On computable numbers with an application to the Entscheidungsproblem. *P. Lond. Math. Soc. (2)* **42** (1936-7) 230–267; A correction, ibid. **43** (1937) 544–546; reprinted in Davis *1965*, pp. 155–222.

1945 *Proposals for Development in the Mathematics Division of an Automatic Computing Engine (ACE),* 1945, Report e882 National Physical Laboratory of Great Britain; reprinted 1972 with a forward by D.W. Davies as National Physical Laboratory Report, *Com. Sci.* **45**, and in Carpenter and Doran *1986*, pp. 20–105.

1947 Lecture to the London Mathematical Society, first published in Carpenter and Doran *1986*, pp. 106–124.

1950 Computing machinery and intelligence. *Mind,* **LIX** No. 236 (1950).

von Neumann, J.

1945 First draft of a report on the EDVAC, Moore School of Electrical Engineering, University of Pennsylvania, unpublished; reprinted in Stern *1981*, pp. 177–246.

1961 *Collected Works,* vol. 1, ed. A.H. Taub. Pergamon Press (1961).

1963 *Collected Works,* vol. 5, ed. A.H. Taub. Pergamon Press (1963).

Wang, H.

1974 *From Mathematics to Philosophy.* Humanities Press (1974).

Webb, J.C.

1980 *Mechanism, Mentalism, and Metamathematics: An Essay on Finitism.* Doordrecht: D. Reidel (1980).

Welchman, G.

> 1982 *The Hut Six Story.* McGraw-Hill (1982).

Woodger, M.

> 1958 The history and present use of digital computers in the National Physical Labora-
> tory. *Proc. Cont. Aut.* (November 1958) 437–442; reprinted with a correction in
> Carpenter and Doran *1986,* pp. 125–140.

Part II

From Universal Turing Machines to Self-Reproduction[*]

Michael A. Arbib

Although Turing in his later years became interested in the use of reaction-diffusion equations to model morphogenesis (cf. Turing *1952*) my task in this essay is to trace the way in which Turing's notion of the universal computer led to a computational theory of organism growth and reproduction. The focus, then, is on the work in the 1960's that was spurred by von Neumann's (*1951, 1966*) theory of self-reproducing automata.

Turing's result that there exists a universal computing machine suggested to von Neumann that there might be a universal construction machine A, which, when furnished with a suitable description I_N of any appropriate automaton N, will construct a copy of N.

If the automaton A has description I_A inserted into it, it will proceed to construct a copy of A. However, A is *not* self-reproducing, for A with appended description I_A produces A without I_A; it is as if a cell had split in two with only one of the daughter cells containing the genetic message. Adding a description of I_A to I_A does not help; now $A + I_A + I_{I_A}$ produces $A + I_A$ and we seem to be in danger of an infinite regress. Such a consideration suggested to von Neumann that the correct strategy might involve "duplication of the genetic material". He thus introduced an automaton B that can make a copy of any instruction I with which it is furnished, I being an aggregate of elementary parts, and B just being a "copier". Next C will insert the copy of I into the automaton constructed by A. Finally, C will separate this construction from the system $A + B + C$ and "turn it loose" as an independent entity.

Let us then denote the total aggregate $A + B + C$ by D. In order to function, the aggregate D must have an instruction I inserted into A. Let I_D be the description of D, and let E be D with I_D inserted into A. Then E is self-reproductive and no

---

[*] This essay is based on Chapter 7, "Automata which compute and construct", of Michael Arbib's *Brains, Machines and Mathematics*, 2nd Edition, Springer-Verlag (1987).

vicious circle is involved, since D exists before we have to define the instructions of I_D.

We thus see that once we can prove the existence of a universal constructor for automata constructed of a given set of components, the logic required to proceed to a self-reproducing automaton is very simple.

1. *Cellular Automata*

To set the stage for an outline of the design of a universal constructor, consider an infinite "chess board," with each square either empty or containing a single component. Each component can be in one of various states, and we think of an organism as represented by a group of cells, collected together somewhere in the plane. We are thus talking of regions in a *cellular automaton*. Any square of the board may be empty or contain some component, say of type $C_j(1 \leq j \leq N)$, in some state, say q_i. We may lump these $N + 1$ alternatives into one supercomponent C, which has one more state than the total number of states of the N components. For mathematical purposes, it is easier to think of there being a copy of one fixed component in every cell, so that rather than study the kinematics of components moving around in the plane we look at a more tractable process of how an array of identical components C, consisting initially of an activated array with the remaining cells in the passive state, passes information to compute and to "construct" new configurations.

Von Neumann was able to show that with a 29-state cell "supercomponent" he could set up a simulation of a complex Turing-type machine which, besides being able to carry out computations on its tape, would also be able to "reproduce" itself. The 29 states could be seen as several states corresponding to an OR gate, several states corresponding to an AND gate, several states corresponding to different types of transmission line, and so forth. Von Neumann's proof was not completed at the time of his death, but the manuscript he left was edited by Arthur Burks, and has since been put out as a book called *The Theory of Self-Reproducing Automata* (von Neumann *1966*). The proof is over 100 pages long. (The price we pay for simple components is a complex program. To take an analogy from computer programming, it is like trying to program in machine language rather than in an appropriate assembly language. In other terminology, or in biological terms, we might say that it is like trying to understand a complicated organism directly in terms of macromolecules, rather than via the intermediary of cellular structure.)

Thatcher *1970* used the same 29-state components as von Neumann but gave a more elegant construction of perhaps half the length. Codd *1965*, with remarkable ingenuity and interaction with a computer, showed that a construction similar to von Neumann's could go through using components with only eight states. Arbib

1966, 1967 showed that the construction could be done with great simplicity, in a matter of eight pages, if one allowed the use of much more complicated components. My rationalization for this use of complex components was that if one wishes to understand complex organisms, one should adopt a hierarchical approach, seeing how the organism is built up from cells, rather than from macromolecules.

Rather than go into any details of my construction I shall just briefly present two pictures which give some idea of the basic notions involved. We are to imagine a *CT* machine (Construction and Turing machine) which under the control of a program in its logic box can read and write on a one-dimensional tape in just the way a Turing machine does, and which can write but not read on a two-dimensional tape. The idea is that the two-dimensional tape is to be thought of as a construction area, and the writing of a symbol is to be thought of as equivalent to the placing of a component (Figure 1).

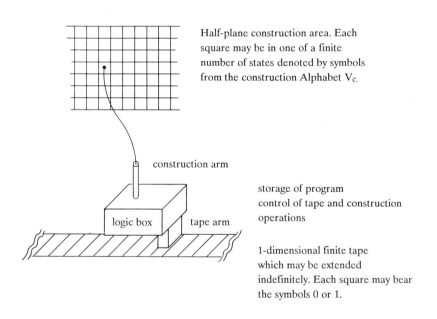

Half-plane construction area. Each square may be in one of a finite number of states denoted by symbols from the construction Alphabet V_c.

construction arm

storage of program
control of tape and construction
operations

logic box   tape arm

1-dimensional finite tape which may be extended indefinitely. Each square may bear the symbols 0 or 1.

Figure 1. A CT-Machine.

Our task is to find a set of components from which we can build tape, logic box, and construction area. Such a component as shown in Figure 2 is a finite-state module that can contain up to 22 instructions from a rather limited instruction set. We are to think of a two-dimensional plane in which these cells are embedded. An automaton is then represented as an activated configuration of these cells. The boxes marked *W* in Figure 2 are *weld registers* which serve to "weld" squares together into a one-dimensional tape in such a way that when any one square of that tape is instructed to move, all cells will move in the indicated direction. The assumption of such a weld operation greatly simplifies our programming.

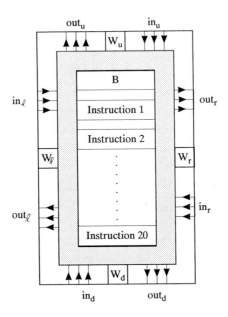

Figure 2. The Basic Module.

In Figure 3 we see an overall plan of an embedded *CT* machine. The logic box has been broken into two pieces, a one dimensional tape which contains the program and two cells which form a control head. The idea is that on activation by the control head, squares of the program tape may either be used to guide the control head in manipulating the computation tape in a Turing machine fashion or else may be used to place selected components in the constructing area and move welded blocks of components around.

A *CT* machine may be completely specified at any stage by a quadruple:

P. its program;
I. the instruction of the program it is executing;
T. the state of the tape (finite support);
C. the state of the construction area (finite support).

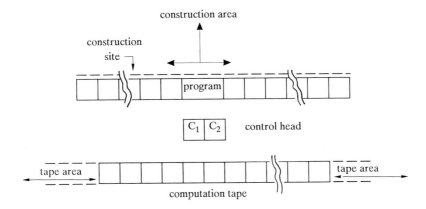

Figure 3. Overall plan of embedded CT-machine.

Arbib (*1969b*, Chapter 10) specifies an instruction code for the modules with which it is not only possible to embed arbitrary Turing machines in the array of those components in the indicated fashion but is also possible to program these machines so that they can construct other such machines in the construction area. This yields the following two results (not proved here):

1 Theorem: *Any* (P, I, T, C) *configuration may be effectively embedded in the tessellation.*

2 Theorem: *There is an effective procedure whereby one can find, for a given CT-automaton A, an embedded CT-automaton c(A) ("constructor of A") which, once started, will proceed to construct a copy of A in the three rows of its constructing area immediately above it, and activate that copy of A.*

Now, my procedure takes perhaps three cells in the program of $c(A)$ to code one cell of A, so it may still seem that any machine is only capable of constructing simpler machines. However, we can effectively enumerate all the (P, I, T, C) configuration with M_n, say, being the nth configuration in this enumeration. It is then a straightforward adaptation of the construction of a Universal Turing machine to prove:

3 Theorem: *There exists a universal constructor M_u with the property that, given the number n, coded in binary form on its tape as I_n, it will construct the nth configuration M_n in its configuration area. Symbolically $I_n : M_u \to M_n$.*

2. *Abstract Theory of Construction*

Why isn't the M_u of Theorem 3 self-reproducing? Because $I_u \to M_u$, but in the "second generation," $M_u \to ?$, and reproduction fails. (A cell which produces a copy of itself minus the genes is not self-reproducing.) To reach self-reproduction we first note a straightforward generalization of **3**:

4 Theorem: *For each recursive function h, there exists a machine M_c (c depends on h) such that*

$$I_n : M_c \to M_{h(n)}$$

Proof. There exists a program $P(h)$ of tape instructions which will convert I_n to the tape-expression $I_{h(n)}$. The machine M_c then has for program $P(h)$ 1 followed by the instructions of the program of M_u.   □

Our use of this result employs Myhill's abstract theory of constructors.

5 Theorem: *For any function g, there exists a machine M_a such that*

$$M_a \to M_{g(a)}$$

Proof. Let $M_{s(x)} = I_x : M_x$, the machine M_x with the description I_x of M_x attached. Then s is a computable function, and so is the composition $g \circ s$. Thus, taking $h = g \circ s$ in **4**, we have for the corresponding M_c that

$$I_n : M_c \to M_{h(n)} = M_{g(s(n))}$$

Setting $n = c$, we obtain

$$M_{s(c)} = I_c : M_c \to M_{g(s(c))}$$

and thus $a = s(c)$ satisfies the theorem.   □

For instance, g could be the function taking M into its mirror image, and so on.

6 Corollary: *There exists a self-reproducing machine.*

Proof. Let $g(x) = x$ in Theorem 5. Then for the corresponding a, $M_a \to M_a$.□

7 Theorem (Myhill *1964*): *Let h(x, y) be a total computable function of two arguments. There is then a machine M_d for which for each a we have that*

$$I_a : M_d \to M_{h(a,d)}$$

Proof. By reasoning similar to the above, we may find r so that M_r, when given a tape comprising I_a followed by I_n (for any a and n) will construct $M_{h(a,t_2(n,n))}$

$$[I_a/I_n] : M_r \to M_{h(a,t_2(n,n))}$$

where $t_2(a, b)$ is the encoding of the machine M_b with instruction tape M_a attached:

$$M_{t_2(a,b)} = I_a : M_b.$$

Now set $d = t_2(r, r)$ so that $M_d = I_r : M_r$. Then we have

$$I_a : M_d = [I_a/I_r] : M_r \to M_{h(a,t_2(r,r))} = M_{h(a,d)} \text{ as claimed.} \qquad \square$$

This result has an amazing corollary: that there exists a machine whose descendants are always "smarter" (in the sense of being able to prove more theorems) than their parent. To see this, we need to recall a few results from recursive function theory:

8 Definition: *A set R is creative if it is recursively enumerable and if there exists a total recursive function f such that*

$$S_n \subset \overline{R} \Rightarrow f(n) \in \overline{R} - S_n.$$

Thus, no creative set is recursive, for if it were, its complement $\overline{R}$ would be recursively enumerable. However, $\overline{R}$ cannot, by the above definition, equal S_n for any n, where S_n is the nth in some effective enumeration of the recursively enumerable sets.

9 Fact: $K = \{n|n \in S_n\}$ *is creative.*

Proof. Since $\overline{K} = \{n|n \notin S_n\}$, $S_n \subset \overline{K}$ implies $n \in \overline{K} - S_n$, and so K is creative with $f(n) = n$. $\qquad \square$

Let, now, Σ be an adequate consistent arithmetical logic. This means that for each r.e. set U there is a predicate $U(n)$ which is provable in Σ, denoted $\Sigma \vdash U(n)$, just in case n belongs to U, while $\{n|\Sigma \vdash \neg U(n)\}$, the set n for which $\neg U(n)$ is a theorem of Σ, is a recursively enumerable subset of $\overline{U}$.

If K is creative, let f be the corresponding function of Definition 8, and let m be such that $S_m = \{n|\Sigma \vdash \neg K(n)\} \subset \overline{K}$. Then $f(m) \in \overline{K}$, but $f(m) \notin S_m$. Hence, $\neg K(f(m))$ represents the true statement $f(m) \in \overline{K}$, but is not provable in Σ. (This gives us Gödel's Incompleteness Theorem.) If we now construct a new logic Σ' by joining $\neg K(f(m))$ as an additional axiom, we have a logic in which the truth $f(m) \in \overline{K}$ is represented by a theorem.

Now, for each logic Σ, we may specify a Turing machine $Z_{h(\Sigma)}$ which, started on a blank tape, proceeds to print out (scratchwork and) an effective enumeration of the theorems of Σ. Further, we may so structure h that the passage from Σ to $h(\Sigma)$ is effective, and given any n, we can tell whether it is an $h(\Sigma)$, and, if so, for which Σ.

Given a Turing machine Z_n, let $g(n) = n$ if Z_n is not $Z_{h(\Sigma)}$ for some logic Σ. If $n = h(\Sigma)$, let $Z_{k(n)}$ be Z_n modified so that it prints out only theorems of the form $\neg K(x)$ for our creative set K, and let $S_{t(n)}$ be the recursively enumerable set of the x so enumerated. But then $\neg K(f(t(n)))$ may be consistently adjoined to the axioms of Σ to yield a new logic Σ'. Let $g(n) = h(\Sigma')$. Clearly, then, g is a total recursive function, and we have

10 Theorem (Myhill *1963*): *There is a total recursive function g such that for the Turing machine $Z_{h(\Sigma)}$ which prints out theorems of the adequate consistent arithmetical logic Σ, the Turing machine $\Sigma_{g(h(\Sigma))}$ is $Z_{h(\Sigma')}$ for an adequate consistent arithmetical logic Σ' with more theorems than Σ.*   □

Thus, while Gödel's incompleteness theorem points to an inevitable limitation of any axiomatization of arithmetic, Myhill's theorem points out the much less well-known fact that this limitation can be *effectively* overcome. And, of course, the process may be iterated mechanically again and again. This has an amusing application due to Myhill *1964* in the study of self-reproducing automata.

Theorem 7 from the abstract theory of constructors states that if $h(x, y)$ is a total recursive function of two arguments, there is then a machine M_d for which always

$$I_a : M_d \rightarrow M_{h(a,d)}.$$

Now let $M_a < M_b$ mean that M_b prints out all the strings that M_a prints and more, and all the strings that M_b prints out are true statements of arithmetic. We shall prove:

11 Theorem: *There exists an infinite sequence of machines $\{M_{z_i}\}$ such that we have simultaneously*

$$M_{z_i} < M_{z_{i+1}}$$

and

$$M_{z_i} \rightarrow M_{z_{i+1}}$$

Proof. Simply take $M_{h(a,d)} = I_{g(a)} : M_d$.   □

We may call each $M_{h(a,d)}$ a machine each of whose descendants "outsmarts" its predecessor. Myhill observes that the theorem is a brutal parody of the growth, in any usable sense, of intelligence, but is of methodological significance in that it suggests the possibility of encoding a potentially infinite number of "directions to posterity" on a finitely long "chromosomal" tape.

3. Simple Self-Reproduction

We next recall the recursion theorem from the theory of computable functions:

9 The Recursion Theorem: *Let h be any total computable function; and let $(\varphi_0, \varphi_1, \varphi_2, \ldots, \varphi_n, \ldots)$ be a fixed effective enumeration of the (partial) computable functions. Then h has an "index fixed point" n_0, i.e., we have*

$$\varphi_{n_0} = \varphi_{h(n_0)}. \qquad \square$$

Smith *1968* showed how to use the recursion theorem to construct self-reproducing configurations in a manner far simpler than that just outlined. We first note that for any Turing machine Z we can define a one-dimensional cellular space $\tilde{Z}$ which simulates Z in real time, with the state of each cell encoding a state-symbol pair of Z, having introduced a new "null state", i.e., the state-space of Z is just $(Q + \{\Lambda\}) \times X$, with (Λ, b) the quiescent state. As shown in Figure 4, the space $\tilde{Z}$ is just an 'image' of the tape, with the square bearing symbol x_0 and scanned by the control box in state q being encoded by $\tilde{Z}$-state (q, x_0), while every other square with symbol x_1 is encoded by (Λ, x_1). It is a straightforward exercise for the reader to write out the nearest-neighbor interaction rules for $\tilde{Z}$ that mimic the behavior of Z. For such a space, we say that Z is wired into $\tilde{Z}$, or that $\tilde{Z}$ has Z wired in.

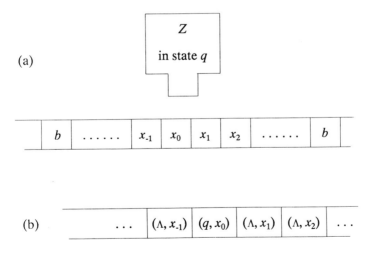

Figure 4. (a) A Turing machine Z. (b) A one-dimension cellular space Z which simulates Z in real time.

Let the wired-in computer be a universal Turing machine U. We use the notation $x \underset{P}{\rightarrow} y$ to mean that Turing machine program P acting on initial tape x halts with y on the tape as its final result, so that we have

$$f(P), x \underset{U}{\rightarrow} y$$

where f is the encoding function which encodes program P and tape x as the tape $f(P)$, x for U, and

$$x \underset{P}{\rightarrow} y.$$

10 Lemma (Smith *1968*; cf. Lee *1963*): *For an arbitrary one to one recursive function f from (program, tape) pairs to tapes and for an arbitrary total recursive function g such that $g(x) = y$, there exists a machine with program P such that*

$$\underset{\uparrow \ P}{x \rightarrow} (f(P), x, y, f\ (P), x).$$
$$\uparrow$$

Proof. Define the function h from programs to programs such that $h(Q)$ is a program which reads arbitrary input tape x, encodes program Q and tape x to get $f(Q)$, x, prints $f(Q)$, x, computes $g(x)$ to get y, prints y, prints $f(Q)$, x again (either by re-encoding $f(Q)$, x or by copying the result of the first encoding), and finally moves the head to the leftmost symbol in the rightmost encoding $f(Q)$, x. That is,

$$\underset{\uparrow \ h(Q)}{x \rightarrow} (f(Q), x, y, f(Q), x). \qquad \square$$

Clearly, h can be chosen total recursive. Thus, by the recursion theorem, there exists P which is a computational fixed point of h such that

$$\underset{\uparrow \ P}{x \rightarrow} (f(P)), x, y, f(P), x).$$

Thus, in cellular space $\tilde{U}$ with U wired in, the following situation can hold:

$$f(P), x \underset{U}{\rightarrow} (f(P), x, y, f(P), x)$$

at some time $T > 0$. If we decree that V is U so modified that on completing such a computation it will backtrack to the rightmost comma, and start anew without changing anything to the left of that coma, we will obtain

$$f(P), x \underset{V}{\rightarrow} (f(P), x, y, f(P), x, y, f(P), x)$$

at some time some later time $T' > T$ and so forth. Hence, we have shown the following:

11 Theorem (Smith *1968*): *Let $\tilde{V}$ be a computation-universal cellular space with a modified universal Turing machine V wired in. Then there exists a configuration* *c = f(P) in $\tilde{V}$ which is both self-reproducing and computes any given total recursive function g on supplied data x.*   □

With this we conclude our celebration of the 50th anniversary of the Universal Turing Machine, in which we have complemented the contributions of other authors by emphasizing the fruits of von Neumann's decision to explore the implications of universal computation for universal construction, and thus for (computationally sophisticated) self-reproduction. For further reading on embryological models, see Ransom *1981*; and for an essay on recent applications of cellular automata, see Hayes *1984*.

References

Arbib, M.A.

    1966 A simple self-reproducing universal automaton. *Inf. Cont.* **9** (1966) 177–89.

    1967 Automata theory and development: part I. *J. Theor. Bio.* **14** (1967) 131–56.

    1969 *Theories of Abstract Automata.* Englewood Cliffs, NJ: Prentice-Hall (1969).

Burks, A.W., ed.

    1970 *Essays on Cellular Automata.* Urbana: University of Illinois Press (1970).

Codd, E.F.

    1965 *Propagation, Computation and Construction in 2-Dimensional Cellular Spaces.* Technical Publications of the University of Michigan (1965). Reprinted as *Cellular Automata.* New York: Academic Press Inc. (1968)

Hayes, B.

    1984 Computer recreations: The cellular automaton offers a model of the world unto itself. *Sci. Am.* **250** (March 1984) 12–21.

Lee, C.H.

    1963 A Turing machine which prints its own code script. In *Proc. Symp. Math. Theory of Automata,* vol. 12, Microwave Research Institutes Symposia Series, pp. 155–64. Brooklyn, NY: Polytechnic Press (1963).

Myhill, J.

    1963 Notes for a series of lectures on recursive functions. In: lecture notes for a summer conference on automata theory, University of Michigan.

    1964 The abstract theory of self-reproduction. In: *Views on General Systems Theory,* ed. M.D. Mesarovic, pp. 106–118. New York: John Wiley and Sons, Inc. (1964).

Ransom R.

>   1981   *Computers and Embryos: Models in Developmental Biology.* Chichester: John
>   Wiley and Sons (1981).

Smith, A.R.

>   1968   Simple computation-universal cellular spaces and self-reproduction. In: *Confer-
>   ence Record IEEE 9th Annual Symposium on Switching and Automata Theory*,
>   pp. 269–77.

Thatcher, J.W.

>   1970   Universality in the von Neumann cellular model. In: *Essays on Cellular Automata*,
>   ed. A.W. Burks, pp. 132–186. Urbana: University of Illinois Press (1970).

Turing, A.M.

>   1936-7   On computable numbers with an application to the Entscheidungsproblem, *P.
>   Lond. Math. Soc. (2)* **42** (1936-7) 230–65; A correction, ibid., **43** (1937) 544–46.

>   1952   The chemical basis of morphogenesis. *Phi. T. Roy. B* **237** (1952) 37–72.

von Neumann, J.

>   1951   The general and logical theory of automata. In: *Cerebral Mechanisms in Behavior:
>   The Hixon Symposium.* New York: John Wiley and Sons, Inc. (1951).

>   1966   *Theory of Self-Reproducing Automata,* edited and completed by A.W. Burks. Ur-
>   bana: University of Illinois Press (1966).

Computerizing Mathematics:
Logic and Computation

Michael J. Beeson

1. Introduction: Mathematics = Logic + Computation

COMPUTATION

The two central aspects of mathematics are logic and computation. Mathematics has been used to compute the orbits of the planets and spaceships, to explain the ice ages, and to design tall buildings and majestic ships. The calculus, whose very name emphasizes calculation, was created in part to solve the problem of the calculation of longitude at sea, whose application to navigation was an important commercial problem of the era. Applied mathematics and engineering rely upon the computational aspect of mathematics.

LOGIC

On the other hand, the logical aspect of mathematics is almost as old as the computational, going back as it does to Apollonius and Euclid. Proofs are often thought of, at least by mathematicians, as the heart of mathematics. The computational achievements of the eighteenth century were followed by logical achievements of the nineteenth and twentieth, as "rigorous foundations" were provided for the theory of real numbers, differential equations were proved to have solutions, even if formulas could not be found for them, and such famous theorems as the Prime Number Theorem (according to which the n-th prime number is approximately $n \log n$) were finally given flawless demonstrations.

CONNECTIONS

Often proofs were sought to explain the results of computations, as in the case of the prime number theorem. Sometimes they were sought to explain the failure to find computations, as in the proofs that the circle could not be squared nor the fifth-degree equation solved by radicals. From the beginning of mathematics there have been intimate connections between logic and computation. The exact nature of these connections is still being investigated. There have been some exciting and surprising discoveries in this century, some even in the past fifteen years. It is one purpose of this paper to describe these discoveries.

Computerizing Mathematics

Today the development of ever-more-powerful computers and programming languages presents us with the possibility of *computerizing mathematics*. The subject of *automated deduction* deals with computerizing the logical aspect of mathematics, while the subject of *symbolic computation* deals with computerizing the computational aspect.[1] The theoretical connections between logic and computation have led to demonstrable computer programs and even to practical applications, and will lead to more in the future. The second purpose of this paper is to describe some of these applications and explain their relations to the theoretical ideas.[2]

Turing's Work

It is a particularly appropriate topic for a volume dedicated to Turing, who was deeply involved in the precise mathematical and philosophical analysis of both the notion of computation and the notion of proof. His first and most important paper (*1936-7*) defined computation using Turing machines and applied the notion to an analysis of the notion of proof. He used these methods to show that Hilbert's *Entscheidungsproblem*, or decision problem for formal logic, is recursively unsolvable.[3] Turing is a major historical figure in the elucidation of the natures of computation and logic and of their relationship. In addition, he was influential in creating the current of ideas that eventually produced the modern computer.

2. *Logic Is Computation*

Leibniz

The idea that logic might be reducible to computation is a very old one. Descartes saw his invention of analytic geometry in this light, reducing the logic needed for Euclidean geometry to algebraic computation. Leibniz imagined a "universal language" of symbols into which any argument could be translated. There would be calculation rules to go with it, so that when two gentlemen disagreed, instead of arguing they would say, "Let us calculate!".[4]

---

1 We shall not discuss purely *numerical* computation in this paper.

2 The equation "mathematics = logic + computation" is surely an overstatement, in that there are other aspects of mathematics besides those two, for instance the visual aspect.

3 The same result was proved by Church independently at about the same time, by showing that the reduction of terms in λ-calculus can be defined in logic. Church's paper is reprinted next to Turing's in Davis *1965*.

4 Leibniz *1679* sketches his vision, but the quoted phrase, in Latin *calculemus*, is not there; it is in Leibniz *1965*, vol. VII, p. 125, p. 200; or Leibniz *1923*, series 1, vol. 4, p. 315.

Boole

Nothing seems to have come of Leibniz's idea, however, and the next major step in turning logic into computation was taken by George Boole, who formulated "Boolean algebra" in 1847. His work and much of his notation survive in what we now call "propositional logic". This is the logic of combining propositions by the operators *and, not, or,* and *implies.* Nowadays Boole's ideas are accessible in modern editions of Boole (1856), *An Investigation of the Laws of Thought.* This work is remarkable in that Boole took Leibniz's idea seriously: he tried to show the fallacies in serious philosophical works by analyzing the arguments in algebraic symbols.[5]

Frege and Peano

What we now think of as "logic" goes beyond propositional logic by allowing variable names x, y, and so on, and the operators *for all* and *there exists,* so that one can say something like, "For every x, if x is a man then x is mortal." Logicians construct formal languages for the expression of logical arguments, much as computer scientists construct formal languages for the expression of computation. Many of these languages are variants of "first order logic", the most common kind of logical language. Such a language was introduced for the first time by Gottlieb Frege (1879), and a notation similar to today's was introduced by G. Peano ten years later (Peano 1889). Peano worked without knowledge of Frege; evidently the time was ripe.

Hilbert

The great mathematician David Hilbert contributed to the development of logic in the first two decades of this century, completing the construction of modern first-order logic by 1928, when his formal systems were set out in Hilbert and Ackerman *1928.* This was the same book, incidentally, in which the Entscheidungsproblem was posed.[6] Hilbert and the other creators of the subject he christened as "proof theory" introduced a precisely defined notion which they believed corresponded to the notion of *proof,* taken in the psychological sense of *convincing argument.* The precisely defined notion in question is that of *formal derivation.* A *formal derivation* of a statement A is an arrangement of symbols on paper obeying certain rules of logic, such that the last line (the "conclusion") is the statement A.

Truth and Symbols

A central question was whether calculations with these symbols could suffice to

---

5 Boole's contemporary Augustus de Morgan should share the credit for the invention of symbolic logic; their books were both published in November of 1847. Boole's notation has survived and de Morgan's has not.

6 A proper history of logic would discuss the contributions of Russell, Löwenheim, Herbrand, and Skolem. The curious reader is directed to van Heijenoort *1967,* Davis *1965,* and to the notes between the papers in Gödel *1986.*

settle any question that could be posed in the formal language. More generally, questions naturally arose about the relationship between *mathematical truth* and formal derivability in some logical language. Formal derivability is a precisely defined concept, having to do with exact relationships of concrete symbols. Mathematical truth, on the other hand, seems to be an abstract, elusive, possibly even subjective concept. Hilbert (and others) wished to reduce abstract mathematical truth to formal derivability. They thought they were on the trail of Leibniz's universal calculus. The mathematical philosophy associated with the attempt to find the meaning of mathematics in symbols is called "formalism".

DEATH OF FORMALISM

The decade of the 1930's brought forth the answers to all these questions. The answers were surprising, and many found them difficult to understand. Most of them were "negative", in the sense that they dispelled the dreams of the formalists. It is, however, not the famous negative results which concern us in this paper, but the connections between logic and computation which developed during that decade.

COMPLETENESS

First came Gödel's completeness theorem (*1930*), according to which every purely logical truth *can* be demonstrated in a fixed logical symbolism. This is a result of central interest: it shows that logic *can*, in one sense at least, be reduced to computation! Every logical truth can be demonstrated, in principle, by searching for a formal derivation.

RESOLUTION

The next major development in the reduction of logic to computation took place three decades later, with the publication of the resolution method by Alan Robinson (*1965*).[7] Resolution has been so influential in the subsequent developments that it is worth some effort to understand exactly what it is. I shall therefore give a short but precise explanation of the method, in terms chosen for scientists not trained in logic.

A *proposition letter* or *Boolean variable* is just a variable p that can take only two values, *true* and *false*. It will be convenient to allow them to take any non-negative integer values, identifying 0 with *false* and any non-zero value with *true*. The logical operation *and* is then just multiplication. The logical operation *or* becomes addition. (Note that this is "inclusive or", not "exclusive or": it means at least one and possibly both of the operands is true.) There is one more operator in logic, negation. We write $\bar{p}$ to mean 0 if p is non-zero, and 1 if p is 0. Since we are only interested in whether values are zero or not, we write $a \cong b$ to mean either both are zero or both are

---

7 I apologize for giving short shrift to G. Gentzen, whose contributions were certainly relevant to the topic and are even today influential in automated deduction, and for omitting to describe the work of Church and Curry on λ-calculus and combinatory logic.

non-zero. The distributive laws $a(b + c) = ab + ac$ and $a + (bc) \cong (a + b)(a + c)$ are the expression in algebraic forms of de Morgan's laws. The second of these laws requires $\cong$ for its statement, as it is a logical law which is not one of the laws of integer arithmetic. Another such law is $p + p \cong p$, which permits us to erase one of a pair of identical terms. Using these laws together with $\overline{\overline{p}} \cong p$, we can express any logical formula as a product of sums, where each term in the sums is a *literal*: that is, either p or $\bar{p}$ for some variable p. Of course, the constants 0 and 1 can also appear. Hence the most general problem in propositional logic is to solve a finite set of linear congruences, i.e., equations $\Sigma a_i \cong 1$, where each a_i is a literal.

In this way logic is reduced to equation-solving: to determine if proposition S follows from axioms $A_1, \ldots, A_n$, we only have to show that the equations $S \cong 0, A_1 \cong 1, \ldots, A_n \cong 1$ have no solution. This much was known to Boole.

Now consider methods of solving such equations. You can use the usual methods for solving equations, except that all the values must be kept non-negative, so you cannot subtract one equation from another if that would result in some negative coefficients. But sometimes it is necessary to create negative coefficients in solving equations, so you must have some additional rule. That is where resolution comes in. The resolution rule can be expressed as the following: if two of our congruences have the form $p + E \cong 1$ and $\bar{p} + G \cong 1$, then we can replace these two congruences with $E + G \cong 1$. The variable p does not have to occur at the beginning of the equation. The resolution rule may thus be thought of as a tool for solving logical equations, a kind of "cancellation", something like Gauss elimination for ordinary equations.

What Robinson showed is that resolution is a completely general method of determining whether logical equations can be solved or not; repeated application of the resolution rule is guaranteed to work. You may use *only* resolution; it is not just an adjunct to adding and subtracting equations, it is enough by itself.

I have oversimplified the matter so far by considering only the propositional case. In general instead of propositional variables p we have formulas like $man(x)$; and we are allowed in addition to the "cancellation" rule, to make substitutions of values for the variables. Resolution proceeds by finding a term such as $man(x)$ and another term such as $\overline{(man(Socrates))}$; we then make the substitution of *Socrates* for x and cancel the terms as above. Robinson's resolution method includes a method known as "unification" for generating the necessary substitutions for the variables.

AUTOMATED DEDUCTION

By 1965 the computer had been born and outgrown its infancy; people were beginning to take seriously the old reduction of logic to computation and ask for a concrete realization of logic on the computer. Early work by Hao Wang, Davis, Putnam, and others demonstrated that something along these lines might be possible, but the methods employed seemed too inefficient. The resolution method was a breakthrough in efficiency. The subject of *automated deduction* was born. Two decades later, it is still flourishing. The achievements and current directions of the field will be discussed in Section 5.[8]

3. *Computation Is Logic*

We have seen that logic can be reduced to computation by the device of formalized first-order logic; and that the reduction can even be made on actual computers, using the resolution method. We shall now take up the connection in the other direction: to what extent can computation be reduced to logic?

It was Alan Turing who first investigated the matter. The manner in which he was led to these investigations is of interest, as it grew out of the reduction of logic to computation. Turing was interested in the *Entscheidungsproblem* (decision problem) posed by Hilbert. That problem can be stated as follows: *Given a collection of axioms A stated in first-order logic, and a statement S, could you decide by some universal computation method whether S does or does not follow from A?*

Such a universal computation method would be called a "decision method". If *S* does indeed follow from *A*, you will, according to the completeness theorem, be able to find a formal derivation of it by searching systematically. But if it does not follow, when will you give up your search? There is no point at which you can be assured that the sought-for proof will *never* be forthcoming. Thus merely searching for a derivation, no matter how efficiently, cannot constitute a decision method.

What Turing did was to prove that no such method can be found. The most difficult part of the proof was to correctly and convincingly define "computational method". It is necessary to give such a definition if you want to prove that no method of such-and-such a kind exists. The hypothetical computing machines now known as "Turing machines" served the purpose admirably.

TURING'S SOLUTION
It is not the definition of Turing machines which concerns us here, however, but what Turing did with them to solve the Entscheidungsproblem. What he did was to show that the step-by-step computations of a computer could be described by formulas of logic. The essence of his idea can be explained as follows, using modern computers in place of Turing machines. We may represent the insides of the computer as an array of pigeonholes, each capable of holding a single binary number, that is, either zero or one. This array will include the memory of the computer, the internal registers of the central processing unit – indeed every place where information can exist in the

---

8 Those interested in the development of the field before 1965 may turn to Siekmann and Wrightson *1983*.

computer. Let us say there are N of these locations. The computer is a completely deterministic machine, and it proceeds in discrete steps, governed by an internal clock; if we start the computer in a certain configuration, the configuration one step later is completely determined. Turing's insight is that the relationship between the configuration now and the configuration one step later can be described *logically*. That is, we can associate one proposition letter (Boolean variable) p_i to each of the N locations in the computer, and then we can give N logical formulas ϕ_k, each with up to N propositional variables p_i, such that ϕ_k expresses how the value at location k depends on the N values p_i at the previous step.

By this device, Turing was able to translate his famous result on the "unsolvability of the halting problem" into a negative solution to the Entscheidungsproblem. The result on the unsolvability of the halting problem is a result purely about computation, expressing the impossibility of computational solutions to certain problems. The Entscheidungsproblem is a hybrid problem, asking about the possibility of computational solutions to logical problems.[9]

PROLOG

Just as the work on formal systems of logic of the 1920's and 30's was eventually realized in practical form on the computer, Turing's work on the logical representation of computation also has its descendants in today's computer science. The practical realization of logic as computation began in 1965 with the resolution method. The practical realization of computation as logic began in the early 1970's with the invention of the programming language Prolog.[10] The basis of Prolog is the realization that the resolution method might be used as the basis of a general-purpose programming language, in which computations would actually be automatically generated formal deductions.

The fundamental unit of which Prolog programs are constructed is the *clause*. For example,

triangle(X) :- polygon(X), sides(X,3).

is a Prolog clause. Prolog notation can be thought of as an alternative notation for logic: the comma means *and*, and the "neck symbol" :- means $\leftarrow$, reverse

---

9 Those familiar with the P=NP problem of modern computer science will recognize the essential idea of Cook's theorem here. Cook's theorem shows that the general question of whether NP computations can be reduced to polynomial-time computations is equivalent to the question whether satisfiability in propositional calculus can be verified in polynomial time. Like Turing's solution to the Entscheidungsproblem, this is a reduction of computation to logic; in fact the proof is essentially the same.

10 Prolog had some predecessors, such as PLANNER; so it is a little difficult to pinpoint the historical moment in question. But Prolog became widely used and its predecessors did not. Prolog was created by Alain Colmerauer at the University of Marseilles.

implication. One reads the neck symbol "if", so that the above clause says X is a triangle if it is a polygon and $sides(X, 3)$ is true.[11]

A clause, like any proposition, expresses (declares) a state of affairs. The *declarative interpretation* of the clause is that this state of affairs is really the case. But the same clause also has a *procedural interpretation*: to determine if X is a triangle, what you do is verify that X is a polygon, and if you succeed, then verify $sides(X, 3)$. The fact that propositions can be interpreted as "problems" in this way was first pointed out by Kolmogorov *1932*; the work of Kolmogorov is one of the earliest papers on the connection between logic and computation.

The procedural interpretation is the natural one for Prolog, and the one which experienced Prolog programmers keep in mind while writing programs. Prolog has turned out to be a very useful language for artificial intelligence programming, because the algorithms built in to implement the resolution method provide easy-to-use matching and searching capabilities. Since matching and searching are fundamental techniques in artificial intelligence, certain programs are easier to write in Prolog than in other languages.

Since Prolog doesn't have negation in the ordinary sense of logic, to express a logical clause of the form $\bar{a} + \bar{b} + c$, you would write c :- a, b. Prolog clauses are required to have only one item in the "head", or part to the left of the neck symbol, which corresponds in logical notation to the restriction that there be at most one non-negated part (literal) in the formula. In other words $\bar{a} + \bar{b} + c$ can't be expressed directly in Prolog. This restriction is what allows an efficient version of the resolution algorithm; without it, Prolog would not be an efficient language.

A Prolog program P is nothing but a list of Prolog clauses. You use the program by asking a question Q, where Q expresses a proposition. The Prolog system uses resolution to attempt to derive Q from the axioms P. If there are some variables in the question Q, and if the derivation is successful, those variables will be given values by the resolution process; this is the mechanism by which Prolog programs can return a value. The key to the success of Prolog is in the restriction on the form of clauses; this permits a very efficient implementation of the resolution algorithm. Turing showed that computations *could* be modeled in logic; Prolog actually did it.

The use of languages in which the fundamental computation method is deduction is known as *logic programming*. Certain authors have dreamed of a language that would permit the programmer to specify *what* was to be done, automatically generating instructions for *how* to do it. You could program declaratively, just writing down the facts of the problem. At first sight logic programming appears a step in this direction, and indeed there are meta-theorems about the equivalence of the declarative and procedural interpretations of programs. But this dream must remain

---

11 In the terminology used above to explain resolution, this clause would be written $triangle(X) + \overline{polygon(X)} + \overline{sides(X, 3)} \cong 1$.

a fantasy, since programs written in this fashion often turn out to be hopelessly inefficient.[12]

PROLOG MACHINES

In the next decade we can expect the influence of Prolog to expand. For example, the Japanese "Fifth Generation" project seeks to build a computer whose hardware contains efficient implementations of Prolog's deduction algorithms – a "Prolog machine". We can expect to see commercial software written in Prolog adding useful capabilities to today's database software, and perhaps we will see computers making limited use of natural languages (that was the interest of the creators of Prolog, incidentally). The progression from Turing's abstract, theoretical work to widespread application has taken only half a century.

The interplay of logic and computation is at its most subtle when a Prolog program runs. The programmer intends to have the computer carry out a computation. The Prolog program reduces the computation to logic. The Prolog interpreter which runs the program reduces the logical deductions to computation. These computations, when reduced to machine code, make use of the digital logic built into the processor and memory chips. The flow of information seems now to be logic, now to be computation. Like waves and particles, logic and computation are metaphors for different aspects of some underlying unity.

4. *The Computerization of Symbolic Computation*

The first uses of the computer were for numerical calculations. When Samuels wrote the first checkers-playing program at IBM in 1948, the output had to be numerically coded because the printers attached to the computer could print only numbers.[13] Only in the fifties was it widely realized that the computer could process symbolic data as well, and only in the late sixties were programs written which could make mathematical computations involving symbols as well as numbers. Quite ambitious plans were laid in the late sixties, but the computing power of the time was not

---

12 See Lloyd *1984* for the basic meta-theorems and the ideal of a usable declarative semantics, e.g., p. 59. It must be pointed out that the logical theory of Prolog is in a very unsatisfactory condition. The meta-theorem on the equivalence of declarative and procedural interpretation holds only for an unrealistically small part of Prolog, the part not mentioning any actions: input, output, or changes to the database, and not using any control operators such as *not* or *cut*. If these operators are included, there is no good, widely accepted definition of the semantics of Prolog programs. This has the consequence that different implementations of Prolog behave differently in certain (admittedly unusual) circumstances.

13 Related in Samuels' speech to the opening session of IJCAI 85.

adequate to realize these plans, and there was little ferment in the field until the eighties.

By *symbolic computation* we mean computations such as are done in algebra, trigonometry, calculus, and some more advanced branches of mathematics. The distinction between a "computation" and a "proof" is familiar to all mathematicians, particularly those who also teach mathematics. A computation proceeds systematically by simple steps, and does not involve any reasoning beyond the correct use of the '=' sign. The rules to be applied at each step are fixed and known. Yet computations can become very lengthy and complicated; and the result is errors. Some surveys show that up to 20% of the values given in certain integral tables are incorrect.

There are seven well-known large programs being used for symbolic computation in 1986. In addition to MACSYMA, there are REDUCE, SMP, MAPLE, SCRATCHPAD II, CAYLEY, and MuMath. REDUCE and SMP are similar in spirit to MACSYMA: they implement the basic algorithms of arithmetic, algebra, and calculus, and offer special-purpose packages to users in certain areas, such as tensor-manipulation facilities for use in general relativity. CAYLEY and SCRATCHPAD II offer in addition "computerized *modern* algebra", i.e., computerized groups, rings, fields, and user-defined algebraic structures. At the risk of inviting contradiction I will venture to guess that every algorithm taught in undergraduate mathematics has been programmed into these systems.[14] MuMath offers more limited capabilities, but it runs on a personal computer.[15]

The subject of symbolic computation has entered a new phase in the last five or ten years, one in which it is no longer a matter of programming what mathematicians have always known. Take integration for example: the first symbolic integration programs simply incorporated the "bag of tricks" taught in calculus courses. If the integral isn't of a memorized form, try to find a substitution. If that doesn't work, try integration by parts, and so on. Modern programs use a systematic integration procedure due to Risch *1970* with superior performance. Moreover, the Risch algorithm is not really very complicated, and may well be taught in calculus classes in the future. Only the very great inertia of the educational system prevents it.

Similar improvements have been found in other mathematical procedures. Even arithmetic is not immune: a faster way to multiply two integers has been discovered!

---

14 To be more specific about the capabilities of at least one of these systems: MACSYMA offers arithmetic on numbers of arbitrary precision, on polynomials and rational functions; it provides an equation solver and facilities for matrix and tensor manipulation; facilities for the manipulation of Taylor, Laurent, and Poisson series; facilities for indefinite and definite integration, the computation of limits, the symbolic solution of ordinary differential equations, and facilities for output in FORTRAN or TEX. MACSYMA was developed at MIT in 1969-82, and is said to involve more than 100 programmer-years of effort.

15 For more information on these systems, see van Hulzen and Calmet *1982*.

The way you were taught in school to multiply two n-digit numbers will require in general about n^2 steps, but it is possible to multiply in only a constant times $n^{log_2 3}$ steps. The constant is large, however, so the new method only becomes faster when multiplying numbers of 40 or 200 digits or more.[16]

REWRITE RULES

An important concept in this work is the idea of *rewrite rule*. A rewrite rule is a "one-way equation". Rewrite rules are written like equations, except with an arrow in place of equality:

$$a(b + c) \rightarrow ab + ac.$$

Rewrite rules may be used only in the indicated direction. Thus we may use the rule above to pass from $x^2(y + yz)$ to $x^2y + x^2yz$, but not in the other direction. A paradigm for calculation may be described as follows: the calculator has a given collection of rewrite rules, and when presented with an expression to simplify, simply attempts to apply the rules one after another, rejecting any whose left-hand sides will not match any part of the current expression, and applying any whose left-hand sides will match. This is called *reducing* the expression. When nothing more is left to do the expression is *normal*.[17]

This way of using rewrite rules has played an important role, both in symbolic computation and in automated deduction. Both applications are of interest in the present connection.

It seems that algebraic simplification, as taught to algebra students, is nothing but the repeated use of rewrite rules. Indeed, the concept of rewrite rule is extremely useful in understanding many symbolic computation processes. However, the matter is not quite as simple as it appears. It is very easy to write a computer program which will reduce expressions according to a given collection of rewrite rules. The result, however, is not an efficient symbolic computation system. Too much time is wasted trying to match the left-hand-sides of rules to different subexpressions. Moreover, serious problems are raised by the commutativity and associativity of addition and multiplication. For instance, the rewrite rule $\sin^2 x + \cos^2 x \rightarrow 1$ ought to be applied even if the terms $\sin^2 x$ and $\cos^2 x$ happen to be separated, as in $\sin^2 x + 4 + \cos^2 x$. One can't expect to use the commutative law as a rule $a+b \rightarrow b+a$, because that will lead to infinite loops immediately. A tremendous literature exists on such theoretical issues,[18] all very interesting, but from the practical point of view, the bottom line is that it is necessary to directly program the fundamental algorithms needed in

---

16 This method, due to Karatsuba, is described in Collins et al. *1982*, p. 194. The exact point at which it becomes more efficient depends on your program and machine. Another method due to Schönhage and Strassen *1971* works in time $n \log n \log \log n$, but with such a large constant as to have no practical impact.

17 This view of calculation was introduced in Markov *1949*; it is the essence of "Markov algorithms".

18 See e.g. Siekmann *1984* for more information and further references.

symbolic computation, in essence specifying the order in which the rules are to be applied and the places where one should look for matches to the left-hand side. The programmers of the most well-known symbolic computation system, MACSYMA, discovered this early on (Fateman *1979*, p. 578), and others have verified it for themselves.[19]

The paradigm of computation as normalizing expressions according to certain rewrite rules is much like logic; the rules are simple and universal. In fact, it can be directly translated into a logical description, in a surprisingly small amount of Prolog code. This program is essentially the translation of computation into logic that Turing used in his solution to the Entscheidungsproblem, but starting from rewrite rules instead of Turing machines.

5. *The Computerization of Logic: Automated Deduction*

Since the early 1960's researchers in "automated deduction" have been attempting to mechanize logic. Leibniz and Boole would have been delighted with their work. Most of this work has been based on Robinson's resolution method, although this is no longer true of current work.[20]

THEOREM PROVERS

A good deal of work in automated deduction has been directed towards the construction of "theorem-provers", whose purpose is to find formal derivations of particular statements that may be presented to the program. A theorem-prover can be applied towards the mechanization of the logical aspect of mathematics, which is the main point of interest in this paper; but in passing it should be mentioned that theorem-provers also can be applied in artificial intelligence. For example, one has been used to construct plans of action for a robot with a certain limited repertoire of actions operating in a predictable environment (Fikes and Nillson *1971*). The successes and limitations of theorem-provers in mathematics will be discussed below.

PROOF CHECKERS

Another branch of the work in automated deduction has been directed toward the computerization of formal theories like those of logic. The result of such a piece of work is a language in which one writes mathematical proofs, much as one writes

---

19 Yet when rewrite rules have been incorporated into resolution theorem provers, as they are for example in Bledsoe's theorem-prover (*1984*), there is no corresponding effort to make the application of the rules efficient. This must await a new generation of programs which will combine the capabilities of today's theorem-provers and today's symbolic-computation programs.

20 There is no space in this paper to review the achievements of theorem-proving prior to about 1980; the interested reader may start with Bledsoe and Loveland *1984*, or with the textbook Wos et al. *1984*.

computer programs in a programming language. If a proof is correct, it will be accepted ("checked") by the system, just as a correct program is accepted by a compiler. A number of such systems have been constructed, beginning perhaps with AUTOMATH, a proof-checker constructed in The Netherlands by de Bruijn. A well-known textbook, Landau's *Grundlagen der Analysis*, was translated into the formal language required by AUTOMATH (van Bentham and Jutting *1979*; de Bruijn *1980*). That textbook was selected because its arguments are already quite detailed and formal, but in the translation the length of the text expanded by a factor of eight to ten.

GRAMMAR OF MATHEMATICS

One of the aims of the current work on proof-checkers is to make a *practical* proof-checker, with which one "programs" a proof with about the same amount of effort which it now takes to translate a mathematical algorithm into a computer program. The problems to be solved in developing such a system must be similar in some respects to the problems that have been solved in the past thirty years of the evolution of programming languages. Perhaps the best system currently in use is NuPrl (Constable et al. *1986*), at Cornell. The research team there is now "programming" several proofs a week. Proofs in NuPrl are written in "top-down" style, using indentation to show the structure of the proof as is common in programming languages. The theorem is stated at the top, the main steps are indented one step, their proofs further indented, and so on. The rules of the language constitute a "grammar of mathematics".[21]

The distinction between a proof-checker and a theorem-prover is not as clearcut as it may seem. In order to be useful, a proof-checker must incorporate a theorem-prover capable of filling in "between the lines". If a proof is to be human-readable, the steps between the lines will in general require several "atomic" logical steps. Each line of such a proof is really a command to the proof-checker: Prove this line from the preceding ones.

BOYER-MOORE

An example of a hybrid system is the theorem-prover described in Boyer and Moore *1979*. This system, which is not based on resolution but on rewrite rules, makes use of previously proved lemmas. When you want it to prove a certain theorem, you supply it with what you consider will be the key lemmas. If it doesn't succeed, you give it more lemmas. In general the proofs it finds are the ones you had in mind all along, and were trying to make it find by supplying the right lemmas; this is more like a proof-checker than a theorem-prover. This theorem-prover represents the state-

---

21 NuPrl and some of its cousins have been heavily influenced by the theoretical work of Per Martin-Löf, who has developed formal logical languages with sufficiently rich type structures to support proof-checkers like NuPrl. See Martin-Löf *1984* or Beeson *1985*, Chapter XI.

of-the-art in the early 1980's. It is able to find proofs by mathematical induction, a skill which has eluded resolution-based theorem provers. Since induction is a natural tool in proving the correctness of computer programs, the Boyer-Moore prover has been applied to several such examples. Its crowning achievement is the formal proof of Gödel's second incompleteness theorem.[22]

Most current developments are *not* in resolution-based theorem-proving. The volume of proceedings Shostak *1984* clearly shows the dominant influence of rewrite rules in the subject at that time. The seminal ideas of Knuth and Bendix *1970* are at the heart of these developments. It is not too difficult to get an appreciation of these ideas, and they are as important as resolution for the subject.

KNUTH-BENDIX

A set R of rewrite rules is called *confluent* if whenever an expression E can be rewritten (by many applications of the rules) in two different ways as E_1 and E_2, then one can find a common rewriting of both E_1 and E_2.[23] For example $(a+b)c(d+e)$ might be rewritten as $(ac + bc)(d + e)$ or as $(a + b)(cd + ce)$, but both of these expressions can be rewritten to the same fully expanded form. Knuth's idea (Bendix was the programmer) was that if a set of rules is *not* confluent, there is a mechanical procedure to generate new rules, thus enlarging the original set of rules. If this procedure is applied repeatedly, under favorable circumstances the process will not generate an infinite supply of rules, but finally stop with a confluent set of rules. The procedure itself is simple: Suppose the expression E can be rewritten as E_1 and also as E_2, but these two expressions have no common rewriting. Then we can take $E_1 \rightarrow E_2$ as a new rule. Of course, we could take $E_2 \rightarrow E_1$ just as well, and most of the technicalities of the method arise from the necessity of making the right choice; the wrong choice may lead to infinite regresses. In order to find expressions E which can be rewritten in two different ways, it suffices to examine those finitely many expressions resulting from matching the left-hand side of one rule to a sub-expression of the left-hand side of another rule. For example, if we take the rules $(xy)z \rightarrow x(yz)$ and $aa^{-1} \rightarrow 1$, with the matching substitution $x = a$, $y = a^{-1}$, we get the expression $(aa^{-1})z$ which can be rewritten as $1z$ and as $a(a^{-1}z)$, so (if the rule $1z \rightarrow z$ is also present) we get the new rule $a(a^{-1}z) \rightarrow z$. This is the first new rule that the Knuth-Bendix algorithm produces when it is given the three axioms of group theory as input. (The final output is a set of ten rules.)[24]

---

22 Although the theorem is of great significance, it is mathematically quite simple; most steps in the proof are simple combinatorial coding tricks.

23 The history of rewrite rules and confluence probably begins with Church's λ-calculus. The famous Church-Rosser theorem asserts that the basic rewrite rules of λ-calculus are confluent.

24 Experts will observe an oversimplification here in that we haven't emphasized the *termination* of rewritings.

The relevance of the Knuth-Bendix technique for automated deduction is this: Suppose we are interested in deductions from axioms which can be stated as equations. (Nearly any axioms can be so stated, in a suitable language.) Then we orient those equations in one direction and consider them as rewrite rules. Applying the Knuth–Bendix procedure then amounts to making deductions from the axioms. The problem of generating millions of irrelevant conclusions does not arise, as the Knuth-Bendix procedure generally finds only a few more rules, most of which are interesting. The difficulty, however, is that systems of interest often involve commutative or associative operations, which create infinite regresses and cause many other technical problems. Some of these problems have been overcome; others are the subject of current research. The best result so far by these means is probably the automatic proof of the following theorem, due to Stickel *1984*: Suppose R is a ring in which the identity $x^3 = x$ holds. Then R is commutative, i.e. satisfies the identity $xy = yx$. The proof consists of 34 equations; in the course of generating them, the computer simplified 9,013 equations.

EQUATIONAL LOGIC

I have suggested as a working definition of computation vs. logic that computations consist essentially of a series of equations derived from each other. Why not base a programming language on "equational logic", thus formalizing this notion of computation? O'Donnell *1985* does just that. A program for a function f consists of a finite set of equations. The interpreter is essentially an implementation of the Knuth-Bendix algorithm; it completes the program to a confluent set of rewrite rules, and execution of the program to compute a value $f(x)$ for a specific x consists in reducing the expression $f(x)$ to normal form using those rules. The distinction between logic and computation, already blurring in the case of Prolog, is very difficult to draw at all in O'Donnell's system. The definition of computation as manipulation of equations will not stand serious scrutiny, since theoretically any statement can be turned into a Boolean-valued equation. O'Donnell's work brings this theoretical observation into programming practice.

WU WEN-TSUN

Some very interesting recent work is based neither on resolution nor on rewrite rules. I refer here to the work of Wu Wen-Tsun and his followers, who have returned to the idea of Descartes by reducing theorem-proving in elementary geometry to calculations with polynomials. What Prof. Wu observed is that if you work in a geometry without *order* relations, then the corresponding polynomial formulation does not involve inequalities but only equations. In this situation, modern efficient algorithms for solving equations can be applied. The concept underlying these algorithms is "Gröbner bases"; for a clear explanation see Winkler *1987*. Wu's simple and straightforward observation led to programs that have proved "various famous theorems in the ordinary geometry: theorems of Keukou, Pappus, Pascal, Simson,

Feuerbach, Morley". For example, Morley's theorem says that the neighboring tri-sectors of the three angles of any triangle intersect at the vertices of an equilateral triangle.[25]

6. *The Computational Core of Mathematics*

I will put forward here a view of the nature of mathematics, which I think is supported by several different lines of evidence and makes sense from several viewpoints. The view is that algorithms, or computational methods, are central in mathematics and the logical aspect of mathematics is "wrapped around" this algorithmic core. There are fundamentally only three kinds of theorems in mathematics:

(1) *Theorems which assert the universal validity in some domain of some equation or inequation, each instance of which can be checked.* For example, Fermat's conjecture that $x^n + y^n = z^n$ has only trivial solutions in integers.

(2) *Theorems which assert that certain algorithms will always yield results for inputs in a certain domain, and that a certain equation or inequation involving symbols for these algorithms is always valid.* For example, the theorem that Euclid's algorithm E will always produce the greatest common divisor of two positive integers m and n can be expressed as

$$divides(k, m) \cdot divides(k, n) \leq divides(k, E(m, n))$$

where *divides* takes values 0 or 1.

(3) *Theorems which assert that a certain object must exist, otherwise a contradiction would result, although the proof does not show how to construct the object in question.*

Proofs of kind (3) are called *non-constructive*. It often happens that a proof of kind (2) can be found to replace a proof of kind (3). For example, one can prove that any two positive integers m and n have a greatest common divisor by showing that the least integer of the form $\lambda m + \mu n$ is the greatest common divisor, where λ and μ are (positive or negative) integers. But since there are infinitely many possibilities for λ and μ, this is a non-constructive proof. In this case, however, we can provide a constructive proof of the theorem using Euclid's algorithm.

The phrase "there exists" has a different meaning if we have the Euclid's algorithm proof in mind than if we have the non-constructive proof in mind. In the one case, it hides an algorithm which for some reason we prefer not to make part of

---

25 The most accessible presentation for non-Chinese readers is Wu *1986*; although the work was done in the late seventies, it has only recently come to the attention of many Western researchers. Morley's theorem actually says more than is stated here. See Kutzler and Stifter *1987* for references to more recent work along the lines opened by Wu.

the statement of the theorem. In the other case, it hides our ignorance of an algorithm. These different uses of "there exists" should be carefully distinguished in the computerization of mathematics.

There is a long history of work on this distinction, undertaken by philosophers, logicians, and mathematicians, and more recently by computer scientists. A central point of the investigation is empirical: how many of the uses of "there exists" in actual mathematics hide algorithms, and how many hide ignorance? And of those which hide ignorance, in how many of the cases can suitable algorithms be supplied?

In other words, how many (and which) of the theorems of mathematics can be "constructivized"? By the constructivization of a theorem, we here understand supplying a correct statement and proof of the theorem in which every use of "there exists", even implicit uses in the definitions of concepts mentioned in the theorem, hides an implicit algorithm rather than ignorance.

This question has to be very carefully formulated before it can be answered meaningfully. Certain theorems can be shown to be hopelessly non-constructive. For example: "for each Turing machine *e*, there exists a number which is 0 if *e* halts at input 0 and 1 otherwise". This theorem cannot hide an algorithm, in view of the unsolvability of the halting problem. Many theorems of mathematics are non-constructive in this sense, for example the theorem that a monotone sequence of real numbers has a least upper bound: in general there is no way to compute the upper bound, given only the rule for generating the sequence. (Specker *1949*; see Beeson *1985*, Chapter IV). On the other hand, if careful attention is paid to the formulation of fundamental definitions, most of the main results of mathematics can be reached without needing the obviously non-constructive theorems. (Bishop *1967*; Bishop and Bridges *1985*). The result of sustained systematic efforts at the constructivization of mathematics is this: most instances in which "there exists" does not hide an implicit algorithm are either obviously non-constructive, like the example above, or can be replaced with suitable constructive versions without inventing substantially new algorithms. The vast majority of mathematics is revealed to consist of algorithms wrapped in logic.

The interesting thing is that there are only *very* few areas of mathematics in which "there exists" seems to hide *genuine* ignorance of algorithms, where so far as we know algorithms may be possible, but they are unknown.[26] It has been shown possible to "unwind" the many complicated concepts of modern analysis and algebra until every important theorem is exposed as having an algorithmic core.[27] Those few

---

26 See Beeson *1985*, Chapter I, for a lengthier discussion.

27 The question of the *efficiency* of these algorithms is another matter, which has not yet been systematically investigated. In the cases where the algorithms are inherently inefficient, we can always look for additional input data, or restrictions on the input, which would make the computation efficient. For example, computing zeroes of a continuous function is in general impossible, but can be

theorems of classical mathematics which are inherently non-constructive, such as the convergence of bounded monotone sequences, turn out not to be important.

The work on symbolic computation has been directed towards the computerization of mathematical algorithms. This is useful when the algorithms are *explicit*, but a proper *computerization of mathematics* will have to provide systematic means of keeping track of *implicit* algorithms. Let me explain: *The logical wrapping on the algorithmic core is an essential part of modern mathematics, because it permits the formulation of concepts leaving only their essential parts visible.* Comprehensible mathematics depends on pyramiding definitions of complicated concepts. These definitions mention "there exists". If we are careful, all these uses of "there exist" hide algorithms; these are the "implicit algorithms" of mathematics.

Some examples will clarify the issues:

Example 1. When we define a real-valued function to be *continuous*, we usually say "for every ϵ there exists a δ such that $\mid x - y \mid < \delta \rightarrow \mid f(x) - f(y) \mid < \epsilon$. This is shorthand for a hidden "modulus of continuity", a function $\omega(\epsilon) = \delta$. This hidden modulus is a "witness" to the continuity of f.[28] When we say "Let f be continuous", we mean "Let f be continuous, and let ω witness the continuity of f". When mathematics is computerized, the computer will have to keep track of "witnesses" of the various properties of the objects being manipulated.

Example 2. The famous theorem that every finite group of odd order is solvable. How should we classify this theorem, as type (1) or type (2)? If we think of the theorem as implying the use of particular (presumably efficient) methods for checking solvability, perhaps even producing the sequence of subgroups needed to "witness" solvability, then it is of type (2). If we "wrap" these methods in logic to achieve a more succinct statement of the theorem, we can think of "solvable" as simply a property which can be checked of a particular group. Then the theorem is of type (1); the computational content of the proof is less explicit; but still present, under the wrapping.

The subject of constructive mathematics has traditionally been associated with strongly-held opinions about the philosophy of mathematics. After long study of the field, the author believes that the essence of the discoveries that have been made can be understood in a more or less objective fashion. Disagreement enters only when one begins to discuss the relative importance and "meaning" of those proofs in which "there exists" does not stand for an implicit algorithm. *That* is a philosophical question. Whatever one thinks about that, the fact that "there exists" is used in different ways, and the fact that mathematics can be done systematically so as to maximize its algorithmic content, are objective. In the past, one had only

---

efficient for polynomials, or more generally, functions given by a power series. This can be viewed either as a restriction on the data or as additional input (the series).

28 The term "witness" was introduced in Feferman *1979*.

philosophical motivations for reworking mathematics more algorithmically; now the possibilities for computerizing mathematics present us with practical reasons for so doing.[29]

I have explained the view of mathematics as consisting of an algorithmic core wrapped in a web of logical definitions of constructs and proofs about them. In computerizing mathematics, it seems to me that this view should be a guiding principle of design. The system will have to keep track of the "hidden witnesses" and be able to call them up when needed, but not burden the user with them. The system NuPrl discussed above is constructed along these lines.

7. *Limits of Computerized Mathematics*

TURING'S LIMIT

Turing's work on the *Entscheidungsproblem* showed the impossibility of ever designing a program to settle any mathematical problem. This result was reinforced by Gödel's incompleteness theorems; together these results place a fundamental theoretical limit on the computerization of mathematics. Mathematics must remain a process of searching for the truth; there will always be problems that are beyond the scope of presently understood methods. Nevertheless, mathematicians are experts like any other experts, and their knowledge is subject to the efforts of the "knowledge engineers", who try to build computer programs to imitate experts. Programs for automated deduction might in principle compete with mathematicians in the finding of new results, even though still subject to Turing's limit.

COMBINATORIAL EXPLOSION

One may well ask why the mathematicians have not been put out of business by computers programmed for automated deduction. The answer, in a nutshell, is this: the computer can search for proofs, but there are millions of proofs of trivial, uninteresting theorems, and the proofs of interesting theorems are very few and far between; it's much worse than looking for a needle in a haystack. The problem afflicts all areas of artificial intelligence, and goes by different names: the "Combinatorial Explosion", or the "Problem of Large Search Spaces". The method of simply systematically searching through all possible formal derivations for a proof of a given theorem has been called the "British Museum algorithm", apparently because it is like looking for a specified object in a museum collection by examining every object in turn. In spite of all efforts to build "intelligent" criteria into programs to restrict the search, the fundamental problem remains.

---

29 See Bishop *1967*, Bishop and Bridges *1985*, and Beeson *1985* for further information about constructive mathematics.

INTELLIGENCE DEFINED

It has even been suggested that intelligence can be *defined* as "the use of knowledge to restrict search". According to this view, every problem is solved by a search through some class of possible solutions ("solution space"), which is usually too large to permit exhaustive search. An intelligent agent (living or machine) must use its knowledge somehow to avoid uselessly searching "up the wrong tree". The author does not endorse this definition, but only sets it out here for consideration.

ROLE OF MEANING

Some evidence for this hypothesis: if people are given problems in symbolic logic, with no meaning attached to the symbols, they don't perform better than computers. Mathematicians perform better than computers at finding proofs of interesting theorems because they understand the meaning of the symbols involved. This understanding calls into play mental mechanisms which are not at all understood. Certain early programs in automated deduction tried to make use of the method of checking hypotheses in a "typical diagram", as is often done while trying to prove geometry theorems. This helped some, and such "semantical methods" are now standard in the art of automated deduction. Nevertheless, as Bledsoe puts the matter, the computer asks, "Which clauses shall I resolve next?", while the human asks, "What shall I *do* next?". The human will reflect, rearrange, reformulate, recall a similar problem, make a broad plan of action, and conjecture subgoals; these are all ways of bringing knowledge to bear to restrict search.

DATA TYPES

Another method of introducing some knowledge about the meaning of terms is the method of "data types". A data type is the answer to a question, "what kind of object is this?". For examples of data types consider: integer, square, real number, group, finite set, automorphism group. One strand of research in automated deduction is the construction of formal languages which allow a rich collection of data types. A test problem here is the theorem of LaGrange in group theory, according to which the cosets of a subgroup of a finite group all have the same number of elements, i.e., are in one-to-one correspondence. The proof of this theorem is very simple by ordinary mathematical standards, yet it seems to be too difficult for automated deduction; and the bottleneck seems to be that several different data types are involved. The mathematician easily keeps them straight. Specifically, the following data types are involved: G, the type of group elements; $P(G)$, the type of subsets of G; C, the type of functions from one subset of G to another (C is needed because one-to-one correspondences between cosets are of type C); and finally, one needs either an operation leading from a subgroup H and element a of G to the coset aG, or one needs an operation leading from H to the type of cosets of H.

COMPUTER CHESS

The problem of playing chess is similar to that of proving theorems, in that there is a large search space of possible moves, most of which are uninteresting. Computers today are much better at chess than they are at mathematics, and it is interesting that they have achieved that ability by sheer brute-force search techniques, rather than by programs involving the representation of knowledge of chess strategies. No program exists today which can be presented with a chess situation and answer the question, "What strategy would you recommend for White?" The search space of chess is just small enough that specially constructed hardware has enabled the machine to compete with very good human players; the same techniques won't work on the game of Go, which has a much larger search space.

DREYFUS BROTHERS

The view that intelligence consists of using knowledge to guide a search is embodied in the design of current "expert systems", as well as in programs for automated deduction. In spite of the enormous amounts of publicity surrounding them, expert systems have not succeeded in equalling the performance of human experts in any field. Stuart and Hubert Dreyfus *1986* attribute the limitations of expert systems to the inadequacy of the view of intelligence they embody, namely the application of rules. They maintain that human experts solve problems by "recognizing" the solution, and may then perform computations (use rules) to verify or refine the solution. The use of rules represents a lesser stage which one passes through on the way to becoming an expert. For example, a master chess player does very little searching, but simply "sees" a good move. The Dreyfus brothers performed an interesting experiment: a master chess player was given arithmetic problems to work mentally. After determining the maximum rate at which he could solve arithmetic problems, he was required to continue solving them while playing five-second-a-move chess against a slightly weaker, but master level, player. He was able to "more than hold his own" under these circumstances.[30] One wonders if there are mathematicians so expert that their performance in finding proofs is not degraded if they are required to solve arithmetic problems at the same time.

---

30 Dreyfus and Dreyfus *1986*, p. 33. That reference does not state that the rate of arithmetic problems was maximum; this claim was made in a lecture by S. Dreyfus.

8. *Discussion*

Cast of characters: Scepticus

Professor Infinity   mathematician

Practicus   engineer

Ordinarius   layperson

Roboticus   researcher in automated deduction

Dr. Lehrmeister   educator

The Author has convened this panel in order to assess the significance of the developments discussed in the paper you just read. He has asked the participants to discuss the topic of logic, calculation, and the computerization of mathematics from their own particular viewpoints, and particularly to share their visions of the future of the subject.

Roboticus: I think this must be one of the most exciting periods of history in which to be alive. Just as machines took over physical work from people in the Industrial Revolution, now machines are starting to take over intellectual work, even mathematics. Machines will write the mathematics of the future and apply it, too. It will free human beings to discover the essence of what it means to be human.

Skepticus: Come now, try to keep your head on your shoulders, will you? The machines have yet to prove a single interesting theorem, let alone create a new subject or make a significant application of mathematics. They day when computers will be more than tools is not even conceivable.

Professor Infinity: So far, they aren't even tools for most of us mathematicians. We teach our classes at the blackboard, we prove our theorems on the back of envelopes. We use the computer only to replace the typewriter when we go to write up our results for publication.

Practicus: Well, that's an overly negative assessment. MACSYMA is in use by 3500 people at over 600 sites today,[31] and is being used to make calculations on significant problems in physics and engineering as well as pure mathematics. It and other computer algebra systems are being used in such areas as high energy physics, celestial mechanics, general relativity, electron optics, and plasma physics.[32] So the computer is certainly at least a tool in today's applied mathematics.

Professor Infinity: It has been used as a tool in pure mathematics, too. The famous four-color theorem was proved in 1976 at the University of Illinois, and the last step of the proof was an examination of a large (but finite) number of cases by computer. Of course, that was a mere calculation; the essential ideas, the logic of the proof, were supplied by humans. No automated deduction was involved. Moreover, that

---

31 Stated in a lecture by Richard Brenner of Symbolics Corp.

32 Calmet and van Hulzen *1982*.

example is just about the only good example you can find of a computer being used to help in a proof: it was the exception, not the rule.

Roboticus: You people remind me of those who scoffed at horseless carriages! Of course, the technology is in its infancy, and most mathematicians are still unaware of the possibilities. There was a time when telephones were an oddity; you would have pointed out that people carried on their business perfectly well by mail and visits. But a hundred years from now, every mathematician will be trained in mathematics using a computer. People will think that a proof *is* something you have gotten accepted by the computer. Students will do their homework on the computer, which will tutor them as well as accepting or rejecting their proof. You won't be able to *publish* your new theorem until it passes the American Mathematical Society's official proof-checker. That is, in a refereed journal – you'll be able to publish anything you want instantly just by putting it in a public area of an electronic bulletin board, from where anyone can print it out at home or in a library.

Ordinarius: What a different world it will be. What will be left for the human mathematicians to do? It sounds to me as if all they will do is ask the questions – the computer will have all the answers. It will find the proofs, do the calculations, even make the applications. You mathematicians will be as obsolete as blacksmiths.

Practicus: I personally don't believe the computer will ever be more than a calculating tool. Possibly it will become a more sophisticated calculating tool; MAC-SYMA has opened my eyes to the possibilities. But mathematics will go on more or less as it is now.

Professor Infinity: As long as *somebody* needs to ask the questions, we mathematicians won't be obsolete. After all, that's really the essence of mathematics, asking the right questions. Calculation is really not the heart of mathematics, nor is logic. Creativity is required. Mathematics is an art form, not just an exercise in logic and calculation as the Author would have you believe.

Skepticus: Let's keep our feet on the ground here. Take a simple, straightforward mathematics book like Hardy and Wright's *Number Theory*. Can you translate that into any of these automated proof-checkers? Not by a long shot. On page 2 you find the convention that p will always stand for a prime number; you can't make that convention in AUTOMATH or NuPrl, and such conventions are built upon pyramidically in mathematics. A few pages later the concept of "greatest common divisor" is discussed, and you find that you need a way of expressing algorithms. You will not be able to use your favorite programming language to do that; you will have to use the unfamiliar and perhaps inconvenient language built into the proof-checker. Because of the lack of convenient typing mechanisms in the formal language, no proof-checker has yet verified a proof of LaGrange's theory in group theory. These difficulties are not going to be overcome any time soon; Roboticus, you are indulging in science-fiction fantasies. And when you stop discussing proof-checkers and start discussing the possibility of machines *discovering* proofs, you move from the fantastic to the ridiculous. If you can't even get them to *verify* proofs produced by humans, you had better forget about getting them to *discover* proofs.

Practicus: Remember, thirty years ago a computer as powerful as the Apple II took up a whole room, needed air-conditioning, and ran slower. It was only fifty years ago that Turing showed us how to program a computer; his Turing machines were the first programming language. Considering how difficult it is to program in Turing machine language, one would never have believed at that time the ease with which complex algorithms can today be encoded in languages like C or Prolog. Having this example of the evolutionary development of languages before us, it is not at all impossible to foresee a similar evolution of powerful and usable languages in which to express mathematics.

Ordinarius: I am certainly willing to believe it. I have seen so many astonishing developments in technology! One year a thing is impossible and the next year it is on sale at Radio Shack. They tell me that expert systems found molybdenum in Idaho in 1980[33] and that they will be able to shoot down hundreds of attacking ICBM's.

Scepticus: Well, the area where that molybdenum was found had been partially explored, and it is difficult for a non-expert to judge the actual contribution of the expert system; I haven't heard of it finding any more deposits in the intervening seven years. And I hope we never have to find out whether an expert system can conduct a nuclear war, but I have serious doubts. Remember, even the most successful of today's expert systems (XCON) gets correct answers only ninety to ninety-five percent of the time.[34] To give another example, there is an expert system, PUFF, which is supposedly expert in the diagnosis of respiratory disease. In the expert system literature you can read glowing reports of it;[35] it is supposedly in daily use at a medical center in San Francisco. Yet in the Sunday newspaper I read an interview with the director of that medical center, who says PUFF is so inaccurate that he now uses it only to get nice printouts of medical information.

Roboticus (a little defensively): Well, medical diagnosis is not mathematics. It's a notoriously imprecise area. Mathematics ought to be much easier. After all, MAC-SYMA doesn't get wrong answers.

Practicus: Every large program has a few bugs, and MACSYMA is no exception. But then again, it replaces tables and human computations, both of which were erroneous much more often than MACSYMA.

Dr. Lehrmeister: You theoreticians amaze me. All this discussion about the computer in mathematics and you never think of the students! Yet it is in education that the computer will find its most important uses. In the future students of all the sci-

---

33 Campbell et al. *1982*.
34 Waterman *1986*, p. 218.
35 Aikins et al. *1983*.

ences, including mathematics, will do much of their work at a computer. They will solve homework problems line by line, with the computer checking and if desired, correcting each line. When they want to learn how to solve the problems, they will ask the computer to show them some examples. Symbolic computation has been adequate for these purposes for some time, and now that the personal computer has acquired enough speed and memory to support these kinds of educational programs, they are just around the corner, not a hundred years in the future like Roboticus' fantasies.

Scepticus: Well, it's not as easy as all that. These programs don't work the way you want students to think, they often have more advanced algorithms. Moreover, they don't print out anything but the answer, which doesn't do the student much good. People have tried to use them in mathematics education, but these problems were stumbling-blocks.

Dr. Lehrmeister: You are right about that, you can't just take MACSYMA or MuMath into the classroom. You have to start over, building *cognitive fidelity* into the system, programming it to print out exemplary solutions as well as answers and to work the way you want the student to work. While you're at it, you should make use of ideas from cognitive science to tailor your teaching to the student. The computer will have a "model" of the student's knowledge and tailor its output accordingly.

Roboticus: Dr. Lehrmeister, I think your point is more general: it is not only mathematicians who stand to benefit from the computerization of mathematics. You pointed out that education will be transformed: but also engineering will be transformed. Of course engineering is already being affected by symbolic computation, but it will also be affected by automated deduction. In particular software engineering will be a different discipline a few decades from now. The problem of reliability in software will be solved once and for all. The way to reliable software is to produce software which has been mathematically proved to meet its specifications. Systems will be built, in fact are being built today, which will support the production of such automatically verified software. The three activities of specifying the task of the software, writing the program to perform that task, and giving a mathematical proof that the program does meet that specification, will not be separate tasks, but will be performed simultaneously. To the programmer it will seem like writing a well-structured, well-commented program. The essential ingredient will be the use of a formalized logical language to replace the informal "comments" in today's programs.

Practicus: Roboticus is right that such systems are under construction today. For example, MITRE Corporation and Oddysey Research Associates both have government contracts for this work. The Department of Defense wants to have all its programs written in ADA and verified to be correct, so it has hired dozens of logicians and computer scientists to arrange an ADA program-proving environment.

In addition, Oddysey is going to provide correctness proofs for engine-control chips for General Motors. They want to know the chips are mathematically proved correct before they put them in every new American automobile.

Scepticus: How will you know the specifications have been correctly written? Remember that Gemini V came down a hundred miles from the planned spot because the orbit program didn't include the effect of the motion of the earth in its orbit around the sun. Couldn't that have happened just as easily in the specification as in the program? And how will you know that the program will always be used according to its specifications? Perhaps the Gemini program's specifications said it was only for short flights. And how will you know that there is no bug in the verifier itself, which might let a crucial program bug pass? It's irresponsible for scientists to pretend that these applications of automated deduction can lead to reliable systems.

Roboticus: On the contrary, it would be irresponsible *not* to apply what knowledge we have to improving the reliability of military programs. I for one find it vital to make sure the programs that control our missiles have as few bugs as possible!

Skepticus: That would be fine if the effort to make them reliable had no side effects. But your work will be cited by generals in testimony to Congress as evidence that these complex systems have been "proved" reliable; so more and better should be built. You cannot responsibly ignore the social and political context of your work.

Dr. Lehrmeister: Ladies and gentlemen, please keep the discussion civil! This is a public place. And try to stay on the subject, which is logic and calculation, and the computerization of mathematics, not social policy.

Ordinarius: It does seem more and more difficult to keep mathematics in the ivory tower where it used to be. I read that even number theory is being used for cryptography these days. I suppose it's not too surprising that logic also has applications. After all, it's supposed to be the science of reasoning, and now that at least simple reasoning can be computerized, you'd expect applied logic to enter society at many places.

Practicus: That's certainly true. Why, that most abstract of subjects, proof theory, has been used as a tool in the optimization of algorithms. The "pruning of a proof tree" translates into a customized algorithm; instead of using the same algorithm to solve every instance of the problem, you customize the algorithm to the input before using it. That work was a Ph.D. thesis, and the student in question saw how to apply the technique to the optimization of algorithms for computer vision. At present the robots on assembly lines are little more than programmable machine tools, that can repeat a sequence of motions originally performed by a human operator. It would help if they could "see" in three dimensions, but existing algorithms are too slow, so they see only in two dimensions. Proof theory is now helping them see in three dimensions.[36]

---

36 Goad *1983*.

Roboticus: Logic is going to be the foundation of artificial intelligence. Today's expert systems are interesting tools, but they are by no means intelligent, because they lack common sense. Eventually, however, we will represent common sense in a suitable logical language. When we find the right formalism, we will put dozens of people to work for a few years typing in all of common sense. After that we will have a large data base which we can make available to any programmer whose program needs to have common sense. That will be the beginning of true artificial intelligence. It may take fifty years; it may take five hundred years, I don't know. But it will happen.[37]

Scepticus: Everyone would be happy if computers could show a little common sense! But even the task of writing down in English the major categories of common sense knowledge, in other words to make a catalogue of what we consider common sense, has yet to be undertaken, and nobody really has a good idea how to apply logic to the problem. It will be closer to five hundred than to fifty years.

Practicus: Leaving these wild speculations aside, let's consider what the next steps should be. On the one hand we have automated reasoning, on the other symbolic computation. Why are these two separate branches of research? Mathematics is computation plus logic; any system that attempts to computerize mathematics should surely incorporate *both* automated deduction and symbolic computation. Yet no such system exists anywhere today. That ought to be a promising direction of research!

Roboticus: I know of one small experiment in that direction,[38] in which a small program with limited capabilities in both automated deduction and symbolic computation was able to automatically derive the equation of motion of a pendulum. Some knowledge of physics was represented in logic, along with the semantical relations between formulas and physical quantities, and statements about the validity of equations were derived. The symbolic computation system could be called upon for the purely mathematical steps in the derivation.

Practicus: Yes, but that small experiment is very far from the kind of integrated system we need. What we need is a combination of MACSYMA and NuPrl, not created just by pushing together two large systems alien to each other, but designed and built as a unit, interacting at many levels. Moreover, the system needs to be able to explain its answers, which MACSYMA cannot do, and to extract usable algorithms in some (or several) common programming languages from correct proofs. It needs a sufficiently elaborate syntax that one could translate Hardy and Wright into it with minimum effort, or any other rather rigorous mathematical text. One must be able to represent scientific or expert-system style knowledge in it as easily as in Prolog, and use that knowledge in symbolic computations.

---

37 John McCarthy, in his invited address to the International Joint Conference on Artificial Intelligence, 1983; but no corresponding paper appears in the proceedings.

38 Beeson *1987*.

Scepticus: That's a very ambitious project. I have my doubts whether such a thing can ever actually be carried out.

Roboticus: There are no theoretical obstacles to the creation of such systems. The very existence of NuPrl and MACSYMA shows that the essential tools are at hand. It's just a matter of putting in enough effort in the right direction.

Dr. Lehrmeister: Systems with the capability to represent scientific knowledge in logical language, and call upon symbolic computation systems for mathematical calculations, will certainly be necessary before intelligent computer-aided instruction in the sciences can come into full flower. For example, H. Graves has envisioned a program in which the student can select components such as weights, rods, and links on a screen and "construct" a simulated physical system. The computer would simultaneously construct a logical representation of the system, derive the equations of motion, and construct a "simulator", a program which could numerically solve those equations of motion and drive a graphics display to animate the student's system. This combination of automated deduction and symbolic computation is presently beyond reach, but only in practice, not in theory.

Practicus: It is easy to see that systems with combined logical and computational capabilities like that, particularly if they also have powerful graphics, will find widespread practical applications in engineering. At present one can use computer-assisted design tools to essentially computerize the drafting process. With the aid of automated deduction and symbolic computation, suitably combined, a great deal more of the engineering design process can be computer-assisted, if not automated entirely. If a student can create model physical systems and have their equations derived, an engineer can create models of the system she is trying to design.

Professor Infinity: I am afraid pure mathematics will never be the same. There was a time when mathematics was pursued only by her lovers; their quiet life together in the ivory tower was seldom interrupted. Now that she is a public figure, that privacy is a thing of the past.

References

Aikins, J. S., J.C. Kunz, and E.H. Shortliffe

    1983 PUFF: an expert system for interpretation of pulmonary function data. *Comp. Bio. M.* **16** (1983) 199–208.

Beeson, M.

    1985 *Foundations of Constructive Mathematics: Metamathematical Studies.* Berlin/ Heidelberg/New York: Springer-Verlag (1985).

    1987 *Automatic Derivation of the Equation of Motion of a Pendulum.* To appear.

van Benthem-Jutting, L.S.

    1979 *Checking Landau's "Grundlagen" in the AUTOMATH System. Math. Cent. T.* **83.** (1979).

Bishop, E.

    1967 *Foundations of Constructive Analysis.* New York: McGraw-Hill (1967).

Bishop, E., and D. Bridges

    1985 *Constructive Analysis.* Berlin/Heidelberg/New York: Springer-Verlag (1985).

Bledsoe, W.W.

    1984 Some automatic proofs in analysis. In: Bledsoe and Loveland *1984,* pp. 89–118.

Bledsoe, W.W., and D.W. Loveland

    1984 *Automated Theorem Proving: After 25 Years,* AMS Contemporary Mathematics Series, vol. 29. Providence, RI: AMS (1984).

Boole, G.

    1847 *The Mathematical Analysis of Logic.* Cambridge (1847); reprinted by Oxford University Press, Oxford, England (1948, 1951).

    1854 *An Investigation into the Laws of Thought.* London (1854); modern edition by Dover, New York (1958).

Boyer, R., and J. Moore

    1979 *A Computational Logic.* New York: Academic Press (1979).

Buchberger, B., G.E. Collins, R. Loos, and R. Albrecht

    1982 *Computer Algebra: Symbolic and Algebraic Computation.* Vienna/New York: Springer-Verlag (1982).

Calmet, J., and J.A. van Hulzen

    1982 Computer algebra applications. In: Buchberger et al. *1982,* pp. 245–258.

Campbell, A.N., V.F. Hollister, R.O. Duda, and P.E. Hart

    1982 Recognition of a hidden mineral deposit by an artificial intelligence program. *Science* **217** (1982) 927–929.

Collins, G.E., M. Mignotte, and F. Winkler

    1982 Arithmetic in basic algebraic domains. In: Buchberger et al. *1982,* pp. 189–220.

Constable, R., et al.

    1986 *Implementing Mathematics with the NuPrl Proof Development System.* Englewood Cliffs, NJ: Prentice-Hall (1986).

Davis, M., ed.

    1965 *The Undecidable.* New York: Raven Press (1965).

de Bruijn, N.G.

1980 A survey of the project AUTOMATH. In: *To H.B. Curry: Essays on Combinatory Logic, Lambda Calculus and Formalism,* eds. J.P. Seldin and J.R. Hindley. New York: Academic Press (1980).

Dreyfus, H., and S. Dreyfus

1986 *Mind over Machine.* New York: The Free Press (1986).

Fateman, R.J.

1979 MACSYMA's general simplifier: philosophy and operation. In: *Proceedings of MACSYMA Users' Conference,* Washington, D.C., June 20–22, 1979, pp. 563–582. Cambridge, MA: MIT Press (1979).

Feferman, S.

1979 Constructive theories of functions and classes. In: *Logic Colloquium '78: Proceedings of the Logic Colloquium at Mons, 1978,* eds. M. Boffa, D. van Dalen, and K. McAloon, pp. 159–224. Amsterdam: North-Holland Publ. Co. (1979).

Fikes, R., and N. Nilsson

1971 STRIPS: A new approach to the application of theorem proving to problem solving. *Artif. Intel.* **2** (1971) 189–208.

Frege, G.

1879 *Begriffsschrift,* a formula language, modeled upon that of arithmetic, for pure thought. In: van Heijenoort *1967,* pp. 1–82.

Goad, C.

1983 Special purpose automatic programming for 3D model-based vision. In: *Proceedings of the DARPA Image Understanding Workshop,* June 1983. Mclean, VA: Science Applications (1983).

Gödel, K.

1930 Die Vollständigkeit der Axiome des logischen Funktionenkalküls, German original with English translation. In: Gödel *1986,* pp. 102–123.

1986 *Collected Works,* vol I, ed. S. Feferman et al. New York: Oxford University Press (1986).

Hilbert, D.

1925 *On the Infinite.* In: van Heijenoort *1967,* pp. 129–138.

Hilbert, D., and W. Ackermann

1928 *Grundzüge der theoretischen Logik.* Berlin: Springer-Verlag (1928).

Knuth, D.E., and P.B. Bendix

1970 Simple word problems in universal algebras. In: *Computational Problems in Abstract Algebra,* ed. J. Leech, pp. 263–297. Pergamon Press (1970).

Kolmogorov, A.N.

    1932 Zur Deutung der intuitionistischen Logik. *Math. Z.* **35** (1932) 58–65.

Kutzler, B, and S. Stifter

    1987 New approaches to computerized proofs of geometry theorems. In: *Proceedings of the Conference on Computers and Mathematics,* Stanford University, July 30– August 1, 1986. To appear.

Leibniz, G.W.

    1679 Two studies in the logical calculus. In: *Philosophical Papers and Letters,* 2nd edition, ed. L.E. Loemker, pp. 235–247. Dordrecht/Boston: Reidel (1969).

    1923 *Sämtliche Schriften und Briefe.* Darmstadt (1923ff.), Leipzig (1938 ff.), Berlin (1950 ff.): Deutsche Akademie der Wissenchaften zu Berlin.

    1965 *Die philosophischen Schriften*, ed. C.I. Gerhardt. 7 vols., Berlin (1857–90); reprinted by Georg Olms Verlag, Darmstadt (1965).

Lloyd, J.W.

    1984 *Foundations of Logic Programming.* Berlin/Heidelberg/New York: Springer-Verlag (1984).

Markov, A.A.

    1949 *Theory of Algorithms.* English translation by Jacques J. Schorr-Kon and PST staff, Academy of Sciences of the USSR, Moscow (1954); there was a German translation of 1954 also.

Martin-Löf, P.

    1984 *Intuitionistic Type Theory.* Naples: Bibliopolis (1984).

McCarthy

    1985 (IJCAI Proceedings)

Moses, J.

    1971 Symbolic integration, the stormy decade. *Comm. ACM* **14** (1971) 548–60.

O'Donnell, M.J.

    1985 Equational Logic as a Programming Language. Cambridge, MA: MIT Press (1985).

Peano, G.

    1889 The principles of arithmetic, presented by a new method. In: van Heijenoort *1967,* pp. 83–97.

Rand, R.H.

    1984 *Computer Algebra in Applied Mathematics: An Introduction to MACSYMA.* London: Pitman (1984).

Risch, R.H.

1970 The solution of the problem of integration in finite terms. *B. Am. Math. Soc.* **76** (1970).

Robinson, A.

1965 A machine-oriented logic based on the resolution principle. *J. As. Comp. Mach.* **12** (1965); reprinted in *Siekmann and Wrightson 1983,* pp. 397–415.

Schönhage, A., and V. Strassen

1971 Schnelle Multiplikation großer Zahlen. *Computing* **7** (1971) 281-292.

Shostak, R., ed.

1984 *7th International Conference on Automated Deduction,* Lectures Notes in Computer Science 170, Berlin/Heidelberg/New York: Springer-Verlag (1984).

Siekmann, J.

1984 Universal unification. In: Shostak *1984,* pp. 1–42.

Siekmann, J., and G. Wrightson, eds.

1983 *Automation of Reasoning,* vol. 1. Berlin/Heidelberg/New York: Springer-Verlag (1983).

Specker, E.

1949 Nicht konstruktiv beweisbare Sätze der Analysis. *J. Symb. Log.* **14** (1949) 145–158.

Stickel, M.

1984 A case study of theorem proving by the Knuth-Bendix method: discovering that $x^3 = x$ implies ring commutativity. In: Shostak *1984,* pp. 248–258.

Turing, A.

1936-7 On computable numbers, with an application to the Entscheidungsproblem. *P. Lond. Math. Soc. (2)* **42** (1936-7) 230–265; A correction, ibid. **43** (1937) 544–546. Reprinted in Davis *1965,* pp. 115–153.

van Heijenoort, J.

1967 *From Frege to Gödel: A Source Book in Mathematical Logic, 1879–1931.* Cambridge, MA: Harvard University Press (1967).

van Hulzen, J.A., and J. Calmet

1982 Computer Algebra Systems. In: Buchberger et al. *1982,* pp. 221–243.

Waterman, D.A.

1986 *A Guide to Expert Systems.* Reading, MA: Addison-Wesley (1986).

Winkler, F.

   1987 Solution of equations I: Polynomial ideals and Gröbner bases. In: *Proceedings of the Conference on Computers and Mathematics,* Stanford University, July 30–August 1, 1986, to appear.

Wos, L., R. Overbeek, E. Lusk, and J. Boyle

   1984 *Automated Reasoning: Introduction and Applications.* Englewood Cliffs, NJ: Prentice-Hall (1984).

Wu Wen-Tsun

   1986 Basic principles of mechanical theorem-proving in elementary geometries. *J. Autom. Reas.* **2** (1986) 221–252.

Logical Depth and Physical Complexity

Charles H. Bennett

Abstract. Some mathematical and natural objects (a random sequence, a sequence of zeros, a perfect crystal, a gas) are intuitively trivial, while others (e.g., the human body, the digits of π) contain internal evidence of a nontrivial causal history.

We formalize this distinction by defining an object's "logical depth" as the time required by a standard universal Turing machine to generate it from an input that is algorithmically random (i.e., Martin-Löf random). This definition of depth is shown to be reasonably machine-independent, as well as obeying a slow-growth law: deep objects cannot be quickly produced from shallow ones by any deterministic process, nor with much probability by a probabilistic process, but can be produced slowly.

Next we apply depth to the physical problem of "self-organization", inquiring in particular under what conditions (e.g., noise, irreversibility, spatial and other symmetries of the initial conditions and equations of motion) statistical-mechanical model systems can imitate computers well enough to undergo unbounded increase of depth in the limit of infinite space and time.

1. Introduction

Persons of Turing's genius do not shy away from asking big questions, and hoping to see, in advance of the eventual slow progress of science, the essential outlines of the answers. "What is intelligence?" is one such question that clearly fascinated Turing. "How do complicated structures arise in nature?" is another. On this latter question, seeds planted by Turing have begun to mature to an extent that I think would please him.

Two outgrowths of computability theory, viz. algorithmic information theory and computational complexity theory, have made it possible to formalize satisfactorily the seemingly vague intuitive notion of complexity. Meanwhile, advances in equilibrium and nonequilibrium statistical mechanics have shed considerable light on the conditions, hinted at in Turing's work on morphogenesis (Turing *1952*), required for a simple, initially homogeneous medium to organize itself into structures capable of holding and processing information in the manner of a computer. We begin by showing how universal Turing machines can be used to formalize intuitive notions of complexity, returning in Section 5 to the question of self-organization in physics.

The ability of the universal Turing machine to simulate any known algorithmic process, and the belief (commonly called Church's thesis but independently proposed by Turing) that all algorithmic processes can be so simulated, has prompted and justified the use of universal Turing machines to define absolute properties of mathematical objects, beginning with the distinction drawn by Turing (*1936-7, 1937*) between computable and uncomputable real numbers.

Church's thesis has been found to hold to a considerable extent for the efficiency, as well as the possibility, of computation. A large variety of computational models (roughly, those with at most polynomially growing parallelism) can be simulated by a universal Turing machine in polynomial time, linear space, and with an additive constant increase in program size. This stronger Church's thesis has prompted the definition, on the one hand, of robust dynamic complexity classes such as *P* and *PSPACE*, and on the other hand of a nearly machine-independent algorithmic theory of information and randomness.

The dynamic complexity class *P*, for example, consists of those 0/1-valued functions of a binary string argument computable in time (i.e., number of machine cycles) bounded by a polynomial in the length of the argument, and includes the same functions regardless of whether the computations are performed by single-tape Turing machines, multi-tape Turing machines, cellular automata, or a wide variety of other models. The class *PSPACE* is defined analogously, but with the bound on space (i.e., number of squares of tape) instead of time. Diagonal constructions similar to that used to prove the existence of uncomputable functions can be used to define provably hard-to-compute functions, requiring for example exponential time and space to compute. A major open question of dynamic complexity theory is the $P = PSPACE$ question: it is widely conjectured, but not known, that *P* is a proper subset of *PSPACE*, i.e., that there exist functions computable in polynomial space but requiring more than polynomial time.

Algorithmic information theory (cf. Kolmogorov *1963, 1965*; Solomonoff *1964*; Zvonkin and Levin *1970*) uses a standard universal Turing machine to define the *Kolmogorov complexity* or *infomation content* of a string *x* as the length of its *minimal program* x^*, the shortest binary input which causes the standard machine to produce exactly *x* as output, and the *information content* of a string is identified with the number of bits in its minimal program. A string is said to be *compressible* by *k* bits if its minimal program is $\geq k$ bits shorter than the string itself. A simple counting argument shows that at most a fraction 2^{-k} of strings of length $\leq n$ bits can have this property. This fact justifies calling strings that are incompressible, or nearly so, *algorithmically random*. Like the majority of strings one might generate by coin tossing, such strings lack internal redundancy that could be exploited to encode them concisely, using the given universal Turing machine as decoding apparatus. Because of the ability of universal Turing machines to simulate one another, the property of algorithmic randomness is approximately machine-independent: a string that is incompressible on one machine cannot be compressed, on another machine, by more than the fixed number of bits required to program the first machine to simulate the second.

The relation between universal computer programs and their outputs has long been regarded (Solomonoff *1964*) as a formal analog of the relation between theory and observation in science, with the minimal-sized program representing the most economical, and therefore a priori most plausible, explanation of its output. In view of this analogy, it would be natural to associate with each finite object the cost in dynamic resources of reconstructing it from its minimal program. A "deep" or dynamically complex object would then be one whose most plausible origin, via an effective process, entails a lengthy computation. Just as the plausibility of a scientific theory depends on the economy of its assumptions, not on the length of the deductive path connecting them with observed phenomena, so a slow execution time is not evident against the plausibility of a program; rather, if there are no comparably concise programs to compute the same output quickly, it is evidence of the nontriviality of that output.

A more careful definition of depth should not depend only on the minimal program, but should take fair account of all programs that compute the given output, for example giving two $k + 1$-bit programs the same weight as one k-bit program. This is analagous in science to the exploration of a phenomenon by appeal to an ensemble of possible causes, individually unlikely but collectively plausible, as in the kinetic theory of gases. Several nearly-equivalent definitions of depth are considered in Section 3; the one finally advocated defines an object's "s-significant depth" as the least time required to compute it by a program that is itself compressible by no more than s bits. This formalizes the notion that any hypothesis of the object's more rapid origin suffers from s bits of redundancy. Such redundancy fairly measures the "ad-hocness" of a hypothesis, the extent to which it contains unexplained, a-priori-unlikely internal correlations that could be explained by deriving the original hypothesis from a more concise, nonredundant hypothesis.

The notion of logical depth developed in the present paper was first described in Chaitin *1977,* and at greater length in Bennett *1982* and Bennett *1985*; similar notions have been independently introduced by Adleman *1979* ("potential"), Levin and V'jugin *1977* ("incomplete sequence"), Levin *1984* ("hitting time"), and Koppel, this volume ("sophistication"). See also Wolfram's work on "computational irreducibility" (Wolfram *1985*) and Hartmanis' work on time- and space-bounded algorithmic information (Hartmanis *1983*).

We propose depth as a formal measure of value. From the earliest days of information theory it has been appreciated that information per se is not a good measure of message value. For example, a typical sequence of coin tosses has high information content but little value; an ephemeris, giving the positions of the moon and planets every day for a hundred years, has no more information than the equations of motion and initial conditions from which it was calculated, but saves its owner the effort of recalculating these positions. The value of a message thus appears to reside not in its information (its absolutely unpredictable parts), nor in its obvious redundancy (verbatim repetitions, unequal digit frequencies), but rather in what might be called its buried redundancy – parts predictable only with difficulty, things the receiver could in principle have figured out without being told, but only at considerable cost in money, time, or computation. In other words, the value of a message is the amount of mathematical or other work plausibly done by its originator, which its receiver is saved from having to repeat.

Of course, the receiver of a message does not know exactly how it originated; it might even have been produced by coin tossing. However, the receiver of an obviously nonrandom message, such as the first million bits of π, would reject this "null" hypothesis, on the grounds that it entails nearly a million bits worth of ad-hoc assumptions, and would favor an alternative hypothesis that the message originated from some mechanism for computing pi. The plausible work involved in creating a message, then, is the amount of work required to derive it from a hypothetical cause involving no unnecessary, ad-hoc assumptions. It is this notion of the message value that depth attempts to formalize.

Depth may be contrasted with other ways of attributing dynamic complexity to finite objects, for example by regarding them as inputs of a universal computer rather than as outputs. An object would be complex in this sense if it caused the computer to embark on a long (but terminating) computation. Such objects might be called "ambitious" because they describe a lengthy computation that may never have been done. A deep object, on the other hand, contains internal evidence that a lengthy computation has already been done. When considering moderate levels of ambition, e.g., exponential rather than busy-beaver in the size of the objects being considered, it is best to define ambition strictly as the actual run time of a string when executed as a program. In this case, of course, ambition is not very robust, since a slight change can convert a slow-running program into a fast-running one, or vice versa. In particular, a deep object need not be ambitious. At high levels of ambition, the notion can be given more robustness and interest (cf. Gács *1983*), by defining an ambitious string more broadly as one which is very slow-running itself or from which a very slow-running program can easily be computed. In this case, a string's ambition measures the amount of information it contains about the halting problem, i.e., about how to compute very large numbers; and, as will be seen later, deep objects are necessarily ambitious but not conversely.

Another kind of complexity associated with an object would be the difficulty, given the object, of finding a plausible hypothesis to explain it. Objects having this kind of complexity might be called "cryptic": to find a plausible origin for the object is like solving a cryptogram. A desirable (but mathematically unproven) property for small-key cryptosystems is that encryption should be easy, but breaking the system (e.g., inferring the key from a sufficient quantity of intercepted ciphertext) should be hard. If satisfactory small-key cryptosystems indeed exist, then typical cryptograms generated from shallow plaintexts are shallow (because they can be generated quickly from shallow input information) but cryptic (because even when the cryptogram contains sufficient information to uniquely determine the plaintext and key, the job of doing so is computationally infeasible).

One might argue that a message's usefulness is a better measure of value than its mathematical replacement cost, but usefulness is probably too anthropocentric a concept to formalize mathematically.

Related ideas appear in the fiction of Borges *1964*, e.g., in the story "Pierre Menard, Author of the *Quixote*" about a man who with great effort reconstructs several chapters of *Don Quixote* without errors and without consulting the original, from an intimate knowledge of the book's historical and literary context. Another story, "The Library of Babel" describes a library whose shelves contain, in seemingly random order, all possible 410-page books on a 25-letter alphabet, and the librarians' attempts to discern which of the books were meaningful.

The tradeoff between conciseness of representation and ease of decoding is illustrated in an extreme form by the information required to solve the halting problem. One standard representation of this information is as an infinite binary sequence K_0 (the characteristic sequence of the halting set) whose i'th bit is 0 or 1 according to whether the i'th program halts. This sequence is clearly redundant, because many instances of the halting problem are easily solvable or reducible to other instances. Indeed, K_0 is far more redundant than this superficial evidence might indicate. Barzdin *1968* showed that this information can be compressed to the logarithm of its original bulk, but no concisely encoded representation of it can be decoded in recursively bounded time.

The most elegantly concise representation of the halting problem is Chaitin's irrational number Ω (Chaitin *1975*), defined as the halting probability of a universal computer programmed by coin tossing (the computer starts in a standard state, and whenever it tries to read a bit of its program, a coin is tossed to supply that bit). Such a randomly programmed computation is like the old notion of a monkey accidentally typing the works of Shakespeare, but now the monkey sits at a computer keyboard instead of a typewriter. The result is still trivial with high probability, but any nontrivial computation also has a finite probability of being performed, inversely proportional to the exponential of the length of its program. The essential features of Ω (Chaitin *1975*, Bennett *1979*) are

i) The first n bits of Ω suffice to decide approximately the first 2^n cases of the halting problem (more precisely, to decide the fate of all computations using

$\leq n$ coin tosses), but there is no faster way of extracting this information than to recursively enumerate halting programs until enough have been found to account for all but 2^{-n} of the total halting probability Ω, a job which requires at least as much time as running the slowest n bit program.

ii) Ω is algorithmically random: like a typical coin toss sequence, its first n bits cannot be encoded in less than $n - O(1)$ bits. (This algorithmic randomness may seem contradictory to, but in fact is a consequence of, Ω's compact encoding of the halting problem. Knowledge of Ω_n (i.e., the first n bits of Ω) is sufficient to decide the halting of all programs shorter than n bits, and therefore to determine which n-bit strings are compressible and which are not, and therefore to find and print out the lexicographically first incompressible n-bit string x. If Ω_n itself were significantly compressible, then the algorithm just described could be combined with the compressed representation of Ω_n to yield a $< n$ bit program to produce x, a supposedly incompressible n-bit string. This contradiction proves the $O(1)$-incompressibility of Ω.)

iii) although it solves unsolvable problems, Ω does not speed up the solution of solvable problems any more than a random coin-toss sequence would. (This uselessness follows from its algorithmic randomness. Any special ability of Ω to speed up the computation of a computable function would set it apart from random sequences in general, and this atypicality could be exploited to encode Ω more concisely, contradicting its randomness.)

In summary, we may say that the Ω contains the same information as K_0, but in such a compressed form that as to be random and useless. Ω is a shallow representation of the deep object K_0.

Descending from the realm of the halting problem to more mundane levels of complexity (e.g., polynomial versus exponential), one is tempted to invoke depth as a measure of complexity for physical objects, but one must proceed cautiously. The applicability of any idea from computational complexity to physics depends on what has been called a "physical Church's thesis", the belief that physical processes can be simulated with acceptable accuracy and efficiency by digital computations. Turing was quite aware of this question, and based his informal justification of his model of computation largely on physical considerations. Granting some form of physical Church's thesis, a nontrivial application of depth to physical systems appears to depend on unproven conjectures at the low end of complexity theory, such as $P \neq PSPACE$. These questions are considered in the last section of the paper.

2. Preliminaries, Algorithmic Information

As usual, a natural number x will be identified with the x'th binary string in lexicographic order ($\Lambda,0,1,00,01,10,11,000...$), and a set X of natural numbers will be identified with its characteristic sequence, and with the real number between 0 and 1 having that sequence as its dyadic expansion. The length of a string x will be denoted $|x|$, the n'th bit of an infinite sequence X will be noted $X(n)$, and the initial n bits of X will be denoted X_n. Concatenation of strings p and q will be denoted pq.

We define the information content (and later the depth) of finite strings using a universal Turing machine U similar to that described by Chaitin *1975*. A universal Turing machine may be viewed as a partial recursive function of two arguments. It is universal in the sense that by varying one argument ("program") any partial recursive function of the other argument ("data") can be obtained. In the usual machine formats, program, data and output are all finite strings, or, equivalently, natural numbers. However, it is not possible to average uniformly over accountably infinite sets. Since we wish to average uniformly over programs we adopt a format (Gács *1974*, Levin *1973*, Chaitin *1975*) in which the program is in effect an *infinite* binary sequence, but data and output are finite strings. Chaitin's universal machine has two tapes: a read-only one-way tape containing the infinite program; and an ordinary two-way read/write tape, which is used for data input, intermediate work, and output, all of which are finite strings. Our machine differs from Chaitin's in having some additional auxiliary storage (e.g., another read/write tape) which is needed only to improve the time efficiency of simulations.

We consider only terminating computations, during which, of course, only a finite portion of the program tape can be read. Therefore, the machine's behavior can still be described by a partial recursive function of two string arguments $U(p, w)$, if we use the first argument to represent that portion of the program that is actually read in the course of a particular computation. The expression $U(p, w) = x$ will be used to indicate that the U machine, started with any infinite sequence beginning with p on its program tape and the finite string w on its data tape, performs a halting computation which reads exactly the initial portion p of the program, and leaves output data x on the data tape at the end of the computation. In all other cases (reading less than p, more than p, or failing to halt), the function $U(p, w)$ is undefined. Wherever $U(p, w)$ is defined we say that p is a *self-delimiting program* to compute x from w, and we use $T(p, w)$ to represent the time (machine cycles) of the computation. Often we will consider computations without input data; in that case we abbreviate $U(p, \Lambda)$ and $T(p, \Lambda)$ as $U(p)$ and $T(p)$ respectively.

The self-delimiting convention for the program tape forces the domain of U and T, for each data input w, to be a *prefix set*, that is, a set of strings no member of which is the extension of any other member. Any prefix set S obeys the Kraft inequality

$$\sum_{p \in S} 2^{-|p|} \leq 1. \tag{1}$$

Besides being self-delimiting with regard to its program tape, the U machine must be *efficiently universal* in the sense of being able to simulate any other machine of its kind (Turing machines with self-delimiting program tape) with at most an additive constant constant increase in program size and a linear increase in execution time.

Without loss of generality we assume that there exists for the U machine a constant prefix r which has the effect of stacking an instruction to restart the computation when it would otherwise end. This gives the machine the ability to concatenate programs to run consecutively: if $U(p, w) = x$ and $U(q, x) = y$, then $U(rpq, w) = y$. Moreover, this concatenation should be efficient in the sense that $T(rpq, w)$ should exceed $T(p, w) + T(q, x)$ by at most $O(1)$. This efficiency of running concatenated programs can be realized with the help of the auxiliary storage to stack the restart instructions.

Sometimes we will generalize U to have access to an "oracle" A, i.e., an infinite look-up table which the machine can consult in the course of its computation. The oracle may be thought of as an arbitrary $0/1$-valued function $A(x)$ which the machine can cause to be evaluated by writing the argument x on a special tape and entering a special state of the finite control unit. In the next machine cycle the oracle responds by sending back the value $A(x)$. The time required to evaluate the function is thus linear in the length of its argument. In particular we consider the case in which the information in the oracle is random, each location of the look-up table having been filled by an independent coin toss. Such a *random oracle* is a function whose values are reproducible, but otherwise unpredictable and uncorrelated.

The following paragraph gives a more formal definition of the functions U and T, which may be skipped by the casual reader.

Let $\{\varphi_i^A(p, w) : i = 0, 1, 2 \ldots\}$ be an acceptable Gödel numbering of A-partial recursive functions of two arguments and $\{\Phi_i^A(p, w)\}$ an associated Blum complexity measure, henceforth referred to as time. An index j is called self-delimiting iff, for all oracles A and all values w of the second argument, the set $\{ x : \varphi_j^A(x, w) \text{ is defined} \}$ is a prefix set. A self-delimiting index has efficient concatenation if there exists a string r such that for all oracles A and all strings w, x, y, p, and q, if $\varphi_j^A(p, w) = x$ and $\varphi_j^A(q, x) = y$, then $\varphi_j^A(rpq, w) = y$ and $\Phi_j^A(rpq, w) = \Phi_j^A(p, w) + \Phi_j^A(q, x) + O(1)$. A self-delimiting index u with efficient concatenation is called efficiently universal iff, for every self-delimiting index j with efficient concantenation, there exists a simulation program s and a linear polynomial L such that for all oracles A and all strings p and w,

$$\varphi_u^A(sp, w) = \varphi_j^A(p, w)$$

and

$$\Phi_u^A(sp, w) \leq L(\Phi_j^A(p, w)).$$

The functions $U^A(p, w)$ and $T^A(p, w)$ are defined respectively as $\varphi_u^A(p, w)$ and $\Phi_u^A(p, w)$, where u is an efficiently universal index.

We now present some definitions and review some elementary facts about algorithmic information.

For any string x, the *minimal program,* denoted x^*, is $\min\{p : U(p) = x\}$, the least self-delimiting program to compute x. For any two strings x and w, the minimal program of x relative to w, denoted $(x/w)^*$, is defined similarly as $\min\{p : U(p, w) = x\}$.

By contrast to its minimal program, any string x also has a *print program,* of length $|x| + O(\log |x|)$, which simply transcribes the string x from a verbatim description of x contained within the program. The print program is logarithmically longer than x because, being self-delimiting, it must indicate the length as well as the contents of x. Because it makes no effort to exploit redundancies to achieve efficient coding, the print program can be made to run quickly (e.g., linear time in $|x|$, in the present formalism).

Extra information w may help, but cannot significantly hinder, the computation of x, since a finite subprogram would suffice to tell U to simply erase w before proceeding. Therefore, a relative minimal program $(x/w)^*$ may be much shorter than the corresponding absolute minimal program x^*, but can only be longer by $O(1)$, independent of x and w.

A string is *compressible* by s bits if its minimal program is shorter by at least s bits than the string itself, i.e., if $|x^*| \leq |x| - s$. Similarly, a string x is said to be compressible by s bits relative to a string w if $|(x/w)^*| \leq |x| - s$.

Regardless of how compressible a string x may be, its minimal program x^* is compressible by at most an additive constant depending on the universal computer but independent of x. (If $(x^*)^*$ were much smaller than x^*, then the role of x^* as minimal program for x would be undercut by a program of the form "execute the result of executing $(x^*)^*$".) Similarly, a relative minimal program $(x/w)^*$ is compressible relative to w by at most a constant number of bits independent of x or w.

The *algorithmic probability* of a string x, denoted $P(x)$, is defined as $\sum\{2^{-|p|} : U(p) = x\}$. This is the probability that the U machine, with a random program chosen by coin tossing and an initially blank data tape, will halt with output x. The *time-bounded algorithmic probability,* $P_t(x)$, is defined similarly, except that the sum is taken only over programs which halt within time t. We use $P(x/w)$ and $P_t(x/w)$ to denote the analogous algorithmic probabilities of one string x relative to another w, i.e., for computations that begin with w on the data tape and halt with x on the data tape.

The *algorithmic entropy* $H(x)$ is defined as the least integer greater than $-\log_2 P(x)$, and the conditional entropy $H(x/w)$ is defined similarly as the least integer greater than $-\log_2 P(x/w)$.

Among the most important properties of the algorithmic entropy is its equality, to within $O(1)$, with the size of the minimal program:

$$\exists c \forall x \forall w H(x/w) \leq |(x/w)^*| \leq H(x/w) + c. \tag{2}$$

The first part of the relation, viz. that algorithmic entropy should be no greater than minimal program size, is obvious, because of the minimal program's own contribution to the algorithmic probability. The second half of the relation is less obvious (cf. Gács *1974*, Chaitin *1975*, and Lemma 2). The approximate equality of algorithmic entropy and minimal program size means that there are few near-minimal programs for any given input/output pair (x/w), and that every string gets an $O(1)$ fraction of its algorithmic probability from its minimal program.

Finite strings, such as minimal programs, which are incompressible or nearly so are called *algorithmically random*. The definition of randomness for finite strings is necessarily a little vague because of the $\pm O(1)$ machine-dependence of $H(x)$ and, in the case of strings other than self-delimiting programs, because of the question of how to count the information encoded in the string's length, as opposed to its bit sequence. Roughly speaking, an n-bit self-delimiting program is considered random (and therefore not ad-hoc as a hypothesis) iff its information content is about n bits, i.e. iff it is incompressible; while an externally delimited n-bit string is considered random iff its information content is about $n + H(n)$ bits, enough to specify both its length and its contents.

For infinite binary sequences (which may be viewed also as real numbers in the unit interval, or as characteristic sequences of sets of natural numbers) randomness can be defined sharply: a sequence X is incompressible, or algorithmically random, if there is an $O(1)$ bound to the compressibility of its initial segments X_n. This class of infinite sequences was first characterized by Martin-Löf *1966*; several equivalent definitions have since been given (Levin *1973*, Chaitin *1977*). Intuitively, an infinite sequence is *random* if it is typical in every way of sequences that might be produced by tossing a fair coin; in other words, if it belongs to no informally definable set of measure zero. *Algorithmically random* sequences constitute a larger class, including sequences such as Ω (Chaitin *1975*) which can be specified by ineffective definitions. Henceforth, the term "random" will be used in the narrow informal sense, and "incompressible", or "algorithmically random", in the broader, exact sense.

We proceed with a few other useful definitions.

The *busy beaver function* $B(n)$ is the greatest number computable by a self-delimiting program of n bits or fewer.

The *halting set* K is $\{x : \varphi_x(x) \text{ converges }\}$. This is the standard representation of the halting problem.

The *self-delimiting halting set* K_0 is the (prefix) set of all self-delimiting programs for the U machine that halt: $\{ p : U(p) \text{ converges }\}$.

K and K_0 are readily computed from one another (e.g., by regarding the self-delimiting programs as a subset of ordinary programs, the first 2^n bits of K_0 can be recovered from the first $2^{n+O(1)}$ bits of K; by encoding each n-bit ordinary program as a self-delimiting program of length $n + O(\log n)$, the first 2^n bits of K can be recovered from the first $2^{n+O(\log n)}$ bits of K_0.)

The *halting probability* Ω is defined as $\sum \{2^{-|p|} : U(p)$ converges $\}$, the probability that the U machine would halt on an infinite input supplied by coin tossing. Ω is thus a real number between 0 and 1.

The first 2^n bits of K_0 can be computed from the first n bits of Ω, by enumerating halting programs until enough have halted to account for all but 2^{-n} of the total halting probability. The time required for this decoding (between $B(n - O(1))$ and $B(n + H(n) + O(1))$ grows faster than any computable function of n. Although K_0 is only slowly computable from Ω, the first n bits of Ω can be rapidly computed from the first $2^{n+H(n)+O(1)}$ bits of K_0, by asking about the halting of programs of the form "enumerate halting programs until (if ever) their cumulative weight exceeds q, then halt", where q is an n-bit rational number.

In the following, we will often be dealing with a prefix set S of strings having some common property, e.g., the set of all self-delimiting programs to compute x in time t. Two useful lemmas relate the *measure* of such a set

$$\mu[S] = \sum_{p \in S} 2^{-|p|}$$

to the compressibility of its individual members.

Lemma 1: *If S is a prefix set of strings whose total measure $\mu[S] = \sum_{x \in S} 2^{-|x|}$ is at least 2^{-m}, and y is an arbitrary string (or the empty string), then at least some members of S must not be compressible by more than m bits relative to y :* $\forall_y \exists_{x \in S} |(x/y)^*| \geq |x| - m$.

Proof. Suppose on the contrary that for some S and y, $\mu[S] \geq 2^{-m}$ and $|(x/y)^*| < |x| - m$ for all $x \in S$. Then the set $\{(x/y)^* : x \in S\}$, also a prefix set, would have measure greater than 1, because each of its members is more than m bits shorter than the corresponding member of S. But the Kraft inequality forbids any prefix set from having measure greater than 1; the lemma follows. $\square$

Lemma 2: *If a prefix set S of strings, having total measure $\mu(S) = \sum_{x \in S} 2^{-|x|}$ less than 2^{-m}, is computable by a self-delimiting program of s bits, then every member of S is compressible by at least $m - s - O(1)$ bits.*

Proof Sketch. This lemma can be proved using, for each S, a special-purpose self-delimiting computer C_S designed to compress each member of S by exactly m bits, in other words, to produce each output in S in response to some m-bit-shorter input which is not the prefix or extension of any other input. This special computer is then simulated by the general purpose U machine, at the cost of expanding all its programs by a constant amount (equal to the number of bits required to describe S and m), to obtain the desired result.

The existence of C_S is guaranteed by a more general result (Chaitin *1975*, theorem 3.2) proving the existence of a special purpose computer C satisfying any consistent, recursively-enumerable list of requirements of the form $< x(k), n(k) >$ ($k = 0, 1, 2 \ldots$), where the k'th requirement $< x(k), n(k) >$ asks that a self-delimiting program of length $n(k)$ be "assigned" to the output string $x(k)$. The special purpose computer may be thought of more abstractly as a partial recursive function whose range is $\{x(k)\}$ and whose domain is a prefix set. The requirements are called consistent if they obey the Kraft inequality $\sum_k 2^{-n(k)} \leq 1$, and the computer is said to satisfy them if there are precisely as many programs of length n with output x as there are pairs $< x, n >$ in the list of requirements. In Chaitin *1975* it is shown in more detail how a straightforward "greedy" allocation rule, which attempts to satisfy each requirement in order of recursive enumeration by the first available string of the requested length that is not the prefix or extension of any previously allocated string, works so well that it only fails if the list of requirements would violate the Kraft inequality.

In the present application, the requirements are for uniform compression of each member of the set S, i.e., $n(k) = |x(k)| - m$. If S is computable by a self-delimiting program of s bits, then the feasible degree of compression m can be computed from the same program that computes S. Thus the U simulating the special purpose C_S machine requires an additional program length of s bits; and each member of S is compressible by $k - s - O(1)$ bits. $\square$

3. Depth of Finite Strings

We begin by considering finite objects (e.g., strings or natural numbers), where the intuitive motivation for depth is clearer, even though mathematically sharper (polynomially or recursively invariant) results can be obtained with infinite sequences.

We consider several candidates for the best definition of depth:

Tentative Definition 0.1: *A string's depth might be defined as the execution time of its minimal program.*

The difficulty with this definition arises in cases where the minimal program is only a few bits smaller than some much faster program, such as a print program, to compute the same output x. In this case, slight changes in x may induce arbitrarily large changes in the run time of the minimal program by changing which of the two competing programs is minimal. Analogous instability manifests itself in translating programs from one universal machine to another. This instability emphasizes the essential role of the quantity of buried redundancy, not as a measure of depth, but as a certifier of depth. In terms of the philosophy-of-science metaphor, an object whose minimal program is only a few bits smaller than its print program is like an observation that points to a nontrivial hypothesis, but with only a low level of statistical confidence.

To adequately characterize a finite string's depth one must therefore consider the amount of buried redundancy as well as the depth of its burial. A string's depth at significance level s might thus be defined as that amount of time complexity which is attested by s bits worth of buried redundancy. This characterization of depth may be formalized in several ways.

Tentative Definition 0.2: *A string's depth at significance level s be defined as the time required to compute the string by a program no more than s bits larger than the minimal program.*

This proposed definition solves the stability problem, but is unsatisfactory in the way it treats multiple programs of the same length. Intuitively, 2^k distinct $(n+k)$-bit programs that compute same output ought to be accorded the same weight as one n-bit program; but, by the present definition, they would be given no more weight than one $(n+k)$-bit program. Multiple programs can be fairly taken into account by the next definition.

Tentative Definition 0.3: *A string's depth at significance level s depth might be defined as the time t required for the string's time-bounded algorithmic probability $P_t(x)$ to rise to within a factor 2^{-s} of its asymptotic time-unbounded value $P(x)$.*

This formalizes the notion that for the string to have originated by an effective process of t steps or fewer is less plausible than for the first s tosses of a fair coin all to come up heads.

It is not known whether there exist strings that are deep according to Definition 0.2 but not Definition 0.3, in other words, strings having no small fast programs, even though they have enough large fast programs to contribute a significant fraction of their algorithmic probability. Such strings might be called deterministically deep but probabilistically shallow, because their chance of being produced quickly in a probabilistic computation (e.g., one where the input bits of U are supplied by coin tossing) is significant compared to their chance of being produced slowly. The question of whether such strings exist is probably hard to answer because it does not relativize uniformly. Deterministic and probabilistic depths are not very different relative to a random coin-toss oracle A (this can be shown by a proof similar to that (Bennett and Gill *1981*) of the equality of random-oracle-relativized deterministic and probabilistic polynomial time complexity classes); but they can be very different relative to an oracle B deliberately designed to hide information from deterministic computations (this parallels Hunt's proof (Hunt *1978*) that deterministic and probabilistic polynomial time are unequal relative to such an oracle).

Although Definition 0.3 satisfactorily captures the informal notion of depth, we propose a slightly stronger definition for the technical reason that it appears to yield a stronger slow growth property (Theorem 1 below).

Definition 1 (Depth of Finite Strings): *Let x and w be strings and s a significance parameter. A string's depth at significance level s, denoted $D_s(x)$, will be defined as $\min\{T(p) : (|p| - |p^*| < s) \wedge (U(p) = x)\}$, the least time required to compute it by a s-incompressible program. At any given significance level, a string will be called t-deep if its depth exceeds t, and t-shallow otherwise.*

The difference between this definition and the previous one is rather subtle philosophically and not very great quantitatively. Philosophically, Definition 1 says that each *individual* hypothesis for the rapid origin of x is implausible at the 2^{-s} confidence level, whereas the previous definition 0.3 requires only that a weighted average of all such hypotheses be implausible. The following lemma shows that the difference between Definition 1 and Definition 0.3 is also small quantitatively.

Lemma 3: *There exist constants c_1 and c_2 such that for any string x, if programs running in time $\leq t$ contribute a fraction between 2^{-s} and 2^{-s+1} of the string's total algorithmic probability, then x has depth at most t at significance level $s + c_1$ and depth at least t at significance level $s - \min\{H(s), H(t)\} - c_2$.*

Proof. The first part follows easily from the fact that any k-compressible self-delimiting program p is associated with a unique, $k - O(1)$ bits shorter, program of the form "execute the result of executing p^*". Therefore there exists a constant c_1 such that if all t-fast programs for x were $s + c_1$-compressible, the associated shorter programs would contribute more than the total algorithmic probability of x. The second part of the lemma follows because, roughly, if fast programs contribute only a small fraction of the algorithmic probability of x, then the property of being a fast program for x is so unusual that no program having that property can be random. More precisely, the t-fast programs for x constitute a finite prefix set, a superset S of which can be computed by a program of size $H(x) + \min\{H(t), H(s)\} + O(1)$ bits. (Given x^* and either t^* or s^*, begin enumerating all self-delimiting programs that compute x, in order of increasing running time, and quit when either the running time exceeds t or the accumulated measure of programs so far enumerated exceeds $2^{-(H(x)-s)}$). Therefore there exists a constant c_2 such that, by Lemma 1, every member of S, and thus every t-fast program for x, is compressible by at least $s - \min\{H(s), H(t)\} - O(1)$ bits.   □

The ability of universal machines to simulate one another efficiently implies a corresponding degree of machine-independence for depth: for any two efficiently universal machines of the sort considered here, there exists a constant c and a linear polynomial L such that for any t, strings whose $(s + c)$-significant depth is at least $L(t)$ on one machine will have s-significant depth at least t on the other.

Depth of one string relative to another may be defined analogously to Definition 1 above, and represents the plausible time required to produce one string, x, from another, w.

Definition 1.1(Relative Depth of Finite Strings): *For any two strings w and x, the depth of x relative to w at significance level s, denoted $D_s(x/w)$, will be defined as $\min\{T(p, w) : (|p| - |(p/w)^*| < s) \wedge (U(p, w) = x)\}$, the least time required to compute x from w by a program that is s-incompressible relative to w.*

Depth of a string relative to its length is a particularly useful notion, allowing us, as it were, to consider the triviality or nontriviality of the "content" of a string (i.e., its bit sequence), independent of its "form" (length). For example, although the infinite sequence 000... is intuitively trivial, its initial segment 0^n is deep whenever n is deep. However, 0^n is always shallow relative to n, as is, with high probability, a random string of length n.

In order to adequately represent the intuitive notion of stored mathematical work, it is necessary that depth obey a "slow growth" law, i.e., that fast deterministic processes be unable to transform a shallow object into a deep one, and that fast probabilistic processes be able to do so only with low probability.

Theorem 1 (Slow Growth Law): *Given any data string x and two significance parameters $s_2 > s_1$, a random program generated by coin tossing has probability less than $2^{-(s_2-s_1)+O(1)}$ of transforming x into an excessively deep output, i.e., one whose s_2-significant depth exceeds the s_1-significant depth of x plus the run time of the transforming program plus $O(1)$. More precisely, there exist positive constants c_1, c_2 such that for all strings x, and all pairs of significance parameters $s_2 > s_1$, the prefix set $\{q : D_{s_2}(U(q, x)) > D_{s_1}(x) + T(q, x) + c_1\}$ has measure less than $2^{-(s_2-s_1)+c_2}$.*

Proof. Let p be a s_1-incompressible program which computes x in time $D_{s_1}(x)$, and let r be the restart prefix mentioned in the definition of the U machine. Let Q be the prefix set $\{q : D_{s_2}(U(q, x)) > T(q, x) + D_{s_1}(x) + c_1\}$, where the constant c_1 is sufficient to cover the time overhead of concatenation. For all $q \in Q$, the program rpq by definition computes some deep result $U(q, x)$ in less time than that result's own s_2-significant depth, and so rpq must be compressible by s_2 bits. The sum of the algorithmic probabilities of strings of the form rpq, where $q \in Q$, is therefore

$$\sum_{q \in Q} P(rpq) < \sum_{q \in Q} 2^{-|rpq|+s_2} = 2^{-|r|-|p|+s_2} \mu(Q). \tag{3}$$

On the other hand, since the self-delimiting program p can be recovered from any string of the form rpq (by deleting r and executing the remainder pq until halting occurs, by which time exactly p will have been read), the algorithmic probability of p is at least as great (within a constant factor) as the sum of the algorithmic probabilities of the strings $\{rpq : q \in Q\}$ considered above:

$$P(p) > \mu(Q) \cdot 2^{-|r|-|p|+s_2-O(1)}.$$

Recalling the fact that minimal program size is equal within a constant factor to the $-\log$ of algorithmic probability, and the s_1-incompressibility of p, we have $P(p) < 2^{-(|p|-s_1+O(1))}$, and therefore finally

$$\mu(Q) < 2^{-(s_2-s_1)+O(1)},$$

which was to be demonstrated. $\square$

An analogous theorem (with analogous proof) also holds for the improbability of rapid growth of depth relative to an arbitrary fixed string w:

Theorem 1.1 (Relative Slow Growth Law): *Given data strings x and w and two significance parameters $s_2 > s_1$, a random program generated by coin tossing has probability less than $2^{-(s_2-s_1)+O(1)}$ of transforming x into an excessively deep output, one whose s_2-significant depth relative to w exceeds the s_1-significant depth of x relative to w plus the run time of the transforming program plus $O(1)$. More precisely, there exist positive constants c_1, c_2 such that for all strings x and w, and all pairs of significance parameters $s_2 > s_1$, the prefix set $\{q : D_{s_2}(U(q, x)/w) > D_{s_1}(x/w) + T(q, x) + c_1\}$ has measure less than $2^{-(s_2-s_1)+c_2}$.*

EXAMPLES OF SHALLOW OBJECTS

The trivial string 000... has already been mentioned. It is shallow in the sense that there exists a low-order polynomial L (linear for the machine model we are using) and a significance parameter s such that for all n, $D_s(0^n/n) < L(n)$. Here s represents the size and $L(n)$ the running time of a fixed program that for any n computes 0^n from n. Similarly, there exist s and L such that for all n and k, a random string x of length n produced by coin tossing satisfies $D_{s+k}(x/n) < L(n)$ with probability greater than $1 - 2^{-k}$. In this case the shallowness of x is shown by a fast near-incompressible program that simply copies x verbatim off the program tape, using the data n to decide when to stop. This shallowness of random sequences applies not only to those generated by coin tossing, but to ineffectively definable sequences such as the halting probability Ω, whose algorithmic randomness implies that there is a significance level s at which $D_s(\Omega_n/n)$ increases only linearly with n.

As emphasized in the introduction, Ω's concentrated information about the halting problem does not make it deep, because the information is encoded so concisely as to appear random.

EXAMPLES OF VERY DEEP OBJECTS

Very deep strings can be constructed by diagonalization, for example, by programs of the form

"Find all n-bit strings whose algorithmic probability, from computations halting within time T (a large number), is greater than 2^{-n}, and print the first string not in this set."

This program runs very slowly, using time about $T \cdot 2^T$ to evaluate the algorithmic probabilities by explicitly simulating all T-bit coin toss sequences, but eventually it outputs a specific string $\chi(n, T)$ guaranteed to have T-fast algorithmic probability less than 2^{-n}, even though the string's relatively concise description via the above algorithm guarantees a slow algorithmic probability of at least $2^{-H(n)-H(T)+O(1)}$. We can then invoke Lemma 3 (relating depth to the rise time of algorithmic probability) to conclude that $\chi(n, T)$ has depth at least T at significance level

$$n - H(T) - \min\{H(n - H(n) - H(T)), H(T)\} - O(1),$$

which (taking into account that in nonvacuous cases $H(T) < n$) is at least $n - H(T) - O(\log n)$.

Because the halting set K has the ability to speed up any slow computation, deep objects such as the diagonal strings considered above are rapidly computable from K. Therefore, by the slow growth law, the halting set must be deep itself. More precisely, by arguments similar to those used in Barzdin's paper (Barzdin *1968*) on the compressibility of the halting problem, it can be shown that for any $c < n$, the initial 2^n bits of K have depth between $B(n - c - O(\log n))$ and $B(n - c + O(\log n))$ at significance level 2^c.

It is not hard to see that the busy beaver function provides an approximate upper bound on the depth of finite strings. Given a string x, its length n, and a value of the significance parameter s, all of which can be encoded in a self-delimiting program of size $n - s + O(\log n)$, one can compute the depth $D_s(x)$, which must therefore be less than $B(n - s + O(\log n))$ by the definition of the busy beaver function.

Very deep objects, because they contain information about their own depth, are necessarily ambitious in the broad sense (cf. introduction) of containing information about how to compute large numbers. On the other hand, ambitious objects need not be deep. For example, Ω is ambitious but shallow.

On the Efficient Generation of Depth

The diagonal method mentioned above for calculating deep strings suffers from exponential overhead, taking more than 2^T time to generate an object of depth T. One naturally wonders whether there are ways of generating depth with an efficiency more closely approaching the maximum allowed by the slow growth law: depth T in time T.

One way of doing so would be simply to generate a string of *length* T, say a string of T zeros. Time $O(T)$ is clearly both necessary and sufficient to generate this string, but it would be more satisfying to find an example of an efficiently generated object deeper than its own bulk. Only then would the object contain evidence of being the visible part of a larger invisible whole.

Unfortunately it appears that nontrivial and efficient production of depth may depend on plausible but unproven assumptions at the low end of complexity theory. Motivated by the finding that many open questions of complexity theory can be easily shown to have the answers one would like them to have in the relativized context of a random oracle (e.g. $P^A \neq NP^A \neq PSPACE^A$), Bennett and Gill *1981* informally conjectured that pseudorandom functions that "behave like" random oracles exist absolutely, and therefore that all "natural" mathematical statements (such as $P \neq NP$) true relative to a random oracle should be true absolutely. Their attempt to formalize the latter idea (by defining a broad class of statements to which it was supposed to apply) was unsuccessful (Kurtz *1983*), but the former idea has been formalized quite successfully in the notion of a cryptographically strong pseudorandom function (CSPSRF) (Blum and Micali *1984*, Yao *1982*, Goldreich, Goldwasser and Micali *1984*, Levin *1985*). A CSPSRF is a 0/1-valued polynomial time computable function G of two variables with the property that, if the first variable (s, "the seed") is chosen randomly from among strings of a given length, the resulting function of the second variable $G_s(x)$ cannot be distinguished from a random oracle $A(x)$ in time polynomial in the length of the seed. "Cannot be distinguished" means that there is no probabilistic or deterministic test by which an adversary, ignorant of the seed s, but knowing the algorithm for G and having the ability to evaluate $G_s(x)$ for arbitrary x, could distinguish the pseudorandom oracle G_s from a truly random oracle A, except with low probability, or by using time more than polynomial in the length of the seed. Exponential time of course would be sufficient to distinguish the two oracles with certainty, by searching exhaustively for a seed which exactly reproduced the behavior of one oracle (the pseudorandom one) but not the other.

Below we show how a random oracle, and hence a cryptographically strong pseudorandom function, would permit deep strings to be generated efficiently.

Given a 0/1-valued random oracle A it is routine to construct a random function $\xi_A(x)$ which maps strings randomly onto strings of the same length. The statistical structure of such random length-preserving functions is well known and facilitates the construction, from a standard starting string such as 0^n, of the deep objects considered below.

Let $\xi_A^k(0^n)$ denote the k'th forward image of 0^n, where $k < 2^{n/2}$, under the length-preserving function derived from random oracle A. This string, the target of a chain of k pointers through the random structure of the ξ function, is readily computed if enough time ($O(k \cdot n^2)$ in our model) is allowed to evaluate the ξ function k times; but, for typical random oracles, if less time is allowed, the probability of finding the correct target is only $O(1/2^n)$, representing pure luck. This statement is true for typical oracles; for a small minority of oracles, of measure $O(k^2/2^n)$, the chain of pointers starting from 0^n would begin to cycle in fewer than k iterations, permitting the target to be found more quickly. Returning to the case of a typical oracle, Lemma 3, relating depth to the time-dependence of algorithmic probability, can be used to show that the target string $\xi_A^k(0^n)$ has, at significance level $n - O(\log n))$, depth proportional to the time $O(k \cdot n^2)$ actually used by the straightforward algorithm for computing it.

This result holds, with high probability, in the relativized world containing a random oracle. If one assumes the existence of CSPSRF, then a similar result holds in the real world: using a pseudorandom length-preserving function ξ_{G_s} derived from a CSPSRF with seed s randomly chosen among strings of length n, one obtains target strings of the form $\xi_{G_s}^k(0^n)$ which can be generated in polynomial time in $k \cdot n$, but with high probability (in the choice of the seed s) have depth exceeding a smaller polynomial in $k \cdot n$. If this were not the case, the pseudorandom G_s could be distinguished from a random A in polynomial time by demonstrating a greater fast algorithmic probability of the target string in the former case.

Another sort of deep object definable with the help of the ξ function are pre-images of 0^n, i.e., members of the set $\{x : \xi(x) = 0^n\}$. The number of pre-images is binomially distributed, approaching a Poisson distribution for large n, so that a given string (such as 0^n) has no pre-images approximately $1/e$ of the time, one pre-image $1/e$ of the time, and m pre-images $e^{-1}/m!$ of the time. Pre-images of 0^n, when they exist, are deep because for random ξ they cannot be found except by exhaustive search. In the terminology of NP problems, a pre-image is "witness" for membership of 0^n in the range of ξ_A, a set which, relative to a random oracle A belongs to $NP^A - P^A$.

REMARKS ON THE TRANSITIVITY OF DEPTH

The slow growth law says that deep objects cannot quickly be produced from shallow ones. It is natural to wonder whether this property can be extended to a transitive law for relative shallowness; in other words, if x is shallow relative to w, and y is shallow relative to x, does it follow that y is shallow relative to w?

The answer is no, as can be seen from the following example: Let w be a random string (produced e.g., by coin tossing), x be the empty string, and y be the bitwise exclusive-or of w with some deep string d. Then y is also random and shallow, and so shallow relative to x, as x is relative to w; however y is deep relative to w, since d can easily be regenerated from y and w. Therefore simple transitivity does not hold.

A more cumulative sort of transitivity can be shown to hold: for all w, x, and y, if x is shallow relative to w, and y is shallow relative to the ordered pair $< w, x >$, then y is indeed shallow relative to w, at least within logarithmic error terms in the significance parameter such as those in Lemma 3. In particular, cumulative transitivity holds when w is empty: if x is shallow and y is shallow relative to x, then y is shallow.

SCALAR MEASURES OF DEPTH

One may well wonder whether, by defining some sort of weighted average run time, a string's depth might be expressed by a single number, unqualified by a significance parameter. This may be done, at the cost of imposing a somewhat arbitrary rate of exchange between the two conceptually very different quantities run time and program size. Proceeding from tentative Definition 0.3 above, one might try to define a string's average depth as the average run time of all computations contributing to its algorithmic probability, but this average diverges because it is dominated by programs that waste arbitrarily much time. To make the average depth of x depend chiefly on the fastest programs of any given size that compute x, one can use the "harmonic mean", or reciprocal mean reciprocal, run time in place of a straight average. The *reciprocal mean reciprocal depth* of a string x may thus be defined as

$$D_{rmr}(x) = \frac{\sum\{2^{-|p|} : (U(p) = x)\}}{\sum\{(2^{-|p|}/T(p)) : (U(p) = x)\}}. \tag{4}$$

In this definition, the various computations that produce x act like parallel resistors, the fast computations in effect short-circuiting the slow ones. (The ratio of rmr depth to algorithmic probability, called "hitting"time, was introduced by Levin *1984* to measure the difficulty of solving NP-type problems by an optimal search algorithm; related ideas are explored in Adleman *1979*.) Due to the short-circuiting of slower programs, no matter how small, by the print program, rmr depth does not allow strings to have depth more than exponential in their length; however, it does provide a simple quantitative measure of a string's nontriviality. Among efficiently universal machines, it is machine-independent to within a polynomial depending on the machines.

The denominator of the above formula for rmr depth is like the numerator, except that it penalizes each program according to the logarithm of its run time. By using a more slowly growing penalty function, the inverse busy beaver function, one obtains another unparameterized depth measure which may be more suitable for very deep objects.

$$D_{bb}(x) = \min\{s + k : D_s(x) < B(k)\}. \tag{5}$$

This depth measure is closely related to the quantity called sophistication by Koppel, this volume.

4. *Depth of Infinite Sequences*

In attempting to extend the notion of depth from finite strings to infinite sequences, one encounters a familiar phenomenon: the definitions become sharper (e.g., recursively invariant), but their intuitive meaning is less clear, because of distinctions (e.g., between infinitely-often and almost-everywhere properties) that do not exist in the finite case. We present a few definitions and results concerning depth of infinite objects.

An infinite sequence X is called *strongly deep* if at every significance level s, and for every recursive function f, all but finitely many initial segments X_n have depth exceeding $f(n)$.

It is necessary to require the initial segments to be deep almost everywhere rather than infinitely often, because even the most trivial sequence has infinitely many deep initial segments X_n (viz. the segments whose *lengths* n are deep numbers).

It is not difficult to show that the property of strong depth is invariant under truth-table equivalence (Rogers *1967*) (this is the same as Turing equivalence in recursively bounded time, or via a total recursive operator), and that the same notion would result if the initial segments were required to be deep in the sense of receiving less than 2^{-s} of their algorithmic probability from $f(n)$-fast programs. The characteristic sequence of the halting set K is an example of a strongly deep sequence.

A weaker definition of depth, also invariant under truth-table equivalence, is perhaps more analogous to that adopted for finite strings:

An infinite sequence X is *weakly deep* if it is not computable in recursively bounded time from any algorithmically random infinite sequence.

As remarked above, computability in recursively bounded time is equivalent to two other properties, viz. truth-table reducibility and reducibility via a total recursive operator. These equivalences are not hard to demonstrate. We will call this reducibility by the traditional name of truth-table reducibility, even though the other two characterizations may be more intuitive.

By contrast to the situation with truth-table reducibility, Gács has recently shown (Gács *1986*) that every sequence is computable from (i.e., Turing reducible to) an algorithmically random sequence if no bound is imposed on the time. This is the infinite analog of the far more obvious fact that every finite string is computable from an algorithmically random string (e.g., its minimal program).

Every strongly deep sequence is weakly deep, but by intermittently padding K with large blocks of zeros, one can construct a weakly deep sequence with infinitely many shallow initial segments.

Truth table reducibility to an algorithmically random sequence is equivalent to the property studied by Levin et al. of being random with respect to some recursive measure. Levin calls sequences with this property "proper" (Zvonkin and Levin *1970*) or "complete" (Levin and V'jugin *1977,* Levin *1984*) sequences (we would call them strongly shallow), and views them as more realistic and interesting than other sequences because they are the typical outcomes of probabilistic or deterministic effective processes operating in recursively bounded time. Deep sequences, requiring more than recursively bounded time to generate, and especially ineffectively defined sequences such as K or Ω, he regards as unnatural or pathological by comparison. We take a somewhat opposing view, regarding objects of recursively unbounded depth as perhaps useful analogs for the less deep objects that may be found in nature.

V'jugin *1976* has shown that weakly deep sequences arise with finite probability when a universal Turing machine (with one-way input and output tapes, so that it can act as a transducer of infinite sequences) is given an infinite coin toss sequence for input. These sequences are necessarily produced very slowly: the time to output the n'th digit being bounded by no recursive function, and the output sequence contains evidence of this slowness. Because they are produced with finite probability, V'jugin sequences can contain only finite information about the halting problem. This contrasts with the finite case, where deep strings necessarily contain information about K. It is not known whether all strongly deep strings contain infinite information about K.

5. *Depth and Complexity in Physics*

Here we argue that logical depth is a suitable measure of subjective complexity for physical as well as mathematical objects, and consider the effect of irreversibility, noise, and spatial symmetries of the equations of motion and initial conditions on the asymptotic depth-generating abilities of model systems. Many of the ideas mentioned here are treated at greater length in Bennett *1987,* Bennett *1986,* and Bennett and Grinstein *1985.*

"Self-organization" suggests a spontaneous increase of complexity occurring in a system with simple, generic (e.g., spatially homogeneous) initial conditions. The increase of complexity attending a computation, by contrast, is less remarkable because it occurs in response to special initial conditions. This distinction has been highlighted recently by the discovery of models (e.g., classical hard spheres moving in an appropriate periodic array of obstacles (Fredkin and Toffoli *1982,* Margolus *1984*) which are computationally universal on a subset of initial conditions, but behave in a quite trivial manner for more general initial conditions.

An important question, which would have interested Turing, is whether self-organization is an asymptotically qualitative phenomenon like phase transitions. In other words, are there physically reasonable models in which complexity, appropriately defined, not only increases, but increases without bound in the limit of infinite space and time? A positive answer to this question would not explain the natural history of our particular finite world, but would suggest that its quantitative complexity can legitimately be viewed as an approximation to a well-defined qualitative property of infinite systems. On the other hand, a negative answer would suggest that our world should be compared to chemical reaction-diffusion systems (e.g., Belousov-Zhabotinsky), which self-organize on a macroscopic, but still finite scale, or to hydrodynamic systems (e.g., Benard) which self-organize on a scale determined by their boundary conditions. A thorough understanding of the physical prerequisites for qualitative self-organization may shed some light on the difficult issue of the extent to which our world's observed complexity is conditioned by the posterior existence of sentient observers.

The suitability of logical depth as a measure of physical complexity depends on the assumed ability ("physical Church's thesis") of Turing machines to simulate physical processes, and to do so with reasonable efficiency. Digital machines cannot of course integrate a continuous system's equations of motion exactly, and even the notion of computability is not very robust in continuous systems (e.g., a computable, differentiable function can have a noncomputable derivative (Myhill *1971*)) but for realistic physical systems, subject throughout their time development to finite perturbations (e.g., electromagnetic and gravitational) from an uncontrolled environment, it is plausible that a finite-precision digital calculation can approximate the motion to within the errors induced by these perturbations. Empirically, many systems have been found amenable to "master equation" treatments in which the dynamics is approximated as a sequence of stochastic transitions among coarse-grained microstates (van Kampen *1962*). Presumably, many mundane hydrodynamic and chemical systems could be efficiently simulated by discrete stochastic models using this approach if the mesh size were made fine enough and the number of states per site large enough.

To see how depth might be used to measure physical complexity, consider an infinite hard sphere gas at equilibrium. The intuitive triviality of this system can be formalized by observing that the coarse-grained state of a typical region in the gas (say a specification, at p digits precision, of the positions and velocities of all the particles in a region of diameter l) has depth bounded by a small polynomial (in lp). Since the gas is at equilibrium, its depth does not increase with time.

Now consider the same gas with a nonequilibrium initial condition, e.g., a periodic modulation of the density. The depth of a local region within the gas would now increase with time, representing the duration of the plausible evolution connecting the present configuration with the significantly nonrandom initial condition.

Now let the evolution of the gas be subject to noise. The regional depth would increase for a while as before, but eventually the significance parameter characterizing this depth would fall to near zero, as the noise gradually obliterated the region's correlation with the system's nonrandom initial condition. Thus, in the realistic presence of noise, the hard sphere gas with nonequilibrium initial condition is not self-organizing.

We do not attempt to survey the vast range of mathematical models used in physics, within which computationally complex or self-organizing behavior might be sought, but instead concentrate somewhat arbitrarily on cellular automata, in the broad sense of discrete lattice models with finitely many states per site, which evolve according to a spatially homogeneous local transition rule that may be deterministic or stochastic, reversible or irreversible, and synchronous (discrete time) or asynchronous (continuous time, master equation). Such models (cf. the recent review Wolfram *1986*) cover the range from evidently computer-like (e.g., deterministic cellular automata) to evidently material-like (e.g., Ising models) with many gradations in between.

In general it appears that reversibility, noise, asynchrony, and spatial reflection-symmetry of the dynamical law hinder computation, whereas their opposite properties facilitate it. Generic values of model parameters (e.g., coupling constants, transition probabilities) and generic initial conditions also tend to hinder computation, whereas special parameters and special initial conditions facilitate it. A further variable is the complexity of the (finite) rule, in particular the number of states per site. Complex rules, of course, give more scope for nontrivial behavior, but are less physically realistic.

Various tradeoffs are possible, and not all the favorable properties need be present at once to obtain nontrivial computation. The billiard ball cellular automaton of Margolus *1984,* for example, though simple, reversible, and reflection-symmetric, is computationally universal. The 3-dimensional error-correcting automaton of Gács and Reif *1985,* on the other hand, is computationally universal in the presence of generic noise, but is irreversible and lacks reflection symmetry. Gács' one-dimensional automaton (Gács *1983*) has similar qualitative properties, but is also enormously complex.

More of the favorable properties need to be invoked to obtain "self-organization", i.e., nontrivial computation from a spatially homogeneous initial condition. In Bennett *1986* we described a rather artificial system (a cellular automaton which is stochastic but noiseless, in the sense that it has the power to make purely deterministic as well as random decisions) which undergoes this sort of self-organization. It does so by allowing the nucleation and growth of domains, within each of which a depth-producing computation begins. When two domains collide, one conquers the other, and uses the conquered territory to continue its own depth-producing computation (a computation constrained to finite space, of course, cannot continue for more than exponential time without repeating itself). To achieve the same sort of self-organization in a truly noisy system appears more difficult, partly because of the conflict between the need to encourage fluctuations that break the system's translational symmetry, while suppressing fluctuations that introduce errors in the computation.

Irreversibility seems to facilitate complex behavior by giving noisy systems the generic ability to correct errors. Only a limited sort of error-correction is possible in microscopically reversible systems such as the canonical kinetic Ising model. Minority fluctuations in a low-temperature ferromagnetic Ising phase in zero field may be viewed as errors, and they are corrected spontaneously because of their potential energy cost. This error correcting ability would be lost in nonzero field, which breaks the symmetry between the two ferromagnetic phases, and even in zero field it gives the Ising system the ability to remember only one bit of information. This limitation of reversible systems is recognized in the Gibbs phase rule, which implies that under generic conditions of the external fields, a thermodynamic system will have a unique stable phase, all others being metastable. Irreversible noisy systems escape this law (Bennett and Grinstein *1985*), being able to store information reliably, and perform reliable computations (Gács and Reif *1985*), even when the noise is biased so as to break all symmetries.

Even in reversible systems, it is not clear why the Gibbs phase rule enforces as much simplicity as it does, since one can design discrete Ising-type systems whose stable phase (ground state) at zero temperature simulates an aperiodic tiling (Robinson *1971*) of the plane, and can even get the aperiodic ground state to incorporate (at low density) the space-time history of a Turing machine computation. Even more remarkably, one can get the structure of the ground state to diagonalize away from all recursive sequences (Myers *1974*). These phenomena have been investigated from a physical viewpoint by Radin *1985* and Radin and Miekisz *1986*; it is not known whether they persist at finite temperature, or in the presence of generic perturbations of the interaction energies.

Instead of inquiring, as earlier, into the asymptotic time-dependence of depth in infinite systems, we can ask how much depth can be generated by a discrete system of size n, given unlimited time. The answer depends on open questions in computational complexity theory. Assuming the existence of cryptographically strong pseudorandom functions which require only polynomial time and space to compute, such a system, if it is capable of universal computation, can generate states exponentially deep in n; the production of greater depth (except accidentally, with low probability) is forbidden by the system's Poincaré recurrence. On the other hand, if the dismal state of affairs $P = PSPACE$ holds, only polynomially deep states can be produced in polynomial space.

Acknowledgements. The development of these ideas is the outcome of thoughtful discussions and suggestions, extending over two decades, from Ray Solomonoff, Gregory Chaitin, Rolf Landauer, Dexter Kozen, Gilles Brassard, Leonid Levin, Peter Gács, Stephen Wolfram, Geoff Grinstein, Tom Toffoli, Norman Margolus, and David Griffeath, among others.

References

Adleman, L

  1979 *Time, Space, and Randomness.* MIT Report LCS/TM-131, April 1979.

Barzdin, J.M.

  1968 Complexity of programs to determine whether natural numbers not greater than n belong to a recursively enumerable set. *Sov. Math. Dokl.* **9** (1968) 1251-1254.

Bennett, C.H.

  1979 *On Random and Hard-to-Describe Numbers,* IBM Report RC 7483 (1979), and Gardner, Martin, Mathematical games, *Sci. Am.* 20-34 (November 1979) based on this report.

  1982 On the logical depth of sequences and their reducibilities to random sequences. Unpublished manuscript (1982).

  1985 Information, dissipation, and the definition of organization. In: *Emerging Syntheses in Science,* ed. D. Pines, pp 297–313. Santa Fe, NM: Santa Fe Institute (1985).

  1986 On the nature and origin of complexity in discrete, homogeneous, locally-interacting systems. *Found. Phys.* **16** (1986) 585–592.

  1987 Information dissipation and the definition of organization, In: *Emerging Synthesis in Science,* ed. D. Pines. Reading, MA: Addison-Wesley (1987).

Bennett, C.H., and J. Gill

  1981 Relative to a random oracle A, $P^A \neq NP^A \neq co\text{-}NP^A$ with probability 1. *SIAM J. Comp.* **10** (1981) 96–113.

Bennett, C.H., and G. Grinstein

  1985 Role of irreversibility in stabilizing complex and nonergodic behavior in locally interacting discrete systems. *Phys. Rev. L.* **55** (1985) 657–660.

Blum, M., and S. Micali

  1984 How to generate cryptographically strong sequences of pseudo random bits. *SIAM J. Comp.* **13** (1984) 850–864.

Borges, J.-L.

  1964 In: *Labyrinths: Selected Stories and Other Writings,* eds. D.A. Yates and J.E. Irby. New York: New Directions (1964).

Chaitin, G.

    1975 A theory of program size formally identical to information theory. *J. ACM* **22** (1975) 329–340.

    1977 Algorithmic information theory. *IBM J. Res.* **21** (1977) 350–359, 496.

    1987 *Algorithmic Information Theory.* Cambridge: Cambridge University Press (1987).

    1987a *Information Randomness and Incompleteness – Papers on Algorithmic Information Theory.* Singapore: World Scientific Press (1987).

Fredkin, E., and T. Toffoli

    1982 Conservative logic. *Int. J. Theor.* **21** (1982) 219.

Gardner, M.

    1979 Mathematical games. *Sci. Am.* 20–34 (November 1979)

Garey, M., and D. Johnson

    1979 *Computers and Intractability, a Guide to NP Completeness.* Freeman (1979).

Gács, P.

    1974 On the symmetry of algorithmic information. *Sov. Math. Dokl.* **15** (1974) 1477.

    1983 Technical Report No. 132, Computer Science Department, University of Rochester (1983), to appear in *J. Comput. Sys.*

    1983a On the relation between descriptional complexity and probability. *Theo. Comp.* **22** (1983) 71–93.

    1986 Every sequence is reducible to a random sequence. *Inf. Contr.* **70** (1986) 186–192.

Gács, P., and J. Reif

    1985 In: *Proceedings of the 17th ACM Symposium on the Theory of Computing,* pp. 388-395 (1985).

Goldreich, O., S. Goldwasser, and S. Micali

    1984 How to construct random functions. In: *Proceedings of the 25th IEEE Symposium on Foundations of Computer Science.* (1984).

Hartmanis, J.

    1983 Generalized Kolmogorov complexity and the structure of feasible computation. In: *Proceedings of the 24th IEEE Symposium on the Foundations of Computer Science* (1983) 439–445.

Hunt, J.W.

    1978 *Topics in Probabilistic Complexity,* Ph.D. dissertation Stanford Univ. Electrical Engineering (1978).

Kolmogorov, A.N.

    1963 On tables of random numbers. *Ind. J. Stat.* **A25** (1963) 369–376.

1965 Three approaches to the quantitative definition of information. *Prob. Inf. Trans.* **1** (1965) 1–7.

Koppel, M.

1988 Structure. This volume.

Kurtz, S.

1983 On the random oracle hypothesis. *Inf. Contr.* (1983).

Levin, L.A.

1973 On the notion of a random sequence. *Sov. Math. Dokl.* **14** (1973) 1413–1416.

1984 Randomness conservation inequalities: Information and independence in mathematical theories. *Inf. Contr.* **61** (1984) 15–37; preliminary draft *MIT Technical Report MIT/LCS/TR-235* (1980).

1985 One-way functions and pseudorandom generators. In: *ACM Symposium on Theory of Computing* (1985).

Levin, L.A., and V.V. V'jugin

1977 *Invariant Properties of Informational Bulks,* Lecture Notes in Computer Science 53, pp. 359–364. Springer-Verlag (1977).

Martin-Löf, P.

1966 Definition of random sequences. *Inf. Contr.* **9** (1966) 602–619.

Miekisz, J., and C. Radin

1986 The unstable chemical structure of quasicrystalline alloys, preprint 1986.

Myers, D.

1974 Nonrecursive tilings of the plane II. *J. Symb. Log.* **2** (1974) 286–284.

Margolus, N.

1984 Physics-like models of computation. *Physica* **10D** (1984) 81–95.

Myhill, J.

1971 A recursive function defined on a compact interval and having a continuous derivative that is not recursive. *Mich. Math J.* **18** (1971) 97–98.

Radin, C.

1985 Tiling, periodicity, and crystals. *J. Math. Phys.* **26** (1985) 1342; Correlations in classical ground states. *J. Stat. Phys.* **43** (1986) 707.

Robinson, R.M.

1971 Undecidability and nonperiodicity for tilings of the plane. *Invent. Math.* **12** (1971) 177–209.

Rogers, H. Jr.

1967 *Theory of Recursive Functions and Effective Computability,* pp. 141, 255. New York: McGraw-Hill (1967).

Schnorr, C.P.

    1973 Process complexity and effective random tests. *J. Comput. Sys.* **7** (1973) 376–388.

Solomonoff, R.J.

    1964 A formal theory of inductive inference. *Inf. Contr.* **7** (1964) 1–22.

Turing, A.M.

    1936-7 On computable numbers, with an application to the Entscheidungsproblem. *P. Lond. Math. Soc. (2)* **42** (1936-7) 230–265; A correction, ibid. **43** (1937) 544–546;

    1937 Computability and λ-definability. *J. Symb. Log.* **2** (1937) 153–163 (1937).

    1952 The chemical basis of morphogenesis. *Phi. T. Roy. B* **237** (1952) 37–72.

van Kampen, N.

    1962 In: *Fund. Prob. in Stat. Mech.,* ed. E.G.D. Cohen. North-Holland Publ. Co. (1962).

V'jugin, V.V.

    1976 On Turing invariant sets. *Sov. Math. Dokl.* 1976.

Wolfram, S.

    1985 Undecidability and intractability in theoretical physics. *Phys. Rev. L.* **54** (1985) 735–738.

    1986 *Theory and Applications of Cellular Automata.* World Scientific Press (1986).

Yao, A.

    1982 Theory and applications of trapdoor functions. In: *Proceedings of the 23rd IEEE Symposium on Foundations of Computer Science,* pp. 80–91 (1982).

Zvonkin, A.K., and L.A. Levin

    1970 The complexity of finite objects and the development of the concepts of information and randomness by means of the theory of algorithms. *Russ. Math. S.* **256** (1970) 83–124.

The Busy Beaver Game
and the Meaning of Life

Allen H. Brady

1. Introduction

The representation of the Universal Computing Machine in the guise of the stored-program digital computer is now well known among serious students of computer science. On the periphery, a modern generation of technologists seems unable to conceive of a time when "computer" meant "a person who computes" and the concept of "programming" was not ubiquitous. For practical reasons, a modern computer is much more complex than is minimally necessary to achieve universality. The original formulation of a Universal Computing Machine (Turing *1936-7*) involved a table based upon 15 symbols and 28 states with no restriction imposed on the length of a sequence of atomic acts ("move" and "print") permitted prior to a change in state. An aboriginal use of *macros* simplified the description.

Briefly stated, the Universal Computing Machine is an active finite-state device of limited size connected to a passive medium of unlimited extent. The active device writes on and reads from the passive medium. Any other finite-state machine of any size or complexity whatsoever, even one connected to its own passive recording medium, may be reduced to nothing more than an abstraction recorded in the medium of the Universal Machine. The Universal Machine without any change in its mechanism then assumes the identity of the machine which has been described to it.

A competition to find the *smallest* Universal Turing Machine received at one time a certain amount of attention (cf. Minsky *1967*). The measure of "smallness" proposed was the size of the state-symbol product in the machine description. A clever construction by Shannon *1956* had demonstrated that with enough symbols any Turing machine can be reduced to only two states. This demonstration, along with his companion construction reducing any machine to a machine with only two symbols, corroborated the notion derived from experience that machine states can be traded for symbols in such a way as to preserve a roughly constant product of the number of states required times the number of symbols required.

Minsky *1962* demonstrated a four-state by seven-symbol Turing machine to simulate a universal Tag system and in a footnote announced that (he and D. Bobrow had determined) there could not be a two-state by two-symbol Universal Turing Machine. In a private conversation (ca. 1964) Professor John McCarthy of Stanford University remarked, "We thought if we were to find the smallest universal machine then we could learn a great deal about computability – of course that wouldn't be so!" And, the last word representing a prevailing feeling is given by Minsky *1967* who suggests that "the question is an intensely tricky puzzle and has essentially *no* serious mathematical interest."

A related path of investigation has looked at simple possibilities for self-reproduction in systems of cellular automata. Codd *1968* and others continued with the pioneering work of von Neumann who used a cellular automaton model suggested by S. M. Ulam (cf. Burks *1970,* and also Arbib, this volume). Codd's work departed from the usual top-down design of functioning systems by virtue of his use of a computer to search for naturally occurring mechanisms which he could employ in his construction of universal cellular systems.

The size of the smallest universal machine remains unknown, however, whether or not it is of any serious mathematical interest. With our minds remaining open to the possibility of practical, scientific, or simply philosophical interest we shall examine the area of simple Turing machines and systems of cellular automata inspired by two games: Rado's Busy Beaver Game and Conway's Game of Life. We shall look at some questions which are hardly explored and some which may be, for reasons beyond our comprehension, virtually unanswerable.

2. *Turing Machine Questions*

THE BUSY BEAVER GAME

The Busy Beaver Game was invented by Tibor Rado *1962*. It is based upon the Kleene *1952* representation of Turing machines in which each sequence of atomic acts consists of printing one symbol and making one move to an adjacent square before changing to the next state. For the Turing machines in his game, Rado added an external *zero* state to denote the act of *halting*. The machines deal with only two symbols, "0" (blank) and "1" (mark). The three-state machine shown in Figure 1 is in the precise form used by Rado as an entry in his game.

The Turing machines of the Busy Beaver Game operate on a two-way infinite tape. The tape is initially blank (all zeros), and the "contest" is among machines having the same number of states. The objective is to find the machine which can print the most marks (ones) on its tape before halting. Not all machines halt, of course, so the problem is not merely one of combinatoric unwieldiness, but one of general undecidability. Rado showed that the maximum number of marks which can

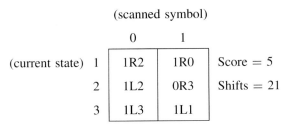

Figure 1. A three-state Turing machine for Rado's Busy Beaver Game.

be placed upon a blank input tape by a machine of k states defines a noncomputable function $\Sigma(k)$. Using essentially a diagonal argument he demonstrated that if f is some computable function then there exists a positive integer n such that

$$\Sigma(k) > f(k) \text{ for all } k > n.$$

Related to the function Σ is another function, the maximum *shift number, $S(k)$,* representing the maximum number of moves or steps that can be taken by a k-state machine which halts after starting on a blank tape. Clearly, $S(k)$ is not computable, else $\Sigma(k)$ could always be computed through a simple process of enumeration once the maximum shift number were known.

The three-state machine shown in Figure 1 will mark five 1's on a blank tape in 21 shifts after beginning in state 1. There are five distinct three-state machines which *score* six, but this is the only machine which will operate for 21 steps (Lin and Rado *1965*). Two machines are shown in Figure 2 which represent those three-state machines (the vast majority) which will never halt. The labels for the two machines describe their classes of nonstopping behavior, and the potential halt entries (unreachable) have been left unspecified.

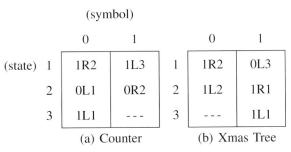

Figure 2. Nonstopping three-state Turing machines (blank input tape).

Experience has shown that a small machine picked at random will very likely never halt. Furthermore, such a machine, started on a blank tape will probably, say,

95% of the time, behave in a trivial way making it obvious that it will never halt. The machines in Figure 2 (from a set of around one percent of the three-state machines considered for the game) represent a slightly more complicated behavior – not the simple looping displayed by a machine which merely "runs away" in one direction down the length of its tape.

While the halting problem for Turing machines with blank input tapes is recursively unsolvable in general, we have no reason to say that given a particular machine we cannot, for logical reasons, declare that the halting problem for that machine on a blank input tape is unsolvable. (Turing demonstrated that there is a *particular* machine, namely the Universal Machine, for which the halting problem is undecidable for *arbitrary* input.) Notwithstanding the fact that very deep mathematical problems such as Fermat's "Last Theorem" and the Goldbach Conjecture each reduce to deciding the halting problem of an individual Turing machine with a blank input tape, we seem intuitively to relegate the individual problems to the combinatoric realm where an ideal computer with sufficient speed and a large enough memory would provide a solution. Rado, on the other hand, wondered out loud about the possibility of there being a *particular* value of n for which the value of $\Sigma(n)$ could not be decided for *logical reasons,* an untenable position in the eyes of his logician colleagues.

With regards to his Busy Beaver Game, Rado *1963* struck a pessimistic note in declaring that "even though skilled mathematicians and experienced programmers attempted to evaluate $\Sigma(3)$ and $S(3)$, there is no evidence that any known approach will yield the answer, even if we avail ourselves of high-speed computers and elaborate programs. As regards $\Sigma(4)$, $S(4)$, the situation seems to be entirely hopeless at present." His pessimism was premature relative to the value for k. At a time when "high-speed" meant a several microsecond cycle time, and integrated circuits were still in the laboratory, high speed computers were not abundant, and their available time was measured and precious. Nevertheless, the value of $\Sigma(3) = 6$ was soon proved (Lin and Rado *1965*), and the value of $\Sigma(4) = 13$ was discovered and the case for $k = 4$ was reduced to manageable proportions (Brady *1966*).

Still, Rado's assessment was only slightly misplaced, because an unreachable lower bound was subsequently demonstrated for $k > 7$. By means of a recursive construction on the states of a machine Green *1964* was able to put what appears to be a nonprimitive recursive lower bound on $\Sigma(k)$ and show that

$$\Sigma(7) \geq 22,961$$
$$\text{and} \quad \Sigma(8) \geq 3 \cdot (7 \cdot 3^{92} - 1)/2.$$

For a small value of k there was then a score to compare with the age of the universe expressed in nanoseconds! If the problem was then tractable for $k = 4$, where did the real difficulties lie?

In 1971, searching for five-state machines with structures similar to those of Green's, D.S. Lynn *1972* first pushed the lower bound for $\Sigma(5)$ to 22 with 435 shifts. Later, when a large amount of low priority computer time became available, he was able to look at what he estimated to be about ten percent of the *tree-normal* five-state machines and extended the limits to $\Sigma(5) \geq 112$ and $S(5) \geq 7,707$ (Lynn *1974*). There the problem sat until a recent flurry of computer searches by several individuals culminated in a nearly *no contest* entry by Uhing *1986* with new lower bounds of

$$\Sigma(5) \geq 1,915 \quad \text{and} \quad S(5) \geq 2,358,063.$$

Uhing used a microprocessor controlling a "hardware Turing-machine simulator. The hardware simulator was constructed using less than \$100 worth of parts and materials, including 32 integrated circuits, sockets and a circuit board." Over several months Uhing simulated about 260,000,000 five-state machines leaving 2,500,000 machines undecided. The high scoring machines are shown in Figure 3.

	0	1			0	1
1	1R2	1L3		1	1R2	1R0
2	0L1	0L4		2	1L3	1R3
3	1L1	1R0		3	0R5	0L4
4	1L2	1R5		4	1L3	0L2
5	0R4	0R2		5	1R4	1R1

Score = 1,915   Score = 1,471

Shifts = 2,133,492   Shifts = 2,358,064

Figure 3: High scoring five-state machines discovered by G. Uhing.

Even though it might appear now that the five-state problem is within grasp, there is a distinct possibility that the limit of practical solvability has in fact been reached. While we can follow Uhing's current champion machines until they halt, it is not clear at all how the machines work. Any cleverness in their construction is not the result of human creation, so there is a conspicuous absence of documentation! In light of Green's results it was easy to accept that the turning point for the Busy Beaver Game might occur at $k = 6$, but such magnitudes as have now been produced for $k = 5$ had never been anticipated. Any hope for solving the problem at this level will require computer programs endowed with a level of intelligence that we have not seen in anything done previously by a machine. Can it be decided by a computer program or will it be necessary to assign one mathematician per unresolved five-state Turing Machine?

THE SMALLEST UNIVERSAL MACHINE

If a particular Turing machine is a Universal Machine then we know there is no (computable) solution to its halting problem for arbitrary input tapes. With respect to saying whether or not a given machine is universal we can only assert that the machine is *not* universal if the halting question can be decided for any input tape. It essentially requires another machine (i.e., a computer program) which will examine the tape and return the correct "YES" or "NO" answer in every case. But going from blank input tapes to arbitrary input tapes is a giant step. This has not yet been done for the set of two-symbol three-state machines (which, without eliminating trivial cases or any sort of symmetry, number nearly two million). Where in the space of *m*-symbol by *n*-state Turing machines might the smallest Universal Machine reside? Perhaps results for the Busy Beaver Game could give us some clues as to where the complex machines lie.

The Busy Beaver Game is naturally open to variations, and the *m*-symbol by *n*-state version is one of them. The traditional scoring rule is that only one symbol other than blank must appear on the final tape (Lee *1963*). This is obviously an arbitrary rule, and it may give us no clue as to when a machine stops. If the maximum shift number is known the matter is moot, and it is really the shift number in the end that counts. Some might argue that the range of tape excursion is the true measure of complexity. But, it is not practical to approach the problem from this standpoint, for a machine which will never stop can spend a very long time inside a restricted region. (The three-state *Counter* machine in Figure 2(a) if left running on a 40 square tape will illustrate this point.) So, it will be the shift number in the end which we deal with in deciding whether or not a particular machine will ever halt while we are watching it run.

On the assumption that the shift number represents a kind of *recursive strength* of a machine, we have tabulated values[1] involving $S(m, n)$ for consideration (Figure 4). Along the diagonal running up to the right from (2,5) to (5,2) the author has no values to supply aside from that of Uhing's for $S(5) = S(2, 5)$. The 3×4 and 4×3 machine spaces are apparently orders of magnitude larger than the 2×5 and 5×2 spaces. One might expect to see results at least comparable to that of $S(2, 5)$. The 3×3 space (with "?" entered) appears to be of an order somewhere between the

---

1 Except for the values for $k = 5$ discovered by Uhing, all lower bound values shown in the paper were produced by computer programs written by the author utilizing a recursive technique involving backtracking to generate distinct machines. While improvements on some of these results no doubt exist, none are known to the author.

sizes of 2×5 and 2×4. Because it is probably at least an order of magnitude greater than 2×4 (the 2×5 space is nearly three orders of magnitude greater than that of the 2×4) no attempt has been made to supply any value.

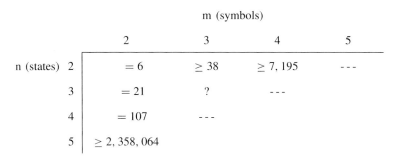

		m (symbols)			
		2	3	4	5
n (states)	2	= 6	≥ 38	≥ 7, 195	- - -
	3	= 21	?	- - -	
	4	= 107	- - -		
	5	≥ 2, 358, 064			

Figure 4. The maximum shift number S for m-symbol by n-state machines.

The state-symbol product as a rule for conservation of computing power seems to hold across the diagonal from 2×3 to 3×2. There it is possible to say that either $S(3, 2) = 38$ or else $S(3, 2) > 15, 000$, and since there are fewer than 3,000 of the 3×2 machines to deal with it seems safe to conclude that $S(3, 2) = 38$. (Someone may well already know this to be a fact – the author does not, but it appears that it should not be too difficult to prove.)

However, what has happened to the conservation rule going across the diagonal where $S(2, 4) = 106$ and $S(4, 2) \geq 7, 195$? (It can be stated that either $S(4, 2) = 7, 195$ or else $S(4, 2) > 15, 000$, although in this case settling for the equality does not seem like a safe bet.) There were almost 400,000 machines generated which is about two thirds the number of machines generated for the 2×4 Busy Beaver problem, but seeing a lower bound for the shift number nearly two orders of magnitude larger than the known value of $S(2, 4)$ we cannot offhandedly say that the two problems are comparable!

Does a universal machine lie anywhere in this matrix? If the shift number indicating relative recursive strength is a gauge of machine complexity, then it would seem a reasonable guess that universal machines should exist in the machine spaces along the diagonal from (2,5) to (5,2). And what about (4,2)? Why does it appear that there is a significant gain in power from using *four* symbols? Does Nature know this already? The four-symbol by two-state machine space deserves careful scrutiny.

3. *Cellular Automata Problems*

GAMES IN LINEAR CELL SPACE

A cellular automata problem equivalent to the Busy Beaver Game was studied by Varshavsky (*1972*) in a linear cell space. An infinite chain of n-state cells all begin in the quiescent state except for one nonquiescent cell which serves as a seed. A cell is taken to be a finite state automaton whose state transition is dependent upon the states of the cells in a three cell neighborhood consisting of itself and the immediately adjacent cells. For fixed value of *n* Varshavsky wished to determine what is the maximum length *L* of active cells possible out of which no further growth of activity may occur in the chain. Varshavsky added the rule that once a cell leaves the quiescent state it must remain active thereafter. The states were therefore numbered 0, 1, ... *n*.

This question ranges over all possible transition tables for the cells, a set of functions in number of the order of $n^{(n^3)}$. Varshavsky reported, "By completely enumerating all possible tables of transition rules it has been shown that for *n* = 3, the maximal length $L(3) = 7$. For *n* = 4 transition rules have been found giving $L(n) = 45$ but this length has not been shown to be maximal". No other information was given on how the "enumeration" was performed, nor in particular exactly how it was determined whether or not a chain of cells becomes stable. For *n* = 3 one is dealing with 3^{27} possible transition functions, though in generating the possibilities from a tree of next-state choices and eliminating symmetry in the process one can readily see how the number of possibilities could be greatly reduced.

Vitanyi *1976*, considering Varshavsky's problem, restricted the flow of information in the chain to one direction, and with a definition of what he called a one directional linear cell space (1 LCS), demonstrated that his space defined a Tag system from which it could be deduced that $L(n)$ diagonalizes the computable functions. But, as Vitanyi points out at the end of his paper, the problem is directly "equivalent to the halting problem for Turing machines by encoding the finite control and the scanned symbol in each cell of the linear cell space".

So, how does one reconcile the flow of information in only *one* direction with a construction embedding the encoding of a Turing machine in each cell when a Turing machine requires information to flow in *both* directions? A constructive solution to this is possible. On alternate cycles let the entire state space shift to the right. (Every cell takes on the state of its left neighbor.) Then, a "left" move by the simulated Turing machine can be executed by allowing the active cell location representing the machine's position to "stand still" as the "tape" moves by. A "right" move is executed during a *nonshift* cycle by letting this active cell location make its actual move to the right. In a sense the embedded machines operate in an "expanding universe" of active cells. In his monograph on cellular automata Codd (*1968*) conjectured "that the existence of unbounded but boundable propagation is a necessary condition for computation universality" in a cellular space. One could interpret the simple construction we have presented here as a means for making a demonstration to the contrary.

The problem of determining $L(n)$ for $n = 4$ appears to be out of reach since the possible number of transition functions climbs to 4^{64}. This represents a rather large space to search. On the other hand, a five cell neighborhood with two-state cells has an interesting 2^{32}. Adding Vitanyi's restriction of the cell space to unidirectional information flow, there would be $n^{(n^2)}$ transition functions with a three cell neighborhood, which is only 3^9 for $n = 3$ and the possible 2^{32} for $n = 4$.

Two Dimensions: The Game of Life

The Game of Life devised by J.H. Conway has had such a run of popularity since its introduction in 1970 that it nearly represents a small industry from magazine articles to books to computer software. It was introduced to the public by Martin Gardner in his Mathematical Recreations column in *Scientific American* (cf. Berlekamp, Conway, and Guy *1982* and Gardner *1983*). It has stimulated both philosophical commentary and science fiction and possesses an entertaining and virtually unending taxonomy of cellular phenomena, containing such combinations as "blinkers", "beehives", "boats", "pulsars", "gliders", and even a "Cheshire cat".

The rules for the game were devised after a certain amount of experimentation. It has a fascinating biological naturalness. Two-state cells ("dead" or "alive") occupy an infinite two-dimensional rectangular space in which only a finite number of cells are initially in the living state. A birth (transition from dead to alive) occurs in a cell at the center of nine squares only if exactly three live neighbors are present. Isolated living cells with no more than one live neighbor will die, and crowded cells with four or more live neighbors will die. Living cells with two or three neighbors will survive.

Conway studied many initial configurations and originally thought that all finite initial patterns eventually degenerated into stable or oscillating patterns (Figure 5) or else disappeared entirely. He offered a $50 prize for anyone finding a pattern which would grow without bound. Such patterns were discovered and led to the creation of an entire system of patterns which could be used to simulate a computer. In other words, with a particular encoding of its cell space, Life becomes a Universal Turing Machine!

In the nine cell Life neighborhood there are 2^{512} possible transition functions (encompassing as well all 2^8 possible neighborhoods involving immediately adjacent cells). Can Conway's Life be the only interesting function? For example, admitting anisotropic transition functions leads to the possibility of directly simulating simple neural nets. Naturally, Life already supports neural nets as a programming layer on top of its simulated computing machine, but the issue here is examination of simple mechanisms functioning directly in a cellular space.

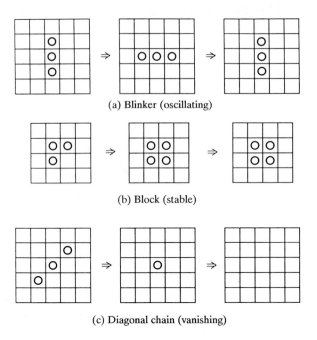

(a) Blinker (oscillating)

(b) Block (stable)

(c) Diagonal chain (vanishing)

Figure 5. Simple cellular examples in Conway's Game of Life.

What about the two-dimensional equivalent to Varshavsky's $L(n)$? We have pointed out that the transition space for $n = 2$ has 2^{512} possibilities. It is difficult to imagine the problem ever being solved for $n = 2$, at least not by any exhaustive enumeration. It would seem to require a theoretical solution.

A cellular universe as an infinite collection of active devices lacks the intuitive appeal of a Turing machine with its infinitely extensible passive medium. However, by restricting the nonquiescent cells to a finite number the two systems are seen to be equivalent. Still, in a very small region of cells with a only a few states little is known which enables us to predict their behavior. These frontiers of Life have barely been explored. It is easy to view anyone's interest in this area as a frivolous preoccupation, because it is difficult to admit that the problems in a simple and seemingly natural domain vastly exceed our present mathematical understanding.

4. The Busy-Beaver Game in Two Dimensions

SIMULATION OF LIFE

If a Turing machine is able to operate in two-dimensions (on a rectangular grid) it is a straightforward matter to design one to simulate the Game of Life. For instance, position the Turing machine in the exact center of a large square border surrounding the active Life cells (Figure 6) where it proceeds in a spiral fashion from the center toward the border leaving a trail as it moves. On the outward pass the Turing machine temporarily marks all those cells which will give birth and also marks all the cells which will die. If any cells give birth while the machine moves around the quiescent cells comprising the border itself, the Turing machine expands the border in all directions by one layer of cells. After completing its trip out to the border, the machine spirals back toward the center marking the new cells as living, deleting the dead cells, and cleaning up its trail. From the center the Turing machine repeats the process for the next cell transition.

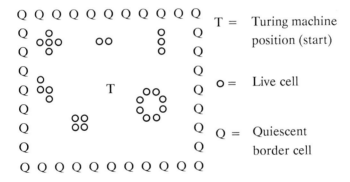

Figure 6. Life simulation by a Turing machine.

Our Turing machine equivalent to the usual Life simulating computer program accommodates Life's hypothetical infinite grid. Since this machine will never halt, it can be very simply stated that Life is a runaway Turing machine!

How does one retain Universality when a Turing machine is constrained never to halt? The simple Turing machine just described is a Universal Machine. It simulates Life, and Life is Universal. The matter of halting comes directly from Conway's original question regarding the stability of Life configurations. We equivalently ask whether or not the border will ever stop expanding. This entire described system (infinite Life grid and simulating Turing machine) can itself be simulated by a Turing machine on its own one-dimensional infinite tape. So the question of border expansion can be tied to the halting problem for the machine with the tape: it merely stops if it has to wait too long for the border to expand. This is not practical, but it is theoretically possible since having enough time to wait is no more of a problem for a Turing machine than having enough tape.

"TurNing" Machines

Moving Turing machines into the two-dimensional domain of Life does nothing to enhance their ultimate computational power. Turing reduced the computing problem from two dimensions to one dimension in the first place in order to get at the simple essence of computer mechanism. However, if we limit ourselves to small machines of approximately the same size, it is reasonable to wonder what degree of recursive strength might be added by removing the one-dimension restriction.

Rather than add the obvious extensions of "Up" and "Down" to "Left" and "Right" why not allow a machine to *turn* as well as move? Let *Left, Right,* and *Back* reorient the machine at the same time it moves, while *Forward* takes it straight ahead. This sacrifices a sense of direction in the external world, but the added power is immediately significant: the Busy Beaver Game improves for *one* state with a score of 4 in 5 shifts. At the risk of breathing new life into an old typographic virus, we shall call these enhanced devices "TurNing" machines. (The ones that never stop can be referred to as "TOuring" machines!)

There is a crystalline flavor in the output of TurNing machines with two states (as one might suspect), and over the machine configuration space of little more than 700 machines there is a likely maximum shift number of 121 and a likely highest score of 37. (The computer enumeration process was restricted to 2,000 moves and machines which fell off the "edge of the earth" 100 squares wide were assumed never to halt.)

An interesting simplification of these machines is obtained by restricting their action to a triangular grid. In the sort of *Chinese checkers* space which results, very small machines seem to find a natural environment where they can spin elaborate webs as they spiral through this simple world. Upon entering a triangular cell through the *base,* a machine will have the choice of moving through the side on its Right, the side on its Left, or Back through the side it entered (reversing direction). The side penetrated becomes the new base. Both the score and shift number improve immediately: a one state machine can turn through six cells! (See Figure 7.)

Among 356 two-state triangular machines examined (which should exhaust the possible configurations) the maximum shift number found was 171, achieved by a machine which also visited the most cells, a total of 62 (Figure 8). A different machine achieved the highest score of 39. If these are not the maximum values then the shift number probably exceeds 2,000.

The number of distinct three-state machines for the game remains quite manageable. Among more than 91,000 machines of the triangular version inspected the

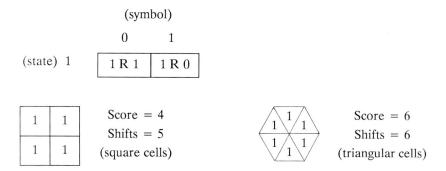

Figure 7. A one-state "TurNing" machine.

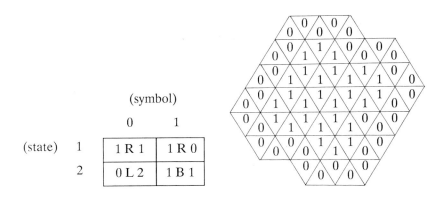

Shifts = 171, Cells visited = 62, Score = 27

Figure 8. A high scoring two-state machine in two dimensions.

highest shift number found was 1,721 and the highest score encountered was 351. Since no machines were watched beyond 2,000 moves, it is not reasonable to predict that either the maximum shift number or the maximum score has been found. It is entirely possible that there are more machines to be inspected, for the generation tree was trimmed on the assumption that any machine which continued past 2,000 moves or reached the edge of a space 50 cells wide did not halt. (It is also possible but not likely that there are more than 356 two-state machines in the previous tree).

The recursive strength of these machines grows rapidly as can be seen from the results of computer runs tabulated in Figure 9. A comparison of the shift number values among the various cases in both one and two dimensions suggests that an attempt to solve the halting problems here would encounter a much higher degree of difficulty than that handled by Brady *1983* for a final proof that $\Sigma(4) = 13$. At least the number of machines to handle would not be unreasonable.

	Triangle cells		Square cells	
	$S(k)$	$\Sigma(k)$	$S(k)$	$\Sigma(k)$
states $k = 1$	$= 7$	$= 6$	$= 5$	$= 4$
2	≥ 171	≥ 39	≥ 121	≥ 37
3	$\geq 1,721$	≥ 351	?	?

Figure 9. Busy Beaver results in two dimensions: "TurNing" machines.

The effect of the addition of more symbols to two-dimensional machines was not examined. A four-symbol by two-state TurNing Machine might well surprise us in its complexity!

5. *Conclusion*

To the more pragmatically minded the Universal Computing Machine is represented on a silicon slate using nor-gates for chalk. The minimal machine in this view is a single nor-gate. It is perhaps just another way of saying that as the number of states is reduced the complexity of the encoding is increased, and in an infinite gate array there is really no distinction between the finite-state machine and its tape. This is not a satisfying view, however. It begs the question.

At another level in the practical world we would experiment with confidence on the 25,000 node neural net representing a snail's brain, or snip off a piece of a chromosome and admire the effect on the offspring of the altered cell. We do this without admitting the limitations of our knowledge of the underlying mechanisms. It is not to say that experimentation is wrong – on the contrary, we have emphasized a problem area involving simple mechanisms wherein we would strongly encourage experimentation. But, we will plead at the same time for more humility on the part of those who are wont to experiment at the high range of the scale.

We have seen how some relatively simple variations in very small mechanisms can have disproportionate and possibly unanticipated effects on their recursive power. How this comes about is not at all clear. Expansion (adding states or symbols) ultimately yields universality in a form which we are able to program in ways consistent with our experience. So, in a strict logical sense, the changes offer nothing new. Nature, however, may have its own methods of programming. Not only should we study the automata mentioned here, but we should be considering many other variations as well. The experience gained could lead to the practical discovery of parallel mechanisms active in natural phenomena. And, furthermore, the mechanized proofs required for rigorous solutions to these problems will give us a sense of the degree to which we can honestly and realistically apply the modifier *artificial* to the term *intelligence*.[2]

Our concluding perspective will be summarized in the following conjectures and predictions proceeding from machines with just one state to machines with six.

Conjecture 1. There exists a *one-state* Universal Machine which operates in two dimensions (i.e., a one-state "TuRning" machine).

Likelihood that this conjecture will be proved: Fair – Nature has probably already proved it, but with time on its side.

Conjecture 2. There does not exist a two-symbol by two-state Universal Machine if *halt* entries are excluded.

Likelihood that this conjecture will be proved: Good, but the characterization of "looping" as a substitution for halting may be an open-ended matter. Halting is useful to make a theoretical point, but in real life survival is the tautological goal. And we no longer build computers which halt – at most they wait in a quiescent state. In any case, solving this problem would be good practice for attacking the following.

Conjecture 3. The halting problem for two-symbol by three-state Turing machines is decidable.

Likelihood that this conjecture will ever be proved: Fair, but this is an extremely challenging problem. Let anyone in doubt pick some machines and try it!

Conjecture 4. There exists a two-symbol by four-state Universal Machine if *halt* entries are excluded. (This is prompted by intuition bolstered by the fact that with four states it is possible to send two signals in each of two directions, and by excluding halt entries some of the recursive strength of the five-state space is assumed.)

Likelihood that this conjecture will ever be disproved: Nil.

Prediction 5. It will never be proved that $\Sigma(5) = 1,915$ and $S(5) = 2,358,064$. (Or, if any larger lower bounds are ever found, the new values may be substituted into the prediction.)

---

2 House and Rado *1963* proposed the study of small Turing machines as a relevant endeavor for the then infant specialty of *artificial intelligence*. They pointed out the theoretical and practical difficulties for problems such as automated searches for optimal machines.

Reason: Nature has probably embedded among the five-state holdout machines one or more problems as illusive as the *Goldbach Conjecture*. Or, in other terms, there will likely be nonstopping recursive patterns which are beyond our powers of recognition.

Prediction 6. From known results for $k = 5$ a six-state machine will be constructed for which it can be "proved" that its shift number (and thus a lower bound for $S(6)$) is an incomprehensibly large value which is in itself difficult to describe.

Reason: It is now clear that determining $\Sigma(6)$ and $S(6)$ is intractable. At this level one can speculate with impunity, and we shall.

Some students of the author were readily convinced after extensive examination and computer testing that Uhing's champion machine for $S(5)$ would never halt. Seeking assurance one student ran her simulator to a point just short of two million moves! From an amusing experience such as this, one is led to consider the possibility that someday a machine of six states (or a just a few more) will be presented by a group of mathematicians along with a "proof" that it will never halt. Suppose then an efficient simulator for the machine were built on the leading but slightly jagged edge of technology and run for an extensive period of time. And then suppose it were to halt! The mathematicians, with solid reasoning to back them up, could make a valid claim that the machine malfunctioned.

But now suppose that instead of building a machine, another group of equivalently qualified thinkers, supported by a great body of mathematical knowledge, countered with a different "proof" that after some unimaginable number of moves the proposed machine would in fact halt. Their number would be so large that building a simulator to check the result would be inconceivable. What then? (It is only speculation, of course!)

For $k = 6$ the problem transcends mechanism. One reaches a point where it becomes impossible to distinguish between the finite and the infinite. Is there a point at which it will transcend logic? Rado's question remains open.

Acknowledgements. The author wishes to thank Luanne Napoli for assistance in editing, Ashraf Kaiser for producing the figures, the American University in Cairo for the use of computer laboratory facilities, and the editor of this volume for his encouragement and patience.

References

Arbib, M.A.

    1977 From Universal Turing Machines to Self-Reproduction, this volume.

Berlekamp, E., J. Conway, and R. Guy

    1982 *Winning Ways for Your Mathematical Plays.* Orlando: Academic Press.

Brady, A.H.

1966 The conjectured highest scoring machines for Rado's $\Sigma(k)$ for the value $k = 4$. *IEEETEC, EC-15* **5** (Oct. 1966) 802–803

1983 The determination of the value of Rado's noncomputable function $\Sigma(k)$ for four-state Turing machines. *Math. Comput.* **40** (April 1983) 647–665.

Burks, A.W. (ed.)

1970 *Essays on Cellular Automata.* Urbana: University of Illinois Press (1970).

Codd, E.F.

1968 *Cellular Automata.* Orlando: Academic Press (1968).

Gardner, M.

1983 *Wheels, Life, and Other Mathematical Amusements.* New York: W.H. Freeman (1983).

Green, M.W.

1964 A lower bound on Rado's sigma function for binary Turing machines. In: *Proceedings of the 5th Annual IEEE Symposium on Switching Circuit Theory and Logical Design*, pp. 91–94 (Nov. 1964).

House, R. W., and T. Rado

1963 An approach to artificial intelligence. IEEE Special Publication S-142 (Jan. 1963).

Kleene, S.C.

1952 *Introduction to Metamathematics.* Princeton: Van Nostrand (1952).

Lee, C.Y.

1963 Lecture Notes from the University of Michigan Engineering Summer Conference, Ann Arbor (1963) 18–21.

Lin, S., and T. Rado

1965 Computer studies of Turing machine problems. *J. ACM* **12** (April 1965).

Lynn, D.S.

1972 New results for Rado's sigma function for binary Turing machines. *IEEETEC C-21***8** (Aug.1972).

1974 Private communication.

Minsky, M.

1962 Size and structure of universal Turing machines using Tag Systems, Recursive Function Theory. *Sym. P. Math.* **5** AMS (1962).

1967 Computation: Finite and Infinite Machines. Englewood Cliffs, NJ: Prentice-Hall (1967).

Rado, T.

1962 On non-computable functions. *Bell Sys. T.* (May 1962) 877–884.

1963 On a simple source for non-computable functions. In: *Proceedings of the Symposium on Mathematical Theory of Automata, Polytechnic Institute of Brooklyn,* pp. 75–81 (April 1963).

Shannon, C.E.

1956 A universal Turing machine with two internal states. In: *Automata Studies,* eds. C. Shannon and J. McCarthy. Princeton: Annals of Math. Studies (1956).

Turing, A.M.

1936-7 On computable numbers with an application to the Entscheidungsproblem. *P. Lond. Math. Soc. (2)* **42** (1936-7) 230–265.

Varshavsky, V.I.

1972 Some effects in the collective behavior of automata. *Mach. Intell.* **7** (1972) 389–403.

Vitanyi, P.M.B.

1976 On a problem in the collective behavior of automata. *Discr. Math.* **14** (1976) 99–101.

Uhing, G.

1986 Unpublished notes, dated Feb. 4. (also in A.K. Dewdney, Mathematical recreations, *Sci. Am.* **252** (April 1985) 20–30.)

An Algebraic Equation for
the Halting Probability

Gregory J. Chaitin

Abstract. We outline our construction of a single equation involving only addition, multiplication, and exponentiation of nonnegative integer constants and variables with the following remarkable property. One of the variables is considered to be a parameter. Take the parameter to be $0, 1, 2, \ldots$ obtaining an infinite series of equations from the original one. Consider the question of whether each of the derived equations has finitely or infinitely many nonnegative integer solutions. The original equation is constructed in such a manner that the answers to these questions about the derived equations are independent mathematical facts that cannot be compressed into any finite set of axioms. To produce this equation, we start with a universal Turing machine in the form of the LISP universal function EVAL written as a register machine program about 300 lines long. Then we "compile" this register machine program into a universal exponential Diophantine equation. The resulting equation is about 200 pages long and has about 17,000 variables. Finally, we substitute for the program variable in the universal Diophantine equation the Gödel number of a LISP program for Ω, the halting probability of a universal Turing machine if n-bit programs have measure 2^{-n}. Full details appear in a book.[1]

More than half a century has passed since the famous papers of Gödel *1931* and Turing *1936-7* that shed so much light on the foundations of mathematics, and that simultaneously promulgated mathematical formalisms for specifying algorithms, in one case via primitive recursive function definitions, and in the other case via Turing machines. The development of computer hardware and software technology during this period has been phenomenal, and as a result we now know much better how to do the high-level functional programming of Gödel, and how to do the low-level machine language programming found in Turing's paper. And we can actually run our programs on machines and debug them, which Gödel and Turing could not do.

I believe that the best way to actually program a universal Turing machine is John McCarthy's universal function EVAL. In his *1960* McCarthy proposed LISP as a new mathematical foundation for the theory of computation. But by a quirk

---

1 This article is the introduction of the book G.J. Chaitin, *Algorithmic Information Theory,* copyright © 1987 by Cambridge University Press, and is reprinted by permission.

of fate LISP has largely been ignored by theoreticians and has instead become the standard programming language for work on artificial intelligence. I believe that pure LISP is in precisely the same role in computational mathematics that set theory is in theoretical mathematics, in that it provides a beautifully elegant and extremely powerful formalism which enables concepts such as that of numbers and functions to be defined from a handful of more primitive notions.

Simultaneously there have been profound theoretical advances. Gödel and Turing's fundamental undecidable proposition, the question of whether an algorithm ever halts, is equivalent to the question of whether it ever produces any output. In another paper (Chaitin *1987*) I have shown that much more devastating undecidable propositions arise if one asks whether an algorithm produces an infinite amount of output or not.

Gödel expended much effort to express his undecidable proposition as an arithmetical fact. Here too there has been considerable progress. In my opinion the most beautiful proof is the recent one of Jones and Matijasevic (*1984*), based on three simple ideas:

1. the observation that $11^0 = 1$, $11^1 = 11$, $11^2 = 121$, $11^3 = 1331$, $11^4 = 14641$ reproduces Pascal's triangle, makes it possible to express binomial coefficients as the digits of powers of 11 written in high enough bases;
2. an appreciation of E. Lucas's hundred-year-old remarkable theorem that the binomial coefficient $\binom{n}{k}$ is odd if and only if each bit in the base-two numeral for k implies the corresponding bit in the base-two numeral for n;
3. the idea of using register machines rather than Turing machines, and of encoding computational histories via variables which are vectors giving the contents of a register as a function of time.

Their work gives a simple straight-forward proof, using almost no number theory, that there is an exponential Diophantine equation with one parameter p which has a solution if and only if the p-th computer program (i.e., the program with Gödel number p) ever halts. Similarly, one can use their method to arithmetize my undecidable proposition. The result is an exponential Diophantine equation with the parameter n and the property that it has infinitely many solutions if and only if the n-th bit of Ω is a 1. Here Ω is the halting probability of a universal Turing machine if an n-bit program has measure 2^{-n} (Chaitin *1985, 1986*). Ω is an algorithmically random real number in the sense that the first N bits of the base-two expansion of Ω cannot be compressed into a program shorter than N bits, from which it follows that the successive bits of Ω cannot be distinguished from the result of independent tosses of a fair coin. It can also be shown that an N-bit program cannot calculate the positions and values of more than N scattered bits of Ω, not just the first N bits (Chaitin *1987*). This implies that there are exponential Diophantine equations with one parameter n which have the property that no formal axiomatic theory can enable one to settle whether the number of solutions of the equation is finite or infinite for more than a finite number of values of the parameter n.

What is gained by asking if there are infinitely many solutions rather than whether or not a solution exists? The question of whether or not an exponential Diophantine equation has a solution is in general undecidable, but the answers to such questions are not independent. Indeed, if one considers such an equation with one parameter k, and asks whether or not there is a solution for $k = 0, 1, 2, \ldots, N - 1$, the N answers to these N questions really only constitute $\log_2 N$ bits of information. The reason for this is that we can in principle determine which equations have a solution if we know how many of them are solvable, for the set of solutions and of solvable equations is r.e. On the other hand, if we ask whether the number of solutions is finite or infinite, then the answers can be independent, if the equation is constructed properly.

In view of the philosophical impact of exhibiting an algebraic equation with the property that the number of solutions jumps from finite to infinite at random as a parameter is varied, I have taken the trouble of explicitly carrying out the construction outlined by Jones and Matijasevic. That is to say, I have encoded the halting probability Ω into an exponential Diophantine equation. To be able to actually do this, one has to start with a program for calculating Ω, and the only language I can think of in which actually writing such a program would not be an excruciating task is pure LISP. It is in fact necessary to go beyond the ideas of McCarthy in three fundamental ways:

1. First of all, we simplify LISP by only allowing atoms to be one character long. (This is similar to McCarthy's "linear LISP".)
2. Secondly, EVAL must not lose control by going into an infinite loop. In other words, we need a safe EVAL that can execute garbage for a limited amount of time, and always results in an error message or a valid value of an expression. This is similar to the notion in modern operating systems that the supervisor should be able to give a user task a time slice of CPU, and that the supervisor should not abort if the user task has an abnormal error termination.
3. Lastly, in order to program such a safe time-limited EVAL, it greatly simplifies matters if we stipulate "permissive" LISP semantics with the property that the only way a syntactically valid LISP expression can fail to have a value is if it loops forever. Thus, for example, the head (CAR) and tail (CDR) of an atom is defined to be the atom itself, and the value of an unbound variable is the variable.

Proceeding in this spirit, we have defined a class of abstract computers which, as in Jones and Matijasevic's treatment, are register machines. However, our machine's finite set of registers each contain a LISP S-expression in the form of a character string with balanced left and right parentheses to delimit the list structure. And we use a small set of machine instructions, instructions for testing, moving, erasing, and setting one character at a time. In order to be able to use subroutines more effectively, we have also added an instruction for jumping to a subroutine after putting into a register the return address, and an indirect branch instruction for returning to the address contained in a register. The complete register machine program for a safe time-limited LISP universal function (interpreter) EVAL is about 300 instructions long. To test this LISP interpreter written for an abstract machine, we have written in 370 machine language a register machine simulator. We have also rewritten this LISP interpreter directly in 370 machine language, representing LISP *S*-expressions by binary trees of pointers rather than as character strings, in the standard manner used in practical LISP implementations. We have then run a large suite of tests through the very slow interpreter on the simulated register machine, and also through the extremely fast 370 machine language interpreter, in order to make sure that identical results are produced by both implementations of the LISP interpreter.

Our version of pure LISP also has the property that in it we can write a short program to calculate Ω in the limit from below. The program for calculating Ω is only a few pages long, and by running it (on the 370 directly, not on the register machine!), we have obtained a lower bound of $127/128$-ths for the particular definition of Ω we have chosen, which depends on our choice of a self-delimiting universal computer.

The final step was to write a compiler that compiles a register machine program into an exponential Diophantine equation. This compiler consists of about 700 lines of code in a very nice and easy to use programming language invented by Mike Cowlishaw called REXX (Cowlishaw *1985*). REXX is a pattern-matching string processing language which is implemented by means of a very efficient interpreter. It takes the compiler only a few minutes to convert the 300-line LISP interpreter into a 200-page 17,000-variable universal exponential Diophantine equation. The resulting equation is a little large, but the ideas used to produce it are simple and few, and the equation results from the straight-forward application of these ideas.

I have published the details of this adventure (but not the full equation!) as a book (Chaitin *1987a*). My hope is that this book will convince mathematicians that randomness not only occurs in nonlinear dynamics and quantum mechanics, but that it even happens in rather elementary branches of number theory.

References

Chaitin, G.J.

    1985 Randomness and Gödel's theorem. *Report RC 11582.* Yorktown Heights, NY: IBM Watson Research Center (1985).

    1986 Information-theoretic computational complexity *and* Gödel's theorem and information. In: *New Directions in the Philosophy of Mathematics,* ed. T. Tymoczko. Boston: Birkhäuser (1986).

1987 Incompleteness theorems for random reals. *Adv. App. Math.* **8** (1987) 119–146.

1987a *Algorithmic Information Theory.* Cambridge, England: Cambridge University Press (1987).

1987b *Information, Randomness and Incompleteness.* Singapore: World Scientific (1987).

Cowlishaw, M.F.

1985 *The REXX Language.* Englewood Cliffs, NJ: Prentice-Hall (1985).

Gödel, K.

1931 On formally undecidable propositions of *Principia mathematica* and related systems I. In: *Kurt Gödel: Collected Works,* vol. I: Publications 1929–1936, ed. S. Feferman. New York: Oxford University Press (1986).

Jones, J.P., and Y.V. Matijasevic

1984 Register machine proof of the theorem on exponential Diophantine representation of enumerable sets. *J. Symb. Log.* **49** (1984) 818–829.

McCarthy, J.

1960 Recursive functions of symbolic expressions and their computation by machine, Part I. *Comm. ACM* **3** (1960) 184–195.

Turing, A.M.

1936-7 On computable numbers, with an application to the Entscheidungsproblem. *P. Lond. Math. Soc. (2)* **42** (1936-7) 230–265; A correction, ibid. **43** (1937) 544–546; reprinted in: *The Undecidable,* ed. M. Davis. Hewlett, NY: Raven Press (1965).

The Price of Programmability

Michael Conrad

1. Introduction

Programmability and computational efficiency are fundamental attributes of computing systems. A third attribute is evolutionary adaptability, the ability of a system to
self-organize through a variation and selection process. The author has previously
proposed that these three attributes of computing are linked by a trade-off principle,
which may be roughly stated thus: *a computing system cannot at the same time
have high programmability, high computational efficiency, and high evolutionary
adaptability* (e.g., Conrad *1972, 1974, 1985*). The purpose of the present paper is
to outline the reasons for the trade-off principle in a manner which, though not
entirely formal, is sufficiently detailed to allow for a well-defined formulation. We
also consider the implications of the principle, first for alternative computer architectures suited to solving problems by methods of evolutionary search and second, for
limits on the capacity of programmable machines to simulate nature and duplicate
intelligence.

2. Programmability

Programmability is difficult to define outside of the context of a specific formalism,
such as Turing machines, recursive functions, or high level computer languages. We
can proceed by defining the concept of a program in terms of a specific formalism,
and rely on the Turing-Church thesis to argue that this definition has an equivalent in
every other formalism. Alternatively, we can make metaobservations about programs
in all these formalisms. In particular we note that programs are rules, or maps, that
feature finiteness of the number of symbols employed, discrete differences among the
symbols, and, in the action of the rule, discreteness in time. The classical definition
of a finite automaton has all three properties, and we can regard the next state

function and output function as comprising the program of the finite automaton. Programmability is the ability to prescriptively communicate a program to an actual system.

Definition 2.1 (Turing machine program): *We recall that a Turing machine is a finite automaton along with a potentially infinite memory tape which it can move and mark. A finite automaton may be defined as a quintuple $< X, Y, Z, \lambda, \delta >$, where X is a finite set of inputs (tape symbols), Y is a finite set of states, Z is a finite set of outputs (tape symbols and moves), λ is the next state function $(X \times Y \rightarrow Y)$, δ is the output function $(X \times Y \rightarrow Z)$, and we assume a discrete time scale. The next state and output functions (the transition functions) comprise the program of the Turing machine (or of the finite automaton).*

Definition 2.2 (simulation): *System S_1 will be said to simulate system S_2 if it executes the same transition function from initial state to final state and output. We do not require that all aspects of S_2's states be represented in S_1's states and we make no assumption about the fidelity of intermediate states. S_1 and S_2 are weakly equivalent if they simulate each other (in this case all aspects of the state description must be included). They would be strongly equivalent if the intermediate states were faithful, that is, if the simulation would hold at every time step. We will say that a class of systems $[S_1]$ is equivalent to a class of systems $[S_2]$ if all the systems in $[S_2]$ can be simulated by some system in $[S_1]$ and conversely. The simulation will be called exact if the final state of S_2 can be precisely ascertained from S_1 and will be called approximate if it can be ascertained to an adequate approximation (any reasonable definition of "adequate" will do).*

Rules (or maps) that generate or describe the behavior of a system may be divided into finite and infinite types. A rule is of the finite type if it can be decomposed into a number of types of distinct operations (to be called primitive operations) that are applied to a finite set of distinct symbols (called primitive symbols). According to the Turing-Church thesis all such finite-type rules are equivalent to Turing machine programs in the strict sense that each of the primitive rules from which they are built can be simulated exactly (that is, without approximation) by Turing machines. The converse is also true if the simulating system is universal. Interpreted in its strongest form, the Turing-Church thesis also asserts that, as far as presently known, all physically realizable maps can in effect be computed by Turing machines, that is, the systems described by these maps can be simulated to an arbitrarily high degree of precision (assuming constraints on space and time bounds are ignored). Thus it should be possible to associate any infinite-type map that is physically realizable, such as the continuous maps of physical mechanics, with rules of the finite type. This distinction motivates the following definition.

Definition 2.3 (program): *A program in the strict sense is a rule that (when embodied) generates the behavior of a system, subject to the condition that the rule is of the finite type. This generalization of Definition 2.1 is justified by the assumption that each distinct element of any finite-type rule can be exactly defined in terms of the Turing machine programs. The maps describing the behavior of systems not of the finite type will not be called programs, but they may be implicitly associated with programs of the finite type (again by virtue of Turing-Church). These associated programs will be called implicit programs.*

We note that the concept of computation can be generalized from the well-defined concept of Turing machine computation in a similar manner. To the extent that all physically realizable dynamics can be simulated by Turing machines it is possible to assign an equivalent amount of Turing machine computation to their behavior. For example, it is possible to use digital simulation to express the amount of *equivalent* digital computation performed by, say, a turbulent fluid. The precise amount of equivalent computational work will depend on the model of computation used for the simulating system and on the efficiency with which the time and processor resources of the simulating system are used. The objection may be raised that computation should be defined in terms of purposes, such as the solution of problems. However, it is clear that Turing machine behavior may serve no useful purpose, aside from generating the particular behavior in question. Moreover, arbitrary continuous dynamical processes could conceivably be incorporated into computing machines as primitive operations in a manner that enhances the utility of these systems for solving problems. Turing machines and models of digital machines generally are particular models of computation. They are particularly useful as reference points for evaluating the amount of computational work performed by arbitrary dynamical systems, not as delimiting the class of behaviors admitted to be forms of computing.

Having defined the concept of program we are in a position to define programmability, and in particular the important concept of structural programmability.

Definition 2.4 (effective programmability): *A real system is (effectively) programmable if it is possible to communicate desired programs to it in an exact manner (without approximation) using a finite set of primitive operations and symbols.*

The finiteness of the set of primitive operations and symbols is tantamount to the requirement that the user's manual be finite. This is what allows the programmer to communicate the rules he conceives exactly, that is, to exert complete control over the rules governing the machine. It is undoubtedly possible to omit the subjective state of the programmer, what he desires to impart to the computer, from the definition. However, our intuitive concept of programmability is strongly connected to our ability to express algorithms directly as programs, using computer *languages*. I have used the term "effective" to emphasize the completeness of the control exerted by the programmer. This sense of complete control is of course also subjective. In fact, it is impossible in general to prove that programs express desired algorithms, and as programs become large it is inevitable that they will be in some measure incorrect (see Avizienis *1983*). A concept of approximate programmability might fit real-world phenomena better than effective, or exact, programmability. However, the concept of exact programmability fits better to the objectives of the majority of programmers.

Programmability can be achieved in three ways. The first is to construct an interpreter, or universal algorithm, that can read and follow any particular program. It is necessary to construct a real system capable of executing this algorithm. The second way is to build a physical realization of the formal system – that is, a machine whose elementary operations and states match the primitive operations and symbols of a programming language. This provides a mechanism for communicating a universal rule to a physical machine. The third way is to use a compiler, or algorithm, which converts the primitive operations and symbols of a psychologically convenient language into primitive operations and symbols that match the operations and components of the machine. The latter two methods are based on the fact that the program, either directly or indirectly, can be explicitly represented in the structure of the physical system, that is, in the state settings and connectivity of its components.

Definition 2.5 (structural programmability): *A physical system is structurally programmable if it is effectively programmable and if its program is mapped by its structure.*

Effective programmability does not necessarily entail structural programmability. In fact it is possible to show that a physical system can be structurally nonprogrammable and at the same time effectively programmable (Conrad *1974a*). To do so it is necessary to construct a structurally nonprogrammable system that is computation universal. If it is computation universal it must be able to support an interpreter. However, all present-day artificial computers are structurally programmable.

Structural programmability may be viewed in terms of the two branches of automata theory – behavior and structure. Behavior theory deals with the capabilities of machines from either a state or language point of view, while structure theory is concerned with either synthesizing systems from primitive components or analyzing them into primitive components. These two branches are logically independent. A combination lock is a finite automaton, but it is not ordinarily decomposable into a base set of elementary-type components that can be reconfigured to simulate an arbitrary physical system. As a consequence it is not structurally programmable, and in this case it is effectively programmable only in the limited sense that its state can be set for achieving a limited class of behaviors. A digital computer used to simulate a combination lock is structurally programmable since the behavior is achieved by synthesizing it from a canonical set of primitive switching components. We shall see that it is usually necessary to pay for this decomposability and universality in terms of the potential efficiency with which the material resources of the system are utilized.

3. Programmability and Efficiency

Computational complexity usually refers to the number of time and processor resources required to solve a problem (Kuck *1978*). I will use the term computational efficiency to refer to the effectiveness with which resources are recruited for solving problems. To compare the ultimate problem-solving power of widely different types of systems, including systems that are structurally nonprogrammable, it is convenient to take the number of processors as the number of particles in the system. We could also consider the costs of composing these processors into an integrated system; however, this cost has negligible impact on our analysis and is in any case highly contingent upon the broader support systems available. Furthermore, we will for simplicity focus on number of particles rather than time, considering the comparative potential computational power of systems over equal intervals of time.

Particles contribute to computation only insofar as they interact with one another. If more interactions are recruited for computing, fewer particles will be necessary. As a consequence we define efficiency in terms of interactions rather than particles, using the laws of physics to translate, when necessary, from number of interactions to number of particles.

Definition 3.1 (computational efficiency and computational complexity): *The computational efficiency of a computing system (denoted by ε) is the number of interactions used for computation relative to the maximum number possible in a system consisting of the same number of particles. The computational power of a computing system is measured by $P = n\varepsilon$, where n is the number of particles it contains. The computational complexity of a problem is the minimum number of resources required to solve it (for the present purposes expressed in terms of number of particles, time, or both).*

According to current force laws, n^2 interactions occur in a system of n particles. A digital computer simulating such a system must calculate all these interactions. In practice the number of interactions among the particles in the digital computer is very much less than n. If the computer is effectively programmable it must operate in a sequential mode, otherwise unanticipated conflicts would always be possible. Roughly speaking a computer consisting of n processors can support at most an n-fold speed-up (e.g., Arbib *1969*). The underlying assumption here is that the definition of a processor in the user's manual should be scale invariant – it should not change as more processors (therefore more particles) are added to the machine. The amount of information processing carried out by a physical system freed from the constraints necessary to support programmability is thus potentially much greater than the potential information processing performed by a system not so constrained.

It may be objected that the nonprogrammable system is not demonstrably processing information. However, we can recall (Section 2) that an equivalence can be established between computing and arbitrary physical dynamics. A digital computer that simulates a nonprogrammable system, such as a turbulent fluid or a DNA molecule, is indubitably processing information. The difference between the digital computer and the simulated system is that the former has a universal simulation capability, whereas the latter may have a limited simulation capability. Restricting the potential number of interactions that can contribute to computing can be justified only if the concept of computing is tied to a particular class of machines. If the equivalence between computing and dynamics is admitted, then n^2 must be the upper limit of interactions that can contribute to computing (Conrad *1984*).

Many constraints are pertinent to the actual computing power achieved by both structurally programmable and structurally nonprogrammable systems. Some of these constraints are connected with restrictions on static and dynamic degrees of freedom of the system (Pattee *1973*). Some are connected with constants of nature, such as the speed of light, Planck's constant, and Boltzmann's constant. When a computing system becomes sufficiently large, the speed of light should be taken into account. Efficiency, ε, would then assume a dependence on size. However, the following model provides a useful first approximation to the comparative computing potential of structurally programmable and structurally nonprogrammable systems (see also Conrad and Hastings *1985*).

Consider a structurally programmable system in which the processors (including both switches and wires) behave according to the specifications in a finite user's manual. As before the system consists of n particles ($1 \leq k \leq n$). It is straightforward to specialize to the probably unrealistic limit in which each manual-defined processor is a single particle. For simplicity we will assume particles are added k at a time (that is, in integer units of processors). The number of interactions that can contribute to computing then increases with the number of particles by at most $C(n/k)k^2$, where n/k is the number of processors and C is a constant representing the number of potential contacts that a processor can have and nevertheless operate according to its definition in the user's manual. As k increases the potential contribution of the processor to computing increases. Usually only a small fraction of the k^2 interactions in a processor will be utilized, however. Furthermore, the factor Cn/k is an upper estimate since effective programmability is in general lost if all processors are used at once even if the machine is effectively programmable when operated in sequential mode. The efficiency of structurally programmable systems thus scales at most as

$\varepsilon = Cnk/n^2 = Ck/n$, and, in the limit of each particle being a manual-defined processor, as $\varepsilon = C/n$. This means that the efficiency of structurally programmable systems decreases as the number of particles in the system increases.

By contrast, structurally nonprogrammable systems can achieve an efficiency of $\varepsilon = 1$. This is the case when none of the interactions are suppressed. However, in this case the system has no flexibility other than that associated with the choice of initial conditions. If constraints are imposed on it to tailor its behavior for a particular task this must reduce the number of interactions. Maximum flexibility is achieved when the allowable number of variations is greatest. According to the binomial theorem this occurs when the number of interactions is $n^2/2$ (Harary and Palmer *1973*). The efficiency of structurally nonprogrammable systems constrained for maximum evolutionary flexibility thus scales as 1/2, independent of the size of the system.

The situation can be summed up in the following definition and theorem:

Definition 3.2 (evolutionary flexibility): *The evolutionary flexibility of a system is the number of possible variations on the pattern of interactions among its constituent particles compatible with the constraints that define the class of systems to which it belongs.*

Theorem 3.1 (efficiency versus programmability and evolutionary flexibility): *The efficiency of a system constrained for structural programmability scales at most as Ck/n, where n is the number of particles, k is the number of particles per manual-defined processor, and C is a constant. The efficiency is less if the system is further constrained to run in a sequential mode, in general a requirement for effective programmability. The efficiency of a structurally nonprogrammable system organized for maximum evolutionary flexibility scales as 1/2, independent of the number of particles it contains. Thus the efficiency of structurally nonprogrammable systems is potentially $n/2Ck$ larger than that of structurally programmable systems when the former are tuned for maximum flexibility and the latter are run in the parallel, least programmable mode. The advantage of structural nonprogrammability increases still further when the programmable systems are run in the sequential mode, and when any of the k^2 interactions with a processor are suppressed. The potential computational power of a structurally programmable system consisting of n particles is a constant, Ck, as compared to n/2 for a maximally flexible, structurally nonprogrammable system.*

To complete the picture, we compare these computational capabilities to the computational resources required for solving problems. We recall that problems are commonly divided into two broad categories, those with polynomial and those with exponential growth rates (Garey and Johnson *1979*). We take polynomial-type problems to be those in which the number of computer resources, here the number of particles, required to solve the problem grow as a small polynomial function of problem size – say as n^2, where n is a measure of problem size. The resources required to solve an exponential-type problem increases combinatorially, say as 2^n.

If the size of a structurally programmable system is increased by 10^{10} and if all of its resources could be recruited in parallel with perfect efficiency, it could potentially increase the size of n^2 type problems it could handle by a factor of 10^5. This is if we consider the computing resource to be the number of manual-defined processors, as is usually done. If we consider the resource to be the number of interactions, we could conceivably take advantage of some of the contacts among processors and some of the interactions within them. Even so, the number of interactions, Ckn, scales as the number of processors, and in the limiting case in which all particles are processors $C = k = 1$. In structurally nonprogrammable systems, however, the number of interactions increases much faster than the number of processors, in fact as the square of this number when the processors are identified with particles. Thus the increase in the computing potential of a structurally nonprogrammable system is in this case 10^5 times better than the increase in computing potential of a structurally programmable system run in a parallel mode. Summing up we have

Theorem 3.2 (structural programmability versus computational complexity): *The size of an n^2-type problem that can be solved by a structurally programmable system consisting of n particles increases by a factor that scales at most as $n^{1/2}$, even if all interactions in the system can be brought to bear. By contrast, the size of the problem solvable by a structurally nonprogrammable system with maximum evolutionary flexibility increases by at most $n/\sqrt{2}$.*

Theorem 3.2 implies that structurally programmable systems cannot keep up with polynomial growth rates in problem size, whereas structurally nonprogrammable systems can in principle keep pace with these growth rates. So far as is known, no system can keep up with exponential growth rates. If a physical system could keep up with exponential growth rates we would have to give up the idea that it is simulatable by digital computers in polynomial time for even the smallest time slices. If it were so simulatable we could solve these exponential time problems in polynomial time by a constant factor speed-up of the digital computer, contradicting the assumption that the problem is of the exponential type. It is of course possible that such nonsimulatable systems exist. This would not contradict the Turing-Church thesis, however. In its strongest form Turing-Church requires all physical processes to be simulatable; but it places no polynomial growth limitation on the number of computational resources required for the simulation.

4. Evolvability and Gradualism

We now investigate necessary and sufficient conditions for evolutionary adaptability. Evolutionary flexibility and gradual transformability (to be defined below) play an important role. In the next section we show that the problem of ascertaining whether a structurally programmable system is gradually transformable is in general unsolvable and that the class of structural changes for which the problem is solvable is much greater for structurally nonprogrammable systems. As a consequence programmable systems are not as effectively structured for evolution as nonprogrammable systems.

Definition 4.1 (evolutionary adaptability): *A system is evolutionarily adaptable (or evolvable) to the extent that it can utilize variation and selection mechanisms to survive in uncertain or unknown environments. Survival means continuation of function at an acceptable level. Evolutionary adaptability may be operationally measured by the uncertainty of the most uncertain environment that a system can tolerate on the basis of variation and selection mechanisms alone. (Entropy measures on the environment transitions, treated as Markov chains, are suitable for this purpose. See Khinchin 1957; Conrad 1983).*

Definition 4.2 (gradual transformability): *A system is gradually transformable if it undergoes small changes in behavior in response to at least one elementary (or noncompound) structure change. The behavior of two systems differs by at most a small amount if they solve approximately the same class of problems, for example, can perform adequately in approximately the same set of environments. The appropriate criteria for two classes of problems to be approximately the same depends on the cost of failure. The weakest reasonable definition is that all systems which perform different but defined computations differ by a small amount as long as their computations terminate. We will say that two systems whose behavior differs from each other in this weak sense of small are weakly transformable into one another if they can be transformed into one another by a sequence of elementary structural changes. Weak transformability can be taken as a necessary condition for gradual transformability.*

In order for a system to be evolvable it must be capable of accepting at least one structural change. A system accepts a structural change if its performance improves or if it is capable of lasting long enough to accept another change, eventually leading to an improvement. A system remains evolvable as long as it satisfies this threshold condition. The reasoning here is that the probability of undergoing a transition to an acceptable structural form becomes negligibly small if simultaneous structural changes (e.g., mutations) are required. If p is the probability of a structural change, the time required for evolution scales as $1/Ap^m$, where m is the number of simultaneous changes that must occur and A is the number of systems in the population. By contrast the time required for evolution to proceed through m single step mutations scales as m/Ap. The evolution time becomes unacceptably large for all values of $m > 1$.

For example, if $p = 10^{-8}$, which is actually large for a biological mutation rate, the rate of evolution would be 10^8 times slower when $m = 2$ than when $m = 1$ for at least one of the possible mutations. Large values of p are not reasonable since this increases the chance of unfavorable structural changes canceling out favorable ones. Detailed calculations which take into account the effects of such canceling mutations, the relative advantage of improved forms, and the number of single step mutations required have been presented elsewhere (Conrad *1983*). These factors, and also the population size, have a negligible impact on the overall picture. Evolution is unfeasible on any reasonable time scale if the evolutionary system is structured in such a way that double or simultaneous structural changes are required for the appearance of an improved form.

The threshold condition can be put into a neat form, analogous to a threshold condition for evolution formulated by Maynard-Smith *1970*. Let N denote the number of single event structural alterations that a system can undergo. In proteins N denotes the number of single mutations or single crossover events. In a computer program N denotes the number of single alterations of code, at any level of hierarchical organization. Let f denote the fraction of these alterations that are acceptable. In order for a system to evolve it must satisfy the threshold condition $fN \geq 1$.

A slight alteration in a system is more likely to be acceptable if it leads to a gradual modification in its behavior. Radically modified behavior might or might not be acceptable, whereas gradually modified behavior is intrinsically acceptable since it entails only small improvement or degradation of performance. Whether gradual transformability is a necessary and sufficient condition for an evolutionarily flexible system to be evolvable or just a sufficient condition depends on the definition of smallness and on the selection pressures imposed. It is possible, though highly implausible, that in some cases similarly structured systems with radically different behaviors could perform satisfactorily in the same environment. However, it can hardly be expected that this will be the case if the difference between the systems is so great that one gives a defined computation and the other does not. Weak transformability can thus be taken as a necessary condition for evolutionary adaptability under the reasonable assumption that mutations from defined to undefined computations are unacceptable.

By itself, however, gradualism does not assure evolutionary adaptability. The gradual transformability must be capable of generating a variety of useful forms. Thus evolutionary adaptability also depends on evolutionary flexibility.

We can summarize the necessary and sufficient conditions for evolutionary adaptability in the following:

Lemma 4.1 (conditions for evolutionary adaptability): *Evolutionary flexibility is a necessary condition for evolutionary adaptability. Gradual transformability is a sufficient condition for an evolutionarily flexible system to be evolutionarily adaptable, and weak gradual transformability is a necessary condition. Gradual transformability increases sharply when $fN \geq 1$, where N is the number of single event structural alterations and f is the fraction of acceptable alterations. It increases moderately with increases in f that push the system further above this threshold condition and decreases sharply when the system falls below this threshold condition.*

Structurally nonprogrammable systems have much higher evolutionary flexibility than structurally programmable systems, assuming that the structural alterations of the latter are constrained to be compatible with structural programmability. The evolutionary flexibility could nevertheless be adequate for performing a wide variety of tasks if the problem of discovering the allowable variants through an evolution process is ignored. Gradual transformability is essential for such discovery, however.

The gradualism property is most likely to be present in systems if they are highly parameterized and highly redundant. It is usually possible to change the parameters of a program over some range without radically altering the execution sequence. The incorporation of continuous dynamics increases parameterization since this increases a system's amenability to continuous deformation. A mouse and an elephant probably share the same rule of development, so far as the overall order of events is concerned, but differ by stretching and compression of events. Gradualism is also increased if it is possible to insert or delete rules that are independent so far as the operation of other rules are concerned, as in a production system type knowledge base of an expert system.

Redundancy serves to buffer the effect of structural change on behavior. Redundancy, taken by itself, confers fault tolerance rather than transformability (Dal Cin *1979*). However, it can serve to enhance transformability in the presence of parameterization and continuous dynamical features. Redundant, noncritical components can serve to absorb some of the impact of a structural change, thereby modulating its effect on components critical to the behavior of the system (Conrad *1983*).

Evolutionary systems can be associated with an adaptive surface which assigns a performance measure to each state of the system. If an evolutionary system fails to satisfy the threshold condition $fN \geq 1$, it becomes trapped on a particular peak on this surface. A system comprising a given number of particles could never be organized to definitely exclude such trapping since whether or not a mutation leads to a small change in behavior depends on the pressures of selection, which in turn changes as the organization of the system changes. However, redundant features can always be added in a stepwise manner without significantly degrading the performance of the system. Mutations of this type serve as f-enhancing mutations. The time required to reach a higher adaptive peak through a series of u f-enhancing mutations followed by m fitness-increasing mutations scales as $(u + m)/Ap$, which is very much faster than $1/Ap^m$, the time required for a double mutation, even when u is large. In an evolutionary system such f-enhancing mutations can hitchhike along with the fitness increasing traits whose evolution they facilitate. The addition of redundancy increases the dimensionality of the space and by doing so opens up *extradimensional* bypasses to higher adaptive peaks (Conrad *1979*).

5. *Programmability and Evolvability*

We now show that structurally programmable systems do not in general meet the requirement for gradual transformability.

Theorem 5.1 (gradual transformability problem): *The problem of ascertaining whether a structurally programmable system will undergo a small change in behavior in response to either single or multiple changes in its structure is unsolvable. It is in general impossible to put any reasonable metric on the amount of behavioral change that is likely to result from a structural change.*

The proof is similar to that of the unsolvability of the halting problem (Turing *1936-7*). Suppose that it is in fact possible to write a program, U, to solve the gradual transformability problem. It is sufficient (as in Definition 4.1) to take the behavior of two systems, P and P', as differing by at most a gradual transformation if they both give defined computations (that is, eventually come to a halt state) and as differing by an unacceptably large amount if one system gives a defined computation and the other does not. We are free to take for P any program that halts. The program U must go to a halt state at least when P' does not. We would then certainly get an answer to the gradual transformability problem since P' would halt if it is similar to P and U would halt if it is not. Suppose that U is itself P'. Then U halts if it does not halt. Since this is a contradiction, the assumption that U is a possible program must be incorrect.

Note that the gradual transformability problem is unsolvable even for the weakest sense of transformability. If it is not possible to ascertain whether an altered system will differ by an infinite amount (as when its computation becomes undefined) it is certainly impossible to solve it when the systems differ by a smaller amount. This is why no reasonable metric can in general be put on the expected amount of structural change. As a consequence structurally programmable systems in general fail to meet both a necessary and a sufficient condition for evolutionary adaptability. For finite systems the notion of unsolvability must be replaced by that of intractability. In the absence of specializing assumptions, it is as intractable as proving program correctness. Experience with digital computers in fact suggests that the vast majority of arbitrary changes in rules lead to undefined or useless behavior.

The unsolvability (or intractability) of the gradual transformability problem in general does not mean that it is unsolvable in all particular cases. We have already observed that rules displaying features of continuity and states displaying features of redundancy allow for gradual deformation of the execution sequence (or, alternatively, realization of a given sequence by a deformed organization). The presence of continuity is incompatible with structural programmability. This is why the unsolvability of the gradual transformability problem does not preclude a reasonable metric for behavior change in structurally nonprogrammable systems. In many instances it is possible to ensure that a structurally nonprogrammable system undergoes acceptably small behavior change by "mutating" it into a higher dimensional space.

Evolution processes in nature have produced a wide variety of powerful biological information processing systems. According to the strong form of the Turing-Church thesis it should be possible to simulate these evolution processes with structurally programmable computers. That is, it should be possible to use structurally programmable computers to build virtual machines capable of supporting evolution. To do so it is necessary to invest sufficient computational resources to pay for the cost of simulating the dynamical features and redundancies that support evolution. If it is not in principle possible to simulate these mechanisms then it must be admitted that an important class of problem solving processes in nature, namely evolution processes, are not simulatable by digital computer.

The computational efficiency of a structurally programmable system is in effect reduced if some of its resources are used to simulate evolution-facilitating features. To properly formulate the relation between evolutionary adaptability and structural programmability we therefore require the following

Definition 5.1 (effective computational efficiency): *A virtual machine is a simulation built on top of a structurally programmable base machine. The effective number of interactions that a virtual machine uses for computing is the number of interactions that the base machine uses for computing minus the number of interactions used for the simulation. The effective computational efficiency of a virtual machine is the effective number of interactions it uses for computing divided by the maximum number possible in a system consisting of the same number of particles as the base machine.*

Theorem 5.1 and Lemma 4.1 combined with this definition may be expressed as

Theorem 5.2 (structural programmability vs. evolutionary adaptability): *Structurally programmable systems fail to satisfy the conditions for evolutionary adaptability when the mutations of structure occur at the level of the base machine. Structurally programmable systems can be used to achieve evolutionary adaptability through*

simulation, in which case the structural mutations occur at the level of the simulation.
Such simulated evolutionary adaptability can only be achieved at the cost of a
decrease in effective computational efficiency.

6.   The Trade-off Principle

The results obtained in the preceding sections can be summarized in

Theorem 6.1 (trade-off principle): *A computing system cannot have all of the fol-*
lowing three properties: structural programmability, high computational efficiency,
and high evolutionary adaptability. Structural programmability and high compu-
tational efficiency are always mutually exclusive. Structural programmability and
evolutionary adaptability are mutually exclusive in the region of maximum effective
computational efficiency (always less than or equal to the computational efficiency).

We may also gather together a number of more specific results:

1. The potential computational efficiency of a structurally programmable system is
   less than that of a structurally nonprogrammable system, the former decreasing
   with the number of particles (or processors) and the latter being independent
   of the number of particles. The potential computational power of a structurally
   programmable system is a constant that depends on the size of the user's manual
   (that is, on the length of the definitions of processors comprising the machine),
   whereas the potential computational power of structurally nonprogrammable systems increases as the square of the number of particles. Furthermore, if a structurally programmable system is operated in a sequential (truly programmable)
   mode its computational power decreases as the number of particles increases
   (Theorem 3.1).
2. The potential evolutionary adaptability of structurally programmable systems is
   less than that of structurally nonprogrammable systems, the former having less
   potential evolutionary flexibility and failing to meet the requirement for gradual
   transformability at the level of the base machine (Lemma 4.1 and Theorems 3.1
   and 5.2).
3. Virtual evolutionary adaptability can be exhibited by structurally programmable
   systems if these systems allocate enough of their time and processor resources to
   simulating the structure-behavior relations that allow for gradual transformability
   and evolutionary flexibility. This allocation of computational resources reduces
   effective computational efficiency (Theorem 5.2 and surrounding discussion).

The trade-off principle does not exclude the coexistence of high evolutionary
adaptability and high computational efficiency. In general evolutionary adaptabil-

ity and high computational efficiency would go together since an evolvable system could learn to use its resources efficiently. This claim could only be satisfactorily demonstrated by construction and demonstration, through computer simulation, that the desired properties are indeed achieved (see next section). However, it is likely that biological organisms provide an instantiation of this claim. As products of evolution, organisms cannot be structurally programmable. Furthermore, they appear to be highly effective for certain types of computations as compared to structurally programmable machines. Similarly, point 3 above has not been given either a constructive or analytical interpretation. However, it is possible to give a fairly general analysis of the organizational features that provide an effective substrate for evolution (Conrad *1983*) and there is nothing about these features which precludes simulation. This does not establish that evolvability comparable to that exhibited by biological systems can be achieved through simulation. It is probably impossible to definitively establish such a claim. However, if the strong form of the Turing-Church thesis is taken as axiomatic it is then possible to assert the simulatability of evolution given sufficient computational resources, or alternatively, to assert that the nonsimulatability of evolution would invalidate the strong form of Turing-Church.

7. Evolutionary Machines

Evolutionary and genetic algorithms have been used for optimization problems (Bremermann *1962*, Holland *1975*) and for adaptive pattern recognition (Conrad et al. *1987*). The trade-off principle implies that such algorithms should be conceived as consisting of two parts. The first is the variation and selection procedure per se. The second is the structure of the substrate on which the search procedure acts. This determines the structure of the adaptive surface on which the evolution is taking place. For evolutionary methods to be effective the substrate should have the evolution facilitating properties of gradualism and flexibility.

The arguments leading to the trade-off principle suggest that it may be possible to advantageously use structurally programmable systems to simulate these evolution-facilitating features. Many of the resources of structurally programmable systems are inefficiently recruited for computing when these systems are run in a programmable mode. *Representing evolution-enhancing features in structurally programmable systems can finance itself to the extent that it increases the effectiveness with which computing resources are used for problem solving.*

Consider first virtual implementations in which the base machine runs in a sequential mode. The computational effort required to simulate an evolution-facilitating substrate entails a decrease in effective computational efficiency. However, this produces a distinctive evolutionary learning capability that may open up problem domains that could not be as effectively addressed by direct use of the base machine.

Problems whose solution demands an adaptive component fall into this domain. Effective programmability is given up at the level of the virtual machine, just as it is in highly parallel machines. As a consequence such machines lend themselves to implementations in which the simulation of the dynamical processes required for evolution are run on many processors in parallel. This simulation, though costly, can be financed by the gain in computational efficiency that results from tapping otherwise dormant processors.

The architecture depicted in Figure 1 illustrates how this could work. The basic modules are cellular automata with input lines and read heads. Each cellular automaton is an array of subcells, each of which is in one of a number of possible states whose transitions are influenced only by neighboring subcells. Such arrays are capable of exhibiting highly elaborate patterns of behavior, ranging from any type of pattern that can be exhibited by partial differential equations to patterns so elaborate that they cannot be predicted in advance except through step-by-step enumeration (Wolfram *1986*). The local rules can be viewed as serving to integrate patterns of input signals in space and time. Particular subcells will thus be activated at particular times in response to different families of patterns. Some of the subcells contain read heads, while others do not. If a read head is located on a subcell that is activated it produces an output signal.

A collection of individual modules with the same cellular automaton dynamics will be called a clone. The clone is the next higher level of organization in the machine. The variation and selection algorithm acts randomly, adding or deleting read heads on each member of the clone and evaluating how well it performs a desired transformation of a training set of input signal patterns to output signal patterns. The configuration of read heads in the best performing module is copied into the other members of the clone, and the process is repeated until the training set is learned. If the local cellular automaton dynamics are such that the training set cannot be learned, then the variation and selection procedure must be applied to the cellular automaton dynamics itself. In this way a variety of differently adapted clones with the same cellular automaton dynamics are harvested after a first round of learning cycles, while clones with different cellular automaton dynamics are harvested after a second round of learning cycles.

If the cellular automaton dynamics causes different input signal patterns to activate the same subcell, it serves to aggregate them into an equivalence class. This is why the machine can generalize from a small training set of signal patterns. However, if such aggregation is admitted it is always possible to define tasks that the dynamics are incapable of performing; this is why the higher level of evolution on the dynamics is necessary. Also note that the cellular automaton dynamics can be chosen so as to be more or less structurally stable (since partial differential equations can be computed via cellular automata). As the dynamics becomes structurally more stable, it becomes more gradually transformable. The cellular automaton mod-

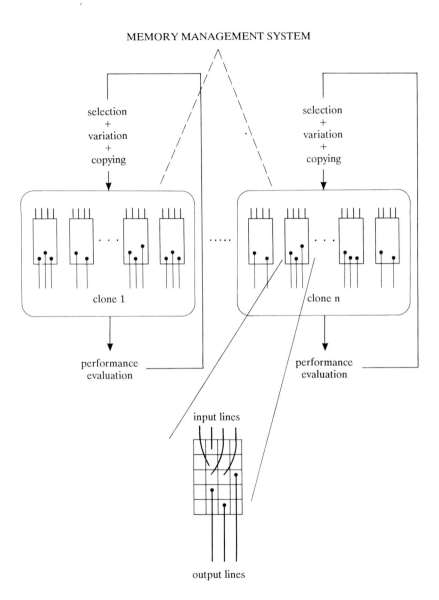

Figure 1. Evolutionary architecture.

ules thus provide both gradual transformability and evolutionary flexibility, the two properties required of a substrate suitable for evolutionary learning.

The top layer of the machine is a memory management system that puts together combinations of differently adapted modules for the performance of complex tasks. Each memory control unit, or reference module, is connected to each of the cellular automaton modules. The activation of a reference module activates a cellular automaton module if its connection to it is facilitated for signal flow. Thus each reference module controls a combination of cellular automaton modules. This organization allows for a higher level of evolutionary learning. However, it is not necessary to describe the operations of this memory level of organization here (see Conrad *1976*).

We have built simulation systems that operate according to the above principles (using modules with differential equation dynamics) and have reported on these elsewhere (Kampfer and Conrad *1983*; Kirby and Conrad *1984, 1986*; Conrad et al. *1987a*). Efficient, hardwired systems for simulating cellular automaton dynamics (cf. Toffoli *1984*) would allow for the efficient utilization of low level parallelism. Such systems are still structurally programmable since the local rules can be effectively prescribed, though in practice it is not possible to prescribe a desired global rule. Continuous analogs of cellular automaton dynamics would be truly structurally nonprogrammable. Biological organisms consisting of molecular components are truly nonprogrammable since it is not in general possible to prescribe desired local rules by means of structural mutations. Nevertheless by linking a structurally nonprogrammable collection of pattern processors together in a suitable way, it is possible to duplicate any computation that could be performed by a structurally programmable machine (Conrad *1974a*). The potential efficiency is greater since each processor can be specifically tailored for the task it performs.

Architectures of the type described could be efficiently implemented using modifications of current silicon technology. The miniaturization of conventional chips is limited by physical side effects, in particular quantum mechanical tunneling of electrons. Tunneling, however, could be exploited to achieve the type of neighbor-neighbor interactions that are necessary for cellular automaton dynamics. Electron diffusion, another phenomenon that must be suppressed in a chip engineered to realize conventional logic, is an excellent type of dynamics for modules in an evolutionary architecture. Macromolecules of the type that occur in organisms would undoubtedly provide the best substrate for evolutionary computing, though. This is due to the fact that molecules such as proteins have highly specific readout capabilities and in addition a high degree of evolutionary gradualism and flexibility.

8. Conclusion

The trade-off principle establishes a link between computing and evolution. Algorithms can be executed in structurally programmable or structurally nonprogrammable systems. The potential computational efficiency is greater in the latter case. Traditional programming, or prescriptive control over the rule obeyed, is given up. However, such systems are well suited for programming by evolutionary means. Evolution is itself a method of problem solving. This form of problem solving can be expressed in an algorithmic framework, but it is different from the algorithms usually implemented on computers in a significant respect. Evolutionary programming involves both a search procedure and a substrate on which this procedure acts. The search procedure can be expressed as an algorithm in a straightforward manner. The substrate can also be described in algorithmic terms, by means of simulation. However, the characterization of the substrate is of such immense importance for the effectiveness of evolution and so closely connected with choice of hardware, that it is probably useful to place evolutionary problem solving in a distinct category.

The recognition of the importance of substrate architecture for evolutionary computing opens up the possibilities of new classes of machines, both virtual and real, that in fundamental respects are more biology-like than present day machines. This class includes machines programmed to simulate the dynamical processes and redundancies that facilitate evolution and, as well, machines directly structured to display these features. The analysis suggests that for some problems the cost of simulating the evolution-enhancing features that make biological materials a good substrate for evolution may be outweighed by the resulting enhancement of adaptive capability and by the more effective use of parallelism.

looseness=1We can finally note that the trade-off principle has a fundamental epistemic implication. Computer scientists commonly assume that it is possible to simulate any mathematically describable process in nature with digital computers, provided that enough time and processor resources are available. It is impossible to prove this strong version of the Turing-Church thesis; but up to the present time there has been no convincing counterexample. On the surface Turing-Church suggests that our understanding of a process, such as a process of human intelligence, can always be crystallized in a computer program and then communicated to an actual machine for execution. The tacit assumption is that this programmability would not itself preclude the computational efficiency necessary for simulating the process in nature. This assumption is incompatible with the trade-off principle. Many processes in nature must be such that we cannot understand them in terms of a computer program and at the same time put our understanding to the test by running the program on a machine. Brain processes of intelligence fall into this category, since the brain is a product of evolution and thus cannot be structurally programmable. Conceivably we could evolve an artificial system to simulate brain-like intelligence; but we would then find it just as difficult to specify and test the program of this artificial system as to specify and test the program that generates the behavior of an organism.

References

Arbib, M.A.

    1969 *Theories of Abstract Automata* Englewood Cliffs, NJ: Prentice-Hall (1969).

Avizienis, A.

    1983 Framework for a taxonomy of fault tolerance attributes in computer systems. In: *IEEE Conference Proceedings of the 10th Annual International Symposium on Computer Architecture,* pp. 16–21 (1983).

Conrad, M.

    1972 Information processing in molecular systems. *Curr. M. Bio.* (now *Biosystems*) **5/1** (1972) 1–14.

    1974 Molecular automata. In: *Physics and Mathematics of the Nervous System,* eds. M. Conrad, W. Güttinger, and M. Dal Cin. Heidelberg: Springer-Verlag (1974) 419–430.

    1974a Molecular information processing in the central nervous system, part I: Selection circuits in the brain. In *Physics and Mathematics of the Nervous System,* eds. M. Conrad, W. Güttinger, and M. Dal Cin, pp. 82–107. Heidelberg: Springer-Verlag (1974).

    1976 Molecular information structures in the brain. *J. Neurosci.* **2** (1976) 233–254.

    1979 Bootstrapping on the adaptive landscape. *Biosystems* **11/2, 3** (1979) 167–180.

    1983 *Adaptability.* New York: Plenum Press (1983).

    1984 Microscopic-macroscopic interface in biological information processing. *Biosystems* **16** (1984) 345–363.

    1985 On design principles for a molecular computer. *Comm. ACM* **28** (1985) 464–480.

Conrad, M., and H.M. Hastings

    1985 Scale change and the emergence of information processing primitives. *J. Theor. Bio.* **112** (1985) 741-755.

Conrad, M., R. Kampfner, and K.B. Kirby

    1987 Simulation of a reaction-diffusion neuron which learns to recognize events (Appendix to: M. Conrad, Rapprochement of artificial intelligence and dynamics). *Eur. J. Oper.* **30** (1987) 280–290.

    1987a Neuronal dynamics and evolutionary learning. To appear in: *Advances in Cognition: Steps Toward Convergence,* eds. M. Kochen and H. Hastings. Washington, D.C.: American Association for the Advancement of Science (1986).

Dal Cin, M.

    1979 *Fehlertolerante Systeme.* Stuttgart: Teubner Studienbücher (1979).

Garey, M., and D. Johnson
1979 *Computers and Intractability.* New York: Freeman (1979).

Harary, F., and E.M. Palmer
1973 *Graphical Enumeration.* New York: Academic Press (1973).

Holland, J.H.
1975 *Adaptation in Natural and Artificial Systems.* Ann Arbor, MI: University of Michigan Press (1975).

Kampfner, R., and M. Conrad
1983 Computational modeling of evolutionary processes in the brain. *B. Math. Biol.* **45** (1983) 931–968.

Khinchin, A.I.
1957 *Mathematical Foundations of Information Theory.* New York: Dover (1957).

Kirby, K.G., and M. Conrad
1984 The enzymatic neuron as a reaction-diffusion network of cyclic nucleotides. *B. Math. Biol.* **46** (1984) 765–783.
1986 Intraneuronal dynamics as a substrate for evolutionary learning. *Physica* **22D** (1986) 205–215.

Kuck, D.
1978 *The Structure of Computers and Computations,* vol. 1. New York: John Wiley and Sons (1978).

Maynard-Smith, J.
1970 Natural selection and the concept of a protein space. *Nature* **225** (1970) 563–564.

Pattee, H.H.
1970 Physical problems of decision-making constraints. In: *Physical Principles of Neuronal and Organismic Behavior,* eds. M. Conrad and M. Magar, pp. 217–225. New York: Gordon and Breach Science Publishers (1970).

Toffoli, T.
1984 CAM: A high-performance cellular automaton machine. *Physica* **10D** (1984) 195–204.

Turing, A.M.
1936-7 On computable numbers, with an application to the Entscheidungsproblem. *P. Lond. Math. Soc. (2)* **42** (1936-7) 230–265.

Wolfram, S.
1986 Approaches to complexity engineering. *Physica* **22D** (1986) 385–399.

Gandy's Principles for Mechanisms as a Model of Parallel Computation

Elias Dahlhaus and Johann A. Makowsky

Abstract. We characterize the classes of Gandy machines which correspond to the class of feasible hardware implementable functions described by NC, the class of functions computable in parallel in polylog time by polynomially many processors.

1. Introduction

R. Gandy, in *1980*, discusses some philosophical aspects of the Church-Turing Thesis related to mechanistic realizability (hardware implementability) of computable functions. Gandy postulates four principles concerning models of computability from which, in contrast to Church's Thesis, it is provable that functions in these models are totally recursive. In other words, whatever is computable by a mechanistic machine (a Gandy machine) is also computable by a Turing machine. He also proves the minimality of those four principles in the sense that no three of them suffice to prove this result. The universe of discourse in Gandy *1980* are the hereditary finite sets with urelements (cf. Barwise *1975*), which also form the background of our Dahlhaus and Makowsky *1986, 1986a, 1987* and Shepherdson *1987*. We assume the reader is familiar with Gandy *1980* and Shepherdson *1973*.

Parallel computation has the aim of speeding up computation. Especially it makes sense to parallelize if one wants to have a real time computation. The current philosophy is that real time means that we can solve a given problem in polylog time (i.e., in time $O(\log^k n)$ where n is the length of the input). A second postulate is that we should need only a "small" number of processors, i.e., at most $O(n^l)$ where l is a constant. The complexity class of decision problems which satisfy these two requirements is called NC (Nick Pippenger's class). Details on the rationale behind this complexity class have been discussed in the pioneering paper Cook *1985*. He used uniform sequences of switching circuits as a model of parallel computation. There are other equivalent models of parallel computation, such as the hardware modification machine of Dymonds and Cook (*1980*), several models of parallel random access machines (called PRAM's), as discussed in Goldschlager *1982*, and Shepherdson's synchronous parallel computation model (Shepherdson *1973*). In Shepherdson *1988* the relationship between Gandy machines and the synchronous parallel computation model is discussed.

It is not very difficult to show that a hardware modification machine can be modeled within the framework of Gandy machines. From this it follows that every function in NC is mechanically realizable by a Gandy machine. The purpose of this paper is to determine which class of Gandy machines captures the class NC. What we actually show is that every machine satisfying Gandy's four principles and whose set of states is first order definable computes a function in NC. This answers a question stated in a side remark in Dahlhaus and Makowsky *1987*.

We would like to thank O. Goldreich, R. Herken, and J.C. Shepherdson for valuable comments on an earlier draft of this paper.

2.   *Definable Gandy Machines*

Recall that a Gandy machine is a pair $< S, F >$ where $S \subset HF$ and $F: S \to S$ is a structural state transition function satisfying the following four postulates (Gandy *1980* and Shepherdson *1987*, section 2):

Principle I: The form of description.
This principle states that a Gandy machine can be described in the language of set theory over the hereditarily finite sets with labels (urelements) in a way which is invariant under permutations of the labels.

Principle II: Limitation of hierarchy.
This principle states that the set theoretic rank of a Gandy machine is bounded.

Principle III: Limitation of hierarchy.
This principle states that each state of a Gandy machine can be uniquely reassembled from parts of bounded size.

Principle IV: Local causation.
This principle states that the next state of a Gandy machine can be uniquely reassembled from its restrictions to overlapping regions and that these restrictions are locally caused. The reassembly can additionally be performed by two functions G_1 and G_2 which are also invariant under permutations of the labels.

Gandy machines $< S, F >$ are models of crystalline automata, or more generally, of hardware components. As Gandy (*1980*, section 6(5)) notes himself, not every recursive function can be realized by an F of a Gandy machine $< S, F >$. He notes that such an F is always bounded by a function in the third level of the Grzegorczyk hierarchy. However, this is not a severe restriction, as the introduction of composition of Gandy machines together with while-statements (or any other similar control structure) allows us to compute arbitrary recursive functions over a fixed finite set of Gandy machines. On the other hand, if one really thinks of Gandy machines as true models of hardware components, then the third level of the Grzegorczyk hierarchy seems too complex to fit today's intuition. We would like to determine those Gandy machines which are feasible from a technological point of view, i.e., which capture the complexity class NC. To this end we add the following postulate:

Principle V: Definability of the set of states.
Let $TC(x) \subset HF$ be the transitive closure with respect to $\in$ of $x \in HF$. We can consider $< TC(x), \in >$ as a first order structure. We additionally require that $S = \{x \in HF :< TC(x), \in > \models \phi\}$ for some first order formula ϕ with $\in$ as the only nonlogical symbol.

Principle V cannot be justified from a mechanistic point of view. However, in most examples one might keep in mind that it can be made true by trivial modifications. We call a Gandy machine satisfying Principles I–V a *definable Gandy machine (DGM)*.

On the other hand we want to relax the notion of a Gandy machine a bit by allowing *infinite* input, i.e., $x \in S$ may be of infinite cardinality provided it is of bounded hierarchy. We call a (definable) Gandy machine with possibly infinite x an *extended (definable) Gandy machine (EGM (EDGM))*.

Theorem: *Let $< S, F >$ be an extended definable Gandy machine and $S^{\#} = S \restriction HF$. Then $F \restriction S^{\#}$ is in NC.*

Remarks: The natural question arises whether the converse of this theorem is also true, especially if one has in mind that NC corresponds to functions computable by uniformly definable sets of circuits. However, if one carries this idea through one sees easily that one cannot always satisfy Principle II. So the best one can get is the following:

Proposition: *Let f be a Boolean-valued function. Then f is in NC if and only if there is an extended definable Gandy machine such that polylog many iterated steps of it implement one application of f.*

3.   Proof of the Main Theorem

In the proof we are going to use a line of reasoning made popular by Immerman (*1983*) which exhibits a relationship between definability of finite structures in some logic (first order, first order augmented with a transitive closure operator) and complexity classes. In particular we use the fact that checking whether a finite structure satisfies a fixed first order sentence can be done in time $O(\log n)$ where n is the size of the structure. We also use the fact that over $< HF, \in >$ first order formulas with bounded quantification really speak about some finite substructure of $< HF, \in >$.

To prove our theorem we first observe some simple steps 1–2. In step 3 we use our additional definability assumption, and in 4 we complete the proof of 1.

Let $< S, F >$ be an extended Gandy machine (not necessarily definable).

1. Given any $x \in S$ we have to compute the pieces of x. Each of the pieces has a transitive closure of bounded cardinality, say q. If the transitive closure $TC(x)$ of x has cardinality n, then we have at most n^q pieces. To be a hereditary subset of any type can be expressed by a formula with bounded quantifiers. Therefore the set of all pieces of x can be computed in $O(\log n)$ time.

2. The transfer functions G_1, G_2 as required in Principle IV compute sets of certain types and have a finite domain. Therefore they can be computed in constant time. The pieces of $F(x)$ are computed using G_1. Then G_2 gives us the information how to past the supports of these pieces together. This can be done in parallel in logarithmic time.

3. The problem which remains is how to reassemble $F(x)$ from its pieces in polylog time by polynomially many processors. The first step is to check for all pieces t_1, t_2 of $F(x)$ with overlapping supports, which $u_1 \in TC(t_1)$ and $u_2 \in TC(t_2)$ are restrictions of a common $y \in TC(F(x))$. For this purpose we consider the first order structure **M** of all elements of the transitive closure of pieces of $F(x)$. Now the set of pairs (u_1, u_2) as above depends only on the first order theory of t_1 and t_2 in **M**. But by compactness considerations we can pick a finite subset of the first order theory which determines this identification. Therefore the set of pairs which are restrictions of a common element can be determined by a first order formula and is therefore in NC. Note that here we use Principle V.

4. Pasting $F(x)$ together can be done by a divide and conquer strategy. We compute a set $U(P)$ of subsets of **M** which represents the elements of $TC(F(x))$. Here P is the set of all pieces of $F(x)$. The set $U(P)$ is computed subsequently as follows:
   We divide P in nearly equal parts P_1 and P_2 and set $U_1 := U(P_1)$ and $U_2 :=$ $U(P_2)$.
   For each $u_1 \in U_1$ and $u_2 \in U_2$ we set $u_1 \cup u_2 \in U(P)$ provided that for each

$t_1 \in P_1$ and $t_2 \in P_2$ with overlapping supports and each $y_1 \in u_1 \cap TC(t_1)$, $y_2 \in u_2 \cap TC(t_2)$, y_1 and y_2 are restrictions of a common element of $F(x)$. This completes the proof of the theorem.

4. Concluding Remarks

This paper more or less describes the location of (definable) Gandy machines in the realm of complexity. Today, what is computable might not be so interesting as it was in Turing's time. Especially in connection with the simulation of physical and technical processes it is more interesting to know what is the real time computable. This paper has shown that a slight modification of Gandy's principles is a model of real time computation. We should not forget that Gandy's starting point was the simulation of crystalline processes.

As Trakhtenbrot, this volume, points out, the Church-Turing thesis can also be interpreted as postulating the interchangeability of hardware and software inside a universal Turing machine. Gandy's analysis of crystalline processes and our analysis of the complexity of functions computable by Gandy machines may then be seen is stating the limit of this interchangeability.

On the other hand this paper may give some hints on how to continue the discussion of the parallel implementation of high-level programming languages such as SETL-like languages (as initiated in Dahlhaus and Makowsky *1986a*) or LISP-like languages (as discussed in Trakhtenbrot, this volume). It seems to us that the development of high level languages with *syntax driven complexity control* is one of the most challenging projects in current software engineering. Theoretical approaches as initiated in Immerman *1983* on one side and Gandy *1980* on the other give us a clue on how this might be possible. Our work in *1986, 1986a* and here shows how this project can be continued.

References

Barwise, K.J.

    1975 *Admissible Sets and Structures: An Approach to Definability Theory.* Berlin: Springer-Verlag (1975)

Cook, S.

    1985 A taxonomy of problems with fast parallel algorithms. Inf. Contr. **64** (1985) 2–22.

Dymonds, P.W., and S. Cook

    1980 Hardware complexity and parallel computation. In: *Proceedings of the 21st Annual Conference on Foundations of Computer Science*, p. 360–372.

Dahlhaus, E., and J.A. Makowsky

    1986 Computable directory queries. In: *Proceedings of CAAP '86,* 11th Colloquium on Trees and Algebra in Programming, Nice, March 24–26, 1986, ed. P. Franchi-Zannettacci, Lecture Notes on Computer Science 214, pp. 254–265. Berlin: Springer-Verlag (1986).

  1986a The choice of programming primitives in SETL-like languages. In: *Proceedings of ESOP '86* (European Symposium on Programming, Saarbrücken, March 17–19, 1986), Lecture Notes on Computer Science 213, pp. 160–172. Berlin: Springer-Verlag (1986).

    1987 Computable directory queries. To appear in the proceedings of the workshop on "Logic and Computer Science: New Trends and Applications", Turin 1986.

Gandy, R.O.

    1980 *Church's Thesis and Principles of Mechanism, The Kleene-Symposium,* eds. J. Barwise et al., pp. 123–148. Amsterdam: North-Holland Publ. Co. (1980).

Goldschlager, L.

    1982 Synchronous parallel computation. *J. ACM* **29** (1982) 1073–1086.

Immerman, N.

    1983 Languages which capture complexity classes. 15th ACM Symposium on Theory of Computing, pp. 347–354 (1983).

Shepherdson, J.C.

    1973 Computation over abstract structures: Serial and parallel procedures and Friedman's effective definitional schemes. In: *Logic Colloquium,* eds. H.E. Rose et al., pp. 445–513. Amsterdam: North Holland (1973).

    1988 Mechanisms for computing over abstract structures. This volume.

Trakhtenbrot, B.

    1988 Comparing the Church and Turing approaches: Two prophetical messages. This volume.

Influences of Mathematical Logic
on Computer Science

Martin Davis

When I was a student, even the topologists regarded mathematical logicians as living in outer space. Today the connections between logic and computers are a matter of engineering practice at every level of computer organization. Companies with names like *Logical Devices* or *Logicsoft* abound. One can walk into a shop and ask for a "logic probe". This is by no means simply a matter of terminology. Issues and notions that first arose in technical investigations by logicians are deeply involved, today, in many aspects of computer science.

To what extent did previous work by logicians actually influence computer scientists and engineers? Did they make use of what had already been done or have they simply "reinvented the wheel" as needed? Alas, it is much easier to point out confluences of ideas than to trace rigorously a path from the original work to the applications. When the ideas in question are truly fundamental, and therefore ultimately simple in the way so many fundamental concepts turn out to be, it is easy for people to just forget how startling the ideas may have been when they were first promulgated. Such ideas can move from the preposterous to the truism in a few years. In Davis *1987* some of these matters are discussed at length in connection with the role of Alan Turing's discovery of the universal computing machine in the actual development of modern electronic computers. In this essay we shall follow the trail of a number of concepts that arose in the work of logicians and have found their way into computer theory and practice.

1. Formal Syntax

One of the first things a novice user of a computer must learn is that computers tend to be totally unforgiving of "minor" lapses in notation. Who has not experienced the frustration of being required to reenter a long line of text simply because a comma somewhere should have been a period? We may say that computer languages (programming languages, operating systems, data base systems, etc.) have a totally prescribed formal syntax. The idea that a specially created language could be useful for extending the range of what could be accomplished by computation goes back to Leibniz (see, for example, Davis *1987*). However, the first actual example of a formal language was presented by Gottlob Frege in his *Begriffsschrift* (1879).

Frege may be regarded as the founder of the *logicist* movement in the foundations of mathematics. Logicism is the thesis that mathematics and logic are literally the same subject: that is, that all entities that occur in mathematics can appropriately be taken to be purely logical constructs and that mathematical proof is simply logical deduction. Thus, a thorough-going logicism must be prepared to give a rigorous account of deductive logic. Frege built on George Boole's earlier discovery that logical deduction can be treated as a branch of mathematics. But – doesn't this lead to a vicious circle? If logic is developed using mathematics, how can it in turn be the underlying foundation for all of mathematics. Formal syntax was Frege's solution to this problem. He created an artificial language, his "Begriffsschrift", literally "concept writing" which Frege characterized as a "Formelsprache des reinen Denkens", a "formal language of pure thought". In this language, deduction is replaced by formal derivation depending solely on symbol manipulation. Thus, the circularity is avoided. Frege's system later turned out to be formally inconsistent (as Frege realized after receiving the famous letter from Bertrand Russell containing the Russell paradox, see below), and therefore his program was certainly not a success. But his achievements were very important. His *Begriffsschrift* contained, for the first time ever, a formulation of what has come to be called first-order logic. But most important of all for computer science was Frege's clear demonstration of how to construct and deal rigorously with a formal language.

Just how remarkable Frege's work was becomes clear on comparing it with that of his successors. Important contributions were made by E. Schröder, G. Peano, and by Bertrand Russell. But this work, mostly done during the early years of the twentieth century, fell far short of Frege's level of rigor. Thus, in Bertrand Russell's *1908* (discussed further below) one finds listed among the "axioms" (Russell calls them "primitive propositions") such purely symbolic expressions as

$$q \supset p \vee q$$

along with the statement in English

A proposition implied by a true premise is true.

Frege would never have permitted such a confusion between the formal language with which he was working and statements in natural language about the formal language. The same formulation appears in *Principia Mathematica,* the three volume edifice developed by A.N. Whitehead and Bertrand Russell as an explicit embodiment of their logicist view.

Emil Post's dissertation (*1921*) marked a return to Frege's standards of rigor. But Post's emphasis was quite different. Unlike the logicists Frege and Russell, Post's main concern wasn't with using formal systems of logic as a "foundation" for mathematics. Rather he saw that, once formulated, these systems could be investigated by ordinary mathematical methods. In particular, Post studied the problem of finding algorithms by which it could be mechanically determined whether particular formulas in the language of *Principia Mathematica* could be derived using the rules of the language. For this purpose Post replaced the dubious "primitive proposition" stated above, with the purely syntactic principle of *modus ponens*:

$$\text{"} \vdash P \text{" and "} \vdash P \supset Q \text{" produce "} \vdash Q \text{"}$$

(which of course is what Russell and Whitehead actually used). From this point of view, the interesting thing about the logicist program, is that it implied that success in finding such algorithms would have led to the mechanization of large parts of mathematics. Post solved only the first part of the problem; he found algorithms for the part of *Principia Mathematica* that we now call the propositional calculus. Post's efforts to extend these results led him to consider formal operations on strings in the most general context, which he called *productions*. As we now know, and as Post himself soon came to realize, the problem in its full generality has no solution: the problem of finding a decision algorithm for the entire system of *Principia Mathematica* is, as we should say today, undecidable. (See Davis *1982a* for an extended discussion of Post's work in the 1920's.)

Post productions are ubiquitous in computer science. Their first application was by Noam Chomsky, who found in them exactly what he needed for his revolutionary theory of the grammars of natural languages. This led Chomsky to his now famous classification or hierarchy of languages based on the specific kind of Post productions permitted in their defining grammars. The connection with computer science became apparent when it turned out that one of Chomsky's classes consisted of just the languages that could be recognized by a finite automaton, and that another, the so-called context-free languages, consisted of the languages recognizable by finite automata equipped with an auxiliary push-down stack. Apparently independently of Chomsky's work, John Backus used Post productions to provide an appropriate syntax for the developing programming language ALGOL 58. And then it turned out that the class of languages that could be described in terms of Backus' syntax were exactly Chomsky's context-free languages!

I was particularly interested to read John Backus' explanation of how he happened to think of his syntax (Backus *1981*, p. 162), because, very much to my surprise, he mentioned my name:

> As to where the idea came from – it came from a class I took from Martin Davis. He was giving it at Atlanta State, talking about Emil Post and the idea of a production.

> It was only in trying to describe ALGOL 58 that I realized that there was trouble about syntax description. It was obvious that Post's productions were just the thing, and I hastily adapted them to that use . . .

My pleasure in having apparently played some role in this matter was somewhat spoiled by my realizing that I had never set foot in "Atlanta State"; in fact, checking with reference librarians made it appear that there has never been such an institution. But I was soon able to clear up the matter to my satisfaction. I had given a course of lectures for IBM during the academic year 1960–61 with John Backus in the audience. The lectures were given at "The Lamb Estate" in Westchester County, New York State, where many of IBM's researchers worked at that time. Surely it was a transcription error. Pleased with my explanation, I published what I thought was a witty little note explaining the error in the *Annals of the History of Computing* (Davis *1982*). Somewhat later I came across Backus *1980* in which Backus once again credited me with being the transmission channel by means of which he became aware of Post's work. Thus it was with some chagrin that, on examining the list of references in this paper, I came to realize that Backus had already given a talk on his "normal form" during the summer of 1959, before he had heard my lectures! Discussions with John Backus in recent years failed to shed any additional light on the matter.

This personal anecdote can serve to underline the difficulty in tracing the path by which ideas move from theory to practice. I have little doubt that Backus' recollection that he was influenced by Post's ideas is correct. But he was not a logician. How did he become aware of Post's work? My book Davis *1958* was a possible source as was Chomsky's early work. There were logicians working with Backus at IBM on the FORTRAN project (see below) – certainly another possibility. But most likely we will never be able to tell.

Of course, what is important and interesting is the fact of the transmission of ideas. Trying to establish the actual path is of real interest only insofar as it helps to establish that the transmission actually did take place. However, this example in which everyone concerned is alive and willing, and there is no acrimony, (conditions which are by no means typical) demonstrates what a difficult undertaking it can be.

2.   Boolean Logic

"True" and "false", "yes" and "no", 0 and 1: it has become a truism that the same dualism suggested by these pairs is responsible for the intimate connection between ordinary propositional logic and binary digital circuitry. The algebra associated with this point of view is that developed by George Boole in the 1840's. Although the idea of representing logical reasoning by a mathematical calculus was already envisioned by Leibniz, it was Boole's work that firmly established this as a reality.

Although Boole's achievement was immense, his work suffered from severe limitations. His vision of logic was limited by the tradition, deriving from Aristotle, of analyzing sentences in subject-predicate form. Thus, his logic amounted to a calculus of predicates or sets. Boole did recognize that the algebraic rules he had discovered also had another "propositional" interpretation in which variables represented sentences or their truth values. However, it remained for Frege, in his *Begriffsschrift* discussed above, to escape from the subject-predicate limitation, and to recognize the primacy of the propositional calculus.

The all-important, but basically simple, observation that every electrical circuit containing switches could be usefully interpreted as a formula of the propositional calculus was made by Claude Shannon in his master's thesis (Shannon *1938*). The idea was just to interpret "true" and "false" as corresponding to a switch being closed or open, respectively. The Boolean connectives "and" and "or" then corresponded to switches being connected in series and parallel, respectively. As Herman Goldstine remarked in his *1972*, "[this] helped change circuit design from an art to a science". For many years, electrical engineers have routinely learned the laws of Boolean logic in courses in "switching theory". Abstract modes of thought were encouraged that made the transition to newer technologies (relays to vacuum tubes to transistors to integrated circuits) natural. In any case, without the systematic analysis that Boolean methods make possible, the construction of modern large-scale digital computers would be virtually unthinkable.

The connections between logic and computer science are two-way. The realization that the logical complexity of a Boolean formula is closely related to the number of elements needed (and hence to the cost) of the corresponding circuits led the logician and philosopher W.V. Quine to work on the difficult combinatorial problem of finding the smallest formula that can represent a given Boolean function (Quine *1952*). Quine's work was taken up and extended by many researchers. Soon this material became a standard textbook topic (see for example, Dietmeyer *1971*). More recently, Boolean circuits are playing an important role in some of the most difficult problems in contemporary theoretical computer science: finding lower bounds on complexity of computation.

3. *Programming Languages*

Programming languages are simply the vehicle by means of which prospective users can state exactly which computational steps they wish to have carried out. In using the early "stored-program" machines of the 1950's, the only programming language available for the user of a particular computer was that machine's own "machine language". Typically the "instructions" of which a program was made up consisted of "orders" to bring binary strings representing numbers from the memory to an "arithmetic" unit where they could be operated on by the operations of ordinary arithmetic before being returned to the memory. The program itself was stored in this same memory in binary coded form, and the instructions were brought sequentially into the arithmetic unit for action. The sequence of arithmetic operations would be interrupted by a "test", typically to determine whether some quantity in the arithmetic unit was positive or negative, which test could result in a "jump" to a next instruction out of sequence.

Although machine languages are typically not so different today from what has just been described, programmers very rarely work in such a stark manner. At worst, they work in an "assembly" language in which mnemonics and symbolic addresses provide at least a minimal human-oriented interface. But most often, programmers work in a "higher-level" language which must be "compiled" (that is, translated once and for all into machine language) or "interpreted" (this means: operated on, one step at a time, by a special program that carries out each step as it is encountered). It is noteworthy that the development of these languages has introduced more and more logical structure. FORTRAN was perhaps the first serious widely used all-purpose programming language. (For a careful discussion of the pre-FORTRAN history of programming languages, see the engaging essay Knuth and Pardo *1980*.) Although the logical facilities available in the initial version of FORTRAN were pretty meager, the team that developed the system was very conscious of the connection between their task and the work of logicians. This can be seen from the fact that several of them attended the now famous Summer Institute for Logic at Cornell University in New York State in 1957, and lectured extensively on the new programming system. Their indebtedness to the formal languages developed by logicians can be seen in the following excerpt from the original FORTRAN manual (quoted in Backus *1981*):

> If E and F are expressions of the same mode, and if the first character of F is not + or
> −, then
>
> $$E + F$$
> $$E - F$$
> $$E * F$$
> $$E / F$$
>
> are expressions of the same mode.

The programming language that university students of computer science usually begin with today is Pascal. Pascal features "Boolean" as one of its basic types: this type consists of the two values "true" and "false". Boolean values can be combined using the operations "and", "or", and "not". Boolean conditions can be used not only to control simple "jumps" by means of "if ... then..." and "if ... then ... else..." constructions, but also to determine the exit point of complex iterated loops using "while ... do" and "repeat ... until" constructions. Facilities of this kind are available in most programming languages in general use today. For example, the form of BASIC provided by the IBM Corporation with its line of personal computers has essentially these same facilities.

Probably the first programming language with extensive logical facilities was John McCarthy's LISP developed in the late 1950's and still in general use (see McCarthy *1981*for an interesting account of LISP's early history). What is special about LISP is that the basic data objects with which it deals are associative lists rather than numbers. Boolean facilities are available in terms of McCarthy's "conditional expressions" the direct ancestor of the "if ... then ... else..." construct mentioned above. Indeed, McCarthy himself was part of the group that developed ALGOL 60, in which the "if ... then ... else..." construct was included, and it was probably the influence of ALGOL that resulted in this construct being included in all serious programming languages. However, from a logical point of view what is most striking about LISP is the explicit use of the LAMBDA operator. This operator transforms an expression into a function or mapping. Thus $\lambda x.x^2$ represents the function mapping a number into its square. It was studied by Alonzo Church, and played a crucial role in the early history of computability theory (see Davis *1982a*, Kleene *1981*, and Rosser *1984*). As McCarthy explained (*1981*, p.176), "To use functions as arguments, one needs a notation for functions, and it seemed natural to use the λ-notation of Church..."

Church's λ-calculus furnishes another example of the continuing cross-fertilization continually taking place between logic and computer science. Church had developed this system in connection with formal systems adequate to formalize classical mathematics. The fact that his original ambitious system turned out to be inconsistent, led Church to retrench and to formulate the λ-calculus as a provably consistent subsystem. After McCarthy had found the λ-operator useful in his LISP, others were attracted to this formalism (notably Corrado Böhm) as providing the basis for a programming language. However, a whole new stimulus to work on the λ-calculus was given by Dana Scott's suggestion that models of this calculus were the appropriate tool for providing programming languages with formal semantics, together with his discovery of an important class of such models.

As a final contribution of logic to programming languages, we mention the notion of *type*. Computer scientists say that Pascal, for example, is a strongly typed language. This means that the programmer must "declare" of each variable occurring in his program to which type it belongs. An attempt to "assign" to such a variable an object not of its type will lead to a compiler error message: TYPE MISMATCH. The idea of strict segregation of objects into types is due to Bertrand Russell *1908*. Russell introduced his theory of types as a response to the contradictions or paradoxes that difficulties with Cantor's transfinite numbers had brought to the fore. Russell himself had analyzed the paradox that results when Cantor's proof that any collection has more subsets than elements is applied to a supposed class of all sets; this analysis produced Russell's famous paradox of the class of all classes that do not contain themselves. Because in this form there is no reference to the transfinite but only to what seemed to be perfectly ordinary logical notions, Russell became convinced that a fundamental repair to our basic principles of logic was called for. The repair he proposed was to develop logic in what today we would call a strongly typed language. Then, there could be no class of all sets – only the class of all sets belonging to some type. And of course, this set would never be of that same type. Although the role of a hierarchy of types has remained important in the foundations of set theory, strong typing has not. It has turned out that one can function quite well with variables that range over sets of whatever type. So, Russell's ultimate contribution was to programming languages!

4. Logic Programming

Computer programming is a demanding task. A problem must be reduced to the carrying out of an explicit sequence of steps, and then these steps must be translated into a programming languages. Logic programming suggests an alternative: the problem will simply be *described* in purely logical terms. The computer will be provided with a logic engine that will carry out all possible logical deductions from this description, and the desired solution will be obtained. The logic engine can be a program for carrying out deductions in first order logic (the logic, first explicitly introduced by Frege, obtained by adding the "quantifiers" *all* and *exists* to the propositional calculus). Such programs, usually called "theorem-provers", have been extensively studied (see Davis *1983* and Loveland *1984* for the history). In his *1980,* Kowalski argued passionately that logic programming is the key to future developments in software.

The principal available logic programming systems are the various dialects of the PROLOG programming language, of which Colmerauer has been the main architect. The well-known inefficiency of theorem-provers (not unrelated to the undecidability of first-order logic) is dealt with in PROLOG by using a brutally minimalist form of first-order logic, the so-called horn clause logic. A *horn clause* is a statement of the form

$$p_1 \wedge p_2 \wedge \cdots \wedge p_n \supset q.$$

Here, $p_1, p_2, \ldots, p_n, q$ may contain variables, but must be atomic, in the sense of containing no other logical operations. (In particular this is a logic without negation.) A program in PROLOG is essentially a list of horn clauses. The PROLOG interpreter is then simply a horn clause theorem-prover. There has been a great deal of interest in PROLOG and in logic programming. In fact, there is now a journal exclusively devoted to this subject. PROLOG interpreters and compilers are available for many computing environments including the ubiquitous IBM-PC/DOS.

Although PROLOG has been used quite effectively, it does not really live up to its billing. The strategy used by the horn clause theorem-prover is so Spartan that results depend critically on the order in which the clauses are given. Even so, various "impurities" need to be added to the logic to have a workable system, particularly the notorious "cut" operation by means of which the programmer prevents the theorem-prover from following unpromising lines of deduction. Nevertheless, PROLOG certainly represents an important and highly visible instance of the association between logic and computing.

5. Conclusions

This volume commemorates the fiftieth anniversary of Turing's discovery of the universal computer in connection with his proof of the undecidability of first order logic. It thus represents, in and of itself, a remarkable instance of the connection we have been discussing. But Turing's paper exhibits the connection in many ways. His universal machine is at the same time a programmed interpreter (for Turing machines conceived as a programming language), thus being the first example of such. The logical interchangeability of software and hardware which this work demonstrated is today very much a fact of life. Even the way Turing associated logical conditions with the separate parts of a program, in his proof that first order logic is undecidable, foreshadowed the use of such conditions in attempts to certify, by mathematical proof, the correctness of a computer program. Who can say what the next fifty years will bring?

References

Backus, John
  1980 Programming in America in the 1950's – Some personal impressions. In: *A History of Computing in the Twentieth Century*, eds. N. Metropolis, J. Howlett, and Gian-Carlo Rota, pp. 125-135. New York: Academic (1980).
  1981 The history of Fortran I, II, and III. Transcript of discussion. In: *History of Programming Languages*, ed. R.L. Wexelblat, pp. 25–73. New York: Academic (1981).

Davis, Martin
  1958 *Computability and Unsolvability.* New York: McGraw-Hill (1958); reprinted by Dover (New York) (1983).
  1982 Lectures at 'Atlanta State'. *Ann. Hist. C.* **4** (1982) 370–371.
  1982a Why Gödel didn't have Church's thesis. *Inf. Contr.* **54** (1982) 3–24.

1983 The prehistory and early history of automated deduction. In: *Automation of Reasoning,* eds. Jörg Siekman and Graham Wrightson, vol. 1, pp. 1–28. Berlin/ Heidelberg/New York: Springer-Verlag (1983).

1987 Mathematical Logic and the Origin of Modern Computers In: *Studies in the History of Mathematics,* pp. 137–165. Washington, D.C.: Mathematical Association of America (1987); reprinted in this volume.

Dietmeyer, Donald L.

1971 *Logical Design of Digital Systems.* Boston: Allyn and Bacon (1971).

Frege, Gottlob

1879 *Begriffsschrift, eine der arithmetischen nach gebildete Formelsprache des reinen Denkens.* Halle (1879); reprinted in: *Begriffsschrift und andere Aufsätze,* ed. Ignacio Angelelli. Hildesheim: Olms (1964); English translation in: van Heijenoort *1967,* pp. 5–82

Goldstine, Herman H.

1972 *The Computer from Pascal to Von Neumann.* Princeton, NJ: Princeton University Press (1972).

Kleene, Stephen C.

1981 Origins of recursive function Theory. *Ann. Hist. C.* **3** (1981) 52–67.

Knuth, Donald E., and Luis Trabb Pardo

1980 The early development of programming languages. In: Metropolis et al. *1980,* pp. 197-273.

Kowalski, R.A.

1980 *Logic for Problem Solving.* New York: North-Holland Publ. Co. (1980).

Loveland, Donald

1984 Automated theorem proving: A quarter century review. *Cont. Math.* **29** (1984) 1–45.

McCarthy, John

1981 History of LISP. Discussion in: Wexelblat *1981,* pp. 173-197.

Metropolis, N., J. Howlett, and Gian-Carlo Rota, eds.

1980 *A History of Computing in the Twentieth Century.* New York: Academic (1980).

Post, Emil L.

1921 Introduction to a general theory of elementary propositions. *Am. J. Math.* **43** (1921) 163–185; reprinted in van Heijenoort *1967,* pp. 264–283.

Quine, W.V.

1952 The problem of simplifying truth functions. *Am. Math. Mo.* **59** (1952) 521–531.

Rosser, J. Barkley

    1984 Highlights of the history of the lambda calculus. *Ann. Hist. C.* **4** (1984) 337–349.

Russell, Betrand

    1908 Mathematical logic as based on the theory of types. *Am. J. Math.* **30** (1908) 222–262; reprinted in van Heijenoort *1967,* pp. 150–182.

Shannon, C.E.

    1938 A symbolic analysis of relay and switching circuits. *Trans. AIEE* **57** (1938) 713ff.

van Heijenoort, Jean (ed.)

    1967 *From Frege to Gödel: A Source Book in Mathematical Logic, 1879–1931.* Cambridge, MA: Harvard University Press (1967).

Wexelblat, Richard L., ed.

    1981 *History of Programming Languages.* New York: Academic (1981).

ADDED IN PROOF: I am informed by B.A. Trakhtenbrot that, independent of Shannon's work, Victor Shestakow had also realized the connection between Boolean algebra and switching circuits. Trakhtenbrot refers to his article in the *Ann. Hist. C.* **6,** No. 4, October 1984.

Language and Computations

Jens Erik Fenstad

C.P. Snow, the physicist and the novelist, on one occasion claimed that P.A. Dirac and F. Crick were the two greatest Englishmen of their generation. Even if he spoke at a banquet celebrating the 70th birthday of Dirac, Snow may be right. This I will not dispute, but I would like to add one more name to this *duo* of eminent Cambridge men – Alan Turing. Physics aiming at understanding the natural world is well represented by Dirac, the life sciences is equally well represented by Crick. And the name of Turing has increasingly become a symbol of the new science of computation and communication, whether it be in "natural" or "artificial" systems.

Turing himself was one of the first who with some insight probed the grey and difficult border area between the natural and the artificial. In a famous paper "Computing Machinery and Intelligence" (*1950*) in the philosophical journal *Mind* he proposed an operational test – *the interrogation game* – for some of the crucial concepts related to "thinking" and "intelligence". He predicted that in about fifty years time a machine could be taught to play the game so well that a human interrogator "will not have more than a 70 per cent chance of making the right identification after five minutes of questioning". And he concluded, "... I believe that at the end of the century the use of words and the general educated opinion will have altered so much that one will be able to speak of machines thinking without expecting to be contradicted".

The Turing test has had an important impact on the new science, giving a sense and a direction to much important research. It has moreover generated an untold number of general and philosophical papers arguing the pros and cons of the thesis. I shall not enter into this debate here. It has been useful as a check on the "naive" researcher – and AI has known several of those, even among its leading practitioners. But the general debate can be carried too far and become counterproductive. Sometimes one has to try to do things, even in the face of a priori "impossibility" arguments. And sometimes one succeeds contra the philosophers.

But there is one aspect of the Turing test that I will draw attention to, the *communication between man and machine.* If we are not to find out the machine, the machine must be able to communicate with us in a "natural way". Turing was no pioneer in "man–machine" communication. It is told that he found it perfectly natural to work directly in a base-32 arithmetic, and if the "machine" so demanded he would not object to writing out the code backwards.

Being able to communicate is part of intelligent behavior. And if we are going to be able to interact with a machine in our way, we need a science of *language and computation* as part of the larger field. Computational linguistics with an emphasis on syntactic processing soon entered the scene. But communication is more than syntax, form must be connected to meaning. This turned out to be more difficult, and progress was slow in coming. We have, however, seen a beginning, and it is my purpose in this paper to give an account of some recent advances in *computational semantics.*

Before we turn to semantics we should fill in some necessary background from formal language theory and computational syntax. N. Chomsky's *Syntactic Structures* (*1957*) marked a theoretical renewal of linguistic science. And with the Chomsky renewal a bond was soon forged with formal or symbolic logic, a part of which is formal language theory. Chomsky himself was one of the active participants in this development. This theoretical development soon joined forces with the emerging computer science, and a vigorous field of computational linguistics was established. This science does not only have important applications to the study of natural languages, it is also an integral part of computer science, e.g., in compiler design.

But this is a study which focuses on linguistic form. "Meaning" was always a recognized part of the chomskian model; however, it was only with the work of R. Montague (*1967*) that a technical adequate meaning component was joined to the syntactic part. Montague's model is based on higher order intentional logic, which – to be kind – is not prima facie very algorithmic in spirit.

Post Chomsky we have seen a number of interesting developments in syntactic theory. Some of these have had a primary linguistic motivation, e.g., the *Lexical-Functional Grammar* developed by J. Bresan and R.M. Kaplan (see Bresan *1982*) and the *Generalized Phase Structure Grammar* developed by G. Gazdar and a number of co-workers (see Gazdar et al. *1985*). Other theories were primarily motivated from the computational point of view, e.g., the *Definite-Clause Grammar* of F.N.C. Pereira and D.H.D. Warren, a formalism tailored for a PROLOG implementation (see Pereira and Warren *1980*) and the *PATR-II* formalism developed at SRI International (see e.g., S.M. Shieber *1986*); for a survey of some of this work see Halvorsen *1987*. At the focal point of many of these developments stands the work of M. Kay on a unification-based approach to grammatical analysis. This work again is related to the "resolution" approach to logic and logic programming.

And an interesting phenomenon has been observed. The "pull" of effective implementation has brought about a confluence of schools. Theoretical linguists have notoriously been a quarreling group (see Newmeyer *1980*). But if you aim at an "applied" science, there must be some coherence and unity to the subject matter. It seems that "unification" and "constraint propagation" can serve as such unifying themes – which, of course, does not mean that there are no more disagreements in the analysis of linguistic phenomena. It means that there is a more unified framework in which to discuss and to compare – a framework which can be used as a basis for implementations and applications.

The stage has now been set and we turn to semantics. Let us recall the overall aim. A system for natural language analysis will provide a framework for relating the *linguistic form* of utterances and their *semantic interpretation.* And the relation between the two must be algorithmic.

Basic to the approach which we present in this paper (see Fenstad et al. *1987*) is an algorithm for converting linguistic form to a format which we call a *situation schema.* The algorithm is in the spirit of current unification-based approaches to grammar and exploit the idea of constraint propagation.

A situation schema has a well-defined (algebraic) structure, suggestive of "logical form". This is a structure different from the standard model-theoretic one; we will argue that it is a structure better adapted for the analysis of the *meaning relation* in natural languages. A situation schema is effectively calculable from the linguistic form, and we believe that it provides a format useful for further processing, e.g., in the construction of a natural language interface with a data base system and also in connection with mechanical translation systems.

In the first part of this chapter we shall provide the necessary background from the semantics of partial information – the *situation semantics* of J. Barwise and J. Perry (see their *1983*). We shall further report briefly on some mathematical investigations into situational logic. The logic is axiomatizable, and we have a well-understood inference mechanism for a many-sorted logic based on a semantics of partial information.

Next we introduce the notion of *situation schema* and explain how this concept is related both to the semantics of partial information and to the syntactic processing of natural language sentences.

In conclusion we discuss possible applications, in particular, situation schemata as a "query level" in question – answering systems. This leads back to the problems raised by the Turing test.

1. The Logic of Situations and Partial Information

Mathematical knowledge is asserted to be complete and independent of time and place. Take any famous unsolved problem, e.g., Fermat's "theorem" or Riemann's hypothesis. Mathematicians agree that they are either true or false. We do not know today which is the case, but that is irrelevant. The truth values are eternal and fixed.

But this is an extreme kind of knowledge. Everyday knowledge tends to be fragmentary; we are usually in a situation of partial information. Take as an example a large scale medical examination where one is testing a certain group of the population for some form of cancer. In many cases the verdict is clear. We have either the "positive" fact that the person examined has this particular kind of cancer or the "negative" fact that he or she does not.

However, in many cases our information is incomplete, and it may require further elaborate testing before a diagnosis can be formed. It may even happen that no verdict will be forthcoming. We also may make mistakes, forcing us to revise our information, e.g., changing a "positive" fact to a "negative" one, or vice versa, at some later stage.

Similarly, in setting up a knowledge base we may at a given stage be able to record some positive facts and some negative facts, but in several cases our information may be incomplete. Of course, at a later stage we may be able to decide further cases; in a certain sense information "grows over time". But changes in the knowledge base may force us to change logical inferences which we drew from the original one. We may thus be engaged in a complicated "updating" of current information.

Standard predicate logic is not well suited to treating partial information. An atomic sentence $P(t_1, \ldots, t_n)$ is either true or false with respect to a model or "knowledge base". We shall in this part describe a semantics better suited to cope with problems concerning partiality.

POSITIVE AND NEGATIVE FACTS

In standard model theory we have a domain of *individuals* D and a set R of *relations* over D. If $r \in R$ is an n-ary relation, we think of r as a set of ordered n-tuples over D, i.e., r is a subset of the Cartesian product D^n. And the basic stipulation is that $a_1, \ldots, a_n \in D$ stand in the relation r iff $< a_1, \ldots, a_n > \in r$, and that r does not hold of $a_1, \ldots, a_n$ iff $< a_1, \ldots, a_n > \notin r$. Information is complete since either $< a_1, \ldots, a_n > \in r$ or $< a_1, \ldots, a_n > \notin r$.

In a situation of partial information we may not be able to assert either the *positive* fact that r holds of $a_1, \ldots, a_n$ or the *negative* fact that r does not hold of $a_1, \ldots, a_n$; the assertion may be undecided. It will prove convenient to change our basic format slightly and write

$$r, a_1, \ldots, a_n; 1$$

to assert that r holds of $a_n, \ldots, a_n$ and to write

$$r, a_1, \ldots, a_n; 0$$

to assert that r does not hold of $a_1, \ldots, a_n$.

Since we may have neither $r, a_n, \ldots, a_n; 1$ nor $r, a_1, \ldots, a_n; 0$ in a given situation, negation may have either a weak or strong sense. The assertion

$$\text{not: } r, a_1, \ldots, a_n; 1$$

may mean that we have $r, a_1, \ldots, a_n; 0$ (*strong negation*) or it may mean that either the assertion is undefined, or if defined that $r, a_1, \ldots, a_n; 0$ (*weak negation*). In a situation of total information the distinction disappears.

LOCATED FACTS

We shall add one more feature to our semantical analysis. A fact may sometimes be *located* at a certain time and place. A medical examination takes place at a certain time and place and the resulting diagonosis makes sense only relative to that particular location. Some facts, e.g., mathematical statements, are better thought of as *unlocated*.

Thus in addition to individuals D and relations R we add the category of *locations* L, and we think of an element $l \in L$ as a connected region of space-time. Thus we may locate our medical examination to the city of Cambridge in the month of January, 1987. The relation in this example is the property of having some specified form of cancer, and the set of individuals may be some specific target group of the population, e.g., testing males aged 50–55 for lung cancer. Note that in this case it is not very intuitive to think of "lung cancer" as a subset of the male population aged 50–55.

Let r be an *n*-ary relation, l a location, and $a_1, \ldots, a_n$ individuals. The format of a basic *located fact* is

$$\text{at } l: r, , \ldots, a_n; 1$$
$$\text{at } l: r, , \ldots, a_n; 0,$$

where the first expresses that at the location l the relation r holds of the individuals $a_1, \ldots, a_n$; the second expresses that it does not hold. The basic format of *unlocated facts* is

$$r, a_1, \ldots, a_n; 1$$
$$r, a_1, \ldots, a_n; 0$$

with the "obvious" reading.

SITUATED FACTS

A knowledge base contains many facts, some located, some unlocated; some positive, some negative. We shall use the word *situation* to describe this state of affairs. Thus a situation s determines a set of facts:

$$\text{in } s: \text{ at } l{:}\, r, a_1, \ldots, a_n; 1$$
$$\text{in } s: \text{ at } l{:}\, r, a_1, \ldots, a_n; 0.$$

The first expresses that in the situation s at the location l, r holds of $a_1, \ldots, a_n$. We leave the reading of the second, as well as the unlocated versions to the reader.

SITUATION SEMANTICS

Situation semantics is grounded in a set of primitives

S	*situations*
L	*locations*
R	*relations*
D	*individuals*

In this paper we shall not worry too much about their ontological status, how they are "sliced out" of reality. We think of a situation as a kind of restricted, partial model which classifies certain basic facts. We assume that the primitives come with some structure. One minimal requirement is that each relation in R is provided with a specificiation of the number of argument slots or roles of that relation. For the purpose of this survey we assume no structure on the set of situations and individuals, although we could speak of one situation as "being part" of another situation. The set L of locations is or represents connected regions of space-time and thus could be endowed with a rich geometric structure (e.g., if we wanted to study shape and structure in connection with vision). Here we shall be much more modest and assume that L comes endowed with two structural relations.

$<$	*temporally precedes*
o	*temporally overlaps*

to account for a simple-minded analysis of past and present tenses.

So much for a background from situation semantics. This theory was created by Jon Barwise and John Perry and we recommend their book, *Situations and Attitudes*, (Barwise and Perry *1983*), for anyone who wants a deeper insight into the theory.

The Formal Logic of Situations

Let us add a few remarks on the *logic* of situation semantics. The model theory or semantics of the logic is – as we explained above – based on a multisorted structure

$$M = < S, L, D, R; In >$$

where *In* is the set of all tuples

$$< s, l, r, a_1, \ldots, a_n, i >$$

$i = 0, 1$, such that

$$\text{in } s: \text{ at } l: r, a_1, \ldots, a_n; i$$

In *Situations, Language and Logic* (Fenstad et al. *1987*) we develop the mathematical study of this class model structures; in particular, we prove several axiomatization theorems, thereby providing a complete inference mechanism for a multisorted logic based on a semantics of partial information.

As a first step we study the logic of a fixed situation. Let *s* be an element of *S* and let

$$M_s = < L, R, D; in_s >$$

where in_s consists of all tuples $< l, r, a_1, \ldots, a_n, i >$ such that $< s, l, r, a_1, \ldots, a_n, i >$ is in *In*. Associated with the class of models M_s we have a two sorted-first order language with atomic formulas of the sort

$$R(l, t_1, \ldots, t_n)$$

where *l* is a location variable or constant, $t_1, \ldots, t_n$ are individual variables or constants, and *R* a two-sorted relation symbol. In addition we have atomic formulae

$$l_1 < l_2$$

to express the relation of temporal precedence on the set *L*. The language will have a *strong* ($\neg$) and a *weak* ($\sim$) negation; recall our discussion above. The basic result is an axiomatization theorem with respect to the class of models M_s. Since the *model theory* of this language seems to be a natural formalism for data base theory, it would be interesting to try to build a PROLOG-style inference mechanism based on the *proof theory* of this logic.

GROWTH OF KNOWLEDGE

We mentioned above problems of nonmonotonicity connected with partial information. When "knowledge grows", i.e., when more facts are decided, we may have to revise inferences drawn from a partial knowledge base. But we would like to know which inferences we may keep. In *Situations, Language and Logic* the following result is proved:

A formula ϕ is called 1-*persistent* if for any two models $M = < L, R, D;\ in >$ and $M' = < L, R, D\ in' >$ where $in \subseteq in'$, if ϕ is true in M then ϕ is true in M'. Notice that M and M' have the same relations, locations and individuals. The inclusion $in \subseteq in'$ means that more facts are decided in M'.

ϕ is called 0-*persistent* if $\neg\phi$ is 1-persistent. ϕ is called *persistent* if it is both 1- and 0-persistent. The notion of persistence is semantical. The following notion is syntactical: a formula ϕ is called *pure* if it does not contain the weak negation symbol $\sim$.

Let $\phi \equiv \psi$ abbreviate the formula $(\phi \supset \psi) \wedge (\psi \supset \phi)$, where $\phi \supset \psi$ is itself an abbreviation for $\sim \phi \vee \psi$. We introduce a strong equivalence by the definition

$$\psi \Leftrightarrow \psi \text{ iff } (\phi \equiv \psi) \wedge (\neg\phi \equiv \neg\psi).$$

Theorem: *A first order formula ϕ is persistent if and only if there exists a pure formula ψ such that $\phi \Leftrightarrow \psi$ is provable.*

2.   Situation Schemata and Natural Language Processing

How is the semantics of partial information to be connected to the analysis of natural language? Let us approach this by way of a simple example. The sentence

John is running

as uttered at a location l_0 (the "discourse location") is true of a situation s if there is some location l in s such that

in s: at l: *run, John;* 1
l o l_0

The sentence

John married Jane

as uttered at l_0 (the discourse location) is true of a situation s if there is some location l in s such that

in s: at l, *marry, John, Jane;* 1
$l < l_0$

The interplay between the utterance situation, the sentence uttered, and the situation described is our central theme. In fact, situation semantics analyzes the *meaning* of a sentence ϕ as a relation between an utterance situation u and a described situation s. We now turn to an explanation of how this is to be done.

SITUATION SCHEMATA INTRODUCED

The *situation schema* is a complex feature structure computable from the linguistic form of utterances and with a choice of features matching the primitives of situation semantics. The basic format is thus

$$
\begin{bmatrix}
\text{REL} & - \\
\text{ARG 1} & - \\
\cdot & \cdot \\
\cdot & \cdot \\
\cdot & \cdot \\
\text{ARG n} & - \\
\text{LOC} & - \\
\text{POL} & -
\end{bmatrix}
$$

Here the features REL, ARG 1 ,..., ARG n and LOC correspond to the primitives *relation, individuals,* and *locations,* POL, abbreviating *polarity,* takes either the value 1 or 0. The values in the schema can either be atomic or themselves complex feature-value structures. The value of the LOC feature is always complex.

Let us return to the sentence

John married Jane.

We shall assign to this sentence the following situation schema. Call the sentence ϕ_1, then SIT.ϕ_1, the situation schema associated to ϕ_1, will be:

$$\text{SIT.}\phi_1 \quad \begin{bmatrix} \text{REL} & marry \\ \text{ARG.1} & John \\ \text{ARG. 2} & Jane \\ \text{LOC} & \begin{bmatrix} \text{IND} & \text{IND.1} \\ \text{COND}_{\text{loc}} & \begin{bmatrix} \text{REL}_{\text{loc}} & precede \\ \text{ARG.'1} & \text{IND.1} \\ \text{ARG.'2} & \text{IND.0} \end{bmatrix} \end{bmatrix} \\ \text{POL} & 1 \end{bmatrix}$$

The interpretation of this schema is relative to an *utterance situation u* and a *described situation s*. The utterance situation decomposes into two parts

    d   *discourse situation*
    c   *the speaker's connection.*

The discourse situation contains information about who the speaker is, who the addressee is, the sentence uttered, and the discourse location. The speaker's connection is a map determining the speaker's meaning of lexical items; this we shall explain in more details below.

A map g defined on the set of indeterminates of $\text{SIT.}\phi_1.\text{LOC}$ and with values in the set L of locations, is an *anchor* on $\text{SIT.}\phi_1.\text{LOC}$ relative to an utterance situation d, c if

$$g(\text{IND.0}) = l_{\text{d}}$$
$$<, g(\text{IND.1}), l_{\text{d}}; 1$$

where l_{d} is the discourse location determined by d and $<$ is the structural relation of temporal precedence on the set L of locations.

We shall use the situation schema $\text{SIT.}\phi_1$ to explain the *meaning* of the sentence ϕ_1 as a relation between the utterance situation d, c and the described situation s. We write the basic meaning relation as

$$d, c[\![\text{SIT.}\phi_1]\!] \; s$$

In our example this relation holds if and only if there exists an anchor g on $\text{SIT.}\phi_1.\text{LOC}$ relative to d, c such that

    in *s*: at $g(\text{IND.1})$: $c(marry)$, $c(John)$, $c(Jane)$; 1.

Note that the speaker's connection is map defined on (morphological) parts of the expression uttered and with values in the appropriate domains of the semantical model, i.e., $c(marry)$ is an element of R, $c(John)$ and $c(Jane)$ elements of D. The latter means that we have in this example adopted a rather simple-minded treatment of names: the speaker's connection picks out a unique referent in the described situation.

COMPLEX ROLES

Roles can be described in more complex ways than by names with unique reference. Our next example is the sentence ϕ_2

> *John married a girl.*

A noun phrase such as *a α* (*an α*) can be given either a generalized quantifier interpretation or a singular NP-reading. In either case our system will generate the following situation schema.

$$
\text{SIT.}\phi_2 \begin{bmatrix} \text{REL} & marry \\ \text{ARG.1} & John \\ \text{ARG.2} & \begin{bmatrix} \text{IND} & \text{IND.1} \\ \text{SPEC} & A \\ \text{COND} & \begin{bmatrix} \text{REL} & girl \\ \text{ARG'.1} & \text{IND.1} \\ \text{POL} & 1 \end{bmatrix} \end{bmatrix} \\ \text{LOC} & \begin{bmatrix} \text{IND} & \text{IND.2} \\ \text{COND}_{\text{loc}} & \begin{bmatrix} \text{REL}_{\text{loc}} & precede \\ \text{ARG'.1} & \text{IND.2} \\ \text{ARG'.2} & \text{IND.0} \end{bmatrix} \end{bmatrix} \\ \text{POL} & 1 \end{bmatrix}
$$

With the singular NP-reading we get

> $d, c[\![\text{SIT}\phi_2]\!]\ s$

if and only if there exists an anchor g on $\text{SIT.}\phi_2.\text{ARG.2}$ relative to d, c and an extension $g' \subset g$ which anchors $\text{SIT.}\phi_2.\text{ARG.2}$ in s, i.e.,

> in s: $c(girl)$, $g'(\text{IND.1});1$

such that

in s: at g(IND.2): c(*marry*), c(*John*), g'(IND.1);1.

Notice that we have used an unlocated fact in the interpretation of ARG.2. This corresponds to the lack of tense markers in the subpart "a girl" in the sentence ϕ_1.

Let us also add a comment on the format of ARG.2. The value of this feature is really a notation for a restricted quantifier

$$\exists\ x\ (\text{girl}(x)\ \wedge\dots)$$

or

$$\exists x \in \text{girl}\ (\dots).$$

We see that ARG.2.IND introduces the quantified variable, ARG.2.SPEC the nature of the quantifier, and ARG.2.COND specifies the domain of variation of the quantifier.

We could at this point have introduced the generalized quantifier reading, our format is well adapted to handle this way of interpreting complex NP's; see Fenstad et al. *1987* for further information on this point.

RESOURCE SITUATIONS AND ANAPHORIC REFERENCE
Let us briefly discuss the pair of sentences

ϕ_3: *The boy married Jane*
ϕ_4: *The boy who married Jane loves her.*

For the first sentence we have the associated schema

$$\text{SIT.}\phi_3\ \begin{bmatrix} \text{REL} & marry \\ \text{ARG.1} & \begin{bmatrix} \text{IND} & \text{IND.1} \\ \text{SPEC} & THE \\ \text{COND} & \begin{bmatrix} \text{SIT} & \text{SIT.1} \\ \text{REL} & boy \\ \text{ARG'.1} & \text{IND.1} \\ \text{POL} & 1 \end{bmatrix} \end{bmatrix} \\ \text{ARG.2} & Jane \\ \text{LOC} & - \\ \text{POL} & 1 \end{bmatrix}$$

We see that the COND of ARG.1 has been expanded with a new feature SIT. In the interpretation of SIT.ϕ_3 the indeterminate SIT.1 will be anchored to some resource situation, which may differ from both the discourse situation and the described situation, and which will be used to determine the domain of variation of the quantifier.

For the sentence ϕ_4 we will generate the following situation schema:

$$
\text{SIT.}\phi_4 \quad
\begin{bmatrix}
\text{REL} & love & & & \\
\text{ARG.1} & \begin{bmatrix}
\text{IND} & \text{IND.1} & \\
\text{SPEC} & \textit{THE} & \\
\text{COND} & \begin{bmatrix} \text{REL} & boy \\ \text{ARG}'.1 & \text{IND.1} \\ \text{POL} & 1 \end{bmatrix} \\
\text{SIT} & \begin{bmatrix} \text{REL} & marry \\ \text{SCHEMA} & \text{ARG.1} & \text{IND.1} \\ \text{ARG.2} & Jane \\ \text{LOC} & - \\ \text{POL} & 1 \end{bmatrix}
\end{bmatrix} \\
\text{ARG.2} & \text{IND.3} \\
\text{LOC} & - \\
\text{POL} & 1
\end{bmatrix}
$$

There are two comments to make. First we notice that the value of ARG.1 has a new feature SIT.SCHEMA to handle relative clauses. Next we see that the schema has a "free indeterminate" IND.3. In many readings we would expect that *Jane* and *her* referred to the same individual. But how do we account for this sameness of reference. Indeterminates must be anchored, so we could have an anchor f on SIT.ϕ_4.ARG.2, and the sameness of reference would be recorded by the fact that $f(\text{IND.3}) = c(\text{Jane})$. But this is an added fact and would not count as a case of anaphoric linking. The latter case would require a further constraint equation

$$(\text{ARG.1.SITSCHEMA.ARG.2}) = (\text{ARG.2}).$$

It is not our purpose on this occasion to argue for a particular theory of anaphoric reference. However, we believe that the format of situation schemata is flexible enough to accommodate various solutions whether obtained by co-indexing, i.e., by using the same indeterminate in different positions, or obtained by an appeal to the mechanism of speaker's connection.

SITUATION SCHEMATA FORMALLY DEFINED

A situation schema is a complex feature-value system. The schemata discussed above can all be generated by the following set of rewriting rules:

SIT.SCHEMA	$\rightarrow$	(SIT)REL^{n}ARG.1 ... ARG.n LOC POL
SIT	$\rightarrow$	*<situation indeterminate>*
REL^n	$\rightarrow$	*<n-ary relation constant>*
ARG.i	$\rightarrow$	{IND$_e$|IND(SPEC COND(SITSCHEMA^n))}
LOC	$\rightarrow$	IND COND$_{loc}$
POL	$\rightarrow$	{1|0}
IND$_e$	$\rightarrow$	*<entity>*
IND	$\rightarrow$	*<indeterminate>*
SPEC	$\rightarrow$	*<quantifier>*
COND	$\rightarrow$	(SIT)REL ARG'.1 POL
COND$_{loc}$	$\rightarrow$	REL$_{loc}$ARG'.1 ARG'.2
ARG'.i	$\rightarrow$	{IND$_e$|IND}.

In the examples discussed above REL$_{loc}$ will expand to the two relations of (temporally) overlapping and preceding. The "lexical expansions" are not included. To insure the correct identification of roles the following constraints are imposed:

$$(ARG.i.IND) = (ARG.i.COND.ARG'.1.IND)$$
$$(LOC.IND) = (LOC.COND_{loc}.ARG'.1.IND).$$

AN EXPANDED FORMALISM

The current format of a situation schema reflects what is computable from the utterance ϕ. However, more information may be necessary in order to spell out the meaning relation

$$u[\![SIT.\phi]\!]s.$$

We remarked that the definite description needs access to a "resource situation" in determining the domain of quantification. We argued that questions of anaphoric reference needed for their resolution further information from the utterance situation. And let us add that problems of quantifier scope cannot be decided from linguistic form alone.

It would be possible to expand the format of a situation schema to include this kind of information. This would give a schema more balanced between the utterance situation and the described situation.

This is a format that we would want to have in a further elaboration of the theory. But it is not at all clear how the extra information, going beyond SIT.ϕ, is to be computed.

FROM UTTERANCE TO SCHEMA

It remains to show how SIT.ϕ can be computed from the sentence ϕ. Our "ideology" is *unification* and *constraint equations,* but this is an ideology which can be implemented in different ways. For purpose of illustration we shall use the LFG-mechanism, but we could equally well have chosen some of the other formalisms mentioned in the introduction.

The computation of SIT.ϕ from ϕ is arranged in two steps. First we assign a simple context-free phrase structure to the given sentence, next we introduce a set of constraint equations which partly are associated with the nodes of the phase-structure tree and partly with the lexical items. We shall not give the full theory, but illustrate the procedure with a simple example. So let use once more return to the sentence

    John married a girl

and see how to compute the situation schema of this sentence as it was given above.

A simple context-free grammar for this sentence is

S	$\rightarrow$	NP	VP
		$(\uparrow ARG.1)=\downarrow$	$\uparrow=\downarrow$
VP	$\rightarrow$	V	(NP)
			$(\uparrow ARG.2)=\downarrow$
NP	$\rightarrow$	DET	N
NP	$\rightarrow$	NPROP	

Disregard for a moment the three equations occurring in the rules above. Using this grammar our sentence can be assigned the following structure:

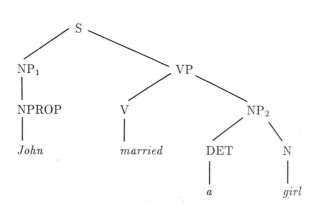

For the context-free part the lexicon is in a certain sense "trivial", we need only list the right classification of the words: *John* is a proper noun, *marry* is a verb, *a* is a determiner, and *girl* is a noun. For the full theory the lexicon must introduce the appropriate constraints (in addition to the "structural" constraints introduced in the context free rules).

For our example we have the following lexicon:

John	NPROP	$(\uparrow IND) = John$
a	Det	$(\uparrow IND) = IND.x$
		$(\uparrow COND) = \downarrow$
		$(\uparrow SPEC) = A$
		$(\downarrow ARG'.1) = (\uparrow IND)$
		$(\downarrow POL) = 1$
girl	N	$(\uparrow COND\ REL) = girl$
married	V	$(\uparrow REL) = marry$
		$(\uparrow LOC) = \downarrow$
		$(\uparrow POL) = 1$
		$(\downarrow IND) = IND.x$
		$(\downarrow COND_{loc}\ REL_{loc}) = precede$
		$(\downarrow COND_{loc}\ ARG'.1) = (\downarrow IND)$
		$(\downarrow COND_{loc}\ ARG'.2) = IND.0$

The equations from the structural rules and the lexicon must be combined to produce the situation schema. The equations are of a general nature and contain a set of "meta-variables" – the "up"- and "down"-arrows – which must be instantiated in each particular application. This is done by the following procedure. Let f_1 be a functional variable assigned to the top node S of the tree. With some of the other nodes of the tree there will be an associated equation coming from the syntax rules, e.g., with the node NP_1 the equation $(\uparrow ARG.1) = \downarrow$ is the associate. The "up"-arrow of this equation points to the "mother" of the node, i.e., to the node S, and the arrow $\uparrow$ shall be replaced by the variable f_1, which was the variable assigned to S. The "down"-arrow points to the node under consideration, i.e., in this case the node NP_1. We must then introduce this node a new functional variable f_2 to replace the arrow $\downarrow$, and we get the equation

$$NP_1: \quad (f_1\ ARG.1) = f_2.$$

In the same way we get the equations

$$\begin{aligned} V: &\quad f_1 = f_3 \\ NP_2: &\quad (f_3\ ARG.2) = f_4 \end{aligned}$$

from the nodes VP and NP_2, respectively. The lexicon contributes a new set of equations.

John: $(f_2 \text{ IND}) = John$

married: $(f_3 \text{ REL}) = marry$
                 $(f_3 \text{ LOC}) = f_5$
                 $(f_3 \text{ POL}) = f_1$
                 $(f_5 \text{ IND}) = \text{IND}.2$
                 $(f_5 \text{ COND}_{loc} \text{ REL}_{loc}) = precede$
                 $(f_5 \text{ COND}_{loc} \text{ ARG}'1) = (f_5 \text{ IND})$
                 $(f_5 \text{ COND}_{loc} \text{ ARG}'2) = \text{IND}.0$

a $(f_4 \text{ IND}) = \text{IND}.1$
                 $(f_4 \text{ COND}) = f_6$
                 $(f_4 \text{ SPEC}) = A$
                 $(f_6 \text{ ARG}'.1) = (f_4 \text{ IND})$
                 $(f_6 \text{ POL}) = 1$

girl: $(f_4 \text{ COND REL}) = girl$

Our equations are "functional equations" corresponding to the fact that the situation schemata are feature-value systems. An equation such as $(f_1 \text{ ARG}.1) = f_2$ means that f_1 is a function which is to be defined on a domain including the feature ARG.1, and the equation tells us that the value of f_1 applied to the argument ARG.1 is another function or feature-value system.

Our set of equations contains six "unknown" functions $f_1, \ldots, f_6$ and we can look upon the equations as imposing certain constraints on these "unknown" functions. When we are seeking a solution to this system of equations we shall always be looking for minimal, consistent solution. *And this minimal, consistent solution is precisely the situation schema associated to the sentence under consideration.*

The reader may verify this in connection with our sample sentence. f_1 which by one of the equations is equal to f_3, is a function with a domain of definition consisting of the features REL, ARG.1, ARG.2, LOC, and POL, and with values that are either "atomic" such as *marry* or complex such as the function or feature-value system f_5 which is the value of f_1 applied to the argument LOC. (We should note one minor notational discrepancy: In the situation schema we have given the value of ARG.1 directly as *John*, whereas the equations will give us as value the function $f_2 = \{< \text{IND}, John >\}$.)

And the style of writing the schema is but one way of exhibiting a finite function in tabular form.

Let us review – bottom up – what we have accomplished. In this last part we have – through one example – sketched an analysis which accepts a sentence from some fragment of a natural language and outputs a situation schema. The situation schema from this point of view is a minimal consistent solution of a set of constraint equations. But situation schemata can also be generated by a set of rewriting rules. And the two are the same: *Any schema generated by the rewriting rules is the schema of some sentence, and, conversely, every "solution" of a grammatical analysis can be generated by the rewriting rules.*

The schema exhibits in a certain sense the "logical form" of the sentence and is used to give the meaning of the sentence as a relation between an utterance situation and a described situation. The semantic foundation for this is a system of "situation semantics", and we started this chapter by giving a brief exposition of this approach to meaning analysis. We also touched the formal study – the model theory of situation semantics is axiomatizable – and we mentioned a persistence result characterizing information invariant under "growth of knowledge".

The situation schema is effectively calculable from surface form and the algorithmic analysis fits well into the current framework based on unification and constraint propagation. But this is not the end of the story: we mentioned the need to take into account further information from the larger background, including the utterance situation. This is a challenge: how is the extra information – going beyond the situation schema – to be tamed and computed?

3. *Towards Applications*

The exposition so far has concentrated on a study of the general features of a natural language system. But we believe that the format which we have isolated, *the situation schema,* has potentialities for applications. The schema is computable from linguistic form, but is independent of surface syntactic structure. It is reminiscent of logical from, but is not a translation into a particular logical formalism. Indeed, from a semantical point of view the situation schema may be incomplete, e.g., leaving questions of quantifier scope undecided. But this is perhaps as it should be. Translation into a logical formalism may force too much structure or "logical form" on the linguistic utterance. And if we look back at the expanded form of a situation schema discussed above, regimentation by one logical formalism is not the goal we should aim at.

However, logic does play an important role as an inference mechanism. But should one, perhaps, try to develop an inference mechanism based directly on situation schemata?

In treating *direct questions* it is also important that a situation schema is not a (declarative) logical formula. The semantics of a direct question has always been an awkward topic. What is the *model-theoretic* meaning of a direct question? Various round-about answers have been tried; the semantics of a question is claimed to be the set of possible answers, each of which has a truth value. But I may understand a question and use it correctly in a communicative act without having to run through in my mind the set of possible answers.

I would like to suggest that a question is an incomplete situation schema which is to be treated as a query with respect to some intended knowledge base. This is an approach which we are pursuing at the moment (and will be reported in the thesis of E. Vestre, *1987*). We are building a small system where the syntactic analysis of direct questions uses a version of Definite-Clause Grammar (see Pereira and Warren *1980*). In this case we use as query language one of the logical formalism developed in PROLOG via a proof-theoretic reduction of the partial logic of one fixed situation to a system of many-sorted predicate logic. An interesting feature is that the algorithm from utterance to schema is reversible, so the system produces an answer in natural language to a question asked in natural language. And this is done in a *principled way*; we are not interested in a "black box" filled with *ad hoc* trickery.

Thus we are back to one aspect of the Turing test, the "natural" communication between man and machine. And it seems that we can report on some success on this particular part of the interrogation game – even if it may be a minor one in the understanding of the larger issues of communication and reasoning in "natural" and "artificial" systems.

References

Barwise, J., and J. Perry

    1983 *Situations and Attitudes.* Cambridge, MA: The MIT Press (1983).

Bresnan, J., and R.M. Kaplan

    1982 Lexical-function grammar: A formal system for grammatical representation. In: *The Mental Representation of Grammatical Relations.* Cambridge, MA: The MIT Press (1982).

Chomsky, N.

    1957 *Syntactic Structures.* Den Haag: Mouton (1957).

Fenstad, J.E., P.K. Halvorsen, T. Langholm, and J. van Benthem

    1987 *Situations, Language and Logic.* Dordrecht: Reidel (1987).

Gazdar, G., E. Klein, G. Pullum, and I. Sag

    1985 *Generalized Phrase Structure Grammar.* Oxford: Basil Blackwell (1985).

Halvorsen, P.-Kr.

    1987 Computer Applications of Linguistic Theory. In: *Linguistics: The Cambridge Survey,* eds. F. Newmeyer and R. Ubell. Cambridge: Cambridge University Press (1987).

Montague, R.

1974 *Formal Philosophy.* New Haven, CN: Yale University Press (1974).

Newmeyer, F.J.

1980 *Linguistic Theory in America.* New York: Academic Press (1980).

Pereira, F.C.N., and D.H.D. Warren.

1980 Definite clause grammars for language analysis. *Artif. Intel.* **13** (1980) 231–278.

Shieber, S.M.

1986 *An Introduction to Unification-Based Approaches to Grammar.* CSLI. Stanford, CA: Stanford University, (1986).

Turing, A.M.

1950 Computing machinery and intelligence. *Mind* **59** (1950) 433–460.

Vestre, E.

1987 *Representasjon av Direkte Spørsmål.* Thesis, University of Oslo (1987). In Norwegian.

Finite Physics

David Finkelstein[*]

1. Introduction

Since I wish to make an offering to the spirit of Alan Turing and at the same time must write about what I do, I propose to examine here how some concepts central to Turing's work in logic and computer theory, such as finiteness, completeness, hierarchy, self-reference, and universality, are expressed in the physics I have been studying lately.

In the two-way traffic between physics and computers, what physics has given computers is mostly hardware, the basic computing elements themselves. The return to physics from computer science has been, in the first place, computational power, creating whole new fields, such as soliton theory, chaotic dynamics, and lattice quantum field theory. But in the second place, since the birth of computer theory there has also been a diffusion of concepts: for example, to mention an extreme case, the concept of the physical world as computer. Turing was involved in all of these transactions, but this paper concerns the conceptual one.

This choice of topic permits me to work up and incorporate a progress report on my own project for a finite quantum theory of nature. An early version of such a theory represents a path in quantum time space as a quantum analogue of a binary sequence, like a code or program for a simple classical machine of fixed topology and Boolean logic. (Hence the name, "space-time code".) The latest permits exact local Lorentz invariance, provides a deeper quantum-logical substructure for time space topology and spin, and allows for many internal dimensions; but the correspondence-limiting process that brings us from quantum time space back to the classical time space continuum has not been found.

---

[*] This paper is based upon work supported in part by National Science Foundation Grant PHY 8410463.

The classical concept of physical universe is suspect in quantum theory: according to Bohr there is no quantum universe. In the last section of this paper I point out that Turing's concept of universal computer suggests an analogous concept of universal quantum system that can be a useful substitute for the possibly nonexistent quantum universe. I propose there a certain extension of the concept of relativity to a third level, counting the level of Newton's and Einstein's relativities as the first and that of Dirac's transformation theory of quantum mechanics as the second. By relativizing the concept of the universe, third-level relativity avoids the special problems of self-reference that arise in quantum theory when we try to treat the entire universe as a physical system. In a third-level relativistic cosmology there is a law of evolution relative to each experimenter but no absolute or constant dynamical law.

2.   Finiteness and Hierarchy

Apparently Turing thought neither the world nor the brain to be a finite-state machine. "Everything really moves continuously", he remarks (Hodges *1983*, p. 419). At the same time, nevertheless, he likened elementary particles to fonts of type, preferred a finite number of dimensions to the customary infinity of quantum theory (ibid., p. 495), and considered brain and presumably world as profitably modeled by finite-state machines for some purposes. Let us consider some pros and cons of such a model, then.

A machine in the simplest sense executes a single unvarying transition rule again and again. So does the world according to physics from Newton until today, where the rule is called the dynamical law. This serial conception of time atomizes the entire process into identical elementary ones, establishing a hierarchy of two levels, the global and the local in time: eternity and the instant. Furthermore, in the simplest models of the brain and in relativistic physics the temporally elementary process is itself composed of still more elementary ones, local in both space and time, which we may call cells.

To model the brain as a finite-state automaton one supposes that each such cell can exist in a finite number of states and can respond in a finite number of ways. One establishes a two-level hierarchy, of brain and cell, and assumes an overall synchronism, an absolute time, making the brain a serial machine.

Such a brain model was already out of date in Turing's time. In the first place, much of the brain is FM, not AM. The inputs and outputs of many nerve cells are best regarded not as solitary impulses but as impulse frequencies, which may also be called impulse currents or impulse probabilities per unit time. Input impulses do not exactly determine output ones; input rates determine output rates. The variables of a brain cell are more conveniently represented as continuous than discrete for many purposes.

Quantum field theory falls somewhere between current brain models and a Turing machine in its discreteness. On the one hand, quantum field theory puts yes-or-no or integer-valued variables into each region, counting the quanta of various kinds, and is in this respect closer to Turing's idealization than to the usual brain theories. On the other hand, in quantum field theory as in the brain, individual input quanta do not causally determine output quanta.

The brain is not serial but highly parallel. If we speak of the logical breadth and depth of a system as the number of processes carried on in parallel and in sequence, then the brain is closer to a logical square than to a line, both numbers being very roughly 10^{12}.

In this sense quantum field theory is more parallel than even the brain, being conspicuously four-dimensional rather than two-dimensional. In the fractal sense, however, the brain has much higher dimensionality than our space, with about 20 dimensions compared to the usual 3. This expresses the fact that any two of the 10^{12} brain cells are only several – say 4 – cells apart, and $10^{12} \approx 4^{20}$, while two of the say 10^{180} cells of space may be 10^{60} apart, and $10^{180} \approx (10^{60})^3$. Time space cells do not have as many dendrites as brain cells.

The time space continuum sharply distinguishes present quantum field theory from both machine and brain. In consequence of this continuum, quantum field theory is not quite consistent, in being infested with divergent integrals, and even when the convergence is improved, as in string theory, not quite operational, in asserting the existence of unattainable points. Below I consider alternatives to the continuum motivated by these problems of practice and principle.

3. Prequantum Locality

Hierarchic structure survives in continuum theories of the world as a principle of locality: the laws of nature are differential equations. In such theories the world is made up of points of time space, and laws of nature govern differential neighborhoods of each point. Here the two levels of hierarchy are the global and the punctual, and locality tells us how the points are connected.

"Science is a differential equation. Religion is a boundary condition", Turing once quipped (ibid., p. 513). This expresses locality as well as a Newtonian conception of God and nature. While Newton was not a strict Newtonian in this sense, he held locality even above his celebrated law of gravity. That law is nonlocal, relating the acceleration of the earth here and now to the position of the sun millions of kilometers away, and so Newton held it to be less than fundamental, despite its grand successes. First Poisson's differential equation, and then some centuries later Einstein's, vindicated Newton's faith in locality, exhibiting Newton's law of gravity as a special solution of a local law which we recognize as more fundamental than

Newton's precisely because of its locality. Thus in one form or another hierarchy and locality existed before mechanics and have outlived it.

The elemental transitions of a nonrelativistic theory of the world (like the Schrödinger equation, for example) occupy a short time but all of space. A decomposition into short times does not force a decomposition into small cells, as Newton's law of gravity illustrates. In a relativistic theory, however, the elemental transitions must be small in order to be truly brief – that is, brief to all observers. If we still wish to speak in terms of cells, rather than transitions or events, we must imagine that these cells appear, accept inputs from their neighbors in the immediate past, send outputs to their neighbors in the immediate future, and vanish. The cells of a brain model or machine have relatively great duration, amounting to a difference in dimensionality. In the kind of picture where world cells are shown as points, a brain cell or machine cell is a line.

In computing machines, there is usually a preferred rest frame determined by the machine cells or hardware, and respected by the machine-level software. Each datum usually has an absolute address. In that sense the world model of relativistic physics, lacking permanent cells, preferred rest frame, and absolute coordinates, has no hardware and might not be called a machine.

4. Experimentality

To be sure, the absolute addresses of the machine may be deliberately hidden from higher-level software by lower-level, and there is no merely logical inconsistency in assuming a similar conspiracy in nature. But an unobservable rest-frame conflicts with what we may call a principle of experimentality: entities that exist can in principle be determined experimentally. Since the theory says what exists and what experiments are possible in principle, experimentality is both a theoretical and an empirical criterion.

One way to build in experimentality is to take the basic theoretical entity to be experimental process itself; much as William James insists that the ultimate units of experience must themselves be experiences. Then one regards each ψ (up to a scalar factor) as standing for a large class of different experimental processes with statistically indistinguishable outcomes. But this seems more appropriate for a phenomenological theory than a fundamental one.

5. Completeness

Quantum theory, suitably formalized, provides an algebra of properties or predicates for a quantum system, a language for experiments on the system, and a rule to determine the truth and falsity of some of its statements: a syntax, a semantics, and a logistics. So does classical mechanics. Neither are complete in the sense of Post, Gödel, or Turing: neither Newton's nor Heisenberg's laws suffice, for example, to tell us the attitude and azimuth of the Moon at noon today. But there has at least been the illusion that Newton's laws are completable with boundary data – if only by God, to return to Turing's aphorism quoted above.

There is no such illusion of completability about quantum theory. Any taking of data invalidates other data, according to the principle of complementarity, and so making some questions about the system decidable makes other questions undecidable. Quantum theory did not invent incompleteness, the stuff that grows at infinity in Newton's mechanics, but it has chopped it fine, sprinkled it throughout all time and space, and rubbed our noses in it.

When Einstein and Bohr debate the incompleteness of quantum theory they use different languages. Einstein's completeness is closer to the logical one used above, but Bohr dubs a quantum theory complete (and von Neumann, maximal) if it merely predicts all that is predictable. Some are distressed by this lower standard, but I find it liberating. Although several founders of the quantum theory expect us not to understand it, few forbid us to enjoy it. And in actuality, despite its incompleteness quantum theory gives us a kind of control and understanding of nature that classical physics never imagined.

6. *Quantum Locality*

The world of quantum theory is less machine-like than classical field theory, not only because quantum processes are indeterminate (indeterminate machines are also useful models) but because complementarity modifies our very concepts of hierarchy and locality. I mention but two such alterations.

In the first place, while in the hierarchies of computing machines and of prequantum physics the upper level reduces to the lower one, in that any full description of the whole is a conjunction of descriptions of its parts, in quantum physics such a reduction is impossible. Determinations of some quantum properties are in principle incompatible with determinations of others: the order in time of such determinations is sometimes material to the results; and the distributive laws of Boolean algebra, relating the logical particles "and" and "or", cannot be assumed for properties of the world. (These statements follow from the quantum principle of superposition.) The quantum whole has more properties than the sum of its parts.

To be sure, a kind of incompatibility occurs in machines: for example, Moore *1956* describes an "uncertainty principle machine". But to account for quantum incompatibility so mechanically is to renounce locality, as illustrated notably in

Bell's Theorem. By accepting quantum theory we once again uphold locality, now even at the price of classical Boolean logic.

Again, to localize a transition in time requires an infinite spectrum of frequencies (this is not peculiar to quantum theory) and therefore of energies (this is). In particular, localization requires both positive energies and negative energies. The positive-energy transitions increase the energy of the system and are therefore inputs, creators or emitters of quanta, and the negative-energy ones decrease the energy and are outputs, destructors or absorbers of quanta. Therefore, while in prequantum theories and machines there may be local inputs and local outputs, in quantum theories what is local cannot definitely be an input but must be a coherent quantum superposition of input and output, creator and destructor, and what is determined to be, say, a creator or input cannot be local.

In this matter, too, in a certain sense, modern quantum field theory is loyal to locality, now even beyond experimentality. We take local processes to be fundamental although our actual experiments involve nonlocal processes of quantum creation and destruction. We manage to account for our actually nonlocal experiments with local laws governing local processes, though our present accounting system has many infinite entries.

The ultimate expression of locality is a principle of gauge invariance, allowing us to change frames independently at different places, and the contemporary successes of the gauge principle in describing gravitational, electroweak and strong interactions are thus ultimate vindications of locality. To be sure, the quantum theory of gravity itself suggests that we cannot localize better than the Planck time, about 10^{-43}s, or the Planck length, 10^{-35}m, so that zero-size points seem to violate experimentality. But this merely modifies the concept of locality without negating it. Perhaps the local entities of nature have finite extent.

That the principle of locality is so old does not mean that it is moribund. It may even be tautologous: perhaps a point interacts only with neighboring points because ultimately we regard points as neighboring just when they interact. This, however, still leaves two remarkable empirical residues to explain:

1. All our forces – gravitational, electromagnetic, and so forth – seem to agree on what points are neighboring.
2. These neighborhoods seem to form a four-dimensional manifold at the macroscopic level.

In a unified theory, where all the interactions are one, the first of these facts becomes clear. The second is the subject of intense research today. The approach to these questions that I describe next is a personal one.

7. *Postquantum Locality*

To make the choice of dimension physical rather than metaphysical, natural rather than supernatural, we must suppose that the topology of the world – the pattern of connections – is not fixed but variable. Such variability is common to many practical machines. Both the Enigma, the code machine that Turing cracked, and the Colossus, the computer with which he cracked it, had plugboards. The Mark IV at Harvard, an early relay computer, switched its connections magnetically. Hand-thrown toggle switches were part of the memory of the MIT Whirlwind I. The brain, too, gradually changes its own topology with experience, by adding and deleting synapses.

Our theories of the world today, however, are not yet as flexible in their topology as our machines or our brains. In special relativity the pattern of connections between time space points is fixed once and for all by, and indeed constitutes, the flat geometry of Minkowski. For the pattern of causal relations among the points of time space defines its topology, and the Minkowski geometry is the unique four-dimensional manifold so defined by causal connections of maximum symmetry; I refer here to famous theorems of A. Alexandrof and C. Zeeman.

General relativity comes slightly closer to a flexible topology when it allows that the world is a four-dimensional manifold. Its pattern of connections sufficiently near any point is like that of Euclidean geometry in four dimensions, but it may have non-Euclidean topology in the large. To be sure, this postulate freezes or hard-wires an infinite amount of local topological information. But it does thaw the global manifold topology. It admits, for instance, various numbers of handles, also called bridges, wormholes, and hyperspace tunnels – and more exotic defects, too. There are many fewer global topological variables, however, than local topological constants.

Quantum theory, applied directly to the continuum theory of the gravitational field, suggests that such global defects proliferate as we experiment with smaller regions and result in what Wheeler calls a space-time foam. Thus the underlying manifold topology that we postulate recedes tantalizingly as we near it, and we can never quite touch it. Assuming such a continuum violates experimentality.

To avoid such a violation, one seeks a physical topology with no underlying time space continuum, but with the continuum as a smoothing approximation useful for long times. In such a topology, the usual continuum of zero-size points is only a workable approximation to a finite but immense network of cells of some nonzero size set by a physical constant **H** with the units of time, and the operative principle of locality is that of connection among these cells. In the extreme form, a pure topological physics, topological variables suffice for all descriptions. Connections are everything.

The problem then is the length and the steepness of the descent from classical time space to quantum topology. For Einstein such a quantization program is "trying to breathe in empty space". But the center of the earth seems an even more appropriate metaphor for our destination, so we shall undertake here a multistage expedition to the core.

At the surface we breathe Einstein's theory of gravity and time space. This supports the first descent, to a Kaluza-Klein-DeWitt hyperspace, whose hypergravity produces not only ordinary gravity but also the electroweak and strong gauge interactions. To go further we notice that gravity and time space are entirely supported by two-component spinors expressing, as Pauli put it, a fundamental quantum two-valuedness, and we suppose that hyperspace is supported exactly in the same way by N-component spinors representing a fundamental quantum N-valuedness. This turns the hyperspace into the Bergmann manifold described below. Beneath this quantum N-valuedness and its associated $SL(N, C)$ invariance group we find a quantum topology, in turn supported by a quantum set theory, all on the back of a quantum logic. (This exposition inverts and straightens out the actual path taken.)

8.   Bergmann Manifold

The idea that time space bears spinors and a spin vector goes back to the youth of quantum theory. At first Pauli and Dirac tack a spin vector σ onto a pre-existing flat time space in order to describe the electron; and then to treat an electron in a gravitational field, Infeld and van der Waerden distribute spin vectors over a curved time space defined by Einstein's fundamental chronometric tensor. Then Bergmann *1957* realizes that the spin vector alone suffices, that σ is deeper than and defines the chronometric and the gravitational field and not conversely.

Spin is the clock and meter-stick of the world, his work suggests. This seems the greatest deepening of our understanding of the nature of geometry since Riemann, who physicalizes geometry, and Einstein, who imbeds it in chronometry. The fruits of such a spinorization of geometry include celebrated works of Penrose and Newman, who use the spinor and its twistor generalization in classical gravity, and Ashtekar, who takes spinor variables into quantum gravity.

Spinorism, the doctrine and program of describing all the fundamental entities of nature solely by spinors, subsumes such spinorization and is somewhat older. By 1957 Penrose was already deep into his theory of spin networks, and Weizsäcker's spinorial theory of fundamental binary quantum alternatives, or urs, was several years old. Their work provides the house of spinorism with two wings. Spinorists like Penrose develop the classical geometric meaning of spinors and seek such meaning for other ψ functions as well, shaping a quantum theory that partakes more of the classical. Spinorists like Weizsäcker regard spinors as describing a fundamental quantum two-valuedness and seek to leave the present quantum theory by the exit facing away from the classical.

If a quantum 2-valuedness gives rise to 4 time space dimensions, a quantum N-valuedness gives rise to N^2, and thus to extra "internal" dimensions that may unify gravity with other interactions in the way explained by Kaluza, Klein, and DeWitt. Such a quantum N-valuedness arises in the quantum topology described below, with the group $SL(N, C)$. This leads at the continuum level to the concept of a *Bergmann manifold* or B_N, an entity of the following parts:

1. A differentiable manifold of yet unspecified dimension n; also called B_N.
2. At each point x of B_N, a linear space $S = S(x)$ of complex N-component vectors called "spinors", the ψ vectors for a local quantum system called *spin*. S represents N-valued quantum experiments on the spin at x in the usual quantum way, although with no Hilbert metric as element of structure. Sometimes spinors of more than two components, and their affiliated constructs, are called *hyper*.
3. In each spin space $S(x)$ a volume element (the *Grassmann form*) $\gamma_{AB...C}(x)$, a skew-symmetric tensor of valence $[^0_N]$, like the γ_{AB} of Bade and Jehle *1953*. This makes S a *Grassmann space* in the sense of Barnabei, Brini, and Rota *1985*. The structural group is therefore $SL(N, C)$.
4. At each x, a real linear isomorphism $\sigma = \sigma(x)$ called the *spin vector* from the linear space of Hermitian forms in S to the tangent space to B_N at x.

It is understood that the Grassmann form and spin vector vary smoothly with x. Because of the spin vector the dimension of a B_N is $n = N^2$, the number of independent Hermitian forms on S. Thus a B_2 has four dimensions and is locally indistinguishable from the original curved time space of Einstein. For the theory of B_2's see Penrose and Rindler *1984-7*.

To reduce confusion, I save the word "metric" for a positive-definite Hilbert metric μ defining quantum probabilities, and use "chronometric" for an indefinite time space pseudometric g defining times. Thus we say for example that time-like differentials are spin images of *metrics* and have positive *chronometric* norm.

The spin vector "solders" to each tangent vector v a Hermitian $N \times N$ matrix. The norm $\|v\|$ is defined as the determinant of that matrix, aside from a purely numerical factor. Thus the fundamental form $\|v\|$ of a B_N is an N-ic form in the components of v, not as in Pythagoras's theorem a quadratic one. Therefore the chronometric tensor g, which had two indices for Einstein and Riemann, now has N indices. A B_N is a Finsler space, and non-Riemannian except for $N = 2$.

One hesitates to leave the known ground of quadratic forms, but it seems possible to do physics with N-ic forms in a way that specializes to the old brand when $N = 2$. For example a Bergmann manifold has a unique *Holm (1986)* connection respecting the spin vector and so also the chronometric, and reducing to Christoffel's for $N = 2$. To be sure, the natural wave equation $\|\partial\|\Phi = 0$ has differential order N, and has

N sheets to its null or characteristic cone. But the differential order with respect to the usual four dimensions remains 2.

This rather complex chronometric structure all ultimately rests on the simple Grassmann form γ. First γ leads to an invariant concept of determinant and volume element, in which the usual Levi-Civita tensor density ϵ is replaced by the true tensor γ. If μ is any probability metric in the spinor space at a time space point x, then σ, the spin vector there, associates the time-like future vector $\delta x = (\delta x^m) = \sigma\mu$ with μ. The determinant of μ defines a density $\Delta = \|\mu\|^{1/N}$ in spinor space, and Δ defines the proper time $\delta\tau$ of δx in seconds by

$$\delta\tau = \mathbf{H}\Delta \tag{1}$$

The physical constant $\mathbf{H}$ thus converts the dimensionless probability density in spinor space into real time, as c translates mass into energy in relativity and h frequency into energy in quantum theory.

The Einstein-Hilbert action R generalizes readily from B_2 to B_N (S.R. Finkelstein *1987*). Since the wave equation has order N, one seeks a quasilinear N-th order action principle. The simplest is

$$R := g^{\{a\}} R_{aa';\{a\}''} := g^{\{a\}} R_{(a)}$$

where $\{a\}$ is the N-adic symmetric collective index $a, a', a_2, \ldots, a_N$ and $\{a\}''$ omits the first two of these N indices. (I omit the gravitational and cosmological constants, which may be added as usual.) This R is the complete contraction of the $(N-2)$-order covariant derivative of the Ricci tensor, and is as close to the Ricci scalar as we get in B_N.

R yields the field equations

$$R^a{}_b - \frac{1}{N} R\delta^a{}_b = T^a{}_b$$

I have contracted $N-1$ of the indices of $R_{(b)}$ with indices of $g^{\{a\}}$ to form the tensor $R^a{}_b$, and have made $T^a{}_b$, which I call the *energy*(-momentum-stress) *tensor* of the source, in the same way from

$$T_{\{a\}} := \frac{\delta L'}{\delta g^{\{a\}}}$$

Finsler geometry or no, the continuity of this theory with Einstein's is evident. There is even a static isotropic "hyper-Einstein" cosmology for each N, that provided by the n-dimensional group manifold $U(N, C)$ (Holm *1987*), and it is amusing that only for $N = 2$ must we add the infamous cosmological term to ρR to legitimize this cosmology.

To see the spin content of the theory, we study small quantum disturbances $\delta\sigma$ in the spin vector σ in normal geodesic coordinates. $\delta\sigma$ is a matrix $\delta\sigma^{A\dot{A}}{}_{\dot{B}B}$ with four spinor indices and each can be "external" $E = 1, 2$ or "internal" $I = 2, \ldots, N$. Each external index contributes spin $\frac{1}{2}$. Thus the modes of $\delta\sigma^{E\dot{E}}{}_{\dot{E}E}$ describe a boson of spin two like the graviton, the modes $\delta\sigma^{E\dot{E}}{}_{iI}$ and $\delta\sigma^{iI}{}_{\dot{E}E}$ describe gauge bosons of spin one like those in Kaluza-Klein-DeWitt theories, and the remaining modes describe gauge quanta of spin $\frac{1}{2}$ and $\frac{3}{2}$ and some described by skewsymmetric tensor fields (S.R. Finkelstein *1987*).

The infinities of the continuum only grow with N. It is time to ask what time space is made of. What is ρ the density of? This is the next stage of our descent.

9. Peano Algebra and Coherent Quantum Logic

When a photon goes through the two-slit diffraction system, may one say that surely it went through slit 1 OR it went through slit 2? Or must one say that it went through slit 1 AND it went through slit 2? Bohr founded quantum logic when he refused to apply the usual AND and OR in such statements about quanta, and von Neumann formalized it when he provided new AND and OR that did apply. According to the von Neumann quantum logic we think of the linear space S as representing a class or predicate $[S]$ of quanta, and should think of $[S]$ as a disjunction of predicates represented by N independent ψ vectors $\alpha, \beta, , \ldots, \zeta$ in S:

$$S = [\alpha] \cup \ldots \cup [\zeta].$$

Here $[\alpha]$ designates the linear space or ray spanned by the vector α, and $\cup$ designates the linear span, a kind of sum. Von Neumann has us add disjoint possibilities, as it were.

Von Neumann's logic is not rich enough for the construction of fundamental quantum physical theories, any more than its classical correspondent, Boolean algebra, is rich enough for classical ones. Set theory is a language adequate for such classical constructions, and I sketch a quantum set theory here. But it calls for a stronger quantum logic than von Neumann's, based on the older ideas of Grassmann.

Grassmann tells us to multiply disjoint possibilities. According to Grassmann, we should represent the linear space S algebraically not as a union or sum but as a kind of *product* of N independent vectors $\alpha, \beta, , \ldots, \zeta$ in it, today called exterior product, written here as

$$\alpha \vee \beta \vee \ldots \zeta := \uparrow$$

There is such a product $\uparrow$ at each point of a B_N. Furthermore, on the Grassmann space S Grassmann would have us build not merely a Grassmann algebra but a *Peano algebra* or double Grassmann algebra, in the following sense of Barnabei, Brini, and Rota *1985*.

A Peano algebra is an algebra with three basic operations $\vee$, $\wedge$, and $+$, each having its own identity element, here written $\downarrow$, $\uparrow$, 0 respectively. $\uparrow$ and $\downarrow$ should be considered as two variants of the number 1. This is the algebra of the *Ausdehnungslehre* of Grassmann, and its two multiplicative operations $\wedge$ and $\vee$, called progressive and regressive products by Grassmann, or *product* and *coproduct* for short here, make it twice as rich as what we call Grassmann algebra today.

It is also our quantum predicate logic. The $\vee$ and $\wedge$ come from classical logic, and the $+$ is quantum superposition. The unit element $\downarrow$ ("bottom") of the $\vee$ operation stands for the null class. That of the $\wedge$ operation, $\uparrow$ ("top"), stands for the universal class. The identity 0 of the $+$ operation is a zero for both $\vee$ and $\wedge$ products and stands for the undefined.

The basic classical logical operations $\vee$ and $\wedge$ that correspond to those of the Peano algebra are not the OR and AND gates $\cup$ and $\cap$, of ordinary computer logic, which lead to the von Neumann lattice quantum logic, but are partial operations, not always defined. The classical operation $A \vee B$ is the disjoint union or *disjoin,* not defined unless A and B are disjoint, and then their union. The disjoin is called arithmetical addition and written $A + B$ by C.S. Peirce (1868). Its quantum version, the product $A \vee B$, defines an exterior algebra $\vee(S)$ with elements of S as first-grade vectors and the complex scalars as 0-grade coefficients. The classical $A \wedge B$, the exhaustive conjunction, is the dual of $A \vee B$, undefined unless A and B are exhaustive (have union $\uparrow$) and then defined as their conjunction or intersection. (This differs from the "arithmetical product" of Peirce (1868).) Its quantum version, the coproduct, again defines an exterior algebra $\wedge(\tilde{S})$, but over $\tilde{S}$, the covectors (grade $N - 1$ elements) of S. There is a complete duality between $\vee(S)$ and $\wedge(\tilde{S})$. The (grade 0) scalars and (grade 1) vectors of each are the (grade N) coscalars and (grade $N - 1$) covectors of the other.

The Peano algebra was from birth regarded as analogous to a logic with its two operations of AND and OR: Peano puts his logical algebra of $\cup$ and $\cap$ and the Peano algebra side by side in the same work. In quantum theory this structural parallel at last becomes a functional one: the Peano algebra not only looks like a logic, it is used as a logic, namely a quantum logic. It is better than the more familiar lattice logic for some purposes:

First, classical logic had a symmetry we took for granted but which common quantum logic lacks. The Boolean algebra of classes or predicates for a system (supposed finite here for simplicity) is isomorphic to the Boolean algebra of sets of such systems. In the lattice quantum logics, this symmetry is broken: the algebra of predicates is a lattice, while the algebra of sets that we use to deal with a plurality of quanta (Fermi-Dirac ensembles) is a Grassmann algebra. Here both algebras are Peano algebras, and the classical isomorphism is restored.

In quantum time space this isomorphism is particularly useful, for two points of time space (for example) define a predicate for a quantum in the time space, and yet at the same time define a set from the point of view of the dynamics of the time space itself, just as the two slits of the Young experiment represent possibilities for the photon and actualities for the experimenter. It seems important for the future development to make the transition from the theory of possibilities to that of actualities as smooth as possible, and with Peano logic this transition is an isomorphism.

Second, the groups of the Peano logics include the $SL(2, C)$ group of special relativity, as well as the unitary groups of particle symmetries in complexified form.

Third, the Peano logic describes more delicate experiments than the lattice one. The old logic assigns to each predicate a definite multiplicity, whose logarithm measures information content. The new one deals with all of the old predicates and also with many more of indeterminate multiplicity. The new experiments permit coherent phase relations between endosystem and exosystem excluded by the Copenhagen quantum theory. To express this I call the new quantum logic *coherent* and the old incoherent. The form of coherent logic given here is that appropriate to odd entities (fermions), the generalizations being clear. An average over the new phase relations of the coherent logic returns us to the incoherent one. Such an average is also involved in the correspondence from quantum topology to classical time space.

One can also build a quantum logic on two group operations, the classical both-or-none (or equivalence) operation Λ and its dual, the either-but-not-both (or inequivalence, or XOR) operation V, with identities ↑ and ↓. Since every point is then a square root of the identity, this leads to a Clifford-algebraic logic. There are thus three fundamentally different kinds of quantum logic, each with its own uses, with the three generating relations:

$$A \cup A = A, \qquad A \cap A = A \qquad \text{(lattice logic)}$$
$$A \, V A = \downarrow, \qquad A \, \Lambda A = \uparrow \qquad \text{(Clifford-algebraic logic)}$$
$$A \vee A = 0, \qquad A \wedge A = 0 \qquad \text{(Peano-algebraic logic)}$$

The lattice logic does not have an isomorphic quantum set theory, the Clifford-algebraic one does not have the group $SL(2, C)$, and the Peano-algebraic used here has both.

10. *Topological Algebra*

For Clifford, Grassmann algebra expresses topology and Clifford algebra metrical geometry. Grassmann would multiply two edge-vectors of a triangle to represent the triangle, for example.

We may instead multiply the three vertices of a triangle to represent the triangle itself, and $N + 1$ elements to represent a higher dimensional generalization of a triangle, an N-dimensional simplex. The first mention of this use of Grassmann algebra I find is by Chevalley *1954*, who associates any simplicial complex with a Grassmann algebra generated by its vertices. The boundary operator is a certain derivative or destructor δ on this algebra. We use the product to build up simplices and the complexes and the coproduct to break them down.

In the classical simplicial theory, a sum of vertices is not a vertex, and the Grassmann algebra of a complex has an absolute basis and metric. In the quantum theory, sums of vertices are again vertices, each experimenter provides a relative basis and metric, and general linear transformations of the basis have meaning.

The exterior algebra of Chevalley does not itself express the topological structure of the complex, leaving this task to an element of the algebra representing the complex, or to a coboundary operator. It is also possible to express such information algebraically by equating to zero those products representing cells that do not belong to the complex. The result, which I call the topological algebra of the complex, is a generalized or "constrained" Peano algebra, a direct sum of Peano algebras. The topological algebra reduces to Chevalley's free Grassmann algebra when we forget the $\wedge$ product only in the case of a simplex, whose topological algebra is generated by the N factors $\alpha, \beta, \ldots, \zeta$ of the above product $\uparrow$.

The topology of a collection of such cells (or a simplicial complex) is determined by their incidence relations. For example, the space with basis vectors α and β is defined to be adjacent to that with basis vectors β and γ because the edges $\alpha \vee \beta$ and $\beta \vee \gamma$ are both incident upon the vertex β.

11. Quantum Complex

Now for the quantum end of the line of theories of time space. I suppose that rather than being attached to otherwise defined time space points, as in a Bergmann manifold, local simplices like $\uparrow(x)$ *are* the cells of time space, defining the topology of the world by their incidence relations as above, that the chronometric density ρ tells the number of such simplices or of their vertices per unit coordinate volume, and that a time-like future vector v^{A^*B} is a Hilbert metric or statistical operator on the local quantum space, as its transformation and positivity suggest. The Bergmann manifold is a smoothing or averaging of this complex, and its spin map relates our empirical coordinate vectors w to the intrinsic vectors $\psi\psi^*$ of the complex by $w = \psi^*\sigma\psi$. Thus the time space light-cone structure, which establishes the distinction between time and space and the local speed of light, is determined by the spin map. In this sense the world is a quantum complex, a complex of cellular quanta, and its chronometry derives from its topology.

Thus the fundamental variables of the quantum complex, as of the already highly developed theories of supersymmetry and supergravity, are Grassmann or anticommuting ones. At the quantum end of our line of time space theories I suppose a quantum-logical theory of a pure quantum topology built entirely on anticommuting variables, a cellular time space structure with relativistic and gauge symmetry at each cell. That quantum theory permits discrete time space structure with continuous symmetry is Hartland Snyder's inspiration.

Kaluza considered hyperspaces with more than four classical dimensions of macroscopic extent, but Einstein and Bergmann found the extra dimension to have quantum size (proportional to Planck's constant). Now this suggests that any extra coordinates have a discrete quantum nature. A trussed roof actually made of three-dimensional cells, tetrahedra, may seem a two-dimensional surface from a distance: its external or fractal dimension is not the same as its internal one. Such a dimensionally heterogeneous structures are inevitable in quantum topology, for even if Ψ is dimensionally homogeneous, $\Psi \vee \Psi^*$ is not. For example, if a plexor Ψ is a superposition of triads $\alpha \vee \beta \vee \gamma$ arising from the tiling of the plane by equilateral triangles, so that the internal and external dimension of Ψ are both two, then $W = \Psi \vee \Psi^*$ has external dimension four (because $2 \times 2 = 4$) but, since cell $\alpha \vee \beta \vee \gamma \vee \alpha^* \vee \beta^* \vee \gamma^*$ is a hexad, internal dimension five. If Ψ is itself dimensionally heterogeneous, then $\Psi \vee \Psi^*$ is more so. Such discrete quantum coordinates can still support the continuous groups of gauge theory.

Whatever the best dynamical principle for the quantum complex, the averaging transition from quantum complex Ψ to classical manifold B_N must yield an effective action for B_N that includes Einstein's if it is to provide a quantum foundation for the Einstein theory of gravity and for the Kaluza-Klein-DeWitt higher-dimensional theory of the other forces. In this theory the gauge group of the forces is the structural group $SL(N, C)$ of the complex.

12. Quantum Set Theory

Polygonal models of nature appear in the *Timaeus,* in molecular chemistry, and in the more recent works of Regge, T.D. Lee, and others. These models, however, still assume a flat metrical or chronometrical continuum within the polygon, exactly what we wish to analyze here. The next step is to strip them of this continuum. This leaves us with purely topological and hierarchical depictions, like family trees, or organizational charts, or LISP programs. The spin networks of Penrose and the causal spaces of Sorkin *1987* are of this deepest class.

Another such theory is the quantum complex as formulated in quantum set theory. Quantum set theory is expressed by a Grassmann algebra SET which has, besides the operations $\vee$ and $+$ and the identity elements $\downarrow$ and 0 already explained, just one additional element of structure, a linear operator ι (Peano's symbol) carrying any set A into the set $\iota(A)$ having A as sole element. ιA is usually written $\{A\}$ nowadays, and the set that is usually written $\{A, B, C\}$ is $\iota A \vee \iota B \vee \iota C$ for Peano (and me), provided A, B, and C are distinct. Because there are an infinite number of finite sets there are no coproduct $\wedge$ and no universal element $\uparrow$ in SET except in an "ideal" sense.

In this model there are elementary experiments, and elementary Grassmann spaces to describe them. The spinor spaces of the Bergmann manifold are the elementary Grassmann spaces that remain after we average the quantum complex into a classical manifold. The constrained topological algebra of the complex corresponds to the algebra differentials on the manifold, which likewise form a constrained Grassmann algebra, since differentials at different points annihilate each other. The elements of this Grassmann algebra play the role for simplex topology that state vectors do for dynamics.

There is no entirely novel mathematical element in quantum set theory. Its $+$ and $\vee$ are well known as quantum superposition and fermionic combination, and $\wedge$ is the Hodge dual of $\vee$.

Its ι has a precedent, too. When we regard vectors as Grassmann products of spinors and antispinors, being second grade, they commute with each other. To form a Grassmann algebra over them, however, we take them to anticommute. We may say that we form anticommuting replicas of them. And that is just what ι does. The first-grade elements of the vectorial algebra are ι-maps of the second-grade elements of the spinorial algebra. Any novelty is in the quantum logical use of these Grassmann and Peano linear operations to express quantum correspondents of the classical predicate operations $\vee$ and $\wedge$.

Conspicuously lacking form this quantum set theory are three prominent quantum theoretic modes of composition: the tensor product, the tensor algebra, and the symmetric tensor algebra. I restrict myself to the Peano algebra in the interest of unity, much as classical finite mathematics builds its universe with the classical counterpart, set algebra: and also for the sake of a finite physics, noting that the two other tensor algebras mentioned are infinite-dimensional and engender risk of divergences. These are clues, not arguments, but one must proceed somehow.

In a quantum set theoretic approach, it seems arbitrary to cut off the hierarchy at two ranks, and indeed almost impossible to do so. If the world is made of points, then in a pure set theory the points must also be made of something. It takes quantum set theory at least five ranks to generate 10^{180} time space points from the null set. Any further contribution of the deeper ranks to physics is unclear.

13. Principle of Fineness

This section concerns unsolved problems. The central one is the correspondence principle that links quantum time space concepts to classical ones as $\mathbf{H} \rightarrow 0$, a limit analogous to $c \rightarrow 0$ and $h \rightarrow 0$. Small systems have high energies and frequencies, and the near-static large-scale gravitational field of classical physics is an expectation value, a statistical parameter describing an average over topological variables that washes out the high frequencies. Thus the correspondence limit $\mathbf{H} \rightarrow 0$ includes an averaging over high-frequency topological variables. Within the Grassmann algebra of the quantum complex it must construct the commutative algebra of classical coordinates, presumably through the quantum law of large numbers, and the generalized Grassmann algebra of differentials of the manifold. This stretch of theory is still in the design stage. Perhaps here where the theory is still most fluid I may illustrate the heuristic principles guiding its development.

The microscopic correspondents of quantum variables like spin and charge are simple enough, but continuum concepts like coordinate, energy, momentum, and mass are even more troublesome here than in general relativity. Einstein thought of coordinates as "arbitrary" functions, in the tradition of 19th century mathematics. Actually they carry all the topological information about time space. We cannot take along all these continuum concepts and relations in our descent into the realm of quantum time. Which do we cache for the return trip, and which do we tote?

I believe we must travel light. My first general rule for the descent is a *principle of fineness*: to distill and carry along the most detailed, most informative, and most local concepts and laws. Thus where classical geometry in the tradition of Euclid sought the weakest, most likely-seeming postulates, aiming for maximum security in its conclusions; quantum chronometry, like quantum physics, seeks the strongest, most unlikely, and most detailed models, courting the highest risk of disaster.

This shapes the quantum topology. For example, since time-like coordinates are more informative than general ones, they must be more detailed, and null coordinates (such as radar coordinates) still more so. But even these are still global, while the numbers they assign to tangent vectors are local, and hence are still more basic. Tangent vectors correspond to edges in an underlying skeleton, and are described by two-index spinors v^{A^*B}, which look like statistical forms or density matrices. This is a mixed concept, and the pure one underlying it must be more basic. So I infer that the fundamental idea underlying the coordinate concept is that of an edge (actually a dual or co-edge) in a graph, described by a spinor, and that the basic spin vector is the identity operator

$$\sigma^m{}_{A^*A} = \delta^{M^*M}{}_{A^*A} = \delta^{M^*}{}_{A^*} \, \delta^M{}_A,$$

of which the usual spin vector is an average.

A significant conclusion follows about scalars or coordinates in quantum topology. In the quantum theory, we do not have a completely given complex with definite edges. Rather, we have a probability for any given edge to belong to any given complex, a matrix element of an occupation number operator. Therefore while in a classical manifold theory a function is either continuous or not, it seems that in quantum topology the ψ vector of a complex defines only a probability for a given function to be continuous.

The fineness principle also shapes the quantum logic. For example, I take the Peano algebra as finer than the lattice because an equation like $A = B \vee C$ is so much more informative than $A \subset B$. (Peirce, too, discusses exactly this difference in information content and, loyal to the Euclidean tradition, chooses as basic not the most informative relation but the most probable one, $A \subset B$.) Again, although set theory is often said to be the theory of the membership relation $A \in B$, to quantize set theory I relinquish $A \in B$ in favor of the more informative algebraic relation $B = \iota A$. The fineness principle biases one towards equations and algebra and away from partial orders and more general relations.

The line of topological, general, and special relativities suggests a natural candidate for a dynamical law of the quantum complex. The action ρR for hypergravity is ineffective in special relativistic theories, because there R vanishes identically. This suggests that a natural action S for topological relativity should be similarly ineffective in general relativity. That is, S should measure departure from being a manifold, just as R measures departure from being flat. And just as the ineffective R gives rise to effective special relativistic actions for small departures from a flat world, one expects the ineffective S to produce effective manifold actions, including R, for small departures from a manifold.

The most conspicuous such variable of the quantum complex is the Euler characteristic χ, the sum for the complex of the numbers of 0-simplices (points), 1-simplices (edges), 2-simplices (triangles), and so forth, with alternating signs. As emphasized by Zumino *1986*, χ is unity for all triangulations of a Euclidean cell, and is ineffective in general relativity. In quantum topology, χ is an alternating sum of occupation number operators. The natural conjecture, then, is that $S = \alpha \chi$, with a possible dimensionless multiplier α.

And why do we see but one time and three space dimensions? Spinorism answers the question of time. A Bergmann manifold of any dimension has but one time-like dimension: it has time-like curves but no time-like surfaces. This follows simply from the existence of an invariant future cone, the positive Hermitian matrices, for if we had two time-like dimensions we could turn any time-like vector continuously into its negative and there could be no future cone. But there still might be $0, 3, 8, 15, \ldots$ spacelike dimensions. Why do we see three?

I do not know. For a higher-dimensional Riemannian manifold, and for most variants of the Kaluza-Klein theory, the structural group is an orthogonal group $SO(n_+, n_-, R)$, and there is no problem in writing down plausible action principles for any n. Indeed, Lovelock *1971* points out, the number of plausible second-order Riemannian actions increases with n. While there are no invariant second-order quasilinear gravitational equations at all for a Bergmann manifold of $N > 2$ (S.R. Finkelstein *1987*) there is a plausible action, the hyper-Hilbert one, of order N.

So spinorism leaves the space dimension problem open for now. The problem first becomes physical in a theory where dimension can change, as in quantum topology, where such changes presumably appear as phase transitions. The world may have had more than four dimensions when it was young and dense, and it may increase its dimensionality again if there is a final compression.

Next comes the entropy question. Suppose there are M quantum complexes corresponding to a given Bergmann manifold: do we superimpose them coherently (entropy zero) or incoherently (entropy log M)? Since light seems to propagate in empty space without line-broadening, I take zero entropy to be a good approximation for time space today. This also raises the issue of superfluidity, as in Nambu's superconducting vacuum. For this we need to pair the odd generators into pairs, and indeed the usual momentum operator ∂_{AB^*} tranforms as a pair destructor, a quasiboson.

I regard it as a further consequence of the fineness principle that one combines the basic quantum elements purely by quantum modes of composition, those tested in the quantum many-body problem, as opposed to classical ones, which are clearly coarser; hence the quantum set theory described above.

14. *Self-reference*

Self-reference is a precondition for life, not a problem. Each living cell carries the plans for itself. But paradoxical self-reference, as in the Cretan paradox and its ilk, is a problem. Paradoxical self-reference seems to threaten when we apply quantum theory to cosmology, as above.

On the one hand we have Bohr's doctrine: there is no quantum universe. (See also Weizsäcker *1985*). Bohr insists that a quantum partition exists separating a quantum part of the universe that is being determined – the endosystem, I call it – from the much vaster part that is determining it – the exosystem. We must use quantum concepts like ψ's and noncommuting operators to describe the endosystem, but the exosystem we must describe with classical language. For Bohr, the quantum partition is a prerequisite for communication: it is on our side of the partition that ordinary language and memory work.

On the other hand, quantum effects must be important when the universe is about the Planck length in size, and when we set out to discuss such quantum effects, it is well-nigh irresistible to write down a wavefunction or ket for them, a cosmic Ψ, despite Bohr's injunction. After all, we already have imperfect communication anyway.

Just as ordinary quantum ψ's describe our view of an endosystem, when we write a cosmic Ψ we say that it describes the universe "from God's viewpoint". This avoids strong self-reference and the apparent violation of the second law of thermodynamics that would occur if we in the universe made a maximal determination of the universe.

But it also recalls Newton's account of Absolute Time as God's Time. It is not that our cosmic Ψ and Newton's Absolute Time result in mathematically inconsistent theories. Nor are they altogether devoid of physical meaning. For example, a cosmic Ψ may limit the partial ψ's used by experimenters living in the universe described by Ψ, with definite experimental implications. But a God's-eye cosmic Ψ is hardly experimental, and experimentality has paid off so magnificently in relativity and quantum theory that it merits respect.

To respect experimentality either we must show maximal self-determination to be approachable in principle, and allow cosmic Ψ's, or show it impossible, and dispense with them. Elsewhere I may use cosmic Ψ's, and then I consider the coherent quantum logic as the one best suited to a quantum universe. Here I present an ultra-relativistic alternative to cosmic Ψ's.

15. Three Relativities

In this century of relativity and quantum theory physicists pay increasing attention to the physics of physicists, for example through transformation theory. By relating one experimenter to another, each transformation theory tells what experimenters are possible, and what information is required to single out one experimenter from all others. A transformation theory is a theory of experimenters.

General relativity and later gauge theories describe the locality of the experimenter as well as the system, allowing transformations of experimenters which vary independently in different parts of the world at different times; and a quantum theory with its unitary transformation theory describes each experimenter by a frame of compatible experiments within the space of all possible ones. Where special relativity is a primitive stick-figure kind of caricature, granting only 10 parameters to an experimenter, general relativity is baroque, ornamenting each frame with four arbitrary smooth functions of four real variables, in infinitely fine detail.

To deal with cosmology without a cosmic Ψ, I propose to supplement relativity and quantum theory with a third and deeper transformation theory, making three levels of transformation theory in all (earlier called c, cq, and q), defined as follows.

An absolute or nonrelativistic physics, such as a solipsistic or totalitarian one, postulates

(1) a universal experimental language,

(2) a universal experiment, and

(3) a universal experimental subject.

The universal experiment is the simultaneous determination of all the variables of the system. The universal subject is the universe itself. Both figure explicitly, for example, in Laplace's fantasy of the intelligence who knows all and does nothing.

In transformation theories of level 1, we deny (1), renounce the universal language with its universal time and space, recognize that different experimenters give different descriptions of the same system, and postulate transformations connecting them. A first-level transformation is like a bilingual dictionary of an inhumanly mechanical kind, giving you a one-word definition in your language for each word in my language.

In a transformation theory of level 2, such as quantum theory, we deny (2), renounce the universal experiment, recognize that experimenters do different experiments, and postulate transformations connecting them. A second-level transformation is like a bilingual dictionary that gives you many possible definitions in your language for each word in my language, each with the probability of its validity. The probabilities sum to unity.

In a transformation of level 3, as in some future post-quantum theory, we deny (3), renounce the universe, admit that experimenters study different experimental systems, and postulate transformations connecting them. A third-level transformation is like a bilingual dictionary where only occasional words in my language translate into yours, and their probabilities do not sum to unity. Some entities exist for me but not for you, and conversely.

Quantum physicists in fact study different endosystems, such as quarks, crystals, and genes. Sometimes we may pay a lip tax to universality and claim that this differentiation and specialization are merely matters of convenience, that all these endosystems are parts of one grand endosystem, the universe. But even in principle there is presently no quantum universe. In field theory, where we deal with the stuff of which everything is made, the Hamiltonians we write down do not describe the entire universe. The exosystem, the part that is left out, is explicitly represented in modern field theory by external sources or currents not subject to variation, by the group parameters entering into symmetry transformations of the theory, and by the infinitesimal variations entering into the dynamical principle.

While Copenhagen quantum theory already renounces the universe, it lacks a transformation theory linking all these different endosystems. The second-level quantum transformation theory of Dirac relates different experimental aspects of one endosystem only.

The simplest third-level transformations merely enlarge or diminish the endosystem. Common (second-level, Copenhagen) quantum theory assumes that the system interface, that between endosystem and exosystem, coincides with that between classical and quantum descriptions. But if the algebra C of a composite system has the form $B \otimes A$, where B and A refer to two parts of the endosystem, it seems permissible to shift the exosystem-endosystem interface to pass through the middle of C, leaving A in the endosystem and absorbing B into the exosystem. After all, there is no fundamental reason why an experimenter cannot have complete quantum information about a small part of herself. Then part of the exosystem is described on the quantum level, and its quantum variables figure as "c numbers" in the theory of the endosystem.

Thus any algebra B that can function as the algebra of some quantum endosystem can by such a shift of interface become part of the algebra of c-numbers. Quaternion quantum mechanics and superquantum mechanics may be regarded as special cases of this general kind.

One may characterize these three levels of relativities by their logics. For the sake of familiarity I consider here the incoherent logics; a similar analysis may be given for the coherent logics. In any of these theories a single experimenter has a repertory of creators forming a complemented Boolean algebra. What varies is the nature of the logic embracing all possible experimenters.

In a first-level transformation theory, this over-all logic is the same as that of any one (complete!) experimenter. In one of the second level it is the ortholattice of projection operators of a Hilbert space. In a third-level one, I propose, it is merely the lattice of subspaces of a linear space – not complemented but complementable.

We may also distinguish these relativities by their algebras of dynamical variables: a commutative *-algebra on level 1, an irreducible *-algebra on level 2, and simply an irreducible algebra on level 3, not starred but starrable.

16.   Beneath Law

Evidently the dynamical law is not an invariant of third-level relativity as it is of first and second. Moreover, the contemporary conception of an a priori action principle and dynamical law is clearly a relic of the medieval concept of the Logos, and I find myself reluctant to perpetuate it. There may well be a natural rather than supernatural origin to the dynamical principle itself, just as the dynamical principle of gravity theory governs the geometry that was previously considered to be a law unto itself.

The kind of substitute for law I envisage is not related to the reciprocity principle of Born, an equation whose solution gives rise to a plurality of possible action principles, supposed to correspond to the plurality of particles. We know better nowadays how one action can give rise to many kinds of quanta.

Rather, it seems to me that what we call dynamical law actually describes the interaction between exosystem and endosystem, and should be regarded as a statistical description of the exosystem from the viewpoint of the endosystem. Thus the hunt for an a priori action principle rests on the still widespread belief in an absolute vacuum and a universal but negligible experimenter.

Indeed in the Feynman path formulation of dynamics, the action principle appears as a ψ vector for a path. Every other kind of ψ vector represents an action of the exosystem on the endosystem, and it may well be that this one does, too. To pursue this concept further requires a quantum theory of the universe, to which I turn next.

17. Universality

Suppose that we always leave ourselves out of the endosystem, lest we be trapped in an infinite hall of dwindling images of images. Then no physicist has or ever will maximally determine the universe, any more than any has ever measured universal time, or simultaneously determined position and momentum.

If I alone cannot determine the universe maximally because of limits to self-reference, then you and I might collaborate in an attempt to do so collectively. We might divide the universe between ourselves into overlapping endosystems. Then each of us may maximally determine his or her own endosystem without problems of self-reference. But if our determinations interfere, we can never fit our partial Ψ's into one valid cosmic Ψ.

This renunciation of the universe is painful at first. Physicists have always felt that the universe was their proper topic. It goes easier with a comforting surrogate made possible by third-level transformation theory, a physical analogue of Turing's universal computer which I call the universal quantum system.

Before Turing, the term "universal computer" might have referred to the computer that the great Laplacian intelligence used in working out the future of the universe. Such a computer would have to represent within itself, with complete precision, the entire universe, including itself. Likely such a universal computer cannot exist in our universe.

Turing's universal computer is one that can emulate or imbed any one actual computer at a time, not all at once. Such a computer can exist, at least as an ideal limit, and Turing was involved in a working approximation to one, the ACE. Similarly, Kron *1939* introduces and applies what may be called a universal electrical network and a universal rotating machine, in much the sense intended here. The transformation of these universal entities into specific ones is of the third level.

Thus even if it turns out that Bohr is right and there can be no quantum universe containing all actual quantum systems, there can still be an ideal universal quantum transforming into any one actual quantum system by a transformation of the third level. The Ψ's of this system, which I call universal Ψ's, are surrogates for cosmic Ψ's. My favorite candidate for the universal quantum system is that described by SET, whose ideal universal element $\uparrow$ may be interpreted either as a set of unconnected points or as an infinite-dimensional simplex: that is, either as a totally disconnected space to which we add connections, or a totally connected space from which we remove connections.

One expects a successor to general relativity and quantum theory to provide a third epistemological constant completing the system of units that begins with c, the limit to communication speed, and h, the limit of control, and thus setting the scale of all the couplings and sizes in the world of particles. Here the constant **H**, the limit to the resolution of our world-picture, plays this role.

We may also expect new unities in a third-level relativity, beyond the mass-energy and frequency-energy unities of the previous two levels. The present theory unites time with probability through Eq. 1, and its transformations unite all physical entities and experimenters.

A transformation theory of the third level does not respect any one dynamical law of the system and may therefore provide a physical basis for all dynamics, in the sense that general relativity does not respect the laws of Euclidean geometry but provides a physical basis for all geometries. Kron realizes this possibility, if only in a limited degree, when he gives the dynamical law for a network in terms of its topology. Similarly, I do not expect an a priori action for the universe.

We began with a nonrelativistic machine. This then evolved into a relativitistic plexus, with active nodes of fleeting transience. But now each node is an elementary experiment, involved in quantum determinations of the properties and laws of the whole process. One wonders what to call this still more relativistic world view. A monadology, perhaps, given its Leibnizian plurality of entelechies; but with such desultory order, not a Cosmos in the old sense, I suppose Turing would agree.

Acknowledgements. I am deeply indebted to Shlomit Ritz Finkelstein, Sarah Flynn, Ronald Fox, Christian Holm, Ernesto Rodriguez, and C.F. von Weizsäcker for discussions.

References

Bade, W.L., and H. Jehle
   1953 An introduction to spinors. *Rev. M. Phys.* **25** (1953) 714–728.

Barnabei, M., A Brini, and G.-C. Rota
   1985 On the exterior calculus of invariant theory. *J. Algebra* **96** (1985) 120–160.

Bergmann, P.G.
   1957 Two-component spinors in general relativity. *Phys. Rev.* **107** (1957) 624–629. I am grateful to R. Penrose for this reference.

Chevalley, C.
   1954 *The Construction and Study of Certain Important Algebras.* Tokyo: Mathematical Society of Japan (1954).

Finkelstein, D.

   1987 Coherent quantum logic. *Int. J. Theor.* **26** (1987) 109.

Finkelstein, S.R.

   1987 Hypergravity Action Principle. *Int. J. Theor.*, in press.

Hodges, A.

   1983 *Alan Turing: The Enigma.* Simon & Schuster (1983).

Holm, C.

   1986 Christoffel formula and geodesic motion in hyperspin manifolds. *Int. J. Theor.* **25** (1986) 1209.

   1987 *Hyperspin Structure of Einstein Universes and Their Neutrino Spectrum.* Ph.D. Thesis, Georgia Institute of Technology (1987).

Kron, G.

   1939 *Tensor Analysis of Networks.* New York: Wiley, New York (1939).

Lovelock, D.

   1971 The Einstein tensor and its generalizations. *J. Math. Phys.* **12** (1971) 498.

Moore, E.F.

   1956 Gedanken experiments on sequential machines. In: *Automata Studies,* eds. C.E. Shannon and J. McCarthy. Princeton, NJ: Princeton University Press (1956).

Peirce, C.S.

   1868a On an improvement in Boole's calculus of logic. *Proc. Am. Ac. Arts Sci.* **7** (1868) 250–261.

   1868b Upon the logics of mathematics. *P. Am. Acad. Arts Sci.* **7** (1868) 402–412.

Penrose R., and W. Rindler

   1984-7 *Spinors and Space-Time,* vol. I., II. Cambridge: Cambridge University Press (1984-7).

Sorkin, R.

   1987 Preprint and personal communication; cf. L. Bombelli, J. Lee, D. Meyer, and R. Sorkin, Space-time as causal set. *Phys. Rev. L.* **59** (1987) 521–524.

Thomas, T.Y.

   1934 *Differential Invariants of Generalized Spaces.* Cambridge: Cambridge University Press (1934).

Weizsäcker, C.F. von

   1985 *Aufbau der Physik.* Munich: Hanser Verlag (1985).

Zumino, B.

   1986 Gravity theories in more than four dimensions. *Phys. Report.* **137** (1986) 109.

Randomness, Interactive Proofs, and Zero-Knowledge – A Survey

Oded Goldreich

Abstract. Recent approaches to the notions of randomness and proofs are surveyed. The new notions differ from the traditional ones in being subjective to the capabilities of the observer rather than reflecting "ideal" entities. The new notion of randomness regards probability distributions as equal if they cannot be told apart by efficient procedures. This notion is constructive and is suited for many applications. The new notion of a proof allows the introduction of the notion of *zero-knowledge proofs*: convincing arguments which yield nothing but the validity of the assertion.

The new approaches to randomness and proofs are based on basic concepts and results from the theory of resource-bounded computation. Elements of this theory are presented only to the extent required for the description of the new approaches.

This survey is not intended to provide an account of the more traditional approaches to randomness (e.g., Kolmogorov Complexity; see also Bennett's account in this volume) and proofs (i.e., traditional logic systems). Whenever these approaches are described it is only in order to confront them with the new approaches.

1. Introduction

When talking about a universal Turing machine, it is implicitly postulated that the input-output relation of the machine (i.e., the function it computes) is the only aspect of the machine that we are interested in. This postulate is naturally imposed upon us when viewing Turing machines as *means* for defining whatever "can be automatically computed". However, when viewing Turing machines (i.e., the computation devices themselves) as the *subject* of an investigation, concentration on their input-output relation should be considered a behavioristic approach.

Recent research trends in computational complexity (i.e., the theory of resource-bounded computation) have even a stronger behavioristic flavor. In these works, one does not consider the input-output relation of a Turing machine, but rather its effect on an (arbitrary) observer (i.e., another Turing machine) with certain natural resource bounds.

Typically, the results we will survey state that the input-output behavior of "ontologically" different machines looks the same to any suitably resource-bounded observer. For example, pseudorandom sequences are defined to be indistinguishable in polynomial-time from truly random sequences, although the first may be generated using substantially fewer coin tosses. Another example is the zero-knowledge proofs, which have the remarkable property of being both convincing and "reproducible" in a slightly weaker sense even by parties which do not have a "real proof".

2. Background: Resource-bounded Computation

We begin this section by recalling the definitions of **P**, **NP**, and **BPP** – the complexity classes corresponding to deterministic, nondeterministic, and probabilistic polynomial-time computations. We continue by presenting the definition of one-way functions, which plays a central role in the construction of pseudorandom generators and in the general results concerning zero-knowledge proofs.

The theory of resource bounded computations is developed in terms of asymptotic behavior. Typically, we will consider the number of steps taken by a Turing machine as a function of its input length, bound this function from above by a "smoother" function, and ask whether the bound is (or can be) a polynomial. This convention allows us to disregard special (short) inputs on which the machine may behave exceptionally good, and helps us concentrate on the "typical" behavior of the machine.

2.1. DETERMINISTIC POLYNOMIAL-TIME COMPUTATIONS

Traditionally in computer science, deterministic polynomial-time computations are associated with efficient computations. (Throughout the article, a *polynomial-time computation* means a computation in which the number of elementary computing steps is bounded by a polynomial in the length of the input.) This association stems from the acceptability of deterministic computing steps and polynomial-time computations as feasible in practice. The preference of deterministic computing steps (over nondeterministic ones) is self-evident. Polynomial-time computations are advocated as efficient due to the moderate growing rate of polynomials and due to the correspondence between problems which are *known* to have "practically efficient" solutions and those *known* to have polynomial-time solutions.

Deterministic polynomial-time computations are captured by the complexity class **P** (*P* stands for **P**olynomial). The complexity class **P** is defined as the set of languages satisfying for each $L \in P$ there exists a Turing machine M and a polynomial $p(\cdot)$ such that the following two conditions hold:

1. On input a bit string x ($x \in \{0, 1\}^*$), machine M halts after at most $p(\|x\|)$ steps, where $\|x\|$ is the length of the string x.
2. On input x, machine M halts in an *accepting state* if and only if $x \in L$. (Otherwise it halts in a "rejecting state".)

2.2. Nondeterministic Polynomial-time Computations

Another interesting complexity class is the set of languages recognizable by nondeterministic[1] polynomial-time Turing machines denoted **NP** (*NP* stands for **N**ondeterministic **P**olynomial-time). Namely, a language L is in NP if there exists a nondeterministic machine M and a polynomial $p(\cdot)$ satisfying:

1. On input $x \in \{0, 1\}^*$, machine M halts after at most $p(\|x\|)$ steps.
2. On input x, there *exists* a computation of M halting in an accepting state if and only if $x \in L$. (Otherwise *all* computations halt in a rejecting state.)

Equivalently, a language L is in **NP**, if there exists a Boolean predicate $P_L: \{0, 1\}^* \times \{0, 1\}^* \rightarrow \{0, 1\}$ such that P_L is computable in (deterministic) polynomial-time and $x \in L$ if and only if there exists a y of length polynomial in $\|x\|$ satisfying $P_L(x, y) = 1$. (The string y can be thought of as encoding the nondeterministic choices of machine M above.) Thus **NP** consists of the set of languages for which there exist short proofs of membership that can be efficiently verified. Both classes (**P** and **NP**) are defined in terms of decision problems (i.e., "is x in the set S?"). Equivalent formulations in terms of search problems (i.e., "given x find y such that (x, y) are in the relation R") are obvious. **P** can be viewed as the class of search problems for which a solution (i.e., a string satisfying the relation with the input) can be found in (deterministic) polynomial-time; while **NP** can be viewed as the class of problems for which a solution, once found, can be verified in (deterministic) polynomial-time.

It is widely believed that $\mathbf{P} \neq \mathbf{NP}$. Settling this conjecture is certainly the most intriguing open problem in theoretical computer science. If true, $\mathbf{P} \neq \mathbf{NP}$ means that there are search problems for which verifying the validity of a solution is substantially easier than coming up with a solution. The vast variety of problems which are **NP**-complete, but are not known to be in **P**, is considered a support to the $\mathbf{P} \neq \mathbf{NP}$ conjecture. A language is **NP**-*complete* if it is in **NP** and every language in **NP** is polynomially-reducible to it (see definition below). Hence, if some **NP**-complete language is in **P** then $\mathbf{NP} = \mathbf{P}$.

---

1 A nondeterministic Turing machine has a transition function which goes from the current local configuration to a (finite) *set of possible* local configurations, rather than going from the current local configuration to the next local configuration.

Definition: *A language L is* polynomially-reducible *to the language L′ if there exists a deterministic polynomial-time Turing machine M so that $x \in L$ if and only if $M(x) \in L'$.*

Traditionally in computer science, **NP**-completeness proofs for languages are considered evidence to the intractability of the corresponding decision problems, since if **P** $\neq$ **NP** then membership in no **NP**-complete language can be recognized in (deterministic) polynomial-time. Among the languages known to be **NP**-complete are *Satisfiability* (of propositional formulae), the *Traveling Salesman Problem* as well as many other optimization problems, and *Graph Colorability* as well as many other combinatorial problems.

2.3. PROBABILISTIC POLYNOMIAL-TIME COMPUTATIONS

Recent trends in computer science regard random computing steps as feasible. Following this approach, we consider computations which can be carried out by *Probabilistic* polynomial-time Turing machines as modeling efficient computations. A probabilistic Turing machine is an "extended" Turing machine that (based on its local configuration) chooses its next move at random (with uniform probability distribution) among a finite number of possibilities. (In a deterministic Turing machine, the next move is determined by the local configuration.) Without loss of generality, we assume that the number of possibilities (for the next local configuration) is either 1 or 2. One can then view the machine as tossing an unbiased coin before each move and determining the next move using the outcome of the coin. On input x, the output of a probabilistic Turing machine M is a random variable defined over the probability space of all possible internal coin tosses. Equivalently, probabilistic Turing machines can be viewed as deterministic machines with two inputs: the ordinary input and an auxiliary "random input". One then considers the probability distributions defined by fixing the first input and letting the auxiliary input assume all possible values with equal probability.

In particular, the complexity class **BPP** (*BPP* stands for **B**ounded-away-error Probabilistic **P**olynomial-time) is defined as the set of languages such that for every $L \in BPP$ there exists a probabilistic polynomial-time Turing machine M satisfying the following two conditions:

1) $Prob(M(x) = 1) > 2/3$ if $x \in L$.
2) $Prob(M(x) = 0) > 2/3$ if $x \notin L$.

It should be stressed that this definition is robust under substitution of $2/3$ by either $1/2 + p(\|x\|)$ or $1 - 2^{-p(\|x\|)}$, where $p(\cdot)$ is an arbitrary positive polynomial. The following thesis underlies all our discussions.

Thesis: *BPP correspond to the set of problems which can be solved "efficiently".*

2.4. ONE-WAY FUNCTIONS

It is generally believed that $\mathbf{P} \neq \mathbf{NP}$, and furthermore that $\mathbf{BPP} \neq \mathbf{NP}$. However, this does not necessarily mean that coming up with hard instances of a "hard" language is easy. (Such instances exist, but may be hard to find.) The reader should note that both $\mathbf{NP}$ and $\mathbf{BPP}$ consider the worst-case complexity of problems, not their average-case complexity. For the results in this article we need a stronger assumption than $\mathbf{NP} \neq \mathbf{BPP}$, namely that there are problems which are hard on most of the instances. This is formulated in terms of the infeasibility of inverting functions, which are easy to evaluate (in the forward direction).

Definition 1. A function $f: \{0, 1\}^* \to \{0, 1\}^*$ is called *one-way* if the following two conditions hold:

1. There exist a (deterministic) polynomial-time Turing machine that on input x outputs $f(x)$.

2. For any probabilistic polynomial-time Turing machine M', any constant $c > 0$, and all sufficiently large n

$$Prob\left(M'(f(x), 1^n) \in f^{-1}(f(x))\right) < \frac{1}{n^c}$$

where the probability is taken over all x's of length n and the internal coin tosses of M', with uniform probability distribution.

Remark: The role of 1^n in the above definition is to allow machine M to run in time polynomial in the length of the preimage it is supposed to find. (Otherwise, any function which shrinks the input more than by a polynomial amount will be considered one-way.)

Motivation to the notion of a negligible fraction: In the definition above, we have required that any machine trying to invert the function will succeed only on a "negligible" fraction of the inputs. We call *negligible* any function $\mu(\cdot)$ that remains smaller than 1 when multiplied by any polynomial (i.e., for every polynomial $p(\cdot)$ the limit of $\mu(n) \cdot p(n)$, when n grows to infinity, is 0). We ignore events which occur with negligible probability (as a function of the input length) since they are unlikely to occur even when repeating the experiment polynomially many times. On the other hand, events which occur with nonnegligible probability will occur with almost certainty when repeating the experiment for a reasonable (i.e., polynomial) number of times. Thus, our notion of an "experiment" with a negligible success probability is robust under polynomial number of repetitions of the experiment.

Motivation for considering infinitely many inputs lengths: The notion of a polynomial-time Turing machine is meaningful only when considering infinitely many input lengths. (Otherwise one can always choose a polynomial which bounds the running time of a machine that halts on all inputs in some finite set.) Furthermore, for any instance length l, there exists a Turing machine M_l which successfully inverts the function f on all instances x of length $\leq l$ within $\|x\| + \|f(x)\|$ steps (machine M_l just incorporates in its transition function the inverses for all instances in this finite set). Both technical difficulties are resolved when considering an infinite set of input lengths.

Assumption: *There exist one-way functions.* Furthermore, there exist one-way 1-1 and onto functions.

The following three number theoretic 1-1 and onto functions are widely believed to be one-way:

- Ex1) **Modular Exponentiation:** In particular, let p be a prime and g be a primitive element of Z_p^* (the multiplicative group modulo p). Define $ME(p, g, x) = (p, g, y)$, where y is the result of reducing g^x modulo p. Inverting ME is known as the *Discrete Logarithm Problem*.

- Ex2) **RSA:** Let p and q be primes, $N = p \cdot q$ and e be relatively prime to $\phi(N) = (p - 1) \cdot (q - 1)$. Define $RSA(N, e, x) = (N, e, y)$, where y equals $x^e \bmod N$.

- Ex3) **Modular Squaring:** In particular, let p and q be primes both congruent to 3 mod 4, and $N = p \cdot q$. Define $MS(N, x) = (N, y)$, where y equals $x^2 \bmod N$. (To make this function one-to-one, restrict x to be a quadratic residue modulo N.) Inverting $MS(N, \cdot)$ is computationally equivalent to factoring N; that is, the problems are reducible to one another through probabilistic polynomial-time transformations.

Remark: The requirement, in condition 2 of Definition 1, that the inverting machine succeeds only on a negligible fraction of the instances of that particular length can be relaxed to requiring that the machine fails on some nonnegligible fraction. Namely,

2') There exists a constant $c > 0$ such that for any probabilistic polynomial-time Turing machine M' and all sufficiently large n

$$Prob\left(M'(f(x)1^n) \notin f^{-1}(f(x))\right) > n^{-c},$$

where the probability is taken as above.

The relaxed form is equivalent to the original Definition 1, both with respect to the existence of arbitrary one-way functions (one-way 1-1 and onto functions), and with respect to the above three particular functions being one-way.

3. Randomness

In this section we survey a recent behavioristic approach to randomness. In this approach a probability distribution is considered "pseudorandom" if no "efficient procedure" can distinguish it from the uniform probability distribution. Remarkably, pseudorandomness so defined is expandable in the sense that (assuming the existence of 1-1 and onto one-way functions) short pseudorandom sequences can be deterministically and efficiently expanded into much longer pseudorandom sequences.

3.1. Definition of Pseudorandom Distributions

A key definition in this approach is that of the infeasibility of distinguishing between two probability distributions. This behavioristic definition views distributions as equal if they cannot be told apart by any probabilistic polynomial-time test. Such a *test* receives as input a single string and outputs some statistics of the input. With no loss of generality, we may assume that the test outputs a single bit, which may be interpreted as a guess of the distribution from which the input was chosen. One considers the probability that, on input taken from the first distribution (resp. second distribution), the test outputs 1. If these two probabilities only differ by a negligible amount, then the corresponding distributions are regarded as indistinguishable by this test.

Preliminaries. (Probability Ensembles): A *probability distribution* is a function π, from strings to nonnegative reals such that $\sum_{\alpha \in \{0,1\}^*} \pi(\alpha) = 1$. A *probability ensemble* indexed by I is a sequence, $\Pi = \{\pi_i\}_{i \in I}$, of probability distributions. Throughout the entire article, we adopt the convention that the probability distributions in an ensemble assign nonzero probability only to strings of length polynomial in the length of the index of the distribution.

Motivation to defining ensembles: Probability ensembles are defined so that we can consider the asymptotic behavior of arbitrary polynomial-time Turing machines on inputs taken from a probability distribution.

Definition 2 (Polynomial Indistinguishability – finite version): Let $\Pi_1 = \{\pi_{1,i}\}_{i \in I}$ and $\Pi_2 = \{\pi_{2,i}\}_{i \in I}$ be two probability ensembles each indexed by elements of I. Let T be a probabilistic polynomial-time Turing machine (hereafter called a *test*). The test gets two inputs: an index i and a string α. Denote by $p_1^T(i)$ the probability that, on input index i and a string α chosen according to the distribution $\pi_{1,i}$, the test T outputs 1 (i.e., $p_1^T(i) = \sum_\alpha \pi_{1,i}(\alpha) \cdot Prob(T(i, \alpha) = 1)$). Similarly, $p_2^T(i)$ denotes the probability that, on input i and a string chosen according to the distribution $\pi_{2,i}$, the test T outputs 1. We say that Π_1 and Π_2 are *polynomially indistinguishable* if for all probabilistic polynomial-time tests T, all constants $c > 0$ and all sufficiently large $i \in I$

$$\| p_1^T(i) - p_2^T(i) \| < \| i \|^{-c}.$$

Motivation to having the index as an auxiliary input to the test: In the above definition, giving the index as auxiliary input to the test is not essential. However, in subsequent definitions presented in this article this convention plays an important technical role. For the sake of uniformity of definitions, we adopt this convention all along.

A probabilistic interpretation of Definition 2: Two probability distributions π_1 and π_2 are equal if they assign identical probability mass to the same string (i.e., if for all $\alpha \in$

$\{0, 1\}^*$, $\pi_1(\alpha) = \pi_2(\alpha)$). Two distributions are *statistically close* if they assign "about the same" mass to the same subsets of strings (i.e., if for all $S \subseteq \{0, 1\}^*$, the absolute difference between the sums $\sum_{\alpha \in S} \pi_1(\alpha)$ and $\sum_{\alpha \in S} \pi_2(\alpha)$ is negligible). Loosely speaking, two distributions are polynomial indistinguishable if they assign "about the same" probability mass to any efficiently recognizable set of strings.

An important special case of indistinguishable ensembles is that of probability ensembles which are polynomially indistinguishable from a uniform probability ensemble. These ensembles are called pseudorandom since, for all practical purposes, they are as good as truly unbiased coin tosses. This is clearly a behavioristic point of view.

Definition 3 (Pseudorandom Distributions – finite version): Let $l: \{0, 1\}^* \to N$ be a *length*) function (mapping strings to integers) $\pi_{0,i}^l$ denote the uniform probability distribution on the set $\{0, 1\}^{l(i)}$, and $\Pi_0^l = \{\pi_{0,i}^l\}_{i \in I}$. Let $\Pi_1 = \{\pi_{1,i}\}_{i \in I}$ be a probability ensemble indexed by l. We say that Π_1 is *pseudorandom* if it is polynomially indistinguishable from Π_0^l, for some length function l.

The above definitions (2 and 3) are titled "finite version" because each of the probability distributions (in the ensembles considered) is a function from *finite* strings to nonnegative reals. The infinite version of these definitions considers instead distributions on *infinite* bit-sequences ("infinite strings"). For the infinite versions, we need to modify the definitions of polynomially indistinguishable distributions, so that the tests run in time polynomial in the length of the first input (the index), while the second input may be infinite. (Here, the first input is essential in order to define the running time of the test!) Another modification, is in defining $\pi_{0,i}$ as a uniform distribution over the set of infinite strings. The technical details are omitted.

3.2. ON THE EXPANDABILITY OF PSEUDORANDOM DISTRIBUTIONS

Having defined pseudorandom ensembles, it is natural to ask whether such ensembles do exist. The answer is trivially affirmative, since the uniform ensemble is pseudorandom (being indistinguishable from itself!) However, this answer is of no interest. We would like to know whether ensembles which are not uniform, and furthermore are not statistically close to uniform, can be pseudorandom. Furthermore, can such ensembles be constructed using less coin tosses than the length of the strings in their support?[2] As we shall see in this section, assuming the existence of one-way 1-1 and onto functions, the answer to both questions is affirmative. A key definitions capturing the second question follows.

---

2 The *support* of a probability distribution is the set of elements which are assigned non-zero probability

Definition 4 (Pseudorandom Generator – finite version): Let $p(\cdot)$ be a polynomial satisfying $p(n) \geq n + 1$. Let G be a deterministic polynomial-time Turing machine that on input of any n-bit string outputs a string of length $p(n)$. Let $\underline{n}$ denote the unary encoding of the integer n. We say that G is a *pseudorandom generator* if the probability ensemble defined by it is pseudorandom. Here, the ensemble defined by G is $\{G_{\underline{n}}\}$ where a string y has probability $m \cdot 2^{-n}$ in the distribution $G_{\underline{n}}$ if there are exactly m strings of length n such that feeding each of them to G yields the output y.

Motivation to the unary encoding of the length: The length of the seed to G (i.e., n) is encoded in unary so that the strings in the support of $G_{\underline{n}}$ have length polynomial in $n(= \|\underline{n}\|)$.

Feeding a pseudorandom generator with seeds taken from a uniform distribution (over $\{0, 1\}^*$), yields a pseudorandom distribution over longer strings. The following theorem states that feeding a pseudorandom generator with seeds taken from a pseudorandom distribution also yields such a pseudorandom distribution.

Theorem 1: *Suppose that* $\Pi_1 = \{\pi_{1,i}\}_{i \in I}$ *is a pseudorandom ensemble, and G is a pseudorandom generator. Let* $\Pi_2 = \{\pi_{2,i}\}_{i \in I}$*, where $\pi_{2,i}$ is defined as the output distribution of G when fed with inputs according to the distribution $\pi_{1,i}$. Then the ensemble Π_2 is pseudorandom.*

Proof's Idea: Assume to the contrary that there exists a (polynomial-time) test T distinguishing between Π_2 and the uniform distribution. Then at least one of the following two statements hold:

1. The test T also distinguishes $\{G_{\underline{n}}\}$ from the uniform distribution (in contradiction to G being a pseudorandom generator).
2. The test T can be modified into a test T' (which first applies G to the tested string and then runs T on the result) so that T' distinguish Π_1 from the uniform distribution (thus contracting the hypothesis that Π_1 is pseudorandom). □

We return to the fundamental question of *whether pseudorandom generators do exist*. We will see that, under the assumption that one-way 1-1 and onto functions exist, the answer is yes. The following definitions and results are used in order to prove this implication. In particular, the equivalence of Definition 3 and Definition 5 plays an important role in proving the pseudorandomness of the construction presented below.

Definition 5 concerns the feasibility of predicting the next bit in a string, which is taken from some distribution. The predictor is given only a prefix of the string. The question is whether there exists an efficient predictor which succeeds with probability nonnegligibly greater than $\frac{1}{2}$.

Definition 5 (Unpredictability): Let $\Pi_1 = \{\pi_{1,i}\}_{i \in I}$ be a probability ensemble indexed by I. Let M be a probabilistic polynomial-time Turing machine that on inputs i and y outputs a single bit (called the *guess*). Let $bit(\alpha, r)$ denote the r-th bit of the string α, and $pref(\alpha, r)$ denote the prefix consisting of the first r bits of the string α (i.e., $pref(\alpha, r) = bit(\alpha, 1) \cdot bit(\alpha, 2) \cdots bit(\alpha, r)$). We say that the machine M *predicts the next bit* of Π_1 if for some $c > 0$ and infinitely many i's.

$$Prob\Big(M(i, pref(\alpha, r)) = bit(\alpha, r+1) \Big) \geq \frac{1}{2} + \|i\|^{-c},$$

where the probability space is that of the string α chosen according to $\pi_{1,i}$, the integer r chosen at random with uniform distribution in $\{0, 1, \ldots, \|\alpha\| - 1\}$ and the internal coin tosses of M. We say that Π_1 is *unpredictable* if there exists *no* probabilistic polynomial-time machine M which predicts the next bit of Π_1.

Definition 5 can be viewed as a special case of Definition 3. Any predictor can be easily converted into a test which outputs 1 if and only if the guess of the predictor is correct. The resulting test will distinguish an ensemble from the uniform ensemble if and only if the original predictor's guesses are nonnegligibly better than "random". Interestingly, the special case is not less powerful. Namely, each successful distinguisher can be converted into a successful predictor.

Theorem 2: *Let Π_1 be a probability ensemble. Then Π_1 is pseudorandom if and only if Π_1 is unpredictable.*

Proof's Idea: Assume that T is a test which distinguishes $\pi_{1,i}$ from the uniform distribution. We consider the behavior of T when fed with strings taken from the *hybrid* distributions $H_i^{(j)}$, where $H_i^{(j)}$ is defined as the distribution resulting by taking the first j bits of a string chosen from $\pi_{1,i}$ and letting the other bits be uniformly distributed. There must be two adjacent hybrids $H_i^{(j)}$ and $H_i^{(j+1)}$, which are distinguishable by T. The $j+$1st bit is predicted using this "gap".   □

The notion of a hard-core predicate (presented below) plays a central role in the construction of pseudorandom generators. Intuitively, a hard-core of a function f is a predicate $(b(x))$ which is easy to evaluate (on input x) but hard to even approximate when given the value of the function $(f(x))$. Recall that f is one-way if it is easy to evaluate (i.e., compute $f(x)$ from x) but hard to invert (i.e., compute x from $f(x)$). Thus, the hard-core maintains in a very strong sense both the easyness (in the forward direction) and the hardness (in the backward direction) of the function.

Definition 6 (hard-core Predicate): Let $f: \{0, 1\}^* \to \{0, 1\}^*$ and $b: \{0, 1\}^* \to \{0, 1\}$. The predicate b is said to be *a hard-core* of the function f if the following two conditions hold:

1. There is a deterministic polynomial-time Turing machine that on input x returns $b(x)$.
2. There is no probabilistic polynomial-time Turing machine M' such that for some $c > 0$ and infinitely many n

$$Prob\Big(M'(f(x)) = b(x)\Big) \geq 1/2 + n^{-c},$$

where the probability is taken over all possible choices of $x \in \{0, 1\}^n$ and the internal coin tosses of M' with uniform probability distribution.

Clearly, if the predicate b is a hard-core of the 1-1 and onto function f then f is hard to invert. Assuming that either of the functions presented in Section 2.4 is one-way, predicates which constitute corresponding hard-cores can be presented. For example, the least significant bit is a hard-core of *RSA* (i.e., given $RSA(N, e, x)$ one cannot efficiently predict the least significant bit of x). In fact, every one-way function f can be "transformed" into a one-way function f' with a corresponding hard-core predicate b'. Thus, unpredictability and computational difficulty play dual roles.

Theorem 3: *If there exists one-way functions (resp. one-way 1-1 and onto function) then there exist one-way functions (resp. one-way 1-1 and onto functions) with a hard-core predicate.*

Proof's Idea: The proof uses the observation that if f is one-way then there must be a bit in its argument x that cannot be efficiently predicted from $f(x)$ with success probability greater than $1 - \frac{1}{\|x\|}$. (Otherwise, with constant probability, all the bits of the argument can be predicted correctly and the argument can be retrieved.) Let $b(i, x)$ denote the ith bit of x. For $\|x_1\| = \|x_2\| = \cdots = \|x_{n^3}\| = n$, define

$$f'(x_1, x_2, \ldots, x_{n^3}) = f(x_2)f(x_2) \cdots f(x_{n^3}),$$

$$b'(x_1, x_2, \ldots, x_{n^3}) = \left[\sum_{i=1}^{n} \sum_{j=1}^{n^2} b(i, x_{i-1) \cdot n^2 + j}) \bmod 2 \right].$$

It can also be shown that the predicate b' is a hard-core of f'. The proof *does not* reduce to showing that a (sufficiently long) sequence of biased and independent 0-1 random variables has sum mod 2 which is almost unibased (since the prediction errors on the various predicates may not be independent random variables)! □

Having a one-way 1-1 and onto function with a hard-core predicate suffices for the following construction of pseudorandom generators.

Construction 1. Let f be a one-way 1-1 and onto function and b a hard-core predicate of f. We define the following polynomial-time Turing machine G. On input x, machine G computes the bits $b_i = b(f^{(i)}(x))$, where $1 \geq i \geq 2\|x\|$ and $f^{(i)}$ denotes the function f iteratively applied i times. Machine G outputs $b_{2\|x\|} \cdots b_2 b_1$.

Lemma 1. Let f, b, and G be as in Construction 1. Then $\{G_n\}$ defined as in Definition 4 is unpredictable.

Proof's Idea: An efficient predictor of the $(k + 1)$-st bit in the sequence defined above can be easily converted into a machine M that on input $f(x)$ guesses $b(x)$ with success probability greater than $\frac{1}{2}$. On input $f(x)$, machine M computes the sequence $b(f^{(k)}(x)), \ldots, b(f^{(2)}(x)), b(f(x))$ and obtains a prediction for $b(x)$. $\square$

Combining Theorem 3, Lemma 1, and Theorem 2, we get

Theorem 4: *If there exist one-way 1-1 and onto functions then there exist pseudorandom generators.*

3.3. DISCUSSION
Before presenting further extensions and applications of the above approach to randomness, let us discuss several conceptual aspects.

Behaviorism versus Ontologism. The behavioristic nature of the above approach to randomness is best demonstrated by confronting this approach with the Kolmogorov-Chaitin approach to randomness. Loosely speaking, a string is *Kolmogorov-random* if its length equals the length of the shortest program producing it. This shortest program may be considered the "true explanation" to the phenomenon described by the string. A Kolmogorov-random string is thus a string which does not have a substantially simpler (i.e., shorter) explanation than itself. Considering the simplest explanation of a phenomenon is certainly an ontological approach. In contrast, considering the effect of phenomena on certain objects, as underlying the definition of pseudorandomness (above), is a behavioristic approach.

Furthermore, assuming the existence of one-way 1-1 and onto functions, there exist probability distributions which are not uniform (and are not even statistically close to a uniform distribution) that nevertheless are indistinguishable from a uniform distribution (by any efficient procedure). Thus, distributions which are ontologically very different, are considered equivalent by the behavioristic point of view taken in the definitions above.

A Realistic View of Randomness. Pseudorandomness is defined above in terms of its observer. It is a distribution which cannot be told apart from a uniform distribution by any efficient (i.e., polynomial-time) observer. Thus, pseudorandomness is subjective to the abilities of the observer. To illustrate this point consider the following *mental experiment.*

Alice and Bob want to play "heads or tails" in one of the following four ways. In all of them Alice flips an unbiased coin and Bob is asked to guess its outcome before the coin rests on the floor. The alternative ways differ by the knowledge Bob has before making his guess. In the first way, Bob has to announce his guess before Alice flips the coin. Clearly, in this way Bob wins with probability ½. In the second way, Bob has to announce his guess while the coin is spinning in the air. Although the outcome is *determined in principle* by the motion of the coin, Bob does not have accurate information on the motion and thus we believe that also in this case Bob wins with probability ½. The third way is similar to the second, except that Bob has at his disposal sophisticated equipment capable of providing accurate *information* on the coin's motion as well as on the environment affecting the outcome. However, Bob cannot process this information in time to improve his guess. In the fourth way, Bob's recording equipment is directly connected to a *powerful computer* programmed to solve the motion equations and output a prediction. It is conceivable that in such a case Bob can improve his guess of the outcome of the coin substantially.

We conclude that the randomness of an event is relative to the information and computing resources at our disposal. Pseudorandom ensembles are unpredictable by probabilistic polynomial-time machines (associated with feasible computations), but may be predictable by infinitely powerful machines (not at our disposal!).

Effectiveness and Applicability. Another interesting property of the above approach to randomness is that pseudorandomness is effective in the following two senses: First, one may construct an efficient (universal) test that distinguishes pseudorandom distributions from ones which are not pseudorandom. In contrast, the problem of determining whether a string is Kolmogorov-random is undecidable. Second, assuming the existence of one-way 1-1 and onto functions, long pseudorandom strings can be efficiently and deterministically generated from much shorter pseudorandom strings. Clearly, this cannot be the case with Kolmogorov-random strings.

   The existence of pseudorandom generators has applications to the construction of efficient probabilistic algorithms (Turing machines). Such algorithms maintain the same performance when substituting their internal coin tosses by pseudorandom sequences. Thus, for every constant $\epsilon > 0$, the number of truly random bits required in a polynomial-time computation on input x can be decreased (from $poly(\|x\|)$) to $\|x\|^\epsilon$.

Randomness and Computational Difficulty. Randomness and computational difficulty play dual roles. This was pointed out already when discussing one-way functions and hard-core predicates. The relationship between pseudorandom generators and one-way computations is even a better illustration of this point. We have shown above that the existence of one-way 1-1 and onto functions implies the existence of pseudorandom generators. On the other hand, one can readily verify that any pseudorandom generator constitutes a one-way function.

3.4. FURTHER EXTENSION: PSEUDORANDOM FUNCTIONS OR EXPERIMENTING WITH THE RANDOM SOURCE

In the previous subsection we have (implicitly) modeled phenomena as single events (bit strings). This model suffices for describing phenomena in which the observer is passive: he can only record the events which occur. A more powerful model allows the observer to conduct experiments, namely, to "feed" the phenomenon with some values and measure the events which correspond to these values. Modeling a phenomenon as a function from events to events (or as a function from environment values to actions) is thus natural and useful. As in the previous subsection, we will present definitions for a pair of indistinguishable phenomena, a pseudorandom phenomenon, and a generator of the latter. In other words, we will present definitions for indistinguishability of functions, pseudorandom functions, and pseudorandom function generators.

For our definition of indistinguishable function ensembles we consider Turing machines with oracles. These machines are able, in addition to the traditional computing steps, to make *oracle queries:* place a string on a special tape and read an "answer" in the next step. Loosely speaking, we will say that two function ensembles are indistinguishable if any polynomial-time oracle Turing machine cannot distinguish the case that its oracle is a function taken from the first ensemble and the case that the oracle is a function taken from the second.

Definition 7. (Indistinguishability of Functions, Pseudorandom Functions and Function Generators): Let $F_1 = \{F_{1,i}\}_{i \in I}$ and $F_2 = \{F_{2,i}\}_{i \in I}$ be two function ensembles, where $F_{j,i}$ is a probability distribution on the functions $f: \{0, 1\}^{\|x\|} \to \{0, 1\}$. We say that F_1 and F_2 are *polynomially indistinguishable* if for every probabilistic polynomial-time oracle machine M, every constant $c > 0$ and all sufficiently large $i \in I$

$$\left\| p_1^M(i) - p_2^M(i) \right\| < \|i\|^{-c},$$

where $p_j^M(i)$ is the probability that M outputs 1 on input i when querying an oracle randomly chosen from the distribution $F_{j,i}$.

The function ensemble $F = \{F_i\}_{i \in I}$ is *pseudorandom* if it is polynomially indistinguishable from the ensemble $H = \{H_i\}_{i \in I}$, where H_i is the uniform probability distribution on the set of functions $f: \{0, 1\}^{\|i\|} \to \{0, 1\}$. We say that $F = \{F_n\}$ is a *pseudorandom function generator* if the following three conditions hold:

1. There exists a probabilistic polynomial-time machine M_1 that, on input $\underline{n}$, randomly selects a function f from the distribution F_n and outputs a (succinct) description of f (denoted $\tilde{f}$).
2. There exists a (deterministic) polynomial-time machine M_2 that, on input $\tilde{f}$ (a description of $f: \{0, 1\}^n \to \{0, 1\}$) and a string $x (\in \{0, 1\}^n)$, outputs $f(x)$. That is, $M_2(\tilde{f}, x) = f(x)$.

3. The ensemble F is pseudorandom.

Similar definitions apply to function ensembles consisting of distributions F_i on functions $f: \{0, 1\}^{\|i\|} \rightarrow \{0, 1\}^{\|i\|}$. Furthermore, one can easily transform ensembles of the first kind to ones of the second type, and vice versa.

As in Section 3.2, we now ask whether there exist nontrivial ensembles of pseudorandom functions, and furthermore whether such ensembles can be efficiently generated. It turns out that this question reduces to the question handled in Section 3.2, namely:

Theorem 5: *Pseudorandom function generators exist if and only if pseudorandom generators exist.*

Proof's Idea: The "only if" direction of Theorem 5 is easy. The generator first uses M_1 to get an $\tilde{f}$ and next uses M_2 to evaluate $f(1), f(2), \ldots$. The "if" direction of the theorem also has a constructive proof. The construction proceeds in two steps: First one uses an arbitrary pseudorandom generator to construct a pseudorandom generator G that doubles the length of its input. Next, G is used to construct a pseudorandom function in the following manner. Let $G_0(x)$ denote the first $\|x\|$ bits output by G on input x, and $G_1(x)$ denote the last $\|x\|$ bits output by G on input x. Extend the above notation so that for every bit σ and bit string α, $G_{\alpha\sigma}(x) = G_\alpha(G_\sigma(x))$. Now let $f_x(y) = G_y(x)$, and F_n is the distribution obtained by picking x uniformly among all n bit strings and using the resulting function f_x. It can be shown that F so defined is a pseudorandom function generator. It is interesting to note that this is not the case if we let $f_x(y) = G_x(y)$. $\square$

Further Discussion. It is interesting to point out the analogy between the above definition of pseudorandom functions and Turing's famous "test of intelligence". (In Turing's test of intelligence, one is interacting arbitrarily with an unknown entity which is either a human or a machine. The machine is said to be (pseudo)intelligent if the tester cannot distinguish the two cases.) In both settings one interacts with an unknown function in order to later determine the "nature" of this function. Failure to determine the "true nature" is interpreted as a proof that the difference in nature is of no importance (as far as functionality goes . . .).

Pseudorandom functions can not be predicted, not even in the following weak sense: any probabilistic polynomial-time oracle Turing machine cannot predict the value of the oracle on an unasked query better than 50-50, when the oracle is a pseudorandom function. This resembles the following quotation of Turing:

> I have set up on a Manchester computer a small programme using only 1000 units of storage, whereby the machine supplied with one sixteen figure number replies with another within two seconds. I would defy anyone to learn from these replies sufficient about the programme to be able to predict any replies to untried values.

3.5. Applications to Cryptography

The most obvious application of pseudorandomness to cryptography is making *one-time pads* a feasible encryption method. One-time pads are the simplest and safest private-key cryptosystem. A cleartext is encrypted by XORing[3] its bits with the currently initial segment of the (*randomly selected*) key, and the resulting ciphertext is decrypted by XORing its bits with the very segment of the key. Each segment of the key is deleted after use, and thus no information about the cleartext can be extracted from the ciphertext. The drawback of one-time pads is that the length of the key in use must at least equal the length of the messages sent. Namely, in order to *secretly* pass a message of length *l* one must exchange secretly another message of length *l*. This is satisfactory from neither a theoretical nor a practical point of view, since the aim is to achieve a high level of security at a much lower "cost". In practice, "pseudorandom sequences" are used instead of the randomly selected key of the one-time pad, *but security can no longer be asserted.* Assuming the existence of pseudorandom bit generators (in the sense discussed in Section 3.2), one can replace the key of the one-way pad by a pseudorandom sequence and prove that the resulting cryptosystem is secure in the following sense: *whatever can be efficiently computed from the ciphertext can be efficiently computed without it.* In other words, as far as polynomial-time computations are concerned, no information about the cleartext is *revealed* from the ciphertext.

Other applications of pseudorandomness to cryptography use the construction of pseudorandom functions (Theorem 5, Section 3.4). For example, it is possible to produce unforgettable *message authentication tags* and *time-stamps*. Suppose that two parties, *A* and *B*, sharing a secret key communicate over a channel tampered by an adversary *C*. The adversary may inject messages onto the channel. The parties would like to be able to verify that the message received has arrived from their counterpart, and not from the adversary. It is suggested that in order to authenticate a message *M*, party *A* just applies the pseudorandom function *f* to *M*, and appends $f(M)$ to *M* as an authentication tag. Party *B* may then verify the validity of this authentication tag, being confident that the message has been sent by *A* (and not injected by *C*). We stress that if *f* is a pseudorandom function then the above scheme is provably secure in the following sense: *even if C gets polynomially many authentication tags to messages of his choice he cannot produce in polynomial-time an authentication tag to any other message.*

---

3 XORing two bit strings means applying exclusive-or (XOR) to each pair of corresponding bits.

4. Interactive Proofs

In this section we survey a recent behavioristic approach to the notion of an efficiently verifiable proof. In this approach, a proof system for proving membership in a language L is a two-party protocol for a *prover* and a *verifier* so that the prover can convince the verifier to accept x (with high probability) if and only if $x \in L$.

4.1. DEFINITION OF INTERACTIVE PROOFS

Before defining the notion of an interactive proof, we define the notion of an interactive pair of Turing machines, which captures the intuitive notion of a two-party protocol.

Definition 8 (Pair of Interactive Turing machines): An *interactive Turing machine* (ITM) is a six-tape deterministic Turing machine with a read-only *input-tape*, a read-only *random-tape*, a read/write *work tape*, a read-only *communication tape*, a write-only *communication tape*, and a write-only *output tape*. The string which appears on the input tape is called the *input*. The contents of the random tape can be thought of as the outcomes of unbiased coin tosses. The string which appears on the output tape when the machine halts is called the *output*. The contents of the write-only communication tape can be thought of as messages sent by the machine, while the contents of the read-only communication tape can be thought of as messages received by the machine.

An *interactive pair of Turing machines* is a pair of ITMs which share their communication tapes so that the read-only communication tape of the first machine is the write-only communication tape of the second machine, and vice versa. Let M_1 and M_2 be an interacting pair of ITMs, then $[M_2(x_2), M_1(x_1)]$ denotes the output of M_1 on input x_1, when M_2 has input x_2.

Intuitively, an interactive proof system for a language L is a two-party protocol for a "powerful" *prover* and a probabilistic polynomial-time *verifier* satisfying the following two conditions with respect to the common input, denoted x. If $x \in L$, then with very high probability the verifier is "convinced" of this fact, when interacting with the prover. If $x \notin L$, then no matter what the prover does, he cannot fool the verifier (into believing that "x is in L"), except for with very low probability. The first condition is referred to as the *completeness* condition, while the second condition is referred to as *soundness*.

Definition 9 (Interactive Proof): An *interactive proof for a language L* is a pair of ITMs P and V satisfying the following conditions:

0) On input x machine V make at most $p(\|x\|)$ steps, where $p(\cdot)$ is a fixed polynomial.

1) *Completeness:* For every constant $c > 0$, and all sufficiently long $x \in L$

$$Prob\Big([P(x), V(x)] = 1\Big) \geq 1 - \|x\|^{-c}.$$

2) *Soundness:* For every ITM P', every constant $c > 0$, all sufficiently long $x \notin L$, and every $y \in \{0, 1\}^*$,

$$Prob\Big([P'(y), V(x)] = 0\Big) \geq 1 - \|x\|^{-c}.$$

Denote by **IP** the class of languages having interactive proofs.

Remarks. Note that it does not suffice to require that the verifier cannot be fooled by the predetermined prover (such a mild condition would have presupposed that the "prover" is a trusted oracle). **NP** is a special case on interactive proofs, where the interaction is trivial and the verifier tosses no coins.

Example of an interactive proof. To illustrate the definition of an interactive proof we present an interactive proof for *Graph Non-Isomorphism.* The input is a pair of graphs G_1 and G_2, and one is required to prove that there exists no 1-1 edge-invariant mapping of the vertices of the first graph to the vertices of the second graph. (A mapping π from the vertices of G_1 to the vertices of G_2 is *edge-invariant* if the nodes v and u are adjacent in G_1 if and only if the nodes $\pi(v)$ and $\pi(u)$ are adjacent in G_2.) It is interesting to note that no short NP-proofs are known for this problem; namely Graph Non-Isomorphism is *not known* to be in **NP**.

The interactive proof proceeds as follows: The verifier chooses at random one of the two input graphs, say $G_\alpha (\alpha \in \{1, 2\})$. The verifier creates a random isomorphic copy of G_α and sends it to the prover, which is supposed to answer with $\beta \in \{1, 2\}$. The verifier interprets $\beta = \alpha$ as evidence that the graphs are not isomorphic; while $\beta \neq \alpha$ leads him to reject. This is repeated several times (with independent random choices!) to collect stochastic evidence. The verifier accepts (the graphs as nonisomorphic) if and only if all of the provers responses are correct.

If the two graphs are not isomorphic, then the prover has no difficulty in always answering correctly (i.e., a β equal to α), and thus the completeness condition is met. If the two graphs are isomorphic it is impossible to distinguish a random isomorphic copy of the first from a random isomorphic copy of the second, and the probability that the prover answers correctly to one "query" is at most $\frac{1}{2}$. the probability that the prover answers correctly all t queries is $\leq 2^{-t}$ and the soundness condition is satisfied.

4.2. Discussion

The terminology of interactive proofs *explicitly* deals with the two fundamental computational tasks related to proof systems: producing a proof and verifying the validity of a proof. For many years **NP** was considered *the formulation* of "whatever can be efficiently verified". This stemmed from the traditional association of deterministic polynomial-time computation with efficient computation. The growing acceptability of probabilistic polynomial-time computations as reflecting efficient computations is the basis of the more recent formalization (namely **IP**) of "whatever can be efficiently verified". As we regard random computing steps as feasible, there is no reason not to allow the verifier to make such steps. Following our convention of disregarding events that occur with negligible probability, we disregard the probability of error in such proofs. (Objections to this approach are discussed in the sequel.) Also, there seems to be no reason to restrict the interaction between the prover and verifier, as long as the verifier remains efficient in terms of (the length of) the claim to be proven. In fact, a hierarchy of complexity classes, parametrized by the number of messages exchanged between the prover and verifier, seems to emerge. Namely, **IP**($f(\cdot)$) denotes the class of languages having interactive proofs in which up to $f(\|x\|)$ messages are exchanged on input x.

The definition of interactive proofs does not meet the traditional notion of a "proof", which disallows the possibility of error. The fact that in an interactive proof an error may occur only with an overwhelmingly small probability does not overrule the objection of a purist (namely that "this is not a proof"). But even the purist must agree that an interactive proof captures the intuitive notion of a convincing argument. Any reasonable person, trusting his own coin tosses, will believe statements "proved" through means of an interactive proof and ignore the overwhelmingly small possibility of an error. Ignoring errors which may occur with overwhelmingly small probability is clearly a behavioristic approach to life. In other words, as far as all practical purposes are concerned, an interactive proof is good as a "real proof" (i.e., an **NP** proof).

Another possible objection to interactive proofs is that they may not be "transferable", but rather convince only a party that either actively participates in them (as a verifier) or believes that the verifier's "random moves" were unpredictable by the prover. (Consider for example the interactive proof for Graph Non-Isomorphism: if the prover can predict the verifier's coin tosses then he can answer correctly even if the graphs are isomorphic.)

In going from **NP** to interactive proofs (i.e., **IP**) we gave away certainty and transferability of the "proof". *Does this buy us anything?* Unless **IP** $\subseteq$ **BPP** (and hence **NP** $\subseteq$ **BPP**), the answer is affirmative as interactive proofs allow us to introduce (nontrivially) the notion of zero-knowledge proofs (see next section). Another possible gain of interactive proofs is that they allow to prove membership in languages *not known* to be in **NP** (e.g., Graph Non-Isomorphism). (*Proving* that **IP** $\neq$ **NP** is way out of the "current state of the art" as it will imply that **NP** is strictly contained in **PSPACE**.)

An interesting question regarding interactive proofs is which ingredients or parameters determine their power. The most important discovery in this direction is that restricting the verifier to send the prover the outcome of his coin tosses does not decrease the power of the proof system. In other words, whatever can be proven with a verifier that "cleverly" chooses his "questions", can be proven with a verifier that chooses his "questions" at random. (The above restricted type of interactive proof is called *an Arthur Merlin game.*) Another result is that increasing the number of interactions by a multiplicative factor does not increase the power of the system (i.e., for every constant $c > 0$ and function f, $\mathbf{IP}(f(\cdot)) = \mathbf{IP}(c \cdot f(\cdot))$). In particular, an interaction in which the verifier gets an answer to one randomly chosen question is as powerful as an interaction consisting of a bounded sequence of questions and answers. Finally, one can show that the error probability in the completeness condition (of Definition 9) is not essential, while the error probability in the soundness condition is essential (unless $\mathbf{NP} = \mathbf{IP}$).

5. Zero-Knowledge Proofs

In this section, we survey the notion of zero-knowledge proofs. To exemplify this notion consider the set (i.e., language) of satisfiable propositional formulae. An easy way to convince a polynomial-time verifier that a formula ψ is satisfiable is to demonstrate a truth assignment which satisfies ψ. This proof, however, reveals much more than the fact that ψ is satisfiable: it yields a satisfying assignment. The reader should note that, unless $\mathbf{P} \neq \mathbf{NP}$, a satisfying assignment is hard to get even if one knows that such an assignment does exist. Thus, the verifier in the above proof obtains (from the prover) knowledge which he could not have computed by himself, even if he believed that ψ is satisfiable. In a zero-knowledge proof, the verifier will be convinced that ψ is satisfiable without getting a satisfying assignment nor obtaining any knowledge about the formula which is not attainable from the formula itself and the fact that it is satisfiable. Thus, the verifier in a zero-knowledge proof is essentially in the same situation as if he has been told by a trusted oracle that ψ is satisfiable.

5.1. DEFINITION OF ZERO-KNOWLEDGE PROOFS

Intuitively, a zero-knowledge proof is a proof which yields nothing but its validity. This means that for all practical purposes "whatever" can be done after interacting with a zero-knowledge prover can be done when just believing that the assertion he claims is indeed valid. (In "whatever" we mean not only the computation of functions but also the generation of probability distributions.) Thus, zero-knowledge is a property of the predetermined prover. It is the robustness of the prover against attempts of the verifier to extract knowledge via interaction. Note that the verifier may deviate arbitrarily (but in polynomial-time) from the predetermined program. This is captured by the formulation sketched below.

Denote by $[P, V^*(x)]$ the probability distribution generated by a machine V^* which interacts with (the prover) P on input $x \in L$. We say that the proof system is *zero-knowledge* if for all probabilistic polynomial-time interactive machines V^*, there exists a probabilistic polynomial-time machine M_{V^*} that on input x produces a probability distribution $M_{V^*}(x)$ such that $\{M_V^*(x)\}_{x\in L}$ and $\{[P, V^*(x)]\}_{x\in L}$ are polynomially-indistinguishable. (We stress that M_{V^*} is an ordinary machine which does not interact with P or any other machine.)

As we argued in Section 3, polynomially-indistinguishable probability distributions should be considered equal for all practical purposes. It follows that the polynomially-indistinguishability of $[P, V^*(x)]$ and $M_V^*(x)$ suffices for saying that nothing substantial is gained by interacting with the prover, except of course conviction in the validity of the assertion $x \in L$.

Example of a zero-knowledge proof. To illustrate the above definition we present a *zero-knowledge* interactive proof for *Graph Isomorphism* (see definition of the problem in Section 4.1). Before doing so let us stress that the isomorphism between the two graphs does constitute a proof for the fact that they are isomorphic, but that this proof is unlikely to be zero-knowledge (as it will imply that the problem of finding an isomorphism between a pair of graphs is in **BPP**).

Our zero-knowledge proof proceeds as follows: The prover creates an random isomorphic copy of the first graph (G_1). The verifier chooses at random one of the two input graphs, and asks the prover to present an isomorphism between the chosen graph and the graph sent by the prover. If the prover fails to present such an isomorphism the verifier rejects. This is repeated several times (with independent random choices!) to collect stochastic evidence. The verifier accepts (the graphs as isomorphic) if and only if all of the provers responses are correct.

If the two graphs are isomorphic, and the prover knows an isomorphism between them, then he has no difficulty in always answering correctly. If the two graphs are not isomorphic it is impossible to answer both possible questions of the verifier. If the verifier chooses his t questions randomly then the probability that the prover answers correctly all of them is $\leq 2^{-t}$.

The following is a very rough sketch of the argument that the above proof is indeed zero-knowledge. One can simulate the conversations between the prover and a verifier by selecting at random one of the two input graphs and creating a random isomorphic copy of it. If the verifier asks to see that isomorphism, we can supply it. Otherwise we repeat the process with new independent random choices. The expected number of repetitions needed to simulate one round of prover-verifier conversation is 2. The distribution created by taking only successful repetitions equals the distribution of the original prover-verifier conversations.

5.2. DISCUSSION

The definition of zero-knowledge presupposes that probabilistic polynomial-time is for "free". Namely, a conversation between the prover and the verifier which can be simulated by the verifier himself in probabilistic polynomial-time contains no knowledge. Thus, implicitly, knowledge is regarded as the result of a computation which is unfeasible for the verifier himself.

Randomness and interaction are essential to the nontriviality of the notion of zero-knowledge. It can be shown that zero-knowledge proofs in which the verifier either tosses no coins or asks no questions exist only for languages in **BPP**. Note that such zero-knowledge proofs are of no interest since every language in **BBP** has a trivial zero-knowledge proof in which the prover sends nothing to the verifier!

The definition of zero-knowledge seems somewhat paradoxical: these proofs yield no knowledge in the sense that they can be constructed by the verifier who believes the statement, and yet these proofs do convince him. The "paradox" is resolved by noting that it is not the text of the conversation that convinces the verifier, but rather the fact that this conversation was held "on line". When constructing such a conversation text by himself the verifier works "off line", concatenating parts of possible conversations while deleting other parts.

5.3. Proving Any NP Statement in Zero-Knowledge

In providing zero-knowledge proofs to any language in **NP**, we use the notion of NP-completeness (see Section 2.2). Namely, we provide a zero-knowledge proof for one **NP**-complete language (denoted L_0) and derive such proofs for any other L ($\in$ **NP**) language by using the polynomial reduction of L to L_0. In presenting a zero-knowledge proof for L_0 we use any secure encryption scheme (in the sense of Goldwasser and Micali *1984*). The existence of such schemes is guaranteed by our assumption that one-way 1-1 functions exist.

As the basis for the construction (i.e., the language L_0), we use the set of 3-colorability graphs. The language *Graph 3-Colorability* consists of the set of graphs, the vertices of which can be colored using three colors such that no two adjacent vertices are assigned the same color.

A Zero-Knowledge Proof for Graph 3-Colorability. The common input to the following protocol is a graph $G(V, E)$, where $V = \{1, 2, \ldots, n\}$ is the vertex set and $E \subseteq V \times V$ is the edge set. In the following protocol, the prover needs only to be a probabilistic polynomial-time machine which gets a proper 3-coloring of G as an auxiliary input. Let us denote this coloring by $\phi(\phi: V \to \{1, 2, 3\})$. Let $m = \|E\|$.

The following four steps are executed m^2 time, each time using independent coin tosses.

1. The prover chooses a random permutation of the 3-coloring, encrypts it, and sends it to the verifier. More specifically, the prover chooses at random a permutation, $\pi \in Sym(\{1, 2, 3\})$, encrypts (separately) each element of the sequence $\pi(\phi(1)), \pi(\phi(2)), \ldots, \pi(\phi(n))$, and sends the resulting encrypted sequence to the verifier.
2. The verifier chooses at random an edge $e \in E$ and sends it to the prover. (Intuitively, the verifier asks to examine the coloring of the endpoints of $e \in E$.)
3. If $e = (u, v) \in E$ then the prover reveals the coloring of u and v and "proves" that they correspond to their encryptions. If $e \notin E$ then the prover stops.
4. The verifier checks the "proof" provided in step (3). Also, the verifier checks that the colors revealed are consistent (i.e., $\pi(\phi(u)) \neq \pi(\phi(v))$, and

$\pi(\phi(u)), \pi(\phi(v)) \in \{1, 2, 3\})$). If either condition is violated the verifier *rejects* and stops. Otherwise the verifier continues to the next iteration.

If the verifier has completed all m^2 iterations then it *accepts*.

The reader can easily verify the following facts: If the graph is 3-colorable and both prover and verifier follow the protocol then the verifier always accepts. If the graph is not 3-colorable and the verifier follows the protocol then no matter how the prover plays, the verifier will reject with probability at least $(1-m^{-1})^{m^2} = \exp(-m)$. Thus, the above protocol constitutes an interactive proof system for 3-colorability. Clearly, this protocol yields no knowledge to the specified verifier, since all he gets is a sequence of random pairs. The proof that the protocol is indeed zero-knowledge (with respect to *any* verifier) is much more complex, and is omitted. We get:

Proposition: *If there exist one-way 1-1 functions then there exist a zero-knowledge interactive proof system for 3-colorability.*

Zero-knowledge proofs for all languages in NP. For every language L in **NP**, there exist an efficient transformation of instances of the language L to instances of 3-colorability. This transformation is called a *reduction,* and is guaranteed by the fact that 3-colorability is **NP**-complete. Incorporating the standard reductions into the protocol for graph 3-colorability, we get

Theorem 6: *If there exist one-way 1-1 functions then every NP language has a zero-knowledge interactive proof system. Furthermore, in this proof system the prover is a probabilistic polynomial-time Turing machine which gets an NP-proof as an auxiliary input.*

Using Theorem 6, one can prove that any language in **IP** has a zero-knowledge proof. Thus, "whatever is efficiently provable" is "efficiently provable in a zero-knowledge manner".

5.4. APPLICATIONS TO CRYPTOGRAPHY

Theorem 6 (above) yields an extremely powerful tool for the design of cryptographic protocols: the *ability to prove any NP statement in a zero-knowledge manner.* To better understand the relevance of this tool, let us briefly discuss the setting in which cryptographic protocols arise.

A cryptographic protocol is a sequence of interactive programs to be executed by *nontrusting* parties. Each party has a local input unknown to the others, and hereafter referred to as *his secret.* Typically, the protocol specifies actions to be taken by each of the parties based on his secret and previous messages. The protocol designer is thus faced with the following problem: how can one party verify that his counterpart has computed the next message in accordance with the protocol? Verification is difficult since the verifier does not know and is not supposed to know the secret of the transmitter. Zero-knowledge proofs for all NP statements are the answer to this problem. The transmitter's claim to having computed his message according to the protocol is an NP statement (and furthermore, the transmitter knows an NP-proof to it). By Theorem 6, this NP claim can be proven without yielding any knowledge of the prover's secret!

The above suggests a powerful methodology for the design of cryptographic protocols. First design your protocol assuming that all parties will follow it properly. Next compile the protocol using zero-knowledge proofs to a protocol which maintains the correctness and privacy of the original protocol even when a minority of the parties display arbitrary adversarial behavior. The details of the compiler are beyond the scope of this survey.

6. Conclusions

The fact that pseudorandom generators and functions exist under a reasonable complexity theoretic assumption (i.e., the existence of one-way 1-1 and onto functions), must be considered at least a plausibility argument. Thus, every reasoning overruling the existence of such generators must incorporate a demonstration that one-way 1-1 and onto functions do not exist. The *possible existence* of pseudorandom generators does not allow us to consider "unbounded" random behavior as necessarily arising from an "unbounded" source of randomness, since a pseudorandom generator may expand a "bounded" amount of randomness to an "unbounded" amount of pseudorandomness. Furthermore, the possible existence of pseudorandom functions implies that a small amount of randomness suffices in order to efficiently determine a random mapping from huge sets into huge sets.

Also, under the same complexity theoretic assumption, the folklore belief that one *necessarily* gains extra insight into a theorem by seeing its proof is seriously shaken. Zero-knowledge proofs are proofs which yield no such insight. Whatever one can efficiently learn from the proof, one can deduce as easily by believing that the theorem is valid.

All the above was discovered through a behavioristic approach to computational notions such as randomness and proofs. We believe that a behavioristic approach is justified when studying computing devices, as much as it is unjustified when studying "thinking beings".

Acknowledgements. First of all, I would like to thank two remarkable people who had a tremendous influence on my professional development. Shimon Even introduced me into theoretical computer science and closely guided me in my first steps. Silvio Micali led my way in the evolving foundations of cryptography and shared with me his efforts to further develop them. Next, I would like to thank my colleagues Benny Chor and Avi Wigderson for their indispensable contribution to our joint research, and for the excitement and pleasure I had when collaborating with them. Special thanks to Leonid Levin for many interesting discussions.

I am grateful to Janos Makowsky and Rolf Herken for suggesting me to write this article and providing guidelines for it. Thanks to Hugo Krawczyk for carefully reading an earlier version of the manuscript, pointing out some errors, and suggesting several improvements. Finally, thanks to Dassi.

Bibliographic Notes. For background on computational complexity consult an appropriate textbook such as Hopcroft and Ullman *1979*, ch. 12-13, and Garey and Johnson *1979*.

The notion of one-way functions was first suggested in Diffie and Hellman *1976*, and the most famous candidate is due to Rivest, Shamir, and Adleman *1978*. A 1-1 and onto function which is one-way, unless factoring is easy appears in Rabin *1979*. Definition 1 (*one-way functions*), however, is a weaker form and is due to Yao *1982*. A special case of Definition 2 (*indistinguishability*) first appeared in Goldwasser and Micali *1984*, the general case is from Yao *1982*. Definitions 3 and 4 (*pseudorandomness*) are due to Yao *1982*, while Definition 5 (*unpredictability*) appears in Blum and Micali *1984*. Theorem 2 (*equivalence of Definitions 3 and 5*) is implicit in Yao *1982*. Definition 6 (*hard-core predicate*), Construction 1 (*pseudorandom generator based on a hard-core predicate*), and Lemma 1 appear in Blum and Micali *1984*. Theorem 3 (*existence of hard-core predicates assuming one-way 1-1 and onto functions*) is implicit in Yao *1982*, where a sketch of the proof of Theorem 4 (*pseudorandom generator based on one-way 1-1 and onto functions*) appears. A finer analysis, which uses a weaker (necessary and sufficient) condition, of Theorems 3 and 4 appears in Levin *1985*. Predicates which are hard-core of the particular number theoretic functions mentioned in Section 2.4, appear in Blum and Micali *1984* and Alexi et al. *1984*. Definition 7 (*pseudorandom functions*) and Theorem 5 (*pseudorandom generators imply pseudorandom function generators*) appear in Goldreich, Goldwasser, and Micali *1986*. Further developments appear in Luby and Rackoff *1986*.

Two different definitions of interactive proofs appear in Goldwasser, Micali, and Rackoff *1985* and Babai *1985*, respectively. In Section 4, we have used the definition of Goldwasser, Micali, and Rackoff *1985*. The definition in Babai *1985* (*Arthur Merlin games*) is a special case of it. The definitions were proven to be equivalent in Goldwasser and Sipser *1986*. The interactive proof for Graph Non-Isomorphism originates from Goldreich, Micali, and Wigderson *1986*.

The definition of a zero-knowledge proof originates from Goldwasser, Micali, and Rackoff *1985*, which contains also a more general definition of the "amount of knowledge contained in a proof". The zero-knowledge proof for Graph Isomorphism as well as Theorem 6 (*all NP languages have zero-knowledge proofs*) appear in Goldreich, Micali, and Wigderson *1986*. (For a definition of secure encryption functions see Goldwasser and Micali *1984*.) The applications mentioned in Section 5.4 appear in Goldreich, Micali, and Wigderson *1986* and *1987*. Further developments appear in Yao *1986* and Goldreich, Micali, and Wigderson *1987*.

References

Alexi, W., B. Chor, O. Goldreich, and C.P. Schnorr

    1984 RSA and Rabin functions: Certain parts are as hard as the whole. In: *Proceedings of the 25th IEEE Symposium on the Foundation of Computer Science*, pp. 449–457. To appear in *SIAM J. Comp.*

Babai, L.

    1985 Trading group theory for randomness. In: *Proceedings of the 17th ACM Symposium on Theory of Computing*, pp. 421–429.

Blum, M., and S. Micali

    1984 How to generate cryptographically strong sequences of pseudo-random bits. *SIAM J. Comp.* **13** (1984) 850–864.

Diffie, W., and M.E. Hellman

    1976 New directions in cryptography. *IEEE Trans. Inf. Th.* **IT-22** (Nov. 1976) 644–654.

Garey, M.R., and D.S. Johnson

    1979 *Computers and Intractability: A Guide to the Theory of NP-Completeness.* New York: W.H. Freeman and Company (1979).

Goldreich, O., S. Goldwasser, and S. Micali

    1986 How to construct random functions. *J. ACM* **33(4)** (1986) 792–807.

Goldreich, O., S. Micali, and A. Widgerson

    1986 Proofs that yield nothing but their validity and a methodology of cryptographic protocol design. In: *Proceedings of 27th Symposium on the Foundation of Computer Science*, pp. 174–187 (1986).

    1987 How to play any mental game or a completeness theorem for protocols with honest majority. In: *Proceedings of the 19th ACM Symposium on Theory of Computing*, pp. 218–229 (1987).

Goldwasser, S., and S. Micali

    1984 Probabilistic encryption. *J. Comput. Sy.* **28(2)** (1984) 270–299.

Goldwasser, S., S. Micali, and C. Rackoff

    1985 Knowledge complexity of interactive proofs. In: *Proceedings of the 17th ACM Symposium on Theory of Computing*, pp. 291–304 (1985). To appear in *SIAM J. Comp.*

Goldwasser, S., and M. Sipser

    1986 Arthur Merlin games versus interactive proof systems. In: *Proceedings of the 18th ACM Symposium on Theory of Computing*, pp. 59–68 (1986).

Hopcroft, J.E., and J.D. Ullman

    1979 *Introduction to Automata Theory, Languages, and Computation.* Reading, MA: Addison Wesley (1979).

Levin, L.A.

    1985 One-way function and pseudorandom generators. In: *Proceedings of the 17th ACM Symposium on Theory of Computing,* pp. 363–365 (1985).

Luby, M., and C. Rackoff

    1986 Pseudorandom permutation generators and DES. In: *Proceedings of the 18th ACM Symposium on Theory of Computing,* pp. 356–363 (1986).

Rabin, M.O.

    1979 Digitalized signatures and public key functions as intractable as factoring. MIT/LCS/TR-212 (1979).

Rivest, R.L., A. Shamir, and L. Adleman

    1978 A method for obtaining digital signatures and public key cryptosystems. *Comm. ACM* **21** (Feb. 1978) 120–126.

Yao, A.C.

    1982 Theory and applications of trapdoor functions. In: *Proceedings of the 23rd IEEE Symposium. on Foundation of Computer Science,* pp. 80–91 (1982).

    1986 How to generate and exchange secrets. In: *Proceedings of the 27rd IEEE Symposium on Foundation of Computer Science,* pp. 162–167 (1986).

Algorithms in the World of
Bounded Resources

Yuri Gurevich[*]

Abstract. Writing an algorithm, one has in mind some addressee (or addressees), i.e., a computing agent, a possible executor of the algorithm. In the classical theory of algorithms, one addresses a computing agent with unbounded resources. We argue in favor of a more realistic theory of algorithms which recognizes the multiplicity of addressees and the fact that their resources may be bounded. We are especially interested in the case when the resources of each relevant computing agent are bounded.

1.   The Assumption of Unbounded Resources

An algorithm is supposed to be given by an exact prescription, defining a computational process. Writing the prescriptions, we make assumptions about the computing agent. The classical theory of algorithms makes a simplifying assumption that the resources of the computing agent are unbounded. Consider for example an algorithm for computing the factorial function in the language of primitive recursive functions:

$$
\begin{aligned}
FACT(0) &= 1, & FACT(x') &= MULT(FACT(x), x'). \\
MULT(x, 0) &= 0, & MULT(x, y') &= SUM(MULT(x, y), x), \\
SUM(x, 0) &= x, & SUM(x, y') &= (SUM(x, y))'.
\end{aligned}
$$

Notice that no provision is made for the case of insufficient computing resources. Let us express the same algorithm as a Pascal program:

```
program FACTORIAL1;
{computes the factorial f of a given natural number n}
    var f, k, n : integer;
begin
    readln(n);
    writeln('n = ', n);
```

---

[*] Partially supported by NSF grants MCS 83-01022 and DCR 85-03275.

```
k := 0;
f := 1;
while k < n do
   begin
   k := k + 1;
   f := f * k
   end;
writeln('n! = ', f)
end.
```

When I tried to run FACTORAL1 with $n = 8$ on my Macintosh, the result was a message stating that the expression $f * k$ is out of range. I guess we are supposed to view our computers as imperfect approximations to an ideal computing agent.

Another, more realistic point of view is that a program may be executed by different computing agents, and some (or even all) of them may have bounded resources. When you write a program (or create a programming language), think about the family of (real or virtual) computing devices that may run your program (or use your language), rather than one ideal computing agent with unbounded resources.

Imagine that FACTORIAL1 is a link in a chain of computations, and its output is the input for another algorithm. To give a standard form to the output of FACTORIAL1, one can elaborate FACTORIAL1 as follows:

```
program FACTORIAL2;
   var f, k, n integer;
begin
   readln(n)
   writeln ('n = ', n)
   k := 0
   f := 1;
   while (k < n) and (f <= maxint /(k + 1)) do
      begin
      k := k + 1;
      f := k * f
      end;
   if k < n then
      writeln ('n! exceeds maxint')
   else
      writeln ('n! = ', f)
end.
```

Is FACTORIAL2 an exact prescription for computing the factorial of an arbitrary natural number n (given in the decimal notation) in the world where resources may be bounded? No, FACTORIAL2 does not ensure the rejection of inputs exceeding maxint in any standard way. Further modification is needed. In particular, we need to change the type of n. Pragmatically speaking, it may be even more important to change the type of f; that would allow us to compute $f = n!$ for many more inputs n. The assumption of unbounded resources simplifies the design of an algorithm. Taking bounded resources into account makes us work harder, but the resulting prescription may be more exact and more useful. Even though this phenomenon is well known, it is not reflected properly in the theory of algorithms.

Sometimes it is easier to explain one's arguments in discussion. To this end, please allow me the liberty of introducing an imaginary opponent, a skeptical graduate student.

Opponent: "In spite of your criticism, I think that the abstraction of unbounded resources is very useful."

It is indeed. Nevertheless its applicability is restricted. It reflects computations where the effort to acquire additional resources—whenever it is necessary—is negligible; after acquiring additional resources, the computation resumes as if it was never interrupted. This mode of computation reflects very well routine computations of human computers. In his celebrated paper (Turing *1936-7*), Turing analyzed a routine computation of a human computer with a pile of sheets of paper before him/her. The pile was abstracted as the Turing tape. It is easy to believe that the supply of paper exceeds the patience of the human computer.

Here is another situation where the abstraction of unbounded resources is justifiable. Consider a relatively small computation process living in a huge resource-sharing environment managed by a flexible operating system. Whenever the process needs more space or another processor or whatever, the operating system quietly and efficiently provides the necessary additional resources.

But, as we saw, it is easy to find situations where the abstraction of unbounded resources is not justifiable. Imagine that your Macintosh runs out of memory during a computation. What can you do?

Opponent: "I can buy a hard disk; it will take care of all of my needs for a while. But I see your point. It is not negligible to acquire additional resources in the case of the Macintosh. What if I keep inserting and removing floppy disks (I can always buy more of those)?"

Then you become essentially a part of a new machine which contains a Macintosh. To make the point, let me consider the following ridiculous example. Imagine that you have bought a personal Turing machine. After a while you may run out of tape. You go and buy some additional tape. Your nearest shop may run out of tape. Then you have to figure out where to go. You use your brain to solve all these problems; the transition table of your Turing machine does not give you all the answers.

2. *Machines With Bounded Resources*

A possible alternative to the assumption of unbounded resources is an assumption that the resources of all relevant computing devices are bounded.

Opponent: "For some purposes you need machines with unbounded resources. For example, no single machine with bounded resources is able to provide an adequate operational semantics for a high-level programming language like Pascal."

It is true that no machine with bounded resources is able to provide an adequate operational semantics for Pascal. But we may consider a family of machines with bounded resources. This creates a situation which is similar to that in logic where the meaning of a first-order formula is given by a family of structures. Each structure of the appropriate signature gives only a local meaning to the formula; the global meaning of a formula is given by all structures of the appropriate signature. A family of Pascal machines with bounded resources is sketched in Gurevich *1987*. Each machine gives a local meaning to a Pascal program, and the global meaning of the program is given by the whole family of Pascal machines. When you write a Pascal program, you may imagine the addressee, i.e., the computing agent, as an ideal machine with unbounded resources. Alternatively, you may think about a machine with bounded resources. This should not make your program implementation dependent. Proving a theorem about finite groups, you may speak about the group of discourse; this does not mean that you prove your theorem for just that one group.

Opponent: "Please, give me a simpler example of a family of machines with bounded resources."

Let T be a Turing machine. Consider bounded-tape versions of T with or without special end-of-tape marks.

Opponent: "The computation of any bounded-tape version of T without the end-of-tape mark is an initial segment of the computation of T, and the cut-off point is irrelevant to the meaning of the program of T. I would prefer to consider the computation of T itself rather then a pretty arbitrary collection of initial segments of the computation."

Yes, in many cases, a machine with unbounded resources gives a cleaner operational semantics. But not always. Machines with bounded resources may know their resources and utilize this knowledge. A program for bounded-tape machines may use the end-of-tape mark for different purposes; it may divide the tape equally into a left and a right part and execute two different processes in a time-sharing fashion. More convincing examples of how machines with bounded resources may use the knowledge of their resources come from real life. Think about operating systems. In particular, think about an operating system which runs on many computers and starts with an inventory of the available resources. Of course, this program (the operating system) can be modeled by a machine with unbounded resources, but this is not necessarily the best way to provide an operational semantics to the program.

Some high level programming languages—such as FORTRAN—do not reflect any information about the resource bounds of the implementation. On the other hand, Pascal has an implementation-defined constant maxint. Implementation-defined constants found in ADA or APL reflect considerable information about the resource bounds. There seems to be a trend to include implementation-defined constants in high level programming languages. We consider this trend symptomatic. It reflects, we believe, the fact that the operational semantics based on machines with bounded resources is closer to the intended meaning of many programs than the operational semantics based on machines with unbounded resources. Enriching a programming language with names for essential resource bounds does not make it implementation-dependent. Pascal with maxint, for example, is not more implementation-dependent than Pascal without maxint.

3. Dynamic Structures

I believe we need a computational model appropriate to deal with machines with bounded resources. To start, it may be reasonable to make some simplifying assumptions. For example, one may suppose the following:

A machine with bounded resources is a finite-state machine which works in discrete time. The states form a legitimate mathematical set. The machine may be nondeterministic and able to do many things at once, but all parts of it obey one clock. The set of states, the input alphabet, the output alphabet and the transition function do not change in time.

Opponent: "Isn't any finite collection a legitimate mathematical set?"

The answer depends on how you define finiteness. But a subcollection of a perfectly good finite set may be not a legitimate mathematical set. For example, the collection of English words is not a legitimate mathematical set; the membership relation takes more than two truth values. Questions like "Is the number of elements even?" or "Is the number of elements prime?" should be meaningful for legitimate finite mathematical sets.

Opponent: "There is already a computational model of machines with bounded resources: the classical theory of finite-state machines?"

Unfortunately, the classical theory of finite-state machines is not adequate to deal with real computers or virtual computing devices like Pascal interpreters. One can view Apple's Macintosh as a finite transducer (or even a finite automaton), but it is unfeasible to write down the full transition table (or a regular expression) for the Macintosh: there are too many states. The problem arises to elaborate the notion of finite-state machines into a notion which is more appropriate to deal with real computers and complicated virtual computing devices. One problem with formalizing machines and some other objects of interest in computer science is their dynamic character.

Opponent: "What other objects?"

Consider for example relational data bases. They are defined often as collections of mathematical relations (predicates). But collections of mathematical relations do not evolve in time, whereas data bases do. Traditional mathematical structures—such as graphs and groups—are static. There is a tendency to formalize dynamic processes by means of static structures; a common trick is to introduce time as another dimension. You can trace that tendency also in the field of foundations. Think about the process of "cleaning" mathematics from constant and variable quantities in favor of functions.

Opponent: "If static structures have served us so well for so long in so many different applications of mathematics, why shouldn't they be appropriate in computer science? What is so special in the applications of mathematics to computer science versus, say, the applications of mathematics to physics?"

The special features of many computer applications include discrete time, the existence and relative simplicity of static configurations and the relative simplicity of transitions from one static configuration to another. (The relative simplicity of static configurations and transitions is often accompanied by an overwhelming number of states and an overwhelming complexity of the whole process.)

Opponent: "I can give you examples of neat mathematical structures of dynamic nature: Turing machines, random-access machines."

I would like to see a more general notion of dynamic structures. Complicated abstract machines, such as virtual Pascal, Ada, and Smalltalk machines, should be covered. I believe we need a theory of dynamic structures. Operational semantics for programming languages would be one application for such a theory.

Opponent: "Plotkin *1981* speaks about transition systems. A transition system is a set of elements (called configurations) with a binary relation (called the transition relation). You were willing to restrict the discussion to the case of discrete time when all parts of the machine obey one discrete clock. The notion of transition systems seems sufficiently general for that case."

I agree. However, transition systems give us static representations of dynamic structures. In the case of a finite-state machine, the transition system is simply the presentation of the machine as a finite automaton, and we know that the theory of finite automata is inadequate to deal with complicated dynamic structures.

It is natural to see configurations of a dynamic structure as static structures. The evolution of a dynamic structure is governed by transition rules. The notion of a family of dynamic structures may be restricted by the requirement that members of the family have the same transition rules (so that there is a single finite formulation of transition rules appropriate to all members of the family).

Different classes of dynamic structures may be defined by imposing syntactical restrictions on transition rules, by allowing or forbidding the evolution of the signature (the language) of the current configuration, by allowing or forbidding the creation of new universes (sorts, types) and the elimination of old ones, and so on.

The notion of dynamic structures is briefly discussed in Gurevich *1987*. The primary example there is a family of Pascal machines. A detailed description of Modula-2 machines will appear in Gurevich and Morris *1987*. The case of Modula-2 is more interesting because of some parallelism allowed there. We are also looking into Ada and Smalltalk from that point of view.

4. *The New Thesis Problem*

The problem is to define a modest class (let us call it U for 'universal') of abstract machines with bounded resources such that every "real" computing device with bounded resources can be closely simulated by an appropriate U-machine of comparable size, and every family of computing devices with bounded resources can be appropriately simulated by a family of U-machines.

It is easy to come up with a cheap solution if one ignores the quality of simulation or the size of the simulating machine (Gurevich *1984*), but a good solution should have important applications. The problem was briefly discussed in Gurevich *1987*, and we intended to take up the issue here but the limitation of time has proved to be too severe. The new thesis problem will be discussed in Blass and Gurevich (in preparation).

By the way, bringing the issue of simulation into the open invites attempts to improve Turing's thesis by imposing a restriction on the allowed simulations and possibly using different machines instead of Turing machines. One may be interested in polynomial-time simulations, linear-time simulations, real-time simulations, etc. There seems to be a consensus among computer scientists that Turing machines are sufficiently good simulators from the point of view of polynomial time. In particular, the polynomial-time version of Turing's thesis seems to be accepted by many. However, one should be a little careful about what is a legitimate computing device in the polynomial version of Turing's thesis. For example, Schönhage *1979* constructed a deterministic random access machine with built-in arithmetical operations which decides the satisfiability of boolean formulas in conjunctive normal form (a known NP complete problem) in polynomial time with respect to the so-called uniform cost criterion. (Under the uniform cost criterion each instruction requires one unit of time whereas a more realistic logarithmic cost criterion takes into account the limited size of machine words in real random access machines; see Aho, Hopcroft, and Ullman *1974*.)

With respect to linear or real time, Turing machines are inadequate simulators. Kolmogorov and Uspenski *1958* introduced more flexible machines. They wrote humbly that the only purpose of their paper was simply to re-examine for themselves Church-Turing's thesis. It is possible nevertheless that they had in mind something more ambitious like the following thesis: Every sequential computing device can be simulated in real time by an appropriate Kolmogorov-Uspenski machine.

The notion of real time simulation is defined as follows in Schönhage *1980*: A machine M' is said to *simulate* another machine M *in real time* if there is a constant c such that for every input sequence x the following holds: if x causes M to read an input symbol, or to print an output symbol, or to halt at time steps

$$0 = t_0 < t_1 < \ldots < t_k,$$

respectively, then x will cause M' to act in the very same way with regard to those externally visible operations at time steps

$$0 = t'_0 < t'_1 < \ldots < t'_k,$$

where $(t'_j - t'_{j-1}) \le c(t_j - t_{j-1})$ for $1 \le j \le k$.

Schönhage *1980* introduced a machine model which is similar to but different from the Kolmogorov-Uspenski model. He "posed the intuitive thesis that this model possesses extreme flexibility and should therefore serve as a basis for an adequate notion of time complexity". A related and blunter thesis is: Every sequential computing device can be simulated in real time by an appropriate Schönhage machine.

To get rid of the restriction to sequential devices, one may use parallel versions of both Kolmogorov-Uspenski machines and Schönhage machines.

Gandy *1980* formulated four assumptions about computing devices, put forward a thesis that every deterministic mechanical device satisfies the four assumptions, and proved that whatever can be calculated by a device satisfying the four assumptions is Turing computable. Much more detailed analysis is needed to argue for or against the polynomial-time version of Turing's thesis or either of the real-time simulation theses. The task does not seem easy. It should involve in particular a closer examination of the assumption of unbounded resources.

Acknowledgements. I am very grateful to Egon Boerger, Bernie Galler, Kit Fine, Saharon Shelah and especially Andreas Blass for enjoyable and useful discussions.

References

Aho, A.V., J.E. Hopcroft, and J.D. Ullman
   1974 *The Design and Analysis of Computer Algorithms.* Reading, MA: Addison-Wesley (1974).

Blass, A., and Y. Gurevich

in prep. A new thesis (tentative title), in preparation.

Church, A.

1936 An unsolvable problem of elementary number theory. *Am. J. Math.* **58** (1936) 345–363.

Gandy, R.

1980 Church's thesis and principles for mechanisms. In: *The Kleene Symposium,* eds. J. Barwise et al., pp. 123–148. North-Holland Publ. Co. (1980).

Gurevich, Y.

1984 Reconsidering Turing's thesis (toward more realistic semantics of programs). Tech. report CRL-TR-36-84, University of Michigan (1984).

1985 A new thesis. AMS Abstracts, Aug. 1985, p. 317.

1987 Logic and the challenge of computer science. In: *Current Trends in Theoretical Computer Science,* ed. E. Börger, pp. 1–57. Rockville, MD: Computer Science Press (1987).

Gurevich, Y., and J.M. Morris

1987 Algebraic operational semantics for Modula-2. Tech. Report CRL-TR-10-87, July 1987, University of Michigan.

Kolmogorov, A.N., and V.A. Uspenski

1958 On the definition of an algorithm. *Uspekhi Mat. Nauk* **13** (1958) 3–28 (Russian); *AMS Translations* **29** (1963) 217–245.

Plotkin, G.D.

1981 A structural approach to operational semantics. Tech. Report DAIMI FN-19, Aarhus University, Aarhus, Denmark.

Schönhage, A.

1979 On the power of random access machines. In: *Automata, Languages and Programming,* ed. H.A. Mauer, pp. 520–529. Berlin: Springer-Verlag (1979).

1980 Storage modification machines. *SIAM J. Comp.* **9** (1979) 490–508.

Turing, A.M.

1936-7 On computable numbers, with an application to the Entscheidungsproblem. *P. Lond. Math. Soc. (2)* **42** (1936-7) 230–236; A correction, ibid. **43** (1937) 544–546.

Beyond the Turing Machine

Brosl Hasslacher

1. Preface

Byzantine complexity, seen in the physical world, contains archetypical features which surface in many disciplines in disguised form – a reflection of the same phenomena in different mirrors. I have extracted those concepts with a rich and natural connection among them and which I think important for designing new kinds of computing strategies for the complexity problem. Many of these ideas go back to von Neumann, but somewhere we dropped the thread in the so-called supercomputer era. Making sense out of extraordinary self organizing complexity belongs to the most difficult problems of this century. Without new kinds of computation models we have no hope of solving it. We must depend on our analytic skills and on our computing machines to act as eyes on this new world. These remarks are about those machines and our interaction with them.

2. Introduction

In the past two decades both mathematicians and physicists have been intensively studying complexity, a problem so difficult that it strains all of our capacity to understand structure in the world. By understand, I mean the construction of models with predictive power which capture the essentials of what we see in nature, and the analysis of such models either with algorithms or by simulation on a computing device. The problem I will mainly discuss is the development of complexity in strongly nonlinear systems with many degrees of freedom, and its self-organization: how much we can understand about such systems and where our power of prediction must be limited because we are beings that use symbols. I shall argue that the study of such systems leads to insights into the construction of new kinds of computing architectures which cannot be efficiently captured by the traditional concept of a Turing machine and which have extraordinary complexity built into their fundamental operation.

Extraordinary complexity is seen in the phase space of chaotic systems, stationary spatial patterns, and ultimately space-time complexity created by the evolution of such systems. I limit myself to the ordinary observable world characterized by energy transitions of the order of magnitude of a few electron volts (e.v.). This does not mean that the complexity problem disappears in the cosmological, relativistic, or quantum world, but it is far less accessible there. For reasons that will become clear, it may also take on different forms in such domains and is not well understood.

So we limit ourselves to patterns of complexity which develop in the few e.v. world of strongly nonlinear classical systems. The phenomena of turbulence in fluids and gases, the so-called "chaotic" behavior of dissipative systems in various dimensions, and the complex behavior that arises under the evolution of simple cellular automata, which are concurrent networks of locally connected few bit finite state machines, are all examples of this. All of them have evaded analytic description except in few special atypical cases.

I will first assume that we have a situation in which we have reached an analytic limit and cannot describe complex structures by an algorithm which takes significantly less time to evolve an outcome than observing a copy of the system itself. This is a strong assumption, and in some cases it will not hold; but it is the usual case, so I will assume it and look at its consequences for analyzing such systems.

Complexity of this order is a subjective concept; it is complexity relative to the capacity of an observer to effectively compute. One can observe the physical world with instruments (i.e., take selective information from it by experiment) or experiment on a model of a physical system. Within a physical computing machine, one designs a virtual machine so that an appropriate symbolic image of the physics defines the transition table in the finite control of an equivalent Turing machine. This virtual world is a model of the physical world. We are interested in computing algorithms defined in this virtual world.

I depart again from what is usually assumed and ask that the observer be a part of the physical world. I will ignore what is in principle possible with hypothetical computation architectures with unlimited resources, and examine only what questions can one ask a physical computing machine, within a certain precision, in a reasonable time. The observer-analyst, which I will call a physicist, although one can substitute mathematician equally well, has to analyze the output of such machines within the constraint of living in the physical world. No one has access to infinite computing resources; this concept is not useful when interacting with the physical world or a faithful model of it, except for technical mathematical reasons, which we will not need.

With such premises, a working definition of a physicist is an information gathering machine, connected to a finite resource computation device, evaluating efficient algorithms. This leads to the idea of polynomial time bounded physicists (P time bounded), and to avoid atypical transient behavior let us bound such a being from above by some smooth polynomial function on the length of the input string. The notion of being P time bounded agrees well with what one can actually do and estimates one's power to analyze in a realistic way. Immediately, one can ask how much analytic power have we lost by making a P time assumption. I shall argue that we have lost nothing essential over an infinite resource machine, and actually gain insight into how much organization and of what complexity we can put on an observed world, whether virtual or real.

All of this will become clearer when we introduce the idea of a standard model for the physical world: the idea of a skeletal game, and fix on what kinds of computational architectures we are allowed to use as analyzers, within P time bounded resources. We will look at the important role of noise in analysis; it induces a spectrum of outcomes from an algorithm which is very close to the physicist's idea of scaling laws for complex systems.

At this point, we remark that, instead of doing an experiment on the physical world or what we think is the physical "outside" world, we can put questions to a simulator of physical phenomena. This is a virtual image within a physical machine of certain algorithmic structures which we believe capture the essential mechanisms of our phenomena. The idea of simulation replaces the more limited concept of computation on an algorithm and is more difficult to deal with. We are now asking a series of questions, not about physics, but about a model of physics.

We can do experiments on this virtual world. We can ask what kinds of questions make no sense to the virtual world, although they may elicit an answer if we performed the same experiment, i.e., asked the same question, on the physical world itself. However, the interrogation of the physical world is equivalent to an oracle whose answer produces no insight. This means that the answer gives no hint of the systematic structure producing it.

We have introduced concepts of computation and simulation which are relative, where we compare the output of two P time bounded devices. We can distinguish complex structures only up to our ability to compute a difference between them which is distinguishable from the outcome of a random coin toss. This is a kind of physical Turing test, and avoids the drastic step of assuming a physical Church's thesis – the idea that the physical world performs computations in the Turing sense. A physical Church's thesis is a misleading viewpoint on the physical world; it injects an algorithmic character into the behavior of the physical world for which there is no evidence. Now we have the idea of the equivalence of virtual physical models, or the complex structures evolved by them, within a class of computational resources, i.e., up to efficient computational procedures.

Note that this is quite different from the original idea of the Turing machine which assumes unbounded resources, in particular infinite time on an infinite rewrite tape over a restricted class of problems – the partially recursive functions. The Turing machine, as originally devised, is an irreducible model for the analytical possibilities of a serial rewrite being, operating a simple finite state machine on a one-dimensional tape, or that can be reduced to such a setup by a diagonalization procedure. It is a model for a number theorist with infinite resources.

In Turing machine theory, and in the complexity analysis of algorithms processed by Turing machines, one concentrates on the properties of output tapes as a function of input tapes – the structure of the Turing machine's finite control is ignored. This is what physicists would call an S matrix theory of computation; the computer is a black box, and we are allowed to look only at the in-out state relations.

We wish to enter the Turing machine and examine the possible structures one can embed in the finite control. Then we look at such computational devices, interacting within P time bounded resources, and at notions of equivalent outputs only under efficient algorithms, ignoring NP problems. The Turing machine is an anthropomorphic model of computation – it may not be a good model for what a physical machine can compute, especially if we look carefully at the possible structures that can be present in the finite state machine (FSM), which makes up the finite control. We will argue that, under well defined conditions, such a FSM can be thought of as encoding a one-way function whose state is impossible to invert under a P time assumption. Such a complex finite control either cannot be reduced to a simpler FSM with a longer input tape or it would take NP resources to do so. This is the basic strategy we adopt to construct machines which are functionally "close" to the structure of complex systems. We will call such machines, extended Turing machines, or Gödel (1) devices, with the conventional Turing machine being Gödel (0) on this scale.

The reason a standard Turing machine is not powerful enough to analyze complex systems in P time lies in the diagonalization procedure so heavily used in Turing machine theory, to prove equivalence to machines endowed with more resources (tapes, heads, tracks, etc.) Diagonalization arguments force a problem onto integral dimension (D) spaces of quite different fundamental scaling structure, for example D either belongs to the natural numbers, N, or to the reals, R. Diagonalization gives no information on what lies metrically between N and R, in the sense of a Hausdorff (H) dimension.

Many nonlinear phenomena, operating at the few e.v. energy level, appear to evolve on spaces of nonintegral H. In the simplest case, they are said to have a fractal dimension between that of N, which has $D = 0$, and R, which has $D = 1$. This is due to the presence of scaling laws in such phenomena and has the consequence that the effective dimension in the sense of scaling in the phase space of such systems is fractional, i.e., $H \neq D$, or $0 < H < 1$.

It is precisely in such scaling regimes, and in their associated fractal spaces, that extraordinary complexity seems to originate. A Turing machine does not naturally inhabit a space of fractional H dimension, since all its operations are countable (elements of N). To repair this problem we will give a Turing machine an FSM, which has all length and time scales built into it, and responds to an input with a spectrum of outputs, controlled by adjustable scaling relations. This is a fractal extension of a Turing machine, which requires at least 2 dimensions to define, and contains essentially NP computational power in the finite control.

To see how one might do this let us first look at the effective model produced when an algorithmic description based on the usual underlying idea of continuous space and time is implemented on a serial digital computer.

First, we remember that we, or P time bounded analysts, are embedded in the physical world. This means we use digital computers whose architecture is given to us in the form of a physical piece of machinery, with all its artificial constraints. We must reduce a continuous algorithmic description to one codable on a device whose fundamental operations are countable, and we do this by various forms of chopping up into pieces, usually called discretization. Using finite differences, elements, or some similar scheme, an algorithm with an operation count belonging to N is constructed and is translated into some high level language. The compiler then further reduces this model to a binary form determined largely by machine constraints.

The outcome is a discrete and synthetic microworld image of the original problem, whose structure is arbitrarily fixed by a differencing scheme and computational architecture chosen at random. The only remnant of the continuum is the use of radix arithmetic, which has the property of weighing bits unequally, and for nonlinear systems is the source of spurious singularities.

This is what we actually do when we compute up a model of the physical world with physical devices. This is not the idealized and serene process that we imagine when usually arguing about the fundamental structures of computation, and very far from Turing machines. Turing machines stand in this process in a similar relation as symbolic logic does to algebraic topology.

How can we build a faithful computational image of the physical world and at the same time a more powerful computational tool? We can construct a simulator which bypasses the continuum and deals from the start with totally discrete variables, a kind of discrete game. Fundamental insight in physics or mathematics does not involve yes-no answers to an algorithm, but rather a search for structures and relationships among structures. These can be abstracted from an appropriately designed iterative game, which evolves an accurate copy of the structures and structural relations of physics. In this sense, discrete games are closer to category theory than to standard classical analysis.

If we define cellular automata as many few bit finite state machines interacting locally with simple rules, then the reduction process from the continuous model to a model that a physical machine can process is just a reduction to a cellular automaton (CA). Instead of mapping from the continuum to an arbitrary CA model, we start with a CA model faithful to the dynamical structures inherent in the problem. This means we invent a standard physical model, built with CA's, whose evolution will be taken as the output to be compared by P time physicists. It also means that the continuum equations, for the strongly nonlinear many degree of freedom system or field theory that ordinarily describes the system, should emerge as macrodynamical equations. These describe large scale collective modes in field theories, and the same equations should describe the collective modes evolved by appropriate CA models or games.

Several CA games are known which reproduce generic qualitative behavior (e.g., the emergence of large scale complexity of correct structure) for quite complex nonlinear field theories (e.g., the equations of fluid dynamics and certain plasma systems). By complex behavior, we mean the generation of turbulence and the development of strong nonlinear instabilities in these systems. For any particular field theory, such as a fluid, there are a large number of possible CA games. Using this freedom one can get arbitrarily good quantitative agreement to continuum models by increasing the number of states per site in the CA model to accurately approximate a number, with a desired degree of accuracy.

At this point, we introduce the idea of a skeletal game for a generic field theory. This is a collision dominated game, having the idea of collisions and allowed classes of them as the central dynamical ingredient, with the minimal structure necessary to reproduce the quantitative features of large systems. The motivation is that evolving such a game captures the core of the solution space which is also generated by the associated partial differential equation that normally describes the system. Doing this, we introduce a virtual discrete microworld within a physical computer, whose structure is fixed by a fundamental collision model for the physics, such as the kinetic theory of fluids, and not by arbitrary machine constraints.

Skeletal games have the property of all games: unless the game is trivial, knowing the rules of the game gives no hint of the outcome of the game. The class of outcomes of nontrivial games can only be found by evolving the game many times: by playing it or simulating it. If the game has many elements and the rules are simple, it is sometimes possible to find a phenomenological description for collective modes, evolved by the game, using the methods of many body theory. But this is just a map to the continuum description of macroscopic modes and leaves us with an intractable partial differential equation describing a certain regime of the game. This is one way to insure that the skeletal game evolves a faithful copy of the physics, but it does not help to analyze the complexity evolved by the game. It is just a reality check.

In general, if a physical process evolves on a time scale t, the skeletal game will also evolve on a time scale t. We need to compare the efficiency of skeletal games to PDE's as algorithmic procedures. If an analytic solution to a PDE is known, then as an algorithm it can be computed roughly in ln t time. This is why analytic solutions are important; one gains exponentially in time over direct simulation of a copy of the system or over a skeletal game. However, in the case of chaotic systems or regimes in a field theory where solutions are highly sensitive to initial conditions, one has to deal with computing exponentially diverging phase space trajectories and defining metrics on the evolving phase space.

This means that although the system is deterministic and any single trajectory can be explicitly followed, two initial points which are metrically close evolve trajectories which diverge from one another very quickly. Chaotic systems are just those whose solution trajectories diverge exponentially in time under a general choice of phase space metrics. In addition, there is inevitable folding and stretching of the manifolds on which these solutions lie, so that the correlations between two nearby points in the phase space of the system appear random.

Since we have a P time analysis constraint, we do not need as strong a definition for random as Kolmogoroff random (K random), where a bit string is also its own shortest algorithmic description. We can allow random to mean evolved by a pseudorandom number generator, specifically one evolved by a 1-way function. With P time bounded resources, pseudo-random and K random are indistinguishable. Later we will see this is closely related to the physics of complex processes.

If the system has dissipation, the phase space can contract down to a manifold of relatively low dimension. But even in this case, the complex structures that evolve on it are, except for specially rigged systems, analytically intractable and computationally NP hard.

Systems which behave this way are said to have positive Lyaponov exponent (L), from the rate of trajectory divergence, usually written in the form exp $(+Lt)$, with L positive. In this case, computing an analytic solution in a parameter regime where L is positive changes a computation taking $\ln t$ to evolve to one taking t. This is the chaotic slow-down phenomenon. In such systems PDE's and CA games take roughly equal time to serially evolve. Simulation and computation become indistinguishable.

In the chaotic sector of nonlinear field theories, the PDE formulation has no computational advantage over skeletal games. The reduction of continuum models to machine form already induces a microworld description, with no attempt made to maintain a faithful copy of the physics within the computing machine. The usual PDE that describes physical continuum behavior is also a phenomenological model for the collective modes of all such models, whether faithful to the physics or not.

So the idea of a skeletal CA game is attractive for two reasons: as a standard backbone model of physics and as the minimal starting point of most compact algorithmic description, provided that the skeletal game is also a minimal game, in the sense of having an irreducible structure.

As tunable parameters in the model are changed, a collision dominated game designed to simulate a strongly nonlinear system will develop prototypical chaotic behavior or generalized turbulence. Chaotic regimes, at some super scale, usually have self-similar and self-organizing behavior and obey scaling laws. By scaling laws, we mean that there exist transformations on the system, consisting of a local coordinate change and a magnification, which brings the system into the same form at $t + dt$ that it had at t. It is form invariant. As the natural scaling interval shrinks down to a point, scaling laws become nowhere differentiable scaling functions.

Examining the behavior of correlation functions in such systems, or its behavior in Fourier space, one often sees power law behavior over many orders of magnitude in the size of the system. Scaling laws are encoding regularity in the dynamical complexity contained in these systems. It is an open question how much of the limit of scaling laws, namely scaling functions, one can measure in chaotic systems without losing the signal in the noise or violating the P time calculator assumption.

To attack this problem, we construct another minimalist game having such complexity built into it, complete with scaling laws and functions. We will need to develop some new concepts.

For continuous dynamical systems with only a few degrees of freedom, an equilibrium state can often be uniquely defined as a state of lowest energy. If a useful potential function description exists, it has at least one minimum. Initialized away from equilibrium, the system evolves by sliding down the potential well until it is trapped in a local or global minimum. If the system is discrete and spatially extended, it often happens that spatially extended islands of marginal energetic stability form and organize into domains, preventing the system from reaching a normal ground state, which comes from the old idea that the formation of domains is some unnatural state. This phenomenon is called frustration. A simple example of a frustrated system is a discrete automaton with a threshold cascade mechanism as the fundamental update rule.

We call the class of such automata "cascade dominated", to distinguish them from ordinary, or collision dominated automata, which have no fundamental cascade mechanism, although they may evolve behavior that acts as if cascade automata were operating. Cascade automata are easy to design and quickly develop extraordinarily complex frustrated states, which have built in spatial and temporal correlation power law behavior.

The simplest example of a cascade dynamics is a granular system containing a tumbling mechanism, under a constant external force, such as gravity. One can try to find the stable state of a heap of dust or sand by pouring it out of a container onto a flat surface. The initial state is far from equilibrium, and at the start the average slope of the heap will be large and the heap unstable. The slope decreases as the heap sheds dust until a critical angle is reached, when the heap ceases motion, at which point we say the heap is critically stable.

Critical stability has a well defined character. As critical stability approaches, stable domains of all sizes appear on the heap, and that continues until all length scales possible in the system are occupied by domains. In an infinitely fine system, this goes on until no length scale is left in the system. The critical state is a basin of attraction for cascade systems initialized far from equilibrium.

The signature of a critically stable state is its behavior under perturbation by noise. In a normal stable state, noise propagates globally through the system. In a critically stable state, noise only propagates locally, until a set of wavelengths present in the noise perturbs a domain with a similar length scale, and the heap redistributes domains by a cascade. Long wavelength noise will excite domains over a wide range of length scales and cause fluctuations in domain lifetimes which have power law behavior over many orders of magnitude. Since one could equally well build up such a system from below a critical slope by sprinkling dust on a plane, critical stability is a boundary regime between a far from equilibrium attractor and one built up slowly from a sequence of equilibrium states. Such a basin of attraction has a fractal character to it, and by definition, a rational Hausdorff dimension.

A simple model of a nontrivial cascade dominated game is a two-dimensional CA with nearest neighbor rules on a square lattice. A site value stands for a field variable, s, and is integer valued. If $n, m = (-1, 0, +1)$ and $b = (1 - |n * m|)$ where $|x|$ is absolute value, $z = (x, y)$ a site in the lattice, and $i = (n, m)$, then the shift invariant site update algorithm is:

$$b * s(z + i) \rightarrow b * [s(z + i) + 1] - 5 * d(n, 0) * d(m, 0)$$

where $d(j, k)$ is the usual Kroeneker function, $d(j, k) = 1$ if $j = k$ and 0 otherwise.

This rule has the effect of reducing the center site field variable by 4 units and incrementing each of its 4 nearest neighbor variables by 1 unit. A threshold value for s is chosen, say C, and since the system should start far from equilibrium, the initial and boundary conditions on $s(z)$ are: random initial values for $s(z) \gg C$ in the interior, and $s(z) = 0$ on the boundary of the automaton (pinned boundaries). The automaton cascades down until all $s(z) < C$, where it halts. The cascade automaton is now in a critical minimally stable state. Pinned boundary conditions are not essential. One can arrange for transport through the system and boundaries, so as to always keep the system fully loaded at criticality. The pinned model is much cleaner for visualizing dynamics.

Such models, which one could call stack automata with tumbling, evade triviality only in dimension 2 or greater. In one dimension, the attractor is trivial, a fixed point. A nonperiodic finite automaton with nontrivial critical stability depending on a cascade cannot be built in 1 dimension, nor can it be related to the composition of one dimensional systems. It is fundamentally a high dimensional structure. Simple automata of this kind evolve behavior similar to many exceedingly complex physical systems, such as spin glasses, percolation phenomena, resistivity networks, and turbulence.

We can use a cascade mechanism as a prototype for a simple kind of generalized turbulence. Many few e.v. systems that are minimally constrained and strongly nonlinear, such as fluids, display cascade phenomena and scale over many orders of magnitude, in space and time correlations, for large regions of parameter space. A critically stable attractor for systems far from equilibrium implies that all length and time scales are present in the system, and has extreme sensitivity to noise perturbation which can propagate only locally. This leads to avalanches or spectral bursts as dynamical outputs.

Cascade automata can be used as natural and robust models for the complex cascade behavior in systems that display generalized turbulence. In the absence of noise, a cascade automaton at a critically stable point will remain there forever. But physical systems always contain random behavior, since they are always at a nonzero temperature. This acts as a spontaneous noise source and causes a cascading of domains with spectral laws. Since cascade automata in the critical state are far from equilibrium attractors, it is very easy to form them in a system, and they are robust in the sense of being independent of tuning parameters. Various turbulence schemes can occur, depending on the physical details that flesh out a skeletal automaton into a physical system.

One way to suppress cascade dynamics is to impose constraints on the system that prevent all but a few length scales from being occupied. These include various exclusion rules, group structures, gauge invariance, and topological or global constraints which have local manifestations, and act as extinguishing mechanisms for cascades. Typically, a strongly nonlinear field theory having extraordinarily complex dynamics is minimally constrained.

The last fundamental characteristic of these simple cascade automata is that they have trivial continuum limits. The cascade model displayed goes over to a simple diffusion equation. The whole essence of the phenomena of critical stability is that it is a discrete cascade.

3. A Cascade Dominated Architecture.

Suppose we take a standard Turing machine and ask that the FSM or finite control be always set to a critically stable state. Such a hypothetical device has at least a 2-dimensional finite control and operates concurrently. It is an ND machine, but with complexity designed into the finite control that is near the complexity of the system or algorithm under study. The scaling laws of the finite control can be adjusted by altering the cascade mechanism, and have an inherently fractal Hausdorff dimension. This is a machine that has an output of ensembles of grammatical structures related by scaling laws rather than algorithmic output in the form of a single bit string.

With a single tape machine, an input string acts as a triggering mechanism for outputs having power law spectra; the power spectrum and correlation functions come from the class of automata obeying the given scaling laws. In this sense, it is an ND machine, performing computational tasks on complex systems in P time, which would take a usual Turing machine NP time to simulate. One can see this by looking at a usual two tape deterministic Turing machine, with one tape containing a program and data, and the second tape holding the output of a critical finite state machine in response to a gaussian noise source. The second tape contains all length and time scales generated from at least a 2-dimensional source and would take exponential time to simulate on a usual deterministic machine.

Given a generalized fractal Turing machine, or Gödel (1) device, we can simulate the behavior of a critically stable system in P time in such a way that its output is indistinguishable from a copy of the physical system to a P time bounded observer.

Cascade dominated automata naturally generate 1-way functions: functions easy to produce algorithmically, but impossible to invert in P time. One-way functions are common in physical phenomena, but instead of producing the usual pseudo-random strings, they produce structures which have correlations fit by power laws. In that sense, physical 1-way functions originating in complex systems, are rich and do not belong to the class of pseudo-random generators, but to a new class, the cascade dominated generators. These are 1-way algorithms which evolve structures that belong to fractal spaces close to those we generally observe in highly complex systems. There are many variations in the structure of the finite control and kinds of information on input tapes one can devise, but I believe this setup is archetypical since any variant scheme can be reduced to the canonical one in P time, with an ordinary Turing machine, by the usual arguments.

Cascade dominated systems are a novel way to examine algorithmic instability. A collision dominated game is equivalent to a cellular automaton lacking a fundamental cascade mechanism. It is an FSM with naive transitions of the type usually considered in Turing machine models. Two and greater dimensional FSM's, operating concurrently, are known to develop frustrated structure, and we can imagine such a Gödel (1) device in operation.

Say that we halt such a machine when it reaches the critically stable point. This can be thought of as a new machine with the finite control set to a critical structure. An input tape is read in and the output is now unstable in that an ensemble of identical copies of such machines will each produce a different output, since the input acts as a noise source, locally exciting the frustrated structure in a chaotic way. The ensemble of outputs will fit a scaling law for correlations and power spectra, but the output of any individual run can be characterized as unpredictable. A deterministic machine can transit dynamically to an ND machine this way, and one of a very special type. Since this whole setup can be modeled (but not in P time) by an algorithm on a usual Turing machine, it is also a model for dynamical chaos in deterministic algorithms.

Critical stability is an energetic phenomenon, so we can attach a meaning to temperature in algorithmic systems. We can define cascade machines set to a critical temperature and so give a meaning to the thermodynamics and statistical mechanics of algorithmic structures, as well as to machines. There exists a natural mapping which takes us from cascade dominated machine and algorithm behavior directly into a statistical mechanical and thermodynamical description of such systems which is quite different from the usual information theory analogies. This is the statistical mechanics of far from equilibrium systems with a frustrated attractor, and it is much closer to the theory of spin glasses than to near equilibrium linear systems. To my knowledge, this has never been explored.

The purpose of these thought machines is not to provide models of phenomena, for there are many obvious alternate setups, but to illustrate the importance of noise on the output of a cascade dominated automaton. In the absence of noise, the critically set finite control is absolutely stable; there are no spontaneous transitions (unless one wants a quantum mechanical computer). Reading any algorithm, even on a one-tape machine, is equivalent to putting noise, taken from the algorithm treated as a probability distribution, onto the finite control. So algorithms will have to be presented to a Gödel (1) machine in a form other than raw bit strings. These devices can serve as models of the development of large scale organization in a locally chaotic system.

There are many kinds of extended Turing machines possible, and a few have been proposed that are quite different from cascade machines, mainly for models of concurrent processing. Our category of extensions, Gödel (1) machines, is defined relative to output complexity, not in the computational complexity sense, but in the dynamical systems sense. The two can be related if the system is simple, but computational complexity is a cumbersome concept for systems with many excited degrees of freedom.

Resource bounded computation – the main theme of computational complexity theory – should be a central consideration for anyone attempting a computational investigation of the nonlinear field theories that describe the various energy regimes of physics. Computing on a model set of equations, especially in the style of analytic stage, then using a computing and display device to explore the regime beyond the present analytic one, and using the results to gain new analytic insight – so-called interactive computation – is engaging in communication, at least with oneself. The idea of P time bounded resources and behavioristic standards of equivalence of structure under such resources, are not engineering concepts but fundamental ones. These ideas are common to any communication between system and analyst or between computational systems.

If analytic schemes are theoretically powerful but painfully slow to evaluate even in P time, then analysis becomes effectively impossible. Complex finite controls, designed to come as close as possible to the kind of behavior under study, is the heart of the present approach. It leads away from the idea of the general purpose or universal machine, which was never conceived to operate in low order polynomial physical time on physically motivated nonlinear problems. A more powerful computational configuration to a physicist is a universal analyzer, or mother machine, parallel in architecture, with high bandwidth ports, to which one can attach very fast, complex, and inexpensive black box devices set up to simulate the structure at hand. This amounts to having an ensemble of disposable FSM's attached to a larger machine containing data reduction and analysis platforms. These are relatively expensive and rarely changed.

Some farsighted computer architects are beginning to design this way, but few in the purely academic mathematical physics community have questioned the utility of the standard pipelined vectorized machines in use today. It is quite clear that, irrespective of their theoretical reduction to a Turing device, parallel architectures will have to be extensively developed as computing objects in a separate class from serial machines. Many architectural experiments will have to be done in the remainder of this century to collect a critical mass of such machines and user experience on them. We can then start a realistic process of evolutionary selection on parallel designs. There is very little deep theory to guide us here. For a while we will have to experiment with architectures, for unlike the prototype of the number theorist, which is the Turing machine, we do not have a similar universal model of the physical or mathematical structures underlying the observable world.

It is encouraging and exciting to find computational complexity theorists, who normally do not concern themselves with the operation of the physical world, coming to many of the same ideas explored here. These are researchers who are involved in developing secure communication between computers in the presence of a hostile computer adversary. Such cryptographic games are a highly simplified model of interaction with a synthetic reality, namely the structures of the playing computers.

The point is that secure communication is by necessity a P time bounded endeavor or it fails to be communication. The degree of complexity is low compared to that experienced in field theories, being limited to pseudorandom numbers instead of cascades, fractal spaces, and scaling relations, but many of the same communication ideas emerge. The notion of a 1-way function and certainly the idea of the functional equivalence of two sequences as effectively random, if they cannot be distinguished by a P time Turing test, were first used in secure communication theory. The intriguing idea of zero-knowledge proofs arose in this context, and has deep connections with the methods of physicists, who search not for the validity of a statement when doing experiments, whether numerical or real, but for structural relations underlying a process. By studying how to attempt to talk to each other with securely encoded messages in P time, we gain insight into the vastly more difficult problem of how to model the physical world with efficient algorithms and machines.

The idea of Gödel (1) devices is important and should be explored. How does one systematically design the structure of an FSM so that it has an intrinsic exponential gain in power that one can then use in P time? Is there a hierarchy of such devices? Do *L* systems, a natural parallel rewrite system, fit into this scheme, and can they be powerfully generalized?

We know of at least one analytic scheme that describes field theory very well and has either a trivial functional fixed point solution or requires a Gödel (1) device to evaluate it in P time, namely the path integral of Feynman. This method searches function space for an optimal system evolution path by a variational principle, essentially searching for an optimum evolution algorithm in parallel. Once found, it has to evaluate this functional algorithm, using functional initial data, also in parallel. Again, analytic solutions or functional fixed points exist only in special cases. Generally, the path integral scheme has a complex functional basin of attraction built into it.

The only methods known to define path integrals then are variations on the scaling functions described above: renormalization group methods. In a sense, the path integral description of the world, which so far is the most compact algorithm for encoding physics, is a kind of model for the analog of formal languages recognized by a Gödel (1) device. The only other method of evaluating a path integral by machine is by a variant of a Monte Carlo scheme – guaranteed to use exponential time resources for any nontrivial problem. Monte Carlo methods are a good example of how serial techniques and machines fail completely when faced with a field theory problem, or at least it is a failure for the present discussion, since they are not efficient or P time procedures.

It is possible that, if one takes a rich representational space for a problem, one can find an analytic solution to a highly constrained field theory system. This is the strategy of relativistic quantum field theory. There is some hope that this will work. However, in that case one could say that the model only contained trivial functional fixed points due to the many constraints imposed by, for example, current superstring theories. That does not help the computational issue in an energy regime where extraordinarily complex structures are known to exist, which are the those of the few e.v. world. And of course it can turn out that at all energy scales the world is extraordinarily complex and no analytic scheme can capture it. The physical world is subtle and has many dimensions to play with. It is unsettling to remember that we are P time bounded symbol beings, dwarfed by the effective computing resources in a thimbleful of water.

Acknowledgements. This work was done during the summer of 1987 at the following: University of California, Los Alamos National Laboratory, Los Alamos, New Mexico, U.S.A.; University of California at San Diego, La Jolla, California, U.S.A.; École Normale Superieure, Paris, France; Institute for Theoretical Physics, Beijing, China. I wish to thank all of these institutions for their hospitality.

References

(1) Cellular Automata:

Wolfram, S. (ed.)
  1986 *Theory and Application of Cellular Automata.* Singapore: World Scientific (1986).

(2) Dynamical Systems and Complexity:

Gukenheimer J., and P. Holmes
  1983 *Non-Linear Oscillations, Dynamical Systems and Bifurcations of Vector Fields.* New York: Springer-Verlag (1983).
  1987 Spatio-temporal coherence and chaos in physical systems. *Physica* **23D** (1987).

(3) Fluid CA Models:

Frisch, U., B. Hasslacher, and Y. Pomeau
  1986 *Phys. Rev. L.* **56** (1986) 1505

Frisch U., D. d'Humieres, B. Hasslacher, P. Lallemand, Y. Pomeau, J.P. Rivet
  1987 Lattice gas hydrodynamics in two and three dimensions. *Comp. Sys.* **1** (1987) 649-707.

(4) Computational Complexity:

Garey, M., and D. Johnson
  1979 Computers and Intractability: A Guide to the Theory of NP Completeness. New York: Freeman (1979).

Hasslacher, B.
  1986 Parallel computation and the limits of predictability in dynamical systems. LANL preprint (1986).

Hopcroft, J.E., and J.D. Ullman
  1979 *Introduction to Automata Theory, Language and Computation.* Reading, MA: Addison-Wesley (1979), and refs. therein.

Wolfram, S.
  1985 Undecidability and intractability in theoretical physics. *Phys. Rev. L. (8)* **54** (1985) 735–738.

(5) Cascade Automata:

Bak, P., C. Tang, and K. Wiesenfeld
  1987 Self-organized criticality: An explanation of $1/f$ noise. *Phys. Rev. L.* **59** (1987) 381–384.

(6) Path Integrals:

Simon, B.
  1979 *Functional Integration and Quantum Physics.* New York: Academic Press (1979).

Feigenbaum, M., and B. Hasslacher
    1982 Irrational decimations and path integrals for external noise. *Phys. Rev. L.* **49** (1982) 605–609.

(7) 1-Way Functions and Pseudo-Random Generators, Zero Knowledge Proofs, etc.:

Goldreich, O.
    1988 Randomness, interactive proofs, and zero-knowledge. This volume.

Goldreich, O., S. Micali, and A. Wigderson
    1986 Proceedings of the 27th Symposium on Foundations of Computer Science. ACM (1986).

Goldwasser, S., and S. Micali
    1984 *J. Comput. Sy. (2)* **28** (1984).

Levin, L.A.
    1985 Proceedings of the 17th ACM Symposium on Theory of Computing. ACM (1985).

Structure

Moshe Koppel

1. Introduction

Although the universal Turing machine's contribution to technology has been immeasurable, its contribution to the formalization and understanding of the most fundamental philosophical ideas may ultimately be regarded as the more profound and significant. According to Church's thesis, the universal Turing machine (UTM) represents the limit of computational power and thus can serve as a formal foundation for theories of computability and complexity.

In this paper we will demonstrate that a reexamination of the original definition of the UTM yields well-defined answers to the following questions:

Given some long finite string
1. What is the structure of the string, that is, what are the projectable properties of the string?
2. What are the likely continuations of the string?
3. What is the sophistication of the string, that is, what is the minimal amount of planning which must have gone into the generation of the string? More picturesquely, if the string is being broadcast by some unknown source, what is the minimum amount of intelligence we must attribute to that source?

Before answering these questions we will show that they are all related to one central concept, the *minimal description* of a string. The minimal description of a string S relative to some UTM is the shortest input to that UTM which results in output S (cf. Chaitin *1975*; Kolmogoroff *1965*; Solomonoff *1964*). Thus we regarded the UTM as an "interpreter" which, given a description, finds the object being described. The length of this minimal description is called the *complexity* of S relative to U.

A simple string such as 1111111... has a short description and thus low complexity; a patternless (random) string can be described only by enumerating it so that its complexity is roughly equal to its length, i.e., its complexity is maximal. Observe, though, that both simple strings and random strings are not sophisticated, in the sense explained in Question 3. Thus sophistication and complexity are orthogonal measures (cf. Bennett, this volume; Chaitin *1979*; Cover *1985*).

The relationship between sophistication and complexity can be made more precise in the following way:

The minimal description of a string consists of two parts. One part is a description of the string's structure, and the other part specifies the string from among the class of strings sharing that structure (Cover *1985*). The sophistication of a string is the size of that part of the description which describes the string's structure. Thus, for example, the description of the structure of a random string is empty and thus, though its complexity is high, its sophistication is low.

Recalling the formal definition of minimal description as minimal input to a UTM, we find that the two parts of a description can be distinguished if the UTM is designed to have not one, but two, input tapes. In fact the original and most natural definition of the UTM provides for two input tapes, one on which is written the program and the other on which is written the data. The universality is then easily expressed as follows: for every partially computable function F, there exists a program P such that for all data D we have $U(P, D) = F(D)$ (where $U(P, D)$ is the output of the UTM given inputs P and D). The realization that both program and data could be regarded as input (data) to a UTM was a major theoretical breakthrough with extraordinary practical consequences. Unfortunately this also led to the merging of the two concepts – program and data – and the loss of a wealth of insights which their distinction affords.

We will use the program-data distinction to answer the three questions posed at the beginning of the paper.

1. The structure of a string is the function computed by the program part of its minimal description.
2. The likely continuations of the strings are those which belong to the range of that program.
3. The sophistication of the string is the size of that program.

The outline of the paper is as follows:

The next section includes the definitions of UTM, minimal description, minimal program, sophistication, extensions of these definitions to infinite strings, and examples.

In Section 3, prediction and probability are defined and the main theorem is provided: the structure of an infinite string is almost uniquely defined and machine-independent.

In Section 4, the relationship between sophistication and "depth" (the running time of minimal descriptions (Bennett, this volume)) is explained, and an algorithm for finding the structure of an infinite string of bounded depth is outlined.

Finally, Section 5 is a sketch of an algebra of string properties.

Many of the ideas in this paper were developed by the author together with H. Atlan and were reported in a joint paper (Koppel and Atlan, *to appear*).

2. *Complexity and Sophistication*

In this section we provide definitions of the fundamental concepts used in this paper. These definitions and the results of the next section were first reported in Koppel and Atlan (to appear).

We will use the following notational conventions:

$|S|$ signifies the length of the string S. S^n signifies the initial segment of length n of the string S. For two strings S_1 and S_2 we say that $S_1 \supseteq S_2$ if S_2 is an initial segment of S_1.

COMPLEXITY

We will use a variant of the classical definition of program-length complexity most closely related to that of Levin *1973*.

We begin by defining a universal Turing machine U.

Let U consist of four tapes: a program tape, a data tape, a work tape, and an output tape.

Given the contents of the cells being scanned in some state, U can switch states, move one of the scanners left or right or print a 0 or 1 in one of the cells being scanned. These instructions are restricted as follows:

1. The program, data and output tapes are scanned left to right only.
2. U writes on the output tapes only if the cell scanned is blank and moves right on the output only if the cell scanned is not blank.
3. The computation halts if and only if a blank is scanned on the data tape.

If beginning in some fixed initial state, with the finite binary string P on the program tape and the (possibly infinite) binary string D on the data tape, and the program and data scanner on the first bits of P and D, respectively, the computation halts with the (possibly empty) binary string S on the output tape, then we say that $U(P, D) = S$. If the computation does not halt but continues printing bits of the infinite binary string α on the output tape then we say that $U(P, D) = \alpha$.

A *process* is a function F from finite binary strings to (finite or infinite) binary strings such that for all S, n we have $F(S^n) \subseteq F(S^{n+1})$. Clearly a function F for which there exists P such that $U(P, D) = F(D)$ is a process. If for every partially computable process F there exists a program P such that $U(P, D) = F(D)$ for all D, then we call U a *universal Turing machine*.

A program P is *total* if $U(P, D)$ is defined for all D. Note that $U(P, D)$ can be finite or infinite. A program is *self-delimiting* if during the course of the computation of $U(P, D)$ the first bit of output is printed while the program scanner reads the last bit of P (cf. Chaitin *1975*, Gács *1974*). We call (P, D) a *description* of a finite or infinite binary string S if P is a total, self-delimiting program such that $U(P, D) \supseteq S$.

Definition. *The complexity of S,* $H(S) = \min\{|P| + |D| \mid (P, D)$ *is a description of S*$\}$.

Having defined complexity in terms of a two-part description, it is easy to sort out the "sophistication" from the complexity.

Definition. *The c-sophistication of S,* $SOPH_c(S) = \min\{|P| \mid \exists D\ (P, D)$ *is a description of S and* $|P| + |D| \leq H(S) + c\}$.

We call a description (P, D) of S, *c-minimal* if $|P| + |D| \leq H(S) + c$. We call a program P a *c-minimal program* for S if P is the shortest program such that for some D, (P, D) is a c-minimal description of S. Thus $SOPH_c(S)$ is the length of a c-minimal program for S.

The c-minimal program for S is that part of the description of S which compresses S – it represents the structure of S. The range of this program constitutes the class of strings which share the structure of S. The class of likely continuation of S is the class of continuations of S which share its structure. This will be discussed further in the next section.

The role of c is as a measure of the "non-ad-hoc-ness" of a minimal program. If a program represents the actual structure of S then it will be c-minimal even for large c. Later in this section we will see an example of the use of c as a measure of confirmation.

SOPHISTICATION OF INFINITE STRINGS
Unlike complexity, sophistication is easily extendable to infinite strings including those which are not recursive. In fact, for infinite strings the parameter c is neatly eliminated.

Let α be an infinite binary string.

Definition: *A program P is called a* weak compression program *for α if there exists some c such that for all n there exists D_n such that (P, D_n) is a c-minimal description of α^n.*

Definition: *A weak compression program is a compression program if the D_n can be chosen such that $D_n \supseteq D_{n-1}$.*

Thus a compression program captures all those properties of a string which reduce the description of the string by an infinite number of bits. If a string has a (weak) compression program we say that it is *(weakly) describable*. A compression program differs from a weak compression program in that it guarantees the existence of a string $D = \lim_{n \to \infty} D_n$ such that (P, D) is a description of α.

Observe that the class of compression programs is independent over the choice of UTM in terms of which they are measured (if we identify programs which compute the same function).

Definition. *The (weak) sophistication of α is $\min\{|P| \mid P$ is a (weak) compression program for $\alpha\}$.*

Let $SOPH'(\alpha)$ signify the weak sophistication of α and let $SOPH(\alpha)$ signify the sophistication of α.

Definition: *α is transcendent if for all c, $\overline{\lim}_{n \to \infty} SOPH_c(\alpha^n) = \infty$.*

Thus if α is transcendent, the more bits of α are revealed the more structure becomes apparent; at no point is the entire structure apparent. This is suggestive of a "living" process constantly transcending any attempt to reduce it to a mechanism. If α is transcendent we define $SOPH'(\alpha) = SOPH(\alpha) = \infty$.

It is not obvious that $SOPH'$ and $SOPH$ are defined for all α, i.e., that every α is either transcendent or (weakly) describable. We have proved (Koppel, to appear) however:

Theorem: *$SOPH'(\alpha)$ is defined for all α.*

We can't prove that $SOPH(\alpha)$ is defined for all α, that is, that all nontranscendent strings have compression programs. We can, however, generalize the notion of compression program somewhat. If α is an infinite string and $U(P, D) = \alpha$ let D_n be the shortest initial segment of D such that $U(P, D_n) \supseteq \alpha^n$.

Definition. *If (P, D) is a description of α and for any other description (P', D') of α, there exists c such that for all n, $|D_n| \leq |D_n'| + c$ then (P, D) is an optimal description of α and P is an optimal program for α.*

Clearly every compression program is an optimal program.

EXAMPLES

We can apply the definitions introduced in this section to three types of strings –

recursive strings, random strings, and images of random strings under recursive processes.

(i) Recursive strings

Let β be the characteristic string of the primes. Consider the following programs each of which generates the primes given appropriate data:

P_1 prints the data
P_2 prints 0 for the even bits (greater than 2) and prints the data for the remaining bits.
P_3 prints 0 for the ith bit if i is a multiple of 3 (greater than 3) and prints the data for the remaining bits.
P_4 prints 1 for the ith bit if and only if i is prime.

Observe that P_4 does not require data at all (except for the mandatory single bit) and it generates an infinite string, namely β.

Now for $j = 1, 2, 3, 4$ let $|P_j| = K_j$ and let U be such that $K_1 \leq K_2 \leq K_3 \leq K_4$. Assume further that if P' is any other program which could produce segments of β then $|P'| > K_4$. Then using the definition of c-minimal programs for *finite* strings we have that the c-minimal program for β^n is

P_1 if $n \leq 2(K_2 - K_1 + c)$
P_2 if $2(K_2 - K_1 + c) < n \leq 2(K_4 - K_2 + c + 1)$
P_4 if $2(K_4 - K_2 + c + 1) < n$

$(P_4, 0)$ is a minimal description of β and P_4 is a compression program for β. If P_4 is the shortest program which outputs the primes then the sophistication of β relative to U is $|P_4|$.

Any minimal program (in any UTM) for β will produce exactly one string, β, given sufficient data. The same of course holds for any recursive string. Thus, using minimal descriptions as a basis for predicting continuations of β^n as n grew, we would successively narrow the class of expected continuations to the ranges of P_1, then P_2, and finally P_4, the range of which consists only of β itself.

Let us consider the progression of c-minimal descriptions we have just seen. As some pattern persists while n gets large, that pattern is incorporated into the minimal program. The precise value of n at which a particular pattern is incorporated into the c-minimal program depends on the relative efficiency of the description versus the next best one and on the parameter c. For example, we never replace P_2 with P_3, since P_3 is larger and less efficient than P_2. We replace P_2 with P_4 precisely when n is large enough that the extra bits of program in P_4 are compensated for by the reduction (compared to P_2) in the number of bits of data used with a net saving of c bits. Thus c can be thought of as the amount of confirmation a description must have before it is accepted.

(ii) Random strings

Without loss of generality but some gain in elegance let us assume that $U(\emptyset, D) = D$ for all D where $\emptyset$ is the empty program. We regard $\emptyset$ as being self-delimiting.

It is natural to define random strings as those which have no structure.

Definition: *An infinite string α is* random *if $(\emptyset, \alpha)$ is a minimal description of α, i.e., if $SOPH(\alpha) = 0$.*

It follows that:

Theorem: *α is random if and only if there exists c such that for all n, $H(\alpha^n) > n - c$.*

The reader is referred to the literature in which the property derived in the theorem has been used extensively as the definition of randomness (Levin *1973, 1984*; Martin-Lof *1971*; Schnorr *1973*; Schnorr and Fuchs *1977*). Different definitions of complexity have given rise to different classes of random strings. The advantage of using minimal processes, as we do, are spelled out in Levin *1973* and Schnorr *1973*.

(iii) Random Pairs

Let γ be an infinite string in which each bit of the random string Θ is doubled (e.g., 0011110011...). If DB is a program which doubles the bits or the data then by definition the description (DB, Θ) is minimal and DB is a compression program.

Observe that DB is not the only possible compression program for γ. Any total, self-delimiting program P such that $U(P, D) = \gamma$ and $D_n \leq (n/2) + c$, is a compression program.

If "structure" is a well-defined concept, then we will have to show that all compression programs for a string are essentially the same. This is the subject of the next section.

3. Structure and Prediction

In this section we compare the various compression programs of a string and show that for our purposes these compression programs are all essentially the same; that is, the structure of a string as represented by its compression program is well-defined.

Let $U(P, D) = \alpha$ for some finite or infinite string α.

Let $I_k(P, D, n) = \{S | S$ is a string of length $n + k$ and there exists x such that
$$U(P, D_n \cdot x) \supseteq S\}$$

(Observe that we choose to define I_k as a function of three variables P, D, n rather than two variables P, D_n because D_n does not uniquely determine n.) $I_k(P, D, n)$ is the set of possible k-bit extensions of α^n, consistent with the description (P, D_n).

For two descriptions of α, (P, D) and (P', D') the extent to which $I_k(P, D, n) = I_k(P', D', n)$ is the extent to which the two descriptions determine the same structure.

Let $A_n((P, D), (P', D')) = \max\{k | I_k(P, D, n) = I_k(P', D', n)\}$

Definition: *A program P is* strong *if for all D, $U(P, D \cdot 0) \not\supseteq U(P, D \cdot 1)$ and $U(P, D \cdot 1) \not\supseteq U(P, D \cdot 0)$. (We call the description (P, D) strong if the program P is strong.)*

Thus, if P is strong, $|U(P, D)| \geq |D|$.

We are now ready to state the

Main Theorem: *If (P, D) and (P', D') are both strong, optimal descriptions of α, then* $\lim_{n \to \infty} A_n((P, D), (P', D')) = \infty$.

The class of optimal descriptions of a string is invariant of choice of UTM. Hence it follows that

Corollary. If (P, D) is a strong, optimal description of α in UTM U and (P', D') is a strong, optimal description of α in UTM U', then $\lim_{n \to \infty} A_n((P, D), (P', D')) = \infty$.

(Observe that the functions $I_k(P, D, n)$ and $I_k(P', D', n)$ which are implicit in A_n are here defined relative to different UTM's – namely, U and U', respectively.)

This corollary means that for any two UTM's, U and U', and any α (the optimal descriptions of which are strong) and any integer k there exists i such that for all $n \geq i$ the predictions we would make for the k bits succeeding α^n would be the same whether we used U or U'.

Proof (Thm). We will show that if there exist two strong, optimal descriptions of α, (P, D) and (P', D'), such that for some k there are arbitrarily high n such that $I_k(P, D, n) \neq I_k(P', D', n)$, then we can construct a description of α, (P^*, D^*), such that for all c there exists n such that $|D_n^*| < |D_n| - c$. This contradicts the optimality of (P, D).

The construction of P^* is as follows:

Run P on all data of length k until k bits of output are generated. Now run P' on all data of length k until k bits of output are generated. If some output of P does not appear as output of P' then call the data of P which resulted in that output, T_1. Now for $i = 1, 2, \ldots$ run P on D_i followed by all data of length k until $1 + k$ bits of output are generated. Since P' is strong, D_i' is uniquely determined by P, D_i and P'. Now run P' on D_i' followed by all data of length k until $1 + k$ bits of data are generated. If some output of P does not appear as output of P' then call the data of P following D_i which resulted in that output, T_i.

By our assumption there exist an infinite number of T_i each of length less than or equal k. The crucial fact is that $D_i \cdot T_i$ is not an initial segment of D since

$U(P, D_i \cdot T_i)$ is not generated by $U(P', D')$ while all initial segments of $U(P, D)$ are generated by $U(P', D')$. There is some smallest $j < k$ and some c' such that for all $i > c'$, if $D_i \cdot T_i^j$ is an initial segment of D then $D_i \cdot T_i^{j+1}$ is not an initial segment of D. If $D_i \cdot T_i^j$ is an initial segment of D call the next bit of D a "determined" bit. Now let P^* simply run P on the data, while finding the T_i as described above. When, for $i > c'$, a T_i is located and the next j bits of data are T_i^j then, since the next bit of data for P is determined (it must be the opposite of the $j + 1^{st}$ bit of T_i^{j+1}), P^* acts as if it were there without reading it. Thus we let D^* be the same as D with the determined bits left out. But by our assumption there are an infinite number of determined bits. Thus for all c there exists n such that $|D_n^*| < |D_n| - c$. This completes the proof of the theorem.

An Example

Recall that γ is an infinite string in which each bit of the random string Θ is doubled. Then by definition the description (DB, Θ) is minimal. It is in fact a strong minimal description. But then by the theorem, if (P, D) is any optimal description of α (even in a different UTM) then for any k there exists i_k such that for all $n \geq i_k$,

$$I_k(P, D, n) = \{S | \alpha^n \text{ is an initial segment of } S, |S| = n + k, S \text{ consists of pairs}\}$$

That is, any optimal description of γ captures precisely the same structure that (DB, Θ) does (over finite extensions). Thus, with k growing with n, the expected k-bit extensions of γ^n are precisely those in the image of DB.

The limitation of this theorem is that i_k depends on k. There exists a strong optimal description of γ, (P, D), such that for arbitrarily large n there exists k such that $I_k(P, D, n) \neq I_k(DB, \Theta, n)$. Let DBE be the program [DB unless in the data a stretch of n consecutive 0's is encountered beginning at the nth bit in which case all subsequent bits are printed but not doubled]. Let $D = \Theta$. Then if Θ does not include any such stretches $U(DBE, \Theta) = \gamma$, so that (DBE, Θ) is a strong optimal description of γ. Nevertheless for any n, $I_{2n+2}(DB, \Theta, n) \neq I_{2n+2}(DBE, \Theta, n)$ since $(\gamma^n \cdot o^{2n+1} \cdot 1) \in I_{2n+2}(DBE, \Theta, n)$ but $(\gamma^n \cdot 0^{2n+1} \cdot 1) \notin I_{2n+2}(DB, \Theta, n)$.

In fact the property that n consecutive 0's are encountered beginning at the nth bit can be replaced by any property which does not hold for Θ. It must therefore not be a property which holds for all random strings, e.g., $\log n$ consecutive 0's follow the nth bit for some n.

This discussion suggests that the "probability" of obtaining an output (e.g. $\gamma^n \cdot 0^{2n+1} \cdot 1$) from one compression program (e.g., DBE) which does not appear in the range of another compression program (e.g., DB) approaches 0 as n (i.e., the known part of the output) gets larger.

In order to formalize this idea we will define "probability".

PROBABILITY

Program-length complexity can be used to define probabilities of strings (cf. Chaitin *1975*; Kolmogoroff *1965*; Solomonoff *1964*; Zvonkin and Levin *1970*). The distinction between program and data in minimal descriptions can be used to distinguish between two different types of probability.

Given a finite initial segment of some string, the probability that some program for that segment is correct for the whole string is inversely proportional to the size of the program. The dependence of the length of programs on the UTM which is used to run them imparts a degree of subjectivity to this probability measure.

On the other hand, the probability that, *given some program for a string,* some initial segment will be extended in a particular way, is determined by a purely logical probability measure.

If program-length complexity is used to define a probability measure, it is important not to artificially merge these two fundamentally different measures.

Now, let an *event* be any set of infinite binary strings. Roughly speaking, the probability of an event E, given some program P, is the proportion of data D such that $U(P, D) = E$.

More formally, we say that the string G is a *generator* of E in P if for all infinite x, $U(P, G \cdot x) \in E$. Then

$PR(E|P) = \sum\{(1/2)^{|D|} \mid D$ is a generator of E but no initial segment of D is a generator of $E\}$.

Thus the $PR(E|P)$ is the probability that $U(P, D)$ is in E if D is determined by coin tossing. We can extend the definition of PR to cases where a description of some finite string α^n is given and we wish to define the probability that the string which succeeds it will be in E.

Let $U(P, D) = \alpha$.

Let $PR(E|P, D, n) = \sum\{(1/2)^{|x|} \mid D_n \cdot x$ is a generator of $\alpha^n E$ but no initial segment of $D_n \cdot x$ is a generator of $\alpha^n E\}$.

Our notation intentionally reflects a parallel with the function $I_k(P, D, n)$.

Conjecture: *If (P, D) and (P', D') are strong, optimal descriptions of α, then* $\lim_{n \to \infty} |PR(E|P, D, n) - PR(E|P', D', n)| = 0$.

(Observe that the conjecture is strong because $PR(E|P, D, n)$ is the probability of E *succeeding* α^n – it is *not* the probability of α^n itself being an initial segment of a string in E.)

The conjecture is motivated by the following argument:

In a one-to-one fashion map each string B to a string B' such that $U(P, D \cdot B) \subseteq U(P, D'_n \cdot B')$. Let $_j\{B_j^{(n)}\}$ be those strings which cannot be thus matched. These are the "deviant" data strings.

The example of DBE suggests that deviant data are those which satisfy some property which does not hold for random strings. Thus it is reasonable to conjecture (cf. Martin-Lof *1971*) that

$$\sum_{n=1}^{\infty} \left(\sum_{j} \left(\frac{1}{2} \right)^{|B_j^{(n)}|} \right) < \infty \qquad (*)$$

But $PR(E|P, D, n) - PR(E|P', D', n) \leq \Sigma \frac{1}{2}^{|B_j^{(n)}|}$ and the conjecture follows immediately.

Observe that $(*)$ implies that $\sum_{n=1}^{\infty} \left(\frac{1}{2} \right)^{A_n} < \infty$ which implies the main theorem.

4. A Structure-finding Procedure

DEPTH

In this section we outline a family of straightforward procedures for determining the structure of an infinite string. Obviously any effective procedure generates hypotheses based on only partial (i.e., finite) information and thus there can be no guarantee that any particular hypothesis is correct. We will show, however, that corresponding to each procedure there is a large class of strings such that for any string in the class the procedure generates an infinite sequence of hypotheses all but a finite number of which are correct.

In order to identify these classes of strings we will need to define the "depth" of strings, first defined by Bennett (Bennett, this volume), and the relevance of which to sophistication has been demonstrated in Schnorr and Fuchs *1977*.

If P is a total program let the function

$ST_P(n)$ = the maximum (over all D) number of steps $U(P, D)$ runs before producing n bits of output.

We say that an infinite string α is weakly F-shallow if there is a (weak) compression program P for α such that for all n, $ST_P(n) \leq F(n)$.

If α is F-shallow then F is an upper bound on the amount of time needed to generate α from a minimal description. We can use the property of F-shallowness to give a *numerical* measure of depth as follows.

For a strictly increasing function F, let γ_F be the characteristic string of the range of F. Let $|F| = \min\{|P| \mid (P, 0)$ is a description of γ_F and P is self-delimiting $\}$.

Definition: *The (weak) depth of $\alpha = \min\{|F| \mid \alpha$ is (weakly) F-shallow $\}$. (If no such F exists then the depth of α is infinite.)*

We will let $D(\alpha)$ signify the depth of α and let $D'(\alpha)$ signify the weak depth of α.

While not translating depth into numerical terms as we do, Bennett has argued persuasively that depth should be regarded as a formal measure of "evolvedness", "meaningful complexity", etc.; that is depth formalizes the same sort of intuitive ideas that sophistication does. It is reassuring, therefore, that we have proved (Koppel, to appear):

Theorem: *There exists c such that for all infinite strings α,*

$$SOPH'(\alpha) - c \le D'(\alpha) \le SOPH'(\alpha) + c.$$

Theorem: *There exists c such that for all describable infinite strings α,*

$$SOPH(\alpha) - c \le D(\alpha) \le SOPH(\alpha) + c.$$

Finding Structure via Converging Hypotheses

We can use the notion of depth to provide an effective procedure for finding the structure of an infinite string. Recall the example of the characteristic string of the primes discussed in Section 2, which illustrated that as more bits of a string become known more structure becomes apparent, thus forcing revisions of hypotheses as to the string's structure. In this section we will formalize that process.

For a finite string S let $H_F(S) = \min\{|P| + |D| \mid (P, D)$ is a description of S and $ST_P(|S|) < F(|S|)\}$.

Let α be some infinite string. For an integer c and a strictly increasing function $F : N \to N$, let $P_{c,F}(\alpha^n)$ be the shortest self-delimiting program P such that for all $i \le n$ there exists $D_i \supseteq D_{i-1}$ such that $U(P, D_i) \supseteq \alpha^i$ and $|P| + |D_i| \le H_F(\alpha^i) + c$ and $ST_P(i) \le F(i)$. Observe that $P_{c,F}$ is not necessarily a c-minimal program for α^n because $P_{c,F}$ need not be total. Since we do not need to check $P_{c,F}$ for totality and since the run-time for $P_{c,F}$ is bounded, $P_{c,F}(\alpha^n)$ is effectively computable from α^n, by exhaustively searching through all programs in order of increasing length, each with all possible data.

For each c, there exists some largest $K_n(c) \le n - c$ such that for all $0 \le i \le K_n(c)$, $P_{c,F}(\alpha^{n-i}) = P_{c,F}(\alpha^n)$.

Definition: *The nth hypothesis (relative to F) for α, is a program $P_F^n(\alpha) = P_{c',F}(\alpha^n)$ where c' is chosen such that for all $c \le n, K(c') \ge K(c)$.*

Clearly, $P_F^n(c)$ is computable from n, F, and α.

Theorem: *If α is an F-shallow infinite string then there exists some K such that $P_F^K(\alpha)$ is a compression program for α and for all $n \geq K$, $P_F^n(\alpha) = P_F^K(\alpha)$.*

That is, for sufficiently high K, the Kth hypothesis for α is a compression program for α (i.e., it is "correct" as a hypothesis for the structure of α) and all subsequent hypotheses are identical to it.

Proof. If α is F-shallow then there exists a program P such that $ST_P \leq F$ and for some constant c and all n, P is a c-minimal program of α^n. But then for all n, $|P_{c,F}(\alpha^n)|$ is defined and bounded by $|P|$. Moreover, $|P_{c,F}(\alpha^n)|$ is nondecreasing with n. Thus, it follows (choosing appropriately among programs of equal length) that there is some smallest constant c' such that there exists K such that for any $i, j \geq K$, $P_{c',F}(\alpha^i) = P_{c',F}(\alpha^j)$.

Then from the definition of P_F^n it follows that for all $n \geq K$, $P_F^n(\alpha) = P_{c',F}(\alpha^n) = P_{c',F}(\alpha^K) = P_F^K(\alpha)$. Moreover, from the definitions of $P_{c',F}$ it follows that for all n, there exists $D_n \supseteq D_{n-1}$ such that $U(P_{c',F}(\alpha^K), D_n) = U(P_{c',F}(\alpha^n), D_n) \supseteq \alpha^n$ and $|P_{c',F}(\alpha^K)| + |D_n| \leq H_F(\alpha^n) + c'$. But since α is F-shallow, there exists c such that for all n, $H_F(\alpha^n) \leq H(\alpha^n) + c$ and therefore for all n, $|P_F^K| + |D_n| = |P_{c',F}(\alpha^K)| + |D_n| \leq H(\alpha^n) + c' + c$. Moreover since $ST_{P_F^K} \leq F$, P_F^K is a total program. Thus P_F^K is a compression program for α and for all $n \geq K$, $P_F^n = P_F^K$ proving the theorem.

Observe that the compression program found by the algorithm is not the smallest compression program for α. The smallest compression program describes only properties of the entire string α – that is, properties which can be used to reduce the size of the description of α by an infinite number of bits. The compression program found by the algorithm describes all properties of initial segments of α which can be used to reduce even slightly the size of all subsequent initial segment.

INDUCTION AND SCIENTIFIC DISCOVERY

It can be argued that the procedure outlined above is a fair representation of the logic of scientific discovery. Using this analogy we can make some interesting observations:

In the model of scientific discovery outlined here, as in other models, the first explanation in some ordering of explanations which is consistent with the observed facts is conjectured as the explanation of the phenomenon. Here the ordering of the explanations is the size order of descriptions and – within descriptions of (almost) equal size – size order of programs.

By using size order of descriptions as the first criterion, we are positing simplicity as the primary determinant of the value of an explanation. By using size order of programs as the second criterion, we are positing parsimony as a secondary determinant of the value of an explanation. This is because the more data a program uses the larger, i.e., less specific, the class of continuations which fall in its range.

Obviously the simplicity of an explanation is relative to the language – indeed the culture – in which that explanation is stated. This is captured here by the dependence on choice of UTM. Significantly, however, while the infinite variety of UTM's allows a certain amount of flexibility in the ordering of descriptions, the invariance results of the previous section demonstrate the limits of this flexibility.

Our model differs from other hierarchal models of scientific discovery in one crucial way. In the standard models a theory (program) is replaced only if it is falsified by subsequent evidence. In our model, however, as more bits of a string are observed, previous minimal programs of the string are changed for one of *two* reasons.

The most straightforward reason for abandoning a program is that subsequent bits of the string are inconsistent with it (do not fall in its range). In this case the program is abandoned in favor of one which is less powerful (shorter, using longer data).

However, another reason for abandoning a program is that more bits of the string make apparent structure not previously apparent – that is, use of a longer, more powerful program results in a shorter description. (See example (i) in Section 2.) Thus it is an integral part of the process of scientific discovery that for the sake of simpler explanations (shorter descriptions), theories (programs) – *even if not falsified* – are replaced by more powerful theories (larger programs) which are less parsimonious and more falsifiable (use less data).

The ideas introduced here allow us to make somewhat more precise different views of what is possible and what is learnable.

A strictly deterministic view of the world could be formalized as asserting that all natural phenomena are recursive. This view is too rigid for current paradigms in physics and biology which assign a role to random phenomena. Therefore strict determinism has been replaced by quasi-determinism, which asserts that all natural phenomena are recursive up to randomness. Quasi-determinism can be formalized as asserting that all natural phenomena are nontranscendent.

A phenomenon is modelable if its structure can be characterized. Formally, a phenomenon (string) is modelable if and only if it is describable, i.e., nontranscendent. As we have seen, these are also the phenomena (strings) which are learnable. Thus quasi-determinism equates the possible with the modelable and the learnable. This equation has been stated explicitly by many proponents of AI.

Observe that according to quasi-determinism, there are phenomena such that even perfect knowledge of their structure allows determination only of the *probabilities* of future instances of the phenomena.

Vitalism can be now formalized as identifying life with transcendence. A being is alive precisely in the sense that it can transcend any reduction of its behavior to recursiveness and randomness.

Indeed, existential psychologists (cf. Becker *1975*) have identified neurosis with nontranscendence; in the absence of transcendence, behavior must be recursive (depressed) or random (schizophrenic) or some combination of the two.

5. *Speculation: An Algebra of String Properties*

SIMPLE UTM'S

While the class of compression programs for α is invariant over choice of UTM, the *smallest* compression program for α (whose length determines $SOPH(\alpha)$) is dependent on U. Thus, recalling the example of Section 3, for all UTM's, the program *DB* (which is natural) and *DBE* (which is messy) are compression programs for U; but for some UTM's *DB* is the shortest compression program for γ, and for others *DBE* is the shortest compression program for γ. Clearly a UTM for which *DBE* is a shorter program that *DB* is pathological. More generally, we can say that a UTM is natural to the extent that its shortest compression programs are natural.

This formalization, of course, begs the question of what is "natural". We propose to answer the question in a way that leads easily to an algebra of structure.

Let $I_U(\alpha)$ be the range of the shortest compression program for α, relative to U. We call a collection of infinite strings I_U a *structure class* relative to U if for some string α, $I_U = I_U(\alpha)$.

We say that a UTM is *natural* if for every string α, $I_U(\alpha) = \cap\{I_U | I_U$ is a structure class and $\alpha \in I_U\}$. That is, U is natural if for any string α the structure of α is precisely the aggregate of all its properties. (Equivalently, let the relation $R_U = \{(\alpha, \beta) | \alpha \in I_U(\beta)\}$; that is $R(\alpha, \beta)$ if α has all the properties of β (and perhaps others); U is natural if and only if R_U is a transitive relation.) We can now easily extend the notion of structure class generated by a string to structure class generated by a set of strings.

Let $I_U(\alpha_1, \ldots, \alpha_n) : \cap\{I_U | I_U$ is a structure class and $\alpha_1, \ldots \alpha_n \in I_U\}$.

We speculate that natural UTM's uniquely determine structure – for natural UTM's, U and U', $I_U(\alpha) = I_{U'}(\alpha)$. This is a strong generalization of the main theorem of Section 3.

AN ALGEBRA OF STRUCTURE CLASSES

From now on we assume that U is a natural UTM and $I(\alpha) = I_U(\alpha)$. We define two relations on structure classes analogous to greatest common divisor and least common multiple, respectively.

Let $I_1 \oplus I_2 = \cap\{I | I$ is a structure class and $I \supseteq (I_1 \cup I_2)\}$.

Observe that if $I_1 = I(\alpha)$ and $I_2 = I(\beta)$ then $I_1 \oplus I_2 = I(\alpha, \beta)$.

Let $I_1 \cdot I_2 = I_1 \cap I_2$

This is the class of strings which have the structure of both α and β.

Let $SOPH(I(\alpha)) = SOPH(\alpha)$. We say I_1 and I_2 are *factors* of I if $I = I_1 \cdot I_2$ and $SOPH(I_1) + SOPH(I_2) \leq SOPH(I)$; that is, the best way to express the property manifested by the class I is as the conjunction of the properties manifested by I_1 and I_2. Observe that for any structure class I, I and $\{0, 1\}^\infty$ are factors of I. We say that I is *prime* if its only factors are I and $\{0, 1\}^\infty$.

Conjecture: *Every structure class has a unique factorization into prime structure classes.*

Recall that in Section 3 we described two types of probability measures – a logical one which measures the probability of an event given the structure of a phenomenon and a partially subjective one which determines the probability that a string has some particular structure. We can use the operations described above to impose certain simple constraints on this second type of probability measure.

Let $PR(I)$ be the probability that a string is in the class I. Then for any structure classes I_1 and I_2 it must be the case that

$$PR(I_1 \cdot I_2) = PR(I_1) \times PR(I_2)/PR(I_1 \oplus I_2).$$

References

Becker, E.

    1975 *Denial of Death.* New York: New York Free Press (1975).

Bennett, C.H.

    1988 Logical Depth and Physical Complexity. This volume.

Chaitin, G.J.

    1975 A theory of program size formally identical to information theory. *J. ACM* **22** (1975) 329–340.

    1979 Toward a mathematical definition of 'life'. In: *The Maximum Entropy Formalism,* eds. R. Levine and M. Tribus, pp. 479–500. MIT Press (1979).

Cover, T.

    1985 Kolmogrov complexity, data compressing, and inference. In: *The Impact of Processing Techniques on Communications,* ed. J.K. Skwirzynski. Martinus Nijhoff Publishers (1985).

Gács, P.

    1974 On the symmetry of algorithmic information. *Sov. Math. Dokl.* **15** (1974) 1474–1480.

Kolmogoroff, A.N.

  1965 Three approaches to the quantitative definition of information. *Probl. Inf. Transm.* **1** (1965) 1–7.

Koppel, M., and H. Atlan

  i.p. Program-length complexity, sophistication and induction. To appear.

Koppel, M.

  i.p. Complexity, Depth and Sophistication. To appear.

Levin, L.A.

  1973 On the notion of a random sequence. *Sov. Math. Dokl.* **14/5** (1973) 1413–1416.

  1984 Randomness conservation inequalities: Information and independence in mathematical theories. *Inf. Contr.* **6/1** (1984) 15–37.

Martin-Lof, P.

  1971 Complexity oscillations in infinite binary sequences. *Z. Warsch. V.* **19** (1971) 225–230.

Schnorr, C.P.

  1973 Process complexity and effective random tests. *J. Comput. Sy.* **7** (1973) 376–388.

Schnorr, C.P., and P. Fuchs

  1977 General random sequences and learnable sequences. *J. Symb. Log.* **42** (1977) 329–340.

Solomonoff, R.J.

  1964 A formal theory of inductive inference. *Inf. Contr.* **7** (1964) 1–22.

Zvonkin, A.K., and L.A. Levin

  1970 The complexity of finite objects. *Russ. Math. S.* **25** (1970) 83–124.

Mental Images and the
Architecture of Concepts

Johann A. Makowsky

In memory of my mother
Marika Erzsébet Makowsky-Deutsch
14.10.1923–9.7.1987

Abstract. We discuss the role of format of data, communication protocols and notational systems in computing. We use the term mental images for the internal data structure of human thinking and formal concept for an externalized data structure associated with it. Format, communication protocol and notation together are called architecture of the concept. It is argued that our current model of computability abstracts too much from the issue of architecture of concepts to provide us with a workable theory of interactive computing and data transfer. Such a future theory would have to take into account aspects of data base theory, cryptographic protocols, probabilistic complexity theory and a theory of learning which extends the statistical theory of estimation of dependencies based on empirical data. We also draw attention to anthropological studies concerning the evolution of mathematical concepts which show that such evolutions are inherently slow. This last aspect serves to dampen our hopes of a quick breakthrough in the evolution of conceptual computing.

1. Introduction

Some fifty years ago a major discussion in the foundations of mathematics seemed settled. The intuitive notion of computability was made precise. It was equated with the mathematical notions of Turing machine computability, recursive functions and λ-definability (Church-Turing Thesis). By now many other models of computation have been suggested, and all of them have been shown to be extensionally equivalent or weaker than any of the above. This is generally taken as evidence that we now understand the notion of computability. But this should not be taken as evidence that we understand the notion of computing as an interactive and distributed activity. Turing based his model of computability on mechanistic aspects of numeric computations and string manipulations of one computing device, be it man or machine. Some of these early models (Turing machines, register machines) were also used for the design of the first hardware. Especially von Neumann advocated the use of universal Turing machines as concepts underlying the physical realization of computers. It was suggested by Trakhtenbrot, this volume, that other early models of computability such as recursive functions or λ-calculus should be viewed as software oriented models. If we push this thought a bit further, then the Church-Turing thesis also stipulates the equivalence and interchangeability of the two notions computable by *hardware* and computable by *software* by a device which has the computational power of a *universal Turing machine*. From Turing's point of view it is fair to say that most issues of computability are today fairly well understood.

However, all these models of computability capture only aspects of computability which reflect the use of string manipulation machines in batch mode. They are inadequate to describe the manipulation and transfer of concepts without explicit reference to the coding of these concepts. They are inadequate to describe various modes of interaction between man and machine and between machines and they are especially inadequate to describe the influence of the choice of particular data structures and data representations on the complexity of computations. The abstraction from these aspects of computability accounts at the same time for the versatility of these models of computability as well as for their uselessness. The present level of computing activities with various user interfaces, virtual machines, networks and the search for new architectures justifies a new look at the basic issues.

The purpose of this paper is to discuss three recent developments in theoretical computer science and to sketch how they possibly contribute to a better understanding of computing as an activity in an interactive context. The three areas discussed are relational data base theory, learnability of concept classes, and the analysis of interaction protocols when concepts are passed from one actor (man or machine) to another.

A system of actors, each equipped with a concept class, is called a cultural system. Such systems are subject of study in anthropology and such an anthropological view of the mathematical activity was initiated by R. Wilder in his book *Mathematics as Cultural System* (Wilder *1981*). In a last section we shall briefly outline how the study of man-machine interaction as a cultural system can further our understanding of computing. The paper is meant to address a wider audience. Technicalities are avoided as much as possible. However, the reader interested in the more technical aspects should be able to fill in all the technical details by consulting the cited references.

2. A Guiding Example

We are all familiar with the game of guessing sets or sequences of natural numbers from a given finite set of examples. Given the numbers $2, 4, 6, \ldots$ we are usually expecting a continuation $8, 10, \ldots$; given $2, 4, 8, \ldots$ we expect $16, 32, \ldots$

etc. Underlying such expectations, however, are concepts such as arithmetical or exponential progression, and it is obviously very easy to find other continuations with very reasonable explanations. The sequence $2, 4, 8, \ldots$ e.g., can be continued with $15, \ldots$ by referring to the three dimensional Euclidean space and the maximal number of regions one gets by cutting it with $1, 2, 3, \ldots$ two dimensional planes. However, a student who will propose $15, \ldots$ as a continuation of $2, 4, 8, \ldots$ in an intelligence test would probably fail, because the author of the test did not include the possibility of explaining the answer. On the other side we would be inclined to consider $15, \ldots$ together with its explanation as a more sophisticated if not more intelligent answer than $16, \ldots$.

When we speak of concepts in our daily usage of the word we might have several things in our mind. What matters here is that having a concept of something means having a set of instances of the concept. In the first example above the instances are $2, 4, 6, \ldots$ and the concept is the set of natural numbers of the form $2n$ for $n > 0$. In the second example the instances are $2, 4, 8, \ldots$ and the concept is the set of numbers of the form 2^n for $n > 0$ or the set of numbers $f(n)$, where $f(n)$ is the maximal number of regions one gets by cutting the three dimensional Euclidean space with n two dimensional planes. The expression $2n$ represents the concept of even numbers, but itself it is again an instance of something. More such instances can be given, such as $3n, 4n, \ldots$, and now the reader would be inclined to continue with $5n, 6n, \ldots$ which can be represented by the expression an, where a and n play different roles. At this stage we might look at the sequence $3, 5, 7, \ldots$ which most people would continue with $9, 11, \ldots$ which cannot be represented by an expression of the form an and we would be lead to expressions of the form $an + b$.

In a similar vein we can also look at pairs of natural numbers, such as $(2, 4), (6, 8), (10, 12) \ldots$ represented by an expression of the form $(2 + 4n, 4 + 4n)$ for $k \geq 0$ or $2 + n - m = 0$ and $m = 4k$ for some $k > 0$. Usually the former type of a solution is considered more desirable for its explicitness. In the case of triples as indicated by $(3, 4, 5,), (5, 12, 13), \ldots$ and represented by all triples such that $a^2 + b^2 = c^2$ finding an explicit representation was considered a major success of mathematics in its period.

The history of mathematics teaches us several lessons concerning the evolution of the concepts of a *set of natural numbers* and a *real function*. For one the concept class "sets of natural numbers" and "real functions" was frequently enlarged. Each such enlargement gave rise to a crisis in mathematics and fierce debates among mathematicians until the new concept class was universally (or almost universally) accepted. For many mathematicians the set of all natural numbers which are Gödel numbers of true formulae of their favorite model of set theory is not a well-defined set of natural numbers. If sets of natural numbers have to be computable this feeling is justified by Gödel's theorem. But still we tend to think that we have a *mental image* of such a set (at least I do) even if the *structure (architecture)* of the underlying *concept* is not clear. I would be hard pressed if I had to describe my favorite model of set theory in an unambiguous way.

3. *Notational Systems*

In the above examples the sequences of numbers are usually given names such as $S_1, S_2, \ldots$ which are treated as atomic names and are not informative at all in the sense that they do not help to think about these sequences. However, it is the *naming* of such sequences which mark considerable progress in our understanding them. Let S_1 be the sequence $2, 4, 6, 8, \ldots$. If instead we name it $2, +2$ we may indicate by this that the sequence starts with 2 and each consecutive element is obtained from the previous one by adding 2. Alternatively we could name it $2n$ indicating by this notation that the elements of the sequence are all the even numbers, but now we have to decide whether 0 is an admissible value for n. Notation clarifies our understanding, but we cannot conclude from the absence of notation a lack of understanding. In Diophantus' work no notation for zero and the negative numbers is present. Philologists concluded from this that the Greeks of Diophantus' time did not have the concepts of zero and the negative numbers. But mathematicians reading Diophantus' work with paper and pencil concluded that he must have been aware of zero and the negative numbers because otherwise the gaps in his arguments could not have been filled. Notation thus makes our reasoning more explicit, and, more important, notation is a *carrier of thought*. The Babylonians wrote $(3 + 5)^2 = 3^2 + 2 \cdot 3 \cdot 5 + 5^2$ and were completely aware of the generality of this statement which would not be clear had they written $(2 + 2)^2 = 2 \cdot 2 + 2 \cdot 2 \cdot 2 + 2 \cdot 2$ since 2 plays three different roles in the latter.

Notation plays a fundamental role in our thinking and in programming machines as well. Notation has inherent limitations which discipline our thinking, but it also has its own temptations which make us go beyond the intended. Very often creativity consists of yielding to these temptations, and stumbling blocks in understanding come from not trusting notation. It took the physicists of the beginning of this century almost twenty years to believe that matrix calculus really furthers the understanding of quantum mechanics, even if not every step has an obvious physical interpretation.

4. *Computing vs. Learning*

Computing deals with the manipulation of strings within a fixed notational system. A class of objects is computable if there exists a notational system such that a Turing machine can generate their names. Computing does not deal with the *creation* of notational systems. Neither does it deal with the mechanism of how we (or machines) assimilate the meaning of a notational system. This rather trivial remark seems to be fundamental when we want to speak of the mechanization of intelligence. When Kepler discovered his laws of planetary movements he had available a host of data. He tried to format these data by making several fundamental decisions. For one, the planetary movement was to be a function of time, the initial position of a planet, its speed and mass. It was not to be dependent on local conditions on earth such as weather or politics, although both weather and politics were considered in his time to be influenced by the position of planets. Once he had a hypothesis concerning the format of the data, he still needed a notational system to represent the planetary movements in form of equations. Contrary to their implicit claim, P. Langley and H. Simon's project "BACON" of mechanization of physical and chemical discoveries (Langley et al. *1983,* in Michalski et al. *1983*) only deals with the last stage of discovery, given the data format and the notational system, how could Kepler determine the actual equations of planetary movements. In the literature this last stage is called learning by example, but I think this name is misleading. It is more precisely described by calling it identification of concepts of a given format by examples. Research in this area was already popular in the Eastern Block countries decades ago, as documented in Hajek and Havranek *1978* and Vapnik *1982.*

5. *Mental Images and the Architecture of Concepts.*

We might use the term "mental images" to describe the data structure of our mental reasoning. We do not have access to this data structure before we are able to externalize it. We do not know its exact format, as we do not even know the format of the data we perceive. I do have a mental image of what a beautiful woman is (or of what artificial intelligence should be) to the extent that I can recognize her, Rachel, (I can recognize a project as *not* being in AI, such as chess playing programs) although I cannot define my concept. I want to introduce here the term *mental image* for concepts as internal data structures of our mind, and *(formal) concept* as the externalized version of the mental image. For a concept to be *formal* it is not enough that we can recognize its instances. A formal concept has a format and a *notational system* which displays this format. The format, however, does not necessarily display all the aspects of a formal concept; it may hide some of them on purpose. A formal concept is an abstraction of a mental image; it is usually a finite dimensional projection of an infinite dimensional image. We shall use the term *architecture* of a concept to mean format and notational system together.

Incidentally, this may also put the rivalry between intuitionism and formalism into a new perspective: the intuitionists insist on displaying all the manipulative aspects of a concept, such that existential statements become constructive, the formalists allow hiding them to an extent that only those aspects needed for a particular argument are on display. In this perspective the axiom of choice is just an encapsulation of a possibly complicated argument. The above discussion can easily be made more concrete in several brief examples.

6.   *The Role of Format and Notation in Music*

Most people have a rather clear mental image of some kind of music relevant to their culture, even without having had explicit musical training. But very few people are aware, even among musicians, to what extent their music is dependent on the existence or absence of notational systems for it. An anecdote attributed to the great linguist R. Jacobson may illustrate the issue. It is said that in Prague Jacobson once set up the following experiment. An excellent flautist of the renowned orchestra, who was particularly known for his capacity to play by ear, was invited to join an African expert in the playing of some tribal wind instrument resembling the flute in many respects. The two were supposed to proceed as follows. The African was to play a short piece of music on his instrument, then the European was to play on his instrument what he thought he had heard. If the African agreed, the session would stop. If not, the African had to play the same piece again, so the European would have a further chance. No verbal communication was to be used, except to express assent or dissent. Here is what supposedly happened. The African played and so did the European. The audience was excited about how well he had repeated the piece, but the African suggested that the European's version was not even similar. He then played what he thought was the same piece, but it was unrecognizable for the European audience, and so they had to continue for many more rounds. Ultimately the two musicians found agreement. It turned out that for the African neither pitch nor rhythm were part of the format which determines the equivalence class of different performances of the same piece, but the only relevant property in common of all the performances he had given was in the quality of the attack of each individual sound. Such music could not have been written in our conventional notational system. But to our ears even the format was wrong. Our ear learns to hear and overhear by a format which is culturally created. No culturally innocent person would consider a Wagner opera played by an orchestra or its piano excerpt version the "same". This very notion of piano excerpt has very much to do with the fact that the piano keyboard has become the "universal Turing machine" of classic and romantic music of western Europe. There are no piano excerpts of Gamelan music! And many new tendencies in twentieth century music have very much to do with the extension of the format of sounds, although the corresponding developments of notation lag behind. In other words the mental image of this new music has not yet been completely converted into a formal concept.

7. *Relational Data Bases*

The above discussion can easily be made more concrete by the example of relational data bases. Relational data bases define a user interface of data in which the data format is well defined and various query languages (sc. notational systems for extracting information from the data bases) can be precisely defined. A very good exposition of the theory of relational data bases may be found in Ullmann *1983*, Maier *1983*, Chandra and Harel *1980,* and the review thereof Makowsky *1987*. The relational approach to data bases might suggest that one can model our mental image of a data base by a formal concept in the sense that exactly the information which can be explicitly extracted is the information we can really obtain. This may be wrong both on the level of the formal concept, inasmuch as implicit information need not be equivalent to explicit information even in the mathematical sense of the notions. A. Zvieli and I have proved (cf. Makowsky *1984*) that the many-sorted implicit queries in the notational system of first order queries are exactly the computable queries in the sense of Chandra and Harel *1980*. But also informally, in the sense that the computable queries over the disjoint union of two data bases form in general a larger class of queries than the union of the computable queries over two data bases. The latter remark is important for the notion of data security and data privacy, e.g., even if we can prove that the data base of the police and the data base of the tax authorities each do not disclose certain information, it may be that software being able to use both may disclose it. More generally, what we can learn by our brain having simultaneous access to several mental images or even formal concepts is different from what we can learn from each alone. This may give us one clue to the different learning capabilities of different humans in otherwise very similar situations. Very important progress in merging ideas of relational data base theory, data structures, boundedness of resources, and programming languages has been made by Y. Gurevich and his school (cf. Gurevich *1987, 1988*).

8. *Communicating Concepts and Communication Protocols*

In contrast to human communication machine communication allows the copying of data in an unambiguous way, if we abstract from the down-to-earth intricacies of the communication channels and communication networks. One can copy data and formal concepts from machine to machine but not mental images. In contrast to machine communication humans can communicate mental images by a mechanism which is not yet completely understood: communication by example. The advantage of this is clear, a finite set of examples is presented to the other and in return the other will form a mental image, which will more or less capture the mental image of the first. The "more or less" is important here, severe misunderstandings are quite frequent in daily life and possibly less frequent in well-defined areas of interest such as legal disputes or mathematics. Another aspect is important here, which was described in the section above on playing music of different cultures: The communication protocol. A riddle familiar among mathematicians illustrates this further:

> A mathematician A meets, after a long time his colleague B. After exchanging formalities A reveals to B that he has three daughters and asks B to guess their ages (integers). As a first hint he says that the product of their ages is 36. B thinks of all the ways of decomposing 36 into three factors and complains to A that there are too many possibilities. A then gives as a next clue that the sum of the ages was equal to B's house number. Now B does some calculations and after a while reports that the ages were still not uniquely determined, whereupon A, as a last clue says, that the eldest very much liked bananas.

The example clearly gives the distinction between computation and hinting by example. The observer C of that story can also guess the ages, even without knowing the house number, but only once he understands that he has to do all the computations to understand the significance of the last clue. B himself also might have modeled the problem first in the integers with multiplication, then also with addition and only at last equipped with a linear ordering. If he did so, then he understood that each clue not only revealed new information on the ages, but also about the format needed to reason about them. Reasoning is not an activity by an individual (man or machine), but a two or rather a many actor game between mind and the external world. It should be noted that the reasoning minds form one or many cultural systems, and their functioning cannot be separated from this. Reasoning consists of introspection and reading of external data. The metaphor of the sole reasoning mind derives from the special case, where the mind assumes that he can simulate all possible reading of external data. This metaphor is similar to the idea of a universal Turing machine. A mathematician can reason about the natural numbers, because the reading of the "empirical" data about them is simulated internally or can be computed on virtual internal workspace, i.e., on paper by pencil. Turing, when proposing his model of computation, had exactly this in mind. But reasoning about external data is different from computing. We can query the external world by measuring certain aspects (not necessarily numerically), and for this we have to invent methods of measurement.

The point I would like to make here is that methods of measurement involve not only determining data about the object to be measured but also a protocol of how these data are to be obtained. In more complex situations this protocol is interactive like in questioning a witness in court (Ericsson and Simon *1984*) or in verification of digital signatures (Goldreich, this volume). This protocol also presupposes something about the computational capacities of the measured object. Measuring the specific weight of the king's crown by measuring the volume involves immersing the crown in water and assuming that the water can "compute" the volume of the crown.

Inventing methods of measurement, inventing the right question, again involves format, notation and computation within this notational system. That something is computable means that there is a notational system in which it can be computed. This hidden existential quantifier in the definition of computability reminds us of the difference between deterministic and nondeterministic computing, e.g., in polynomial time. Computing relates to thinking like P to NP. But this analogy is misleading, because in the former we can convince ourselves that the hierarchy is proper and unbounded. If we wanted to equate reasoning with first order reasoning, then this hidden existential quantifier can be made explicit by passing to second order logic. The problem which remains, is that the domain of this hidden quantifier remains always vague. We do not have, so far, a general definition of format, notation, protocol, and worse, had we one, it could be easily transcended by some kind of a diagonal argument.

9. Cultural Systems

The evolution of notational systems for number systems was studied from an anthropological point of view by R. Wilder (Wilder *1950, 1968, 1981*). His studies deserve attention especially when one has in mind the evolution and development of programming languages, operating systems, user interfaces, and other paradigms of computing. His studies clearly show several phenomena: that the evolution of concepts to widely accepted norms of practice takes much longer and need more than just the availability of such concepts; that the evolution of concepts is not due to individuals but is embedded in a (or several competing) cultural systems which are themselves embedded in host cultural systems; that nevertheless the fame and prestige of the protagonists of science and scientific progress do play an important, possibly also counterproductive, role; that cultural stress and cultural lag play a crucial role in the evolution of concepts; that periods of turmoil are followed by periods of consolidation, after which concepts will stabilize; that diffusion between different fields usually will lead to new concepts and accelerated growth of science; that environmental stresses created by the host culture and its subcultures will elicit observable response from the scientific culture in question; and, finally, that revolutions may occur in the metaphysics, symbolism, and methodology of computing science, but not in the core of computing itself. Wilder has developed (Wilder *1981*) a general theory of "Laws" governing the evolution of mathematics, from which I have adapted the above statements. It remains a vast research project to assimilate Wilder's theory to our context, but it is an indispensable project, if we want to adjust our expectation of progress in computing science to realistic hopes. Wilder's work also sheds some light onto the real problems underlying the so-called "software crisis": The cultural lag of programming practice behind computing science and the absence of various cultural stresses may account for the abundance of programming paradigms without the evolution of rigorous standards of conceptual specifications.

10. Computing vs. Intelligence

Our generally accepted model of computability is rather robust, but still fixes format, notation and protocol. Format and notation are abstracted from by coding into strings (numbers, etc.) and the protocol reduces to putting data into registers (writing it on tape), performing the fixed operations and reading the data. It allows us to analyze the resources needed for a computation and to prove negative results. It allows us to analyze what is computable within fixed formal concepts. It also can lead us in extending the notion of computability to other data types, as in Shepherdson, this volume, Chandra and Harel *1980,* and Dahlhaus and Makowsky *1986.* But it does not give us a means to discuss the *creation of data,* the passage from *mental images to formal concepts,* i.e., what Turing (Turing *1939*; cf. Feferman, this volume) called intuition and ingenuity.

The last twenty years of research in computing (rather than computer) science seem to indicate growing awareness of the role of the choice of data structures and different protocols of data creation and data exchange between machines and between man and machine. Especially research in mathematical cryptography, complexity theory, and learning theory (Goldreich, this volume; Valiant *1984*; Blumer et al. *1986*; Benedek *1988*) all invoke statistical concepts to model aspects of computing which are abstracted from in Turing's model. An anthropological look at the history of mathematics and the evolution of mathematical concepts teaches us caution in our expectation of a breakthrough in our expectations of intelligent machine performance. It may well be that within a couple of hundred years the cumulative experience of man-machine interaction will lead to an integration of man-machine intelligence. It may well be that programming by example will evolve as the ultimate mode of man-machine communication. But as much as children need ten years of constant exposure to language and writing to reach minor literacy, and our cultural system needed several hundred years to reach it, the computer population will need a similar stretch of time to overcome the problems of its infancy. And it is safe to say that the conceptual (both practical and theoretical) groundwork for intelligent man-machine interaction has just started. I hope this paper will stimulate research which will integrate these various lines of thought into a global theory of conceptual vs. computational programming.

Acknowledgments. I wish to thank J. Tal for all he has taught me in music during our joint research on notational systems for electro-acoustic music, R. Herken, J.B. Neidhart, and O. Wiener, for their continuous challenge making me think about these problems, and to

R. Herken, J.B. Neidhart and R. Zucker for encouraging me to write that paper and come to an end with it, though I am fully aware that much more could and should be said.

References

Blumer, A, A. Ehrenfeucht, D. Haussler, and M. Warmuth

    1986 Classifying learnable geometric concepts with the Vapnik-Chervonenkis dimension. In: *Proceedings of ACM-STOCS 1986*, pp. 273-282 (1986).

Benedek, G., and A. Itai

    1988 Nonuniform Learnability. To appear in: *Proceedings of ICALP 1988*, Lecture Notes in Computer Science. Springer-Verlag (1988).

Chandra, A.

    1988 Theory of data base queries. To appear in: *Proceedings of PODS '88*.

Chandra, A., and D. Harel

    1980 Computable queries for relational data bases. *J. Comput. Sy.* **21/2** (1980) 156–178.

Dahlhaus, E., and J.A. Makowsky

    1986 Computable directory queries. In: *Proceedings of CAAP '86*, Lecture Notes on Computer Science, vol. 214, pp. 254-265. Springer-Verlag (1986).

Ericsson, K.A., and H.A. Simon

    1984 *Protocol Analysis: Verbal Reports as Data.* Cambridge, MA: MIT Press (1984).

Feferman, S.

    1988 Turing in the land of O(z). This volume.

Goldreich, O.

    1988 Randomness, interactive proofs, and zero-knowledge. This volume.

Gurevich, Y.

    1987 Logic and the challenge of xomputer science. In: *Current Trends in Theoretical Computer Science*, ed. E. Börger. Computer Science Press (1987).

    1988 Algorithms in the world of bounded resources. This volume.

Hajek, P., and T. Havranek

    1978 *Mechanizing Hypothesis Formation (Mathematical Foundations for a Formal Theory).* Springer-Verlag (1978).

Langley, P., G. Bradshaw, and H. Simon

    1983 Rediscovering chemistry with the Bacon system. In: Michalski et al. *1983*, pp.303-329.

Maier, D.

    1983 *The Theory of Relational Databases.* Computer Science Press (1983).

Makowsky, J.A.

    1984 Model theoretic issues in theoretical computer science, part I: Relational data bases and abstract data types. In: *Logic Colloquium '82*, eds. G. Lolli, G. Longo, and A. Marcja, pp. 303–343. North Holland Publ. Co. (1984).

    1987 Review of Ullmann *1983*, Maier *1983*, and Chandra and Harel *1980*. *J. Symb. Log.* **51/4** (1987) 1079–1084.

Michalski, R., J. Carbonell, and T. Mitchel (eds.)

    1983 *Machine Learning*. Tioga Publishing (1983).

Shepherdson, J.

    1988 Mechanisms for computing over arbitrary structures. This volume.

Trakhtenbrot, B.

    1988 Comparing the Church and Turing approaches: Two prophetical messages. This volume.

Turing, A.M.

    1939 System of logic based on ordinals. *P. Lond. Math. Soc. (2)* **45** (1939) 161–228.

Ullman, J.

    1983 *Principles of Database Systems,* 2nd ed. Computer Science Press (1983).

Valiant, L.

    1984 A theory of the learnable. *Comm. ACM* **27/11** (1984) 1134–1142.

Vapnik, V.

    1982 *Estimation of Dependencies Based on Empirical Data.* Springer-Verlag (1982).

Wilder, R.L

    1950 The cultural basis of mathematics. In: *Proceedings of the International Congress of Mathematicians,* vol. 1, pp. 258–271 (1950).

    1968 *Evolution of Mathematical Concepts. An Elementary Study,* John Wiley (1968).

    1981 *Mathematics as a Cultural System.* Pergamon Press (1981).

The Fifth Generation's Unbridged Gap

Donald Michie

One can imagine that after the machine had been in operation for some time, the instructions would have been altered out of recognition, but nevertheless still be such that one would have to admit that the machine was still doing very worthwhile calculations. Possibly it might still be getting results of the type desired when the machine was first set up, but in a much more efficient manner. In such a case one would have to admit that the progress of the machine had not been foreseen when its original instructions were put in. It would be like a pupil who had learnt much from his master, but had added much more by his own work. (A.M. Turing, "The Automatic Computing Engine", Lecture to the London Mathematical Society, 1947)

Old programs do not learn. They simply fade away. (H.A. Simon, "Why should machines learn?", in *Machine Learning: An Artificial Intelligence Approach,* 1983)

1. Introduction

The "Fifth Generation" planning decisions (JIPDEC 1981) of 1979-80 by the Japan Information Processing Development Centre (JIPDEC) embodied a number of primary themes, subsequently institutionalized in the ICOT laboratory in Tokyo. In addition to improvements to the general run of man-machine communication aids, such as visual and speech input-output, sensory robot interfaces, natural language database query and the like, cognitive compatibility between the new ultra-powerful machines and their users was to be ensured by use of knowledge-based programming techniques. Taken together with radical approaches to the design of computing hardware, the Fifth Generation plans exceed present-day machines in power, instructability, and task-oriented intelligence by margins which leave the reader of the Japanese material somewhat dazed.

The Report of the Alvey Committee (*1982*) issued a year or two later by the British government highlighted a number of implications for industry. I shall here follow a different path, oriented more towards cognitive science than technology.

But I shall also argue that neglect of one specialized theme and of its scientific corollaries has jeopardized Japanese industrial goals and those of other countries.

The theme can be found lying rather inconspicuously in the main Japanese planning documents and is known as computer induction. It is concerned with machine acquisition from data of machine executable rules. When such rules are recognizably human in style we say that the machine is forming concepts. By generalizing over examples, whether supplied tutorially by the user or sampled from real-time monitoring equipment, or dredged from large database files, the machine induces descriptive rules and stores them for later execution by the same or other machines or by humans. Given only that such synthetic rules satisfy certain human-like structural constraints so as to qualify them as concepts, then they can serve as contributions not only to our stock of software but *also to our stock of knowledge.* This was clearly demonstrated by Michalski, Jacobsen, and Chilausky (Chilausky et al. *1976*) for diagnostic knowledge about soybean diseases and prefigured by Michie and Chambers' experiment (*1969*) with automatic acquisition of control rules) (see Michie *1983*). A more elaborate achievement in the Michalski-Chilausky style was recorded in the same year for the program Meta-DENDRAL. This is the inductive inference module of the celebrated DENDRAL expert system developed by Feigenbaum and members of the Heuristic Programming Project at Stanford University. By generalizing over mass spectroscopy data for a particular class of compounds, namely the mono- and poly-keto-androstanes, Meta-DENDRAL was able to infer a set of rules for deducing the molecular structures of these substances from their mass spectra (Buchanan et al. *1976*). The fruits of the program's work were in due course published in the chemical literature as an original contribution to knowledge.

A more recent case from the Space Shuttle program may help to clarify, in miniature so to speak, the steps by which a computer program may convert a collection of facts into a piece of operational knowledge. NASA's requirement was a decision procedure able to monitor real-time variables relevant to the question: At each given instant should the spacecraft be under the control of the autolander or under manual control? Moreover on the basis of these measurements, e.g., atmospheric turbulence, wind direction, altitude of craft, stability, etc., the procedure is required to give real-time answers to this critical decision question.

The autolander's chief designer Mr. Roger Burke, together with members of his design team, attempted the required programming task. At first the team was frustrated by a difficulty universally experienced by expert practitioners of highly tuned mental skills, namely inability to introspect their own decision rules with sufficient clarity and completeness to be able to supply a programmable account of them. The same experts, however, can normally articulate the same skills if allowed to express them in a *language of examples.* The upper part of Figure 1 shows a solution to the Shuttle problem expressed in such a language. Burke and his team arrived at it in interaction with an induction program known commercially as RuleMaster. The program converts example sets into machine-executable rules in a form illustrated in the lower half of the Figure. In addition to being a solution to NASA's problem, such a rule can be viewed as a miniature operational theory – a grain of knowledge, so to speak, to be added to one or another of the Himalayan heaps which the Fifth Generation envisaged building.

Readers of the original Japanese plan might excusably have paused to consider certain preconditional questions, such as:

1. Can the Himalayas be constructed without use of power tools? In 1981 the Japanese set out without such tools. Six years later their induction tool-kit remains unstocked and the knowledge mountain unbuilt.
2. If rule induction tools are to be used, can quality be assured and certificated in the new mass product?
3. Is it prudent to approach the building of such gigantic knowledge structures without any formulated theory of what knowledge is, how its forms are to be discriminated from other processable information, and what means exist for its quantitative measurement?

It seems that the development and testing of machine representations of concepts, even of the somewhat trivial kind illustrated in Figure 1, cannot be done in a satisfactory fashion except as part of a larger enterprise of a scientific character. This enterprise is nothing less than the development of a sound and well quantified theory of knowledge. Since Newton's time the practical bridge builder has had a scientific theory of matter and motion, known as mechanics; the steam engineer has Carnot's thermodynamics; the aerodesigner has fluid dynamics; the agricultural breeder has statistical genetics; the communications engineer has Shannon's theory of information. Before we plunge further into the scenario held out by Japan's Fifth Generation, it seems prudent to ask the question, "What then do the knowledge engineers have for characterizing and scientifically measuring this new resource which they are building – the automated knowledge resource?" The question has a succinct answer: rather little. I shall devote part of this paper to sketching some of the questions for which the missing science is needed, questions of what can and cannot be computed within "the human window". But first let me set JIPDEC's main planning ideas into focus.

2. Vision of Three in One

From page 10 and again from page 31 of JIPDEC's Preliminary Report a diagram is reproduced (Figure 2). What does it mean? Clearly there are three subsystems here, in some way overlapping or interlocking. To the left lies the cognitive machinery of the user, rich in culture, poor in store and processor. To the right is deployed the

MODULE: shuttle

STATE: one

ACTIONS:
noauto [advise "Don't use auto land"]
auto [advise "Use auto land"]

CONDITIONS:
stab [ask "stable?" "stab, xstab"] {stab xstab}
error [ask "errors?" "XL, LX, MM, SS"] {XL LX MM SS}
sign [ask "sign?" "pp, nn"] {pp nn}
wind [ask "winds?" "head, tail"] {head tail}
mag [ask "magnitude?" "L, M, S, O" {L M S O}
vis [ask "visibility?" "y, n"] {y n}

EXAMPLES

–	–	–	–	–	n	⇒	(auto, goal)
xstab	–	–	–	–	y	⇒	(noauto, goal)
stab	LX	–	–	–	y	⇒	(noauto, goal)
stab	XL	–	–	–	y	⇒	(noauto, goal)
stab	MM	nn	tail	–	y	⇒	(noauto, goal)
–	–	–	–	O	y	⇒	(noauto, goal)
stab	SS	–	–	L	y	⇒	(auto, goal)
stab	SS	–	–	M	y	⇒	(auto, goal)
stab	SS	–	–	S	y	⇒	(auto, goal)
stab	MM	pp	head	L	y	⇒	(auto, goal)
stab	MM	pp	head	M	y	⇒	(auto, goal)
stab	MM	pp	tail	L	y	⇒	(auto, goal)
stab	MM	pp	tail	M	y	⇒	(auto, goal)
stab	MM	pp	head	S	y	⇒	(noauto, goal)
stab	MM	pp	tail	S	y	⇒	(auto, goal)

Figure 1a. The final form of a training set of examples for the autolander problem.

```
MODULE shuttle IS
  State: one
    IF (ask "visibility?" "y, n") IS
      "y": IF (ask "errors?" "XL, LX, MM, SS") IS
        "XL" : (advise "Don't use auto land", goal)
        "LX" : (advise "Don't use auto land", goal)
        "MM" : IF (ask "stable?" "stab, xstab") IS
          "stab": IF (ask "sign?" "pp, nn") IS
            "pp" : IF (ask "magnitude?" "L, M, S, O") IS
              "L" : (advise "Use auto land", goal)
              "M" : (advise "Use auto land", goal)
              "S" : IF (asks "winds?" "head, tail") IS
                "head" : (advise "Don't use auto land", goal)
                ELSE (advise "Use auto land", goal)
              ELSE (advise "Don't use auto land", goal)
            ELSE (advise "Don't use auto land", goal)
          ELSE (advise "Dont't use auto land", goal)
        ELSE IF (ask "stable?" "stab, xstab) IS
          "stab" : IF (ask "magnitude?" "L, M, S, O") IS
            "L" : (advise "Use auto land", goal)
            "M" : (advise "Use auto land", goal)
            "S" : (advise "Use auto land", goal)
            ELSE (advise "Don't use auto land", goal)
          ELSE (advise "Don't use auto land", goal)
      ELSE (advise "Use auto land", goal)
  GOAL OF shuttle
```

Figure 1b. The decision rule.

might of tomorrow's gigaflop processors and gigabyte memories. What then is the role of the subsystem in the middle?

It is not at all easy to work out from this diagram alone just what is being proposed. My own attempts at decipherment were aided by confirmation from the diagram's author, Dr. Kaichi Furukawa of Japan's ICOT Laboratory.

The *general* notion is clear enough, namely that the central circle is to contain an artificial intelligence charged with a bridging function – to mediate between our richly endowed but slow and muddled human brains and the new race of super-powerful and super-opaque computing architectures with which we shall have to live. More fine-grained scrutiny of the diagram reveals in the layout of the central circle some landmarks familiar to practitioners of the software art known as "expert systems". The systems here envisaged, however, are a good deal more sophisticated

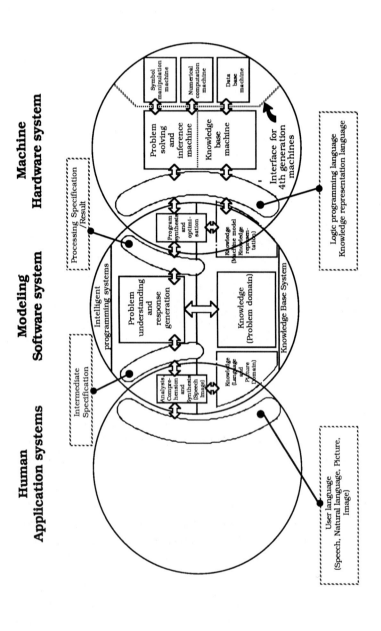

Figure 2. The target product of Japan's Fifth Generation plan for research and development is summarized in the above diagram. The central circle encloses a new AI-based technology intended to mediate between two rapidly diverging technical cultures—between human arts, crafts, and sciences on the one hand and computational "brute force" on the other.

than those that pass as expert systems today. This sophistication stems from the Fifth Generation idea of an expert system:

- such a system is able, on the basis of a representation coming in from the left-hand circle, to understand exactly what the problem is that the user wants to have solved;
- it can formulate a solution as a machine program written in an ultra-high level language, optimized to run efficiently on the knowledge machine with its complex of supercomputer slaves in the right-hand circle;
- it is able to operate in the inverse direction of the above two steps, that is, it can translate the output obtained from the supercomputer complex into a form acceptable by the user as an intelligible answer to his query.

To do all this, the system in the central circle must manipulate models not only of the problem domain itself but also of what goes on inside the two circles on either side of it. So the circle labeled "modeling software system" must model both the problem domain *and* the user's mental and linguistic processes. Moreover, it must also model the right-hand circle's interlinked inference module and knowledge-base (which above I collectively denote as the "knowledge machine"), whereby the answers are to be extracted, organized, packaged, and handed back.

This knowledge machine must in turn manipulate models of the more conventional fourth generation slave resources depicted inside the dotted lines on the extreme right. We can follow the complete life history of a user query through a sequence of stages by stylizing each to the form of an input-operation-output triple (Figure 3).

The task associated with the central circle is thus new. Previous generations of designers gave us computing systems programmable to construct *solutions* within defined problem areas. The Fifth Generation's task is to design systems able to construct *solution strategies*. The central circle's right-hand part must express these strategies as logic programs and associated knowledge structures to be shipped over to the inference and knowledge-base complex in the right-hand circle. In complex domains the construction of strategies – for concreteness I shall call them "calculative plans" – is a quintessential human activity. But whenever a calculative plan is brought into being by human intelligence, the question presents itself as to who or what is to execute it. For execution the human equipment is rather weak, as indicated in Figure 4. Doubtless this is why, ever since the invention of the abacus and the arts of counting and writing, mankind has sought by every possible means to mechanize the tedious business of *executing* plans, as opposed to constructing them. Today's 100-mips machines convey the news that at last this incremental enterprise has won through. Any feasible and precisely formulated calculative plan can now be executed with stunning accuracy and speed. So where next? I shall discuss the Japanese answer

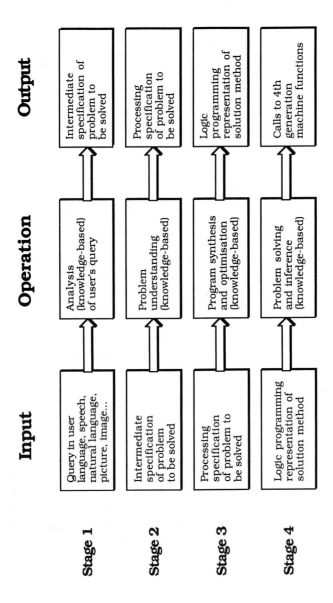

Figure 3. Stages leading from a user query (top left) to a solution (bottom right). An equivalent sequence must be traversed back to the user in turning this into an answer deliverable in acceptable form.

1.	Rate of information transmission along any input or output channel	30 bits per second
2.	Maximum amount of information explicitly storable by the age of 50	10^{10} bits
3.	Number of mental discriminations per second during intellectual work	18
4.	Number of addresses which can be held in short-term memory	7
5.	Time to access an addressable "chunk" in long-term memory	2 seconds
6.	Rate of transfer from long-term to short-term memory of successive elements of one "chunk"	3 elements per second

Figure 4. Numerical values of some key parameters of the brain as problem-solver. As with biological measures generally large variation between individuals is encountered, so that the figures shown in general should be credited with a twofold spread in each direction. The size of short-term memory, however, is not believed to vary much above the memory shown.

to this question with the aid of an example which may seem like a digression. In my view, however, it contains the heart of the matter.

3. Calculating Prodigies Do Not Calculate

From time to time accounts circulate of extraordinary mental feats performed by rare individuals known as calculating prodigies. Possibly the most extraordinary, and certainly the best documented, was Alexander Aitken, for many years Professor of Mathematics at the University of Edinburgh. The following passage is taken from one of a number of scientific papers by Ian Hunter (*1962*), then at Edinburgh, and subsequently Professor of Psychology at Keele University, on the subject of Aitken's specialized mental skill. Hunter speaks of

> ... a large multiplication problem presented to Professor Aitken some years ago by his own children. Problem 12. Multiply 987,654,321 by 123,456,789. "I saw in a flash that 987,654,321 by 81 equals 80,000,000,001; and so I multiplied 123,456,789 by this, a simple matter, and divided the answer by 81. Answer: 121,932,631,112,635,269. The whole thing can hardly have taken more than half a minute."

Does this mean, then, that Alexander Aitken was the possessor of unusual powers of calculation, in the sense of super-fast execution of calculative plans? Hunter eliminated this possibility. Aitken's response to a problem posed to him consisted of two phases. The first phase would last anything from a few seconds to a minute, depending on the difficulty of the problem. During this phase Aitken was sitting still, apparently doing nothing. What he was actually doing was mentally rummaging in a huge internally-stored library of useful facts and tricks concerning the number system so as to put together a *calculative plan.* During the second phase Aitken executed the plan. In other words he was busy calculating. But, however busy he may have been, the digits came out no faster then if the same plan had been handed to a student with the request to execute it. Hence Aitken's prodigious power of lightning calculation was like every other prodigious mental power which has ever been investigated. It was knowledge-based, not calculation-based. Suppose that Aitken were fortunate enough to have a student gifted with a speed and accuracy of brute force calculation as superior to Aitken's as Aitken's "how-to-do-it" knowledge-base was superior to that of ordinary mortals, a Superstudent in fact. Aitken and Superstudent could then form a partnership of far greater problem solving power than possessed by Aitken alone. An even more powerful consortium would be created by enlisting Hunter for the task of generating an explanatory report for each arithmetical feat. Let us call this hypothetical team the Hunter-Aitken-Superstudent system.

To dramatize the relevance to the Fifth Generation of this apparent digression, Figure 5 reproduces Figure 2 together with a companion diagram of the Hunter-Aitken-Superstudent system for fast solution of instant problems in arithmetic. If the survival of planet Earth depended upon improved operation of this system, how would we proceed? We would probably first try to speed up Superstudent. One could say that this is precisely what computing technology has been attempting to do for its first 30 years of life.

For the hypothetical student, our first step perhaps would just be to invest in a range of hand calculators for him to use. I say a range, rather than just one, because for fixed cost there is a tradeoff in calculator design between working memory and running speed. Hence, we envisage Superstudent looking at the calculative plan handed him by the master and saying, "Aha, I see that when I use this plan on this particular problem I will not need to have more than two memory registers in use at any one time. I'll take calculator n°· 8". Notice that Superstudent is already functioning as more than a mere plan executor: he is using his knowledge of calculative plans and of the properties of alternative calculator architectures to optimize his choice.

So to speed up Superstudent yet further our second step might be to write a computer simulation of this limited knowledge-based behavior, and substitute it in the right-hand circle of the three-in-one diagram as a program for calculator selection and control. Aitken must now of course couch his calculative plans in some formal

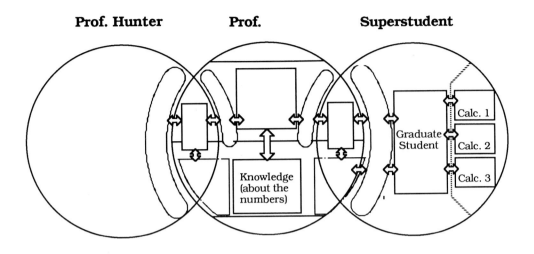

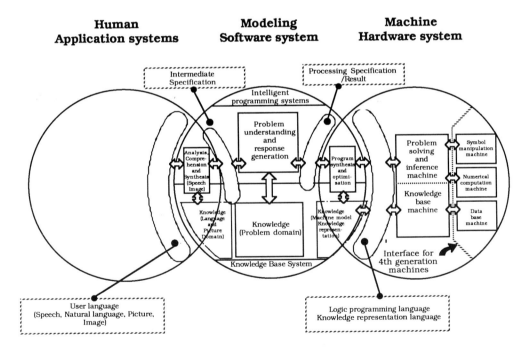

Figure 5. The Hunter-Aitken-Superstudent system as a parable of JIPDEC's three-in-one vision (see text).

language rather than in whatever semiformal mix of mathematical notation and English would suffice for the original Superstudent. A next logical step would be to replace the various hand calculators by high-speed supercomputers specialized to efficient execution of the various kinds of problems and programs submitted to them by the now automated Superstudent. This step finally closes the gap, and takes us back to the right-hand circle of Figure 2. Note some intriguing consequences. First, all of the actions and most of the branches and loops which were originally executed by Superstudent directly from the plan can now advantageously be done on the other side of the interface, inside the dotted lines. But Superstudent must now undertake some more-than-primitive problem solving, in order to effectively command the slave supercomputers. Secondly, the burden thrown upon the knowledge-base component which constitutes Superstudent's lower half is greatly increased. Now it must model not a range of calculators but a heterogeneous collection of supercomputers.

Eventually the shoe begins to pinch in a different place, namely in the middle circle, labeled Alexander Aitken. Can he be mechanized? A design choice presents itself.

1. Discard Aitken and let Hunter try to talk directly to Superstudent. This is the trend of supercomputer design today. If we are not careful, this is how tomorrow's brute force computer technology will evolve. For associated dangers, see Michie and Kopec *1983*.
2. Discard Hunter and build a machine-optimized Super-Aitken, to drive Superstudent: even more dangerous than 1, since a *machine*-optimized agent will be inescapably Martian in mentality, opaque to man.
3. Build a Super-Aitken in the human mould, i.e., as a *brain*-optimized program.

Options 1 and 2 are mirrored in various attempts today to emulate chess-master play by computer (Michie *1982*). The predictable result will be the eventual attainment of a program which the world champion cannot beat, but which he cannot talk to either, whether for purposes of teaching, learning, or trouble-shooting.

Option 3 typifies the long distance goals of machine intelligence.

An engineer whose field of work is in its prescientific phase makes his professional decisions by the seat of his pants, eked out with a little codified lore. Doubtless this is what the Japanese engineers did in opting for option 3 – yet they concluded rightly. Option 3 not only says that the central circle, the Super-Aitken circle, must exist, but it also entails implementation by the software methods of artificial intelligence, the cornerstone of Japan's planned information technology of the 1990's. I began by calling for scientific foundations for the new craft of brain-optimized computation. To convey a flavor of the kind of work to be done, let us consider the three-way design choice posed above.

4. Beyond Intuition

For complex tasks, is it true, as the Japanese have concluded, that only rule-based programming disciplines, geared to mimic the representations used by the expert brain, can provide the needed bridge function? Can a "human window" be defined? Are standard programming styles doomed for some reason to miss this window and so to generate only opaque products, unfriendly and therefore unwelcome and unsafe?

Imagine some computing problem which by 1990 will be run-of-the-mill, of the order of complexity, say, of large problems of air traffic control, or of missile early warning, or of the monitoring of nuclear power stations. For establishing and analyzing the principles at issue the domain could equally well be a tough problem of group theory or chess or some Rubik hyper-cube yet to be invented. For definiteness I have arbitrarily drawn Figure 6 for a size of combinatorial space (the set of distinct configurations of some hyper-Rubik for example) such as would require 10^{50} bits to store explicitly. Clearly a machine to answer questions in such a problem domain cannot in practice do it by dictionary lookup. In theory, though, given a machine specification we may estimate a least compute time for worst-case lookup, and assuming a minimally coded lockup table this identifies a point on a plot time vs. space. The projection of this point onto the space axis (E in Figure 6) marks the given function's information content (see Michie *1976, 1982*).

At the other end of the plot lies a point corresponding, when projected onto the horizontal axis (A in Figure 6), to the shortest program which can compute the function at all, possibly by some simple-minded exhaustive search. For complex domains this type of solution is also ruled out, on grounds of astronomical running time. Now augment the memory alowance little by little. At each stage put a point on the graph for the program with the least worst-case evaluation time achievable within the given ration of program size. Obviously such a least time always exists and has a definite value, whether or not we can measure or estimate it. So we proceed with our hypothetical function, let us call it *h*, until the entire curve is mapped.

Let us now place boundary marks on the two axes, denoting limits to what we expect from the hardware technology of this century. In the diagram marks have been placed (somewhat arbitrarily) at 10^{15} bits of random access memory and 10^{10} primitive calculational steps as a maximum acceptable waiting time β in Figure 6. The two then define what can be termed a rectangle of feasibility. For a function to have a solution program in any practical sense, its tradeoff curve must intersect this rectangle. Points lying in the intersection area correspond to feasibly executable programs. If for all machines which exist or could exist, *h* fails this criterion, then *h* is what is technically called "hard". Feasible in principle but not in practice, its evaluation is of academic interest only.

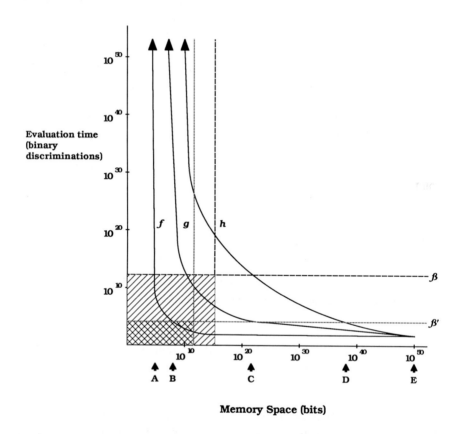

Figure 6. Three hypothetical functions are depicted which have identically the same resource requirements as each other either if expressed as giant lookup tables or alternatively as compact programs for evaluation on demand by calculation. Both forms of representation are ruled out for the functions here envisaged, for reasons of availability of sufficient store, or sufficient computing time, respectively. This is characteristic of problems which are "complex". It means that if a computed solution is to be feasible it must exploit the tradeoff between memory cost and calculation cost. Each function has its own characteristic tradeoff curve which sets bounds to feasibility, as illustrated in the figure for three hypothetical functions f, g, and h. See Note at end of this article for further detail.

Suppose, however, that the problem is like *g*. Then a feasibly executable set of programs exists of which the least-time subset corresponds to those points on *g*'s curve which lie within the feasibility rectangle. Moreover we can rank them in an order of cost-preference by noting that the horizontal and vertical axes are both scaled logarithmically. Hence the least-cost program, measuring cost as store multiplied by CPU-time, is represented by that point which is nearest to the tangent made with the curve by a straight line inclined at 45°. Note, however, that *g*'s curve, although intersecting the larger rectangle, fails to intersect the smaller rectangle. The latter defines the zone of feasibility for the "brain machine", that is, a computing system subject to the same bounds on various information-processing parameters as those known to characterize the human brain. The bounds in the diagram of the smaller "rectangle of brain feasibility" are a little generous. For the logic of this argument it is sufficient simply that they be less in each case than the correspoinding bounds for existing computing machines.

We are now in a position to consider function *f*. Five alternative categories of solution program are in this case possible (Figure 7).

1. Too calculation-intensive to be machine-feasible

2. Too store-intensive to be machine-feasible

3. Machine-feasible: too calculation-intensive to be brain-feasible

4. Machine-feasible: too store-intensive to be brain-feasible

5. Machine-feasible, and potentially brain feasible

Figure 7. Five categories of solution program for one and the same complex problem domain. Following the store-time curve of function *f* (Fig. 6) the five segments of the curve correspond to the five categories shown.

Dropping (1) and (2) from further consideration, first observe that knowledge engineering is the art of writing programs to fall into category (5) or, to speak more precisely, to fall into that subcategory of category (5) which is not only by reason of store-time tradeoff properties potentially brain-feasible, but also satisfies certain necessary structural properties (see below). Note that for complex domains this category is effectively disjoint from all other categories of programs which have characterized the past 30 years of computing. Only programs of category (5) can have the property of being sufficiently store-sparing to be memorizable and sufficiently processor-sparing to be mentally checkable. They are candidates, in fact, for lying within the "human window". I shall show that membership of the lower intersection area depicted as black on the graph, although a necessary condition for brain feasibility, is not a sufficient condition. But first let us revisit the designer's dilemma from which we started, and enquire whether designing for machine optimality can perhaps be combined with the human window criterion. Returning to the 45° tangent test of machine optimality, we can see that whether or not the point thereby distinguished lies within the human window is determined by the form taken by the tradeoff curve of the given function. Thus absolutely no guarantee can be offered the designer that for any given function the two properties can be combined. We, the knowledge engineers, who demand brain feasibility, are actually in worse case still, for an additional and quite different reason: even programs sufficiently processor-sparing and store-sparing to lie in the "human window" intersection area of the graph may be (and typically are) disqualified by their syntactic structure from achieving true brain compatibility. I shall illustrate, from a recently demonstrated form of computer program, the superprogram.

5.   Machine-made Programs: Opaque or Transparent

A superprogram is a program that runs at least twice as fast as the most efficient that could possibly be written by a human programmer. Figure 8 shows execution costs on the Cyber 175 for three different solution programs written for the same difficult problem. The first implements a standard algorithmic approach. The second was the fastest pattern-coded solution which J.R. Quinlan (*1982*), an outstanding programmer and computer scientist, could achieve after many months of work and study. The third was generated by Quinlan's induction program ID3 on the Cyber 175, equipped with a file of primitive descriptors and a file of example data. This last product clearly qualifies as a superprogram. There was no reason to place it outside the rectangle of brain-feasibility depicted in the previous Figure, for it was both compact and fast. Yet to human scrutiny it was totally and obstinately opaque. Why?

The answer lies in the topology of the decision trees (or equivalently in the nesting structure of the corresonding conditional expressions) which ID3 generates. Large branching tree-structures cannot be processed by a device such as the brain (see Figure 4) with no more than seven locations in short-term memory. Responding to this, Shapiro and Niblett (*1982*) developed a method known as structured induction adapted from the discipline of structured programming. The programs which are output by their program-synthesizing method take the form of decision structures humanly recognizable as concepts. We now appreciate that the structural features which hold the balance between humanly transparent and humanly opaque are directly related to known facts about the brain's working memory.

This relationship, coupled with recent techniques for endowing machine-made programs with run-time self-commenting facilities, makes possible the implementation of a "superarticulacy" feature. An expert system may be enabled fully and intelligibly to document a mental skill for which its human possessor can only achieve

Lost 3-ply Experiments

(CDC Cyber 72 at Sydney University)

49 attributes (mixture of 35 pattern based
5 pattern based
10 high-level predicate)

715 distinct instances
177 node decision tree found in 34 seconds

Classification method	CPU time (msec)
Minimax search	285.0
Specialized search	17.5
Using decision tree	3.4

Figure 8. Quinlan's "superprogram discovery". The decision tree was inductively extracted from a file of 715 instances, each consisting of a list of 49 attribute values paired with one or other of two decision classes, lost 3-ply and not lost 3-ply. Of the three machine methods shown for classifying the file, the induced decision tree executed fastest and qualifies as a superprogram.

a fragmentary and fallible articulation (Michie *1986*). This can even be done when the expert system has been synthesized by computer induction rather than hand crafted (Shapiro and Michie *1986*), and raises the possibility not only of generating new knowledge by machine but also of its *human acquisition* from machine constructed knowledge bases. For experimental investigation, problem domains are required which combine machine tractability with impenetrability to the unaided intellect, unaided, that is, by tutorial access to the machine-synthesized knowledge. The use of such "ultra-complex" problem material is discussed elsewhere (Bratko and Michie *1987*).

It is precisely in this area of machine acquisition of user-transparent knowledge that the Fifth Generation project in Japan, as also its Alvey and Esprit followers in Europe, left a critical gap unbridged. A warning of this was publicly uttered by Kazukiwo Kawanobe as early as 1984 in a keynote address in Tokyo to ICOT's first international Conference (Kawanobe *1984*). In his "Current Status and Future Plans" Kawanobe made it clear that ICOT was more than up to target over most of its chosen front. Progress had been formidable in Japanese acquisition and mastery of the latest programming and design disciplines, techniques, and tools, including in particular those of functional and logic programming. But when Kawanobe turned to the *knowledge problem* a different picture was disclosed.

Kawanobe reviewed the goals originally set for attainment at the end of the 3-year "Initial Stage" of the plan. In a passage set out in Figure 9 I have added the word "transparent" in brackets in a way that I believe that Dr. Kawanobe would accept, so as to retain the constraint emphasized by the three-circle diagram. This gives us Figure 9, in which the prime concerns of the knowledge engineer are clearly set out.

Problem Solving and Inference Software

Goal in the initial state (1981–1984)

The final goal of this research involves a number of techniques that have not yet been achieved, such as:

1. deductive inference function which performs completely (transparent) logical inference,

2. the inductive inference function, which involves guess-work based on incomplete knowledge, and

3. the cooperative problem solving function by which systems cooperate with each other to solve a given problem.

Figure 9. K. Kawanobe at the November 1984 FGCS Conference in Tokyo: "Current Status and Future Plans".

Focussing on the specific, but critical, problem of knowledge acquisition, we read of the need for

... clarification of techniques to implement the knowledge base management function, which involves a number of unknown factors, such as the knowledge acquisition mechanism. (Kawanobe: *Intermediate Stage* 1985-89)

and again:

For the basic high level inference software, attempts will be made to develop high level artificial intelligence including inductive inference, analogy and learning functions. (Kawanobe: *Intermediate Stage* 1985-89)

and again:

For the knowledge acquisition basic software, a tool to acquire knowledge from experts will be developed. At the same time attempts will be made to develop techniques to make rules by means of inductive inference. (Kawanobe: *Intermediate Stage* 1985-89)

On secondment to ICOT at the time, Dr. Kawanobe returned soon afterwards to rejoin his own company. Visiting ICOT in 1986, I found continuing awareness of the gap but no remedial plans.

Attribute	*Definition*
knowledge-content	amount of information immediately realizable from brain-feasible programs
advantage	ability to realize more than is stored
penetration	logarithm of advantage
grasp	increase of penetration
rote learning	acquisition of grasp via prestorage of individual facts
parameter learning	acquisition of grasp via numerical adjustment of prestored descriptions
inductive learning	acquisition of grasp via formation and storage of new descriptions
concept learning	acquisition of grasp via formation and storage of brain-feasible descriptions
difficulty of a problem domain	length of shortest brain-feasible programs able to compute solutions to all problems of the domain

Figure 10. Program attributes with well-understood intuitive meanings (left-hand column) paired in each case with an informal definition based on the theory of knowledge-measurement(see Michie *1976, 1982*).

Having missed its declared aim, Japan's Fifth Generation should not be judged as lacking in constructive contribution. A promotional umbrella was created under which ICOT's small center of excellence in logic programming has been able to pursue abstract enquiries in relative peace, and to disseminate the fruits to industrial as well as academic laboratories. Moreover, the Fifth Generation planning documents, as yet without practical prospect of fulfillment, constitute a well researched characterization of an inspiring objective. Other laboratories have meanwhile accumulated the requisite methods of knowledge synthesis and acquisition which correspond to the Fifth Generation's unbridged gap. These laboratories are now at liberty to pick up Japan's abandoned plans and to implement them in earnest. If and when they do, there will be immediate need for a sound system of definition and measurement. A knowledge engineer who cannot measure the knowledge in a patttern-based rule is like a hydraulic engineer who does not know how to measure the amount of liquid in a vessel. For programs of the new type, we may also need schemes like that of

Figure 10, adapted from a first report (Michie *1976*) of work in progress (Michie *1982a*).

Acknowledgement. In preparing this paper I have incorporated materials from a study entitled "Mismatch between machine representations and human concepts: dangers and remedies" by Daniel Kopec and myself, commissioned by the European Economic Community (Forecasting and Assessment of Science and Technology Programme).

Note on Figure 6. Store-time tradeoff curves for hypothetical finite functions f, g, and h. Each has the same information content (10^{50} bits) and the same Chaitin complexity (10^4 bits) (see Chaitin *1975*). The time feasibility (β) limit for functions evaluated by machine has somewhat arbitrarily been placed at 10^{12} time bits. Space feasibility has been set at 10^{15} store bits. Time feasibility for human evaluation of the same functions is set at 10^4 time bits indicated by β' and space feasibility at 10^{10} store bits. The hatched rectangle is the 'zone of machine feasibilitiy', through which curves f and g pass. Only f passes through the cross-hatched 'zone of human feasibility'. The five upward arrows at the baseline mark, respectively: the Chaitin complexity of f, g, and h; the β'-complexity of g; the β'-complexity of h; the information content of f, g, and h. f is semitractable; it is also reducible, since its β'-complexity is less than the human store-bound. This property offers, without guaranteeing, the possibility of human master of f as a problem-solving domain domain. g is also semitractable, but not reducible. h is intractable, and a fortiori not reducible. A useful quantity to keep in mind is the maximum bit rate of human mental calculation, equivalent to about 20 binary discriminations per second. In setting machine time feasibility equal to 10^{12} we assume the availability of machines capable of calculating a hundred million times faster than the brain, say 2000 million binary discriminations per second. On this basis it takes the same length of time to wait for a machine to execute 10^{12} binary discriminations as it does for a human solver to execute 10^4, i.e., 500 secs using the above bit rate estimates for the two devices.

Explanation of the notations used in the above note can be found in Michie *1982*.

References

Alvey Committee

> 1982 *A Programme for Advanced Information Technology: The Report of the Alvey Committee.* London: Her Majesty's Stationery Office (1982).

Bratko, I., and D. Michie

> 1987 Ideas on knowledge synthesis stemming from the KBBKN chess endgame. *ICCA Journal* **10/1** (1987) 3–13.

Buchanan, B.G., D.H. Smith, W.C. White, R.J. Gritter, E.A. Feigenbaum, J. Lederberg, and C. Djerassi

> 1976 Applications of Artificial Intelligence for Chemical Inference XXII. Automatic rule formation in mass spectroscopy by means of the Meta-DENDRAL program. *J. Am. Chem. S.* **98** (1976) 6198.

Chaitin, G.

> 1975 Randomness and mathematical proof. *Sci. Am.* **232** (1975) 47–52.

Chilausky, R.L., B. Jacobsen, and R.S. Michalski

    1976 *Proceedings of the Sixth International Symposium on Multi-variable Logic (UTAH)* (1976).

Hunter, I.M.L.

    1962 An exceptional talent for calculative thinking. *Br. J. Psycho.* **53** (1962) 243–258.

JIPDEC Fifth Generation Computer Committee

    1981 *Preliminary Report on Study and Research on Fifth-Generation Computers in 1979–1980.* Tokyo: Japan Information Processing Development Centre (1981).

Kawanobe, K.

    1984 Current status and future plans of the Fifth Generation Computer Systems project. In: *FGCS 84, Proceedings of the International Conference on Fifth Generation Computer Systems 1984,* Tokyo, Japan, Nov. 6–9, 1984, pp. 3–17 (1984).

Michie, D.

    1976 Measuring the knowledge of content of programs. In: *Technical Report UIUCDCS-R-76-786.* Urbana, IL: University of Illinois, Department of Computer Science (1976).

    1982 Computer chess and the humanisation of technology. *Nature* **299/5882** (1982) 31–394.

    1982a Measuring the knowledge-content of expert programs. *B. Ins. Math. App.* **18** (1982) 216–220. Also reproduced as Chapter 18 of *On Machine Intelligence,* 2nd edition 1986, pp. 219–230. Chichester: Ellis Horwood, and New York: Halstead Press (1986).

    1983 The state of the art in machine learning. In: *Introductory Readings in Expert Systems,* ed. D. Michie, pp. 208–229. New York: Gordon and Breach (1983).

    1986 The superarticulacy phenomenon in the context of software manufacture. *P. Roy. Soc. (A)* **405** (1986) 185–212.

Michie, D., and R.A. Chambers

    1969 Man-machine co-operation on a learning task. In: *Computer Graphics: Techniques and Applications,* eds. R. Parslow, R. Prowse, and R. Eliot Green, pp. 79–186. London: Plenum (1969).

Michie, D., and D. Kopec

    1983 Mismatch between machine representations and human concepts: dangers and remedies. *FAST SERIES No 9.* Brussels: Commission of the European Communities (1983).

Quinlan, J.R.

    1982 Semi-autonomous acquisition of pattern-based knowledge. In: *Machine Intelligence 10,* eds. J.E. Hayes, D. Michie, and Y.-H. Pao. Chichester: Ellis Horwood, and New York: Halsted Press. Also in: *Introductory Readings in Expert Systems,* ed. D. Michie. London/Paris/New York: Gordon and Breach (1982).

Shapiro, A.D., and D. Michie

>   1986 *Advances in Computer Chess 4,* ed. D.F. Beal, pp. 147–165. Oxford: Pergamon
>   (1986).

Shapiro, A.D., and T.B. Niblett

>   1982 *Advances in Computer Chess 3,* ed. M.R.B. Clarke, pp. 73–91. Oxford: Pergamon
>   (1982).

On the Physics and Mathematics of Thought

Roger Penrose

1. Introduction

The questions that I shall touch upon here are closely related to those that Alan Turing concerned himself with in the seminal article that this volume commemorates (Turing *1936-7*). What are the essential physical actions that take place when a person thinks? Is a human brain, indeed, effectively a universal Turing machine? Is the precise material that constitutes a thinking device (such as a brain) of any real consequence? Does it much matter what physical laws are called upon in the action of such a device?

Regrettably I have not time, as I write this, to develop fully the (rather incomplete) ideas that I would wish to express; nor is there really the necessary space for an adequate discussion. As it happens, I am in the process of writing a semi-popular but lengthy account of those ideas, which will be published in due course by Oxford University Press (under the title: *The Emperor's New Mind*). Instead of attempting to summarize the intended contents of that work here, I shall present an abbreviated account based largely on a few excerpts from portions that are already written, albeit in their present preliminary state. (I am grateful to Oxford University Press for their permission to proceed in this way.)

2. Is Mental Activity Merely the Enacting of an Algorithm?

A certain view seems to have gained prominence in recent years, namely that a human brain is, indeed, effectively some kind of computer, and that it differs from a universal Turing machine only in that its storage capacity is finite, that its activity is curtailed in time, that it may involve some random elements, and that it sometimes makes mistakes. This, indeed, seems to have been Turing's own view, more or less (Turing *1950*), and his general concept of what could constitute a thinking device – in particular, a brain – led him originally to his powerful Turing machine concept. On this view, it does not *in principle* matter how the device is physically constructed. Cogs, wires and pulleys, water pipes, or electromagnetic waves would each do as well as electronics, or the electro-chemical activity of neurones. Though the details of the physical action might be different in each case, this would be of no overall consequence – apart from the (perhaps very great) differences in capacity, speed, and reliability.

Must we accede to this view? Indeed, is it necessarily the case that any physical object whatever – where I include a functioning human brain as a physical object – must act by rules that are in accordance with the running of some Turing machine program? Or may it be that some suitably delicately organized (perhaps biological) construction can perform in some fundamentally different, and perhaps even superior, way? My own opinion, for what it is worth, is that this latter possibility is very likely to be true of, in particular, the functioning of (parts of) our brains, and that there is some noncomputable element present in physical laws which may well be being harnessed in the physical basis of our thoughts.

This is, I believe, a distinctly unconventional view. Turing had, after all, presented a very persuasive case that any mechanism, in the ordinary sense of the word, would have to follow the general procedures that he laid down for a Turing machine. I am not disputing his line of argument, which I, also, find persuasive. What I am attempting to dispute is the idea that a physical object need actually behave as a "mechanism" in this ordinary sense. As part of my argument I shall need to explain what I mean by a physical object not behaving as a mechanism. For the moment, let us simply say that such a physical "nonmechanism" might have a behavior which is completely governed by precise physical laws, yet governed in a way that cannot be derived from any algorithm.

My reasons for adopting such a viewpoint come from several different directions. One of these has to do with our conscious perceptions. To investigate this, let us first consider a profoundly opposed viewpoint – that of "strong AI" – which asserts that our thinking consists merely of the enacting of some algorithm. According to this view, consciousness must be some quality that can be actually possessed by algorithms. Somehow, this quality would have to become manifest whenever that algorithm is enacted. However, John Searle (*1980*) has argued strongly that the mere enacting of an algorithm cannot, in itself, evoke the mental qualities that that algorithm is intended to provide. He presents his example of a "Chinese room", in which a human subject manipulates counters with Chinese symbols marked on them – these manipulations following the action of some known algorithm – but without experiencing the mental quality, in this case "understanding", that the algo-rithm is supposed to provide. Searle's arguments are, in my opinion, interesting and fairly persuasive, but by no means definitive.

There is another fundamental problem for the strong AI position: what does it actually *mean* for an algorithm to be "enacted"? Does "enacting" necessarily involve the movement of some material bodies (e.g., of electrons along wires, or of neurological chemicals) in some spatio-temporal configuration which represents the *running* of the algorithm? Or can "mental qualities" be achieved merely by the algorithm's specification having been written out on some page (or in the magnetic fields of some computer disc)? At what stage would "conscious awareness" be conjured into being, according to this view? If that conscious awareness requires the physical movement of bodies, then it is *not* merely a quality possessed by algorithms: conscious thinking would indeed depend upon the *physical action* of the "thinking device" that is involved. If the mere specification of an algorithm can conjure up "awareness", then why does it need physical embodiment at all? Perhaps the algorithm's mere Platonic existence as a mathematical object is, in itself, sufficient to give it conscious feelings!

This dilemma lies behind the scenes of an argument put forward by Douglas Hofstadter (*1981*) – himself a major proponent of the strong-AI view – in a dialogue entitled "A Conversation with Einstein's Brain". Hofstadter envisages a book, of absurdly monstrous proportions, which is supposed to contain a complete description of the brain of Albert Einstein. Any question that one might care to put to Einstein can be answered just as the living Einstein would have, simply by leafing through the book and carefully following all the detailed instructions it provides. Of course "simply" is an utter misnomer, as Hofstadter is careful to point out. But his claim is that *in principle* the book is completely equivalent, in the operational sense of a "Turing test", to a ridiculously slowed down version of Einstein himself. (Recall Turings's idea of a test whereby an interrogator has to decide, merely from the answers to a succession of questions, which of two subjects is a human being and which is a computer. If she consistently cannot tell, then the computer passes, and is deemed to possess the appropriate "mental qualities". We imagine a corresponding criterion for deciding, in effect, whether something – here a book – "is" or "is not" Einstein!) Thus, according to the contentions of strong AI, the book would think, feel, understand, be aware – just as though it were Einstein himself, but no doubt living at a monstrously slowed-down rate (so that to the book-Einstein the world outside would seem to flash by at a ridiculously speeded-up rate). Indeed, since the book is supposed to be merely a particular embodiment of the algorithm which constitutes Einstein's "self", it would actually *be* Einstein.

But now a difficulty presents itself. The book might never be opened, or might be continually pored over by innumerable students and searchers after truth. How would the book "know" the difference? Perhaps the book would not need to be opened, its information being retrieved by means of X-ray tomography, or some other technological wizardry. Would Einstein's awareness be enacted only when the book is being so examined? Would he be aware twice over if two people chose to ask the book the same question at two completely different times? Or would that entail two separate and temporally distinct instances of the *same* internal state of Einstein's awareness? Perhaps his awareness would be manifested only if the book is *changed*? After all, normally when we are aware of something we receive information from the outside world which affects our memories – and the states of our minds are indeed slightly changed. If so, does this mean that it is (suitable) *changes* in algorithms (and here I am including the memory store as part of the algorithm) which are to be associated with mental events rather than (or perhaps in addition to) the *activation* of algorithms? Or would the book-Einstein remain completely self-aware even if it were never examined or disturbed by anyone or anything? Hofstadter touches on (some of) these questions, but he does not really attempt to answer them, nor even properly to come to terms with most of them.

What does it mean to activate an algorithm, or to embody it in physical form? Is changing an algorithm different in any sense from merely discarding one algorithm and replacing it with another? What on earth does any of this have to do with our feelings of conscious awareness? The reader (unless himself or herself a supporter of strong AI) may be wondering why I have devoted this much space to such a patently absurd idea. In fact, I do *not* regard the idea as intrinsically an absurd one – mainly just wrong!

3. Is All Physical Action Algorithmic?

There is, indeed, a certain force in the reasoning behind strong AI which must be reckoned with. If, as appears to be the common belief, it is actually true that the action of *any* physical object is equivalent to the running of some Turing machine, then in particular it must be the case that the activity of a human brain can, in principle, be exactly modeled by a general-purpose computer (i.e., in effect, by a universal Turing machine). Any question that one might put to the human could, in principle, be answered in an identical manner by the computer. Thus, if one adopts Turings's highly *operational* viewpoint – as described according to his "Turing test" – one must accept that the computer, whenever it is (in an appropriate sense) enacting the appropriate algorithm, must have the same mental experiences as the human would.

It may be argued that some such operational procedure is, after all, the way that one judges the presence of mental qualities in human beings other than oneself. The strong-AI viewpoint merely extends this type of judgement from biological structures such as ourselves to other (e.g., electronic) kinds of device. In my own opinion, there is a certain rationale to this. I do not myself see why biological systems should be singled out as the only objects capable of evoking mental qualities. However, I do feel that there is a *different* unjustified assumption being invoked here, namely that all objects in the physical world must necessarily behave algorithmically.

Even Searle, in some of his discussions, seems to be implicitly accepting that electronic computers of the present-day type, but with considerably enhanced speed of action and size of rapid-access store (and possibly parallel action) may well be able to pass the Turing test proper, in the not-too-distant future. He is prepared to accept the contention of strong AI (and of most other "scientific" viewpoints) that "we are the instantiations of any number of computer programs" (Searle *1980,* in Hofstadter and Dennett *1981,* p. 368); moreover, he succumbs to: "Of course the brain is a digital computer. Since everything is a digital computer, brains are too" (ibid., p. 372). Searle maintains that the distinction between the functioning of human brains (which can have minds) and of electronic computers (which, he has argued, cannot) both of which might be executing the same algorithm, lies solely in the material construction of each. He claims, but for reasons he is not able to explain, that the biological objects (brains) can have "intentionality" and "semantics", which he regards as defining characteristics of mental activity, whereas the electronic ones cannot. Just in itself this does not seem to me to point the way towards any helpful scientific theory of mind. What is so special about biological systems, apart perhaps from the "historical" way in which they have evolved (and the fact that *we* happen to be such), which sets them apart as the objects allowed to achieve intentionality or semantics? The claim looks to me suspiciously like a dogmatic assertion, perhaps no less dogmatic, even, than those assertions of strong AI which maintain that the mere enacting of an algorithm can conjure up a state of conscious awareness!

In my opinion Searle, and a great many other people, have been led astray by the computer people. And they, in turn, have been led astray by the physicists. (It is not the physicists' fault. Even *they* don't know everything!) The belief seems to be widespread that, indeed, "everything is a digital computer". It is my intention, eventually, to try to show why, and perhaps how, this need *not* be the case. I shall not be able to develop the full arguments here, however. These must await the completion of my more extended account, but I shall present some indication of them later in this article. For the moment, suppose we accept the commonly believed view that all physical action is indeed algorithmic. We shall see why the strong-AI viewpoint is in some ways a natural development from this.

4. Hardware and Software

In the jargon of computer science, the term *hardware* is used to denote the actual machinery involved in a computer (printed circuits, transistors, wires, etc.), including the complete specification for the way in which everything is connected up. Correspondingly, the term *software* refers to the various programs which can be run on the machine. It was one of Turing's remarkable discoveries that, in effect, any machine whatever, for which the hardware has achieved a certain definite degree of complication and flexibility – that of a *universal* Turing machine – is *equivalent* to any other such machine. This equivalence is to be taken in the sense that for any two such machines A and B there would be a specific piece of software which if given first to machine A would make it act precisely as though it were machine B and, likewise, there would be another piece of software which would make machine B act precisely like machine A. I am using the word "precisely" here to refer to the actual output of the machines for any given input (fed in after the converting software is fed in) and *not* to the *time* that each machine might take to produce that output. I am also allowing that if either machine at any stage runs out of storage space for its calculations then it can call upon some (in principle unlimited) external supply of blank "rough paper" – which could take the form of magnetic tape, discs, drums, or whatever. In fact, the difference in time taken over some task, between the machines A and B, might well be a very serious consideration. It might be the case, for example, that A is much more than a thousand times faster at performing a particular task than is B. It might also be the case that, for the very same machines, there is some other task for which B is over a thousand times faster than A. More-over, these timings could depend very greatly on the particular choices of converting software that are used. This is very much an "in principle" discussion, where one is not really concerned with such practical matters as achieving one's calculations in a reasonable time.

Since, in effect, all modern general purpose computers are universal Turing ma-chines, they are equivalent to one another in the above sense: the differences between them can be entirely subsumed in the software – provided that we are not concerned about differences in the resulting speed of operation and possible limitations on stor-age size. Indeed, modern technology has enabled computers to perform so swiftly and with such vast storage capacities that, for a good many "everyday" purposes, neither of these practical considerations actually represents any serious limitation to what is normally needed, so this effective theoretical equivalence between comput-ers can be seen also at the practical level. Technology has, it seems, transformed entirely academic discussions concerning idealized computing devices into matters which directly affect all of our lives!

As far as I can make out, one of the most important factors underlying the strong-AI philosophy is this equivalence between physical computing devices. The hardware is seen as being relatively unimportant (perhaps even totally unimportant) and the software, i.e., the program, or the algorithm, is taken to be the one vital ingredient. However, it seems to me that there are also other important underlying factors lending some support to this view, these coming more from the direction of physics. I shall next try to give some indication of all this, as I see it.

What is it that gives a particular person his (or her) individual identity? Is it, to some extent, the very atoms that compose his body? Is his identity dependent upon the particular choice of electrons, protons, and other particles that compose those atoms? There are at least two reasons which tell us that this cannot be so. In the first place, there is a continual turnover in the material of any living person's body. This applies in particular to the cells in a person's brain – despite the fact that no new actual brain cells are produced after birth. The vast majority of atoms in each living cell (including each brain cell) are continually being replaced.

The second reason comes from quantum physics – and by a strange irony is, strictly speaking, in contradiction with the first! According to quantum mechanics, any two electrons must necessarily be completely identical with one another, and the same holds for any two protons and for any two particles whatever, of any one particular kind. This is not merely to say that there is no way of telling the particles apart. The statement is considerably stronger than that. If an electron in a person's brain were to be exchanged with an electron in a brick, then the state of the system would be *exactly the same state* as it was before, not merely indistinguishable from it! (The knowledgeable reader who is worrying about an unimportant *sign* difference should imagine that one of the electrons is rotated completely through 360° as we make the interchange!) The same holds for protons and for any other kind of particle, and for whole atoms, molecules, etc. If the entire material content of a person were to be exchanged with corresponding particles in the bricks of his house then, in a strong sense, nothing would have happened whatsoever. What distinguishes the person from his house is the *pattern* of how his constituents are arranged, not the individuality of the constituents themselves.

There is perhaps an analogue of this at an everyday level which is independent of quantum mechanics, but made particularly manifest to me as I write this, by the electronic technology which enables me to type at a word processor. If I desire to change a word, say to transform "make" into "made" I may do this by simply replacing the "k" with "d", or I may choose instead to type out the whole word again. If I do the latter, is the "m" the same "m" which was there before, or have I replaced it with another identical one? What about the "e"? Even if I do simply replace the "k" with "d", rather than retype the word, there is a moment just between the disappearance of "k" and reappearance of "d" when the gap closes and there is (or, at least, sometimes is) a wave of realignment down the page as the placement of every succeeding letter (including the "e") is recalculated, and then re-recalculated as the "d" is inserted. (Oh, the cheapness of mindless calculation in this modern age!) In any case, *all* the letters that I see before me on the screen are mere gaps in the track of an electron beam as the whole screen is scanned some fifty times each second. If I take any letter whatever and replace it with an identical one, is the situation the *same* after the replacement, or merely indistinguishable from it? To try to adopt the second viewpoint (i.e., "merely indistinguishable") as being distinct from the first (i.e., "the same") seems footling. At least, it seems reasonable to call the situation the same when the letters are the same. And so it is with the quantum mechanics of identical particles. To replace one particle by an identical one is actually to have done nothing to the state at all.

The earlier remarks above concerning the continual turnover of atoms in a person's body were made in the context of classical rather than quantum physics. The remarks were worded as though it might be meaningful to maintain the individuality of each atom. In fact classical physics is adequate for this, and we do not go badly wrong, at this level of description, by regarding atoms as individual objects. Provided that the atoms are reasonably well separated from their identical counterparts as they move about, one *can* consistently refer to them as maintaining their individual identities since each atom can be, in effect, tracked continuously, so that one could envisage keeping a tab on each separately. From the point of view of quantum mechanics it would be a convenience of speech only to refer to the individuality of the atoms, but it is a consistent enough description at the level just considered.

Let us accept that a person's individuality has nothing to do with any individuality that one might try to assign to his material constituents. Instead, it must have to do with the *configuration,* in some sense, of those constituents – let us say the configuration in space or in space-time. But the supporters of strong AI go further than this. If the information content of such a configuration can be translated into another form from which the original can again be recovered then, they claim, the person's individuality must remain intact. It is like the sequences of letters I have just typed and now see displayed on the screen of my word processor. If I move them off the screen, they remain coded in the form of certain tiny displacements of electric charge, in some configuration in no clear way geometrically resembling the letters I have just typed. Yet, at any time I can move them back on the screen, and there they are, just as though no transformation had taken place. If I choose to save the part I have just written, then I can transfer the information of the sequences of letters into configurations of magnetization on a disc which I can later remove, and by switching off the machine I neutralize all the (relevant) tiny charge displacements in it. Tomorrow, I can reinsert the disc, reinstate the little charge displacements, and display the letter sequences again on the screen, just as though nothing had happened. To the strong-AI supporters, it is "clear" that a person's individuality can be treated in just the same way. Like the sequences of letters on my display screen, they would claim, nothing is lost of a person's individuality – indeed nothing would really have happened to it at all – if his (or her) physical form were to be translated into something quite different, say into fields of magnetization in a block of iron. They are even driven to claim that the person's conscious awareness would persist while the person's "information" is in this other form. In this view, a "person" is to be taken, in effect, as a piece of software, and his particular manifestation as a material human being is to be taken as the operation of this software by the hardware of his body.

It seems that the reason for these claims is that, whatever material form the hardware takes – for example some electronic device – one could always "ask" the software questions (in the manner of a Turing test), and assuming that the hardware performs satisfactorily in computing the replies to these questions, these replies would be identical to those that the person would make whilst in his normal state. ("How are you feeling this morning?" "Oh, fairly well, thank you, though I have a slightly bothersome headache." "You don't feel, then, that there's ... er ... anything odd about your personal identity ... or something?" "No; why do you say that? It seems to be a rather strange question to be asking." "Then you feel yourself to be the same person that you were yesterday?" "Of course I do!")

5. Mathematical Platonism

Let us try to go along, for a while, with the strong-AI view that the algorithm (or software) is all that counts. Do algorithms have some kind of existence on their own as mathematical objects, independent of any particular physical form? This question returns us to the matter of the Platonic existence of mathematical objects – in particular, of algorithms – and the issue of whether such "existence" might, according to the strong-AI philosophy, be in itself sufficient to give an algorithm conscious feelings! One is led, here, into questions of the philosophy of mathematics. Do mathematical concepts have any (Platonic) existence beyond their mere presence in the minds of mathematicians? If mathematical concepts could exist only in minds, then it would not be of much use attempting to explain minds in terms of algorithms – using a concept of an algorithm which can itself only exist in a mind! One might take the view that mathematical concepts could also exist in terms of some other physical embodiment, such as on the printed page or in a computer memory. But there is the difficulty here that such physical embodiments depend upon the language in which they are written. This language must be "known" in order that the mathematical concept being described be actually well defined – but "known" by whom? It seems that one is led back to the necessary presence of a mind in order to do the "knowing", and one is no better off than before.

To avoid such circularity, it seems that strong-AI supporters must take a more Platonic view of mathematics than that. Let us now address this question, and try to examine how plausible the Platonic view is. How "real", indeed, *are* the objects of the mathematician's world? From one point of view it seems that there can be nothing real about them at all. Mathematical objects are just concepts; they are the mental idealizations that mathematicians make, often stimulated by the appearance and seeming order of aspects of the world about us, but mental idealizations nevertheless. Can they be other than mere arbitrary constructions of the human mind? On the other hand, there often does appear to be some profound reality about these mathematical concepts, going quite beyond the mental deliberations of any particular mathematician. It is as though human thought is, instead, being guided towards some eternal external truth – a truth which has a reality of its own, and which is revealed only partially to any one of us.

The familiar Mandelbrot set provides a striking example. Its wonderfully elaborate structure was not the invention of any one person – nor was it the design of a team of mathematicians. Benoit Mandelbrot himself, who first studied the set, had no real prior conception of the fantastic elaboration inherent in it, although he certainly knew that he was on the track of something very interesting. Indeed, when his first computer pictures began to emerge, he was under the impression that the fuzzy structures that he was seeing were the result of a computer malfunction (Mandelbrot *1986*)! Only later did he become convinced that they were really there in the set itself. Moreover, the complete details of the complication of the structure of Mandelbrot's set cannot really be fully comprehended by any one of us, nor can it be fully revealed by any computer. It would seem that this structure is not just part of our minds, but that it has a reality of its own. Whichever mathematician or computer buff chooses to examine the set, approximations to the *same* fundamental mathematical structure will be found. It makes no real difference which computer is used for performing calculations (provided that the computer is in accurate working order), apart from the fact that differences in computer speed and storage, and graphic display capabilities, may lead to differences in the amount of fine detail that will be revealed and in the speed with which that detail is produced. The computer is being used in essentially the same way that the experimental physicist uses a piece of experimental apparatus to explore the structure of the physical world. The Mandelbrot set is not an invention of the human mind. It was a discovery. Like Mount Everest, the Mandelbrot set is just *there*!

Likewise, the very system of complex numbers, in terms of which that set is described, has a profound and timeless reality which goes quite beyond the mental constructions of any particular mathematician. The beginnings of an appreciation of complex numbers came about in the sixteenth century with the work of Gerolamo Cardano. In this he put forward the first complete expression for the solution (in terms of surds) of a general cubic equation. He noticed, however, that in a certain class of cases – the ones referred to as "irreducible", where the equation has three real solutions – he was forced to take, at a certain stage in his expression, the *square root of a negative number*. Although this was puzzling to him, he realized that if he allowed himself to *take* such square roots, and *only* if so, then he could express the full *real* answer!

While at first it may seem that the introduction of such roots of negative numbers was just a device – a mathematical invention designed to achieve a specific purpose –

it later becomes clear that these objects are achieving far more than that for which they were originally designed. Although the original purpose of introducing complex numbers was to enable square roots to be taken with impunity, by introducing such numbers we find that we get, completely free, the potentiality for taking any other kind of root or of solving any algebraic equation whatever. Later we find many other magical properties that these complex numbers possess – properties that we had no inkling about at first. These properties are just *there*. They were not put there by Cardano – nor by Bombelli, nor Wallis, nor Cotes, nor Euler, nor Wessel, nor Gauss, despite the undoubted farsightedness of these and other great mathematicians – but such magic was inherent in the very structure that they gradually uncovered. When introducing his complex numbers, Cardano could have had no inkling of the many other magical properties which were to follow – properties which go under various names, such as: the Cauchy integral formula, the Riemann mapping theorem, the Lewy extension property. These, and many other equally remarkable facts, are properties of the very numbers, with no additional modifications whatever, that Cardano had first encountered in about 1539.

All this suggests that the finding of mathematical results is a matter of *discovery* and not invention. Yet the matter is perhaps not quite so straight-forward as this. As I have said, there are things in mathematics for which the term "discovery" is indeed much more appropriate than "invention", such as the examples just cited. These are the cases where much more comes out of the structure than is put into it in the first place. One may take the view that in such cases the mathematicians have stumbled upon "works of God". However, there are other cases where the mathematical structure does not have such a compelling uniqueness, such as when in the midst of a proof of some result, the mathematician may find the need to introduce some complicated and far from unique construction in order to achieve one very specific end. In such cases no more is likely to come out of the construction than was put into it in the first place, and the word "invention" seems more appropriate than "discovery". These are indeed just "works of man". On this view, the true mathematical discoveries would, in a general way, be regarded as greater achievements or aspirations than would the "mere" inventions.

Such categorizations are not entirely dissimilar from those that one might use in the arts or in engineering. Great works of art are indeed "closer to God" than are lesser ones. It is a feeling not uncommon amongst artists that in their greatest works they are revealing eternal truths which have some kind of prior ethereal existence, whilst their lesser works might be more arbitrary, of the nature of mere mortal constructions. Likewise, an engineering innovation with a beautiful economy, where a great deal is achieved in the scope of the application of some simple unexpected idea, might appropriately be described as a discovery rather than an invention.

Having made these points, however, I cannot help feeling that, with mathematics, the case for believing in some kind of external existence, at least for the more profound mathematical concepts, is a good deal stronger than in those other cases. There is a compelling uniqueness and universality in such mathematical ideas which seems to be of quite a different order from that which one could expect in the arts or engineering.

How does this relate to the issues of strong AI, according to which mental phenomena are supposed to find their existence within the mathematical idea of an algorithm? In my opinion, the actual concept of an algorithm is indeed a profound and "God-given" notion. Does this viewpoint lend some credence to the strong AI point of view, by providing the possibility of an ethereal type of existence for mental phenomena? Just conceivably so; but if mental phenomena can indeed find a home of this general kind, I do not believe that it can be with the concept of an algorithm. For one thing, strong AI attributes mental significance to *specific* algorithms rather than to be the more "God-given" concept of a *general* algorithm. What would be needed would be something very much more subtle than specific algorithms. The fact that algorithmic things indeed constitute a very narrow and limited part of mathematics will be an important aspect of the later discussions.

6. *Gödel's Theorem*

Are there more clear-cut reasons than these for doubting that conscious thought is merely the enacting of an algorithm? I believe that there are. In particular, it would appear that our (conscious) perception of mathematical truth goes demonstrably beyond anything that can be formulated in algorithmic terms. Gödel's theorem seems to show this, since Turing effectively established that the algorithmic decidability of an area of mathematics is equivalent to its axiomatizability, according to Hilbert's program. Gödel's result tells us that our perceptions of mathematical truth indeed cannot be axiomatized. (Compare Lucas *1961.* Counterarguments to Lucas's line of reasoning have often been put forward, e.g., Benacerraf *1967,* Good *1969,* Bowie *1982,* Hofstadter *1979.* In a general way, I am in sympathy with the Lucas type of viewpoint, though my own viewpoint is perhaps a little different. I do not think that the arguments are conclusive either way. Some of Lucas's critics seem to make the assumption that the behavior of any physical object – e.g., Lucas himself – must be governed by *some* algorithm. It is this assumption that I am disputing!)

Let us briefly recall the Hilbert program: each clear-cut body of mathematical thought is to be encapsulated in a system of axioms and rules of procedure. The axioms are to be expressed as strings of symbols belonging to some precisely defined alphabet, the rules of procedure being some clear set of allowed ways of deriving new such strings of symbols from the ones already obtained. But the hopes of David Hilbert and his followers were dashed by Kurt Gödel in 1931. Gödel showed that any such precise ("formal") mathematical system of axioms and rules of procedure *whatever,* provided that it is broad enough to contain descriptions of simple arithmetical propositions (like "Fermat's last theorem") and provided that it is free from contradiction, must contain some statements which are neither provable nor disprovable by the means allowed within the system. It will be important for us to understand the nature of this "undecidability" and see why Gödel's argument cut to the very core of the Hilbert program. We must see how Gödel's argument indeed enables us, by the use of conscious contemplation and insight, to go beyond the limitations of any particular formalized mathematical system.

According to Hilbert's program, it is in principle possible to check, within such a formalized system, whether any mathematical proof has been correctly given. The hope had been that such a system could be *complete,* and this would have enabled one in principle to decide the truth or falsity of any mathematical statement that could be formulated within the system. Thus for any correctly formed proposition, say, P, one should be able to prove either P or else $\sim P$ (where "$\sim P$" stands for "not P"), depending upon whether P is true or false. In fact, this would even enable us to dispense with worrying about what the propositions *mean* altogether! P would just be a string of symbols, correctly assembled according to certain prescribed ("syntactical") rules (i.e., brackets paired off correctly, etc.). The string of symbols P would be assigned the truth-value *"true"* if P is a theorem (i.e., if P is provable within the system) and it would be assigned the truth-value *"false"* if, on the other hand, $\sim P$ is a theorem. For this to make sense, we require *consistency* in addition to completeness. That is to say, there must be no string of symbols P for which *both* of P and $\sim P$ are theorems. Otherwise P could be *"true"* and *"false"* at the same time.

The point of view that one can dispense with the meanings of mathematical statements, and just regard such statements as being nothing but strings of symbols in some formal mathematical system, is the mathematical standpoint of *formalism.* Some people like that idea. However, it is not one that appeals to me at all. Fortunately, Gödel dealt it a devastating blow! Let us run through his argument to see how he did this.

Part of Gödel's argument was very detailed and complicated, but it is not necessary for us to examine the intricacies of that part. The central idea, on the other hand, was simple, beautiful, and profound. That part we shall be able to appreciate. The complicated part (which also contained much ingenuity) was to show in detail how one may actually code the individual rules of procedure of the formal system, and also the use of its various axioms, into *arithmetical operations.* (It was an aspect of the profound part, though, to realize that this was a fruitful thing to do!) In order to carry out this coding we need to find some convenient way of labeling propositions with natural numbers. One way would be simply to use some kind of "alphabetical" ordering for all the strings of symbols of the formal system for each specific length, where there is an overall ordering according to the length of the string. (Thus, the strings of length one could be alphabetically ordered, followed by the strings of length two, alphabetically ordered, followed by the strings of symbols of length three, etc.) This amounts to using the notation for the strings of symbols in the formal system as a notation for the *natural numbers* expressed in a "base" given by the number of symbols of the system. (In fact Gödel originally used a rather more complicated system.) We shall be particularly concerned with (syntactically correctly formulated) *propositional functions* (propositions dependent on numerical variables) which are dependent on a *single* variable. Let the *n*-th such propositional function (in the chosen ordering of strings of symbols), applied to w, be

$$P_n(w).$$

Then $P_n(w)$ will be some perfectly well-defined particular arithmetical statement concerning the two natural numbers n and w. Precisely *which* arithmetical statement it is will depend on the details of the particular numbering system that has been chosen. That belongs to the complicated part of the argument and will not concern us here. The strings of propositions which constitute a *proof* of some theorem in the system can also be labeled by natural numbers using the chosen ordering scheme. Let

$$\Pi_n$$

denote the *n*-th proof.

Now consider the following propositional function, which depends on the natural number w (where "$\exists$" means "there exists ..."):

$$\sim \exists x [\Pi_x \text{ proves } P_w(w)].$$

The statement in the square bracket is given partly in words, but it is a perfectly precisely defined statement. It asserts that the x-th proof is actually a proof of that proposition which is $P_w(\ )$ applied to the value w itself. In fact, because of the translations into arithmetic that we are supposing have been carried out, it is actually some *arithmetical* statement concerning the two natural numbers x and w. As before, precisely *which* arithmetical statement it is will depend on the details of the numbering systems, and it will depend very much on the detailed structure of the axioms and rules of our formal system. Since all that belongs to the complicated part of the argument, the details of it will not concern us here. Outside the square bracket the existential quantifier ($\exists$) serves to remove one of the variables, so we end up with an arithmetical propositional function which depends on only the one variable w. We have numbered all propositional functions which depend on a single variable, so the one we have just written down must have been assigned a number. Let us suppose that this number is k. Then we have

$$\sim \exists x [\Pi_x \text{ proves } P_w(w)] = P_k(w).$$

Now examine this for the particular w-value: $w = k$. We get

$$\sim \exists x [\Pi_x \text{ proves } P_k(k)] = P_k(k).$$

The specific proposition $P_k(k)$ is a perfectly well-defined arithmetical statement. Does it have a proof within our formal system? Does its negation $\sim P_k(k)$ have a proof? The answer to both of these questions must be "no". We can see this by examining the *meaning* underlying the Gödel procedure. Although $P_k(k)$ is just an arithmetical proposition, we have constructed it so that it asserts what has been written on the left-hand side: "there is no proof, within the system, of the proposition $P_k(k)$". If we have been careful in laying down our axioms and rules of procedure, and assuming that we have done our numbering right, then it cannot be possible to find a proof of this $P_k(k)$ within the system. For if it were, then the "meaning" of the statement that $P_k(k)$ actually asserts would be false, so $P_k(k)$ would have to be false as an arithmetical proposition. Our formal system should not be so badly constructed that it actually allows false propositions to be proved! Thus, it must be the case that there is in fact *no* proof of $P_k(k)$. But this is precisely what $P_k(k)$ is trying to tell us. What $P_k(k)$ asserts must therefore be a *true* statement, so $P_k(k)$ must be true as an arithmetical proposition. It follows that we had also better not be able to find a proof of its negation $\sim P_k(k)$. We are not supposed to be able to prove false propositions within our formal system, and we have now seen that $\sim P_k(k)$ is certainly false! Thus, neither $P_k(k)$ nor $\sim P_k(k)$ is provable within our formal system. This establishes Gödel's theorem.

7. Mathematical Insight

Notice that something very remarkable has happened here. People often think of Gödel's theorem as something negative – showing the necessary limitations of formalized mathematical reasoning. No matter how comprehensive we think we have been, there will always be some propositions which escape the net. But should the particular proposition $P_k(k)$ worry us? As an essential ingredient of the above argument, we have actually established that $P_k(k)$ is a *true* statement! Somehow we have managed to *see* that $P_k(k)$ is true despite the fact that it is not formally provable within the system. The strict mathematical formalists *should* indeed be worried, because by this very reasoning we have established that the formalist's notion of "truth" must be necessarily incomplete. *Whatever* formal system is used, there are statements that we can see are true but which do not get assigned the truth-value *"true"* by the formalist's proposed procedure, as described above. The way that a strict formalist might try to get around this would perhaps be not to talk about the concept of truth at all but merely refer to *provability* within some fixed formal system. However, this seems very limiting. (One could not even frame the particular form of the Gödel argument as given above, using that point of view, since parts of that argument made use of reasoning about what is actually true and what is not true.) Some formalists take a more "pragmatic" view, claiming not to be worried by statements such as $P_k(k)$ because they are extremely complicated and uninteresting as propositions of arithmetic. Such people would assert:

"Yes, there is the odd statement, such as $P_k(k)$, for which my notion of provability or *"truth"* does not coincide with your instinctive notion of truth, but those statements will never come up in serious mathematics (at least not in the kind I am interested in) because such statements are absurdly complicated and unnatural as mathematics."

It is indeed the case that the statement $P_k(k)$ would be an extremely cumbersome and odd-looking mathematical proposition when written out in full. But there are closely related types of Gödel statement for which this is not so (cf. Paris and Harrington *1977*). The above seems to me to be a very strange point of view to adopt for a philosophy of mathematics. Moreover, it is not really all that pragmatic. When mathematicians carry out their forms of reasoning, they do not want to have to be continually checking to see whether or not their arguments can be formulated in terms of the axioms and rules of procedure of some formal system. They only need to be sure that their arguments are valid ways of ascertaining truth. The Gödel argument is another such valid procedure, so it seems to me that $P_k(k)$ is just as good a mathematical truth as any that can be obtained more conventionally using the axioms and rules of procedure that can be laid down beforehand.

Perhaps there are numerous perfectly acceptable results in the mathematical literature whose proofs require insights equally far from the rules and axioms of the original standard formal systems for arithmetic. Mathematical insights – by which I mean the means by which the mathematician arrives at his or her judgements of truth – are not something amenable to the procedures of formalization, and it would seem, therefore, that they are not the results of algorithmic operation.

Be that as it may, it is a clear consequence of the Gödel argument that mathematical truth is *not* an algorithmic matter; and it is something that goes beyond mere formalism. This is perhaps clear even without Gödel's theorem. For how are we to decide what axioms or rules of procedure to adopt in any case when trying to set up a formal system? Our guide in deciding on the rules to adopt must always be our intuitive understanding of what is "self-evidently true", given the "meanings" of the symbols of the system. How are we to decide which formal systems are sensible ones to adopt – i.e., in accordance with our intuitive feelings about "self-evidence" and "meaning" – and which are not? The notion of self-consistency of the system is certainly not adequate for this. One can have many self-consistent systems which are not "sensible" in this sense. "Self-evidence" and "meaning" are concepts which would still be needed, even without Gödel's theorem.

However, without Gödel's theorem it might have been possible to imagine that the intuitive notions of "self-evidence" and "meaning" could have been employed just once and for all, merely to set up the formal system in the first place, and thereafter dispensed with as part of clear mathematical argument for determining truth. Perhaps these "vague" intuitive notions could have roles to play as part of the mathematician's preliminary thinking, as a guide towards finding the appropriate formal argument; but they would play no part in the actual demonstration of mathematical truth. Gödel's theorem shows that this point of view is not really a tenable one in a fundamental philosophy of mathematics. The notion of mathematical truth goes beyond the whole concept of formalism. There is something absolute and "God-given" about mathematical truth. This is what mathematical Platonism is about. Any particular formal system has a provisional and "man-made" quality about it. Such systems indeed have very valuable roles to play in mathematical discussions, but they can supply only a partial (or approximate) guide to truth. Real mathematical truth goes beyond these man-made constructions.

8. *Some Examples of Nonalgorithmic Mathematics*

There are very many areas of mathematics where problems arise which are not algorithmic. Thus, we may be presented with a class of problems to which the answer for each individual problem is either "yes" or "no", but for which no general algorithm exists for deciding which of these two answers is actually valid in each case. Some of these classes of problem are remarkably simple-looking.

First, consider the problem of finding integer solutions of systems of algebraic equations with integer coefficients. Such a set of equations might be

$$z^3 - y - 1 = 0, \quad yz^2 - 2x - 2 = 0, \quad y^2 - 2xz + z + 1 = 0$$

and the problem is to decide whether or not they can be solved for *integer* values of x, y, and z. In fact, in this particular case they can, a solution being given by

$$x = 13, \quad y = 7, \quad z = 2.$$

However, there is no algorithm for deciding this question for an arbitrary set of Diophantine equations. Diophantine arithmetic, despite the elementary nature of its ingredients, is part of nonalgorithmic mathematics!

As a somewhat less elementary-looking example, consider the question of topological equivalence of manifolds. Now, for ordinary surfaces – of *two* dimensions (say, closed and orientable) – there is, in fact, an algorithm for deciding whether or not two of them are topologically identical. This amounts, in effect, to counting the number of "handles" that each surface has. If the number of handles is the same in each case, then the two surfaces are topologically the same. Otherwise the surfaces are topologically different.

Now, suppose that we ask the same question for "surfaces" of *three* dimensions. Disconcertingly, the answer to this question is not known (at the time of writing). Let us pass on to the case of manifolds of *four* dimensions: is there an algorithm for 4-manifolds? Now the answer *is* known! There is *no* such algorithm. There is no algorithm for 5-manifolds either, nor for 6-manifolds, nor for any higher dimension. The question on the case of 4-manifolds is conceivably of some relevance to physics, since according to Einstein's general relativity, space and time together constitute a 4-manifold, and it has been suggested that this nonalgorithmic property might have relevance to "quantum gravity" (Geroch and Hartle *1986*). The fact that there is no algorithm for deciding whether two 4-manifolds are the same or not entails that, in a certain sense, 4-manifolds cannot be completely systematically classified. There is an unending variety of different kinds of structure that can occur.

Let us consider a different type of problem – one which is rather simpler to appreciate, namely the *word problem* (for semigroups). Suppose that we have some alphabet of symbols, and we consider various strings of these symbols, referred to as *words*. The words need not in themselves have any meaning, but we shall be given a certain (finite) list of "equalities" between them which we are allowed to use in order to derive further such "equalities". This is done by making substitutions of words from the initial list into other (normally longer) words which contain them as portions. Each such portion may be replaced by another portion which is deemed to be equal to it according to the list. The problem is then to decide, for some given pair of words, whether or not they are "equal" according to these rules.

As an example, we might have, for our initial list:

$$EAT = AT$$
$$ATE = A$$
$$LATER = LOW$$
$$PAN = PILLOW$$
$$CARP = ME$$

From these we can derive, for example,

$$LAP = LEAP$$

by use of successive substitutions from the second, the first, and again the second of the relations from the initial list:

$$LAP = LATEP = LEATEP = LEAP.$$

The problem now is, given some pair of words, can we get from one to the other simply using such substitutions? Can we, for example, get from CATERPILLAR

to MAN, or, say, from CARPET to MEAT? The answer in the first case happens
to be "yes", while in the second it is "no". When the answer is "yes", the normal
way to show this would be simply to exhibit a string of equalities where each word
is obtained from the preceding one by use of an allowed relation. Thus (indicating
the letters about to be changed in bold type, and the letters which have just been
changed in italics):

$$\text{CA\textbf{T}ERPILLAR}= \text{CA}\textit{R}\text{PILLAR} = \text{CARPIL}\textbf{\textit{LATER}} = \text{CARP}\textbf{ILLOW}$$
$$= \textbf{CARP}\textit{AN} = \textit{ME}\text{AN} = \text{ME}\textbf{AT}\textit{E}\text{N} = \text{M}\textit{AT}\text{EN}$$
$$= \text{MAN}.$$

How can we tell that it is impossible to get from CARPET to MEAT by means
of the allowed rules? For this, we need to think a little more, but it is not too hard
to see, in a variety of different ways. The simplest appears to be the following: in
every "equality" in our initial list, the number of A's plus the number of W's plus
the number of M's is the same on each side. Thus the total number of A's, W's and
M's cannot change throughout any succession of allowed substitutions. However,
for CARPET this number is 1 whereas for MEAT it is 2. Consequently, there is no
way of getting from CARPET to MEAT by allowed substitutions.

Notice that when the two words are "equal" we can show this simply by exhibiting an allowed formal string of symbols, using the rules that we had been given;
whereas in the case where they are "unequal", we had to resort to arguments *about*
the rules that we had been given. There is a clear algorithm that we can use to
establish "equality" between words whenever the words *are* in fact "equal". All
we need do is to make some appropriate "alphabetical" listing of all the possible
sequences of words of given length – and then successively increasing the length,
as above; and then strike from this list any such string for which there is a pair of
consecutive words where the second does not follow from the first by an allowed
rule. The remaining sequences will provide all the sought-for "equalities" between
words. However, there is no such obvious algorithm, in general, for deciding when
two words are *not* "equal", and we may have to resort to "intelligence" in order
to establish that fact. (Intelligence, incidentally, is useful also for establishing the
existence of an "equality", but in that case it is not *necessary*!)

In fact, for the particular list of five "equalities" that constitute the initial list in
this particular case, it is not unduly difficult to provide an algorithm for ascertaining
that two words are "unequal" when they are indeed "unequal". However, in order
to *find* the algorithm that works in this case we should have to exercise a certain
amount of intelligence! For it turns out that there is no general algorithm which
can be used universally for *all* possible choices of initial list. In this sense there is
no algorithm for solving the word problem. The general word problem belongs to
nonalgorithmic mathematics!

There are even certain *particular* selections of initial list for which there is no algorithm for deciding when two words are unequal. One such is given by

$$AH = HA$$
$$HE = EH$$
$$AT = TA$$
$$TE = ET$$
$$TAI = IT$$
$$HEI = IH$$
$$THAT = ITHT$$

(This list is adapted from one given by Tseitin and Scott *1955*.) Thus this *particular* word problem belongs to nonalgorithmic mathematics – in the sense that using this particular initial list we cannot algorithmically decide whether or not two given words are "equal". This type of word problem arose from considerations of formalized mathematical logic ("formal systems", etc.). The initial list plays the role of an axiom system and the substitution rule for words, the role of the formal rules or procedure. The proof that the word problem is nonalgorithmic arises from such considerations.

As a final example of a problem in mathematics which is nonalgorithmic, let us consider the question of covering the Euclidean plane with polygonal shapes, where we are given a finite number of different such shapes and we ask whether it is possible to cover the plane completely, without gaps or overlaps, just using these shapes and no others. Such an arrangement of shapes is called a *tiling* of the plane (Grünbaum and Shephard *1986*). We are all familiar with the fact that such tilings are possible using just squares, or just equilateral triangles, or just regular hexagons. There are many other single shapes which will tile the plane, some of which are quite elaborate. With a *pair* of shapes, the tilings can become more elaborate still. In fact, various pairs of shapes are known which will tile the plane only nonperiodically. For a *general* set of tiles there is no decision procedure (Berger *1966*, cf. also Robinson *1971*): the tiling problem, *also*, is part of nonalgorithmic mathematics! It is perhaps remarkable that such an apparently "trivial" area of mathematics – namely covering the plane with congruent shapes – which seems almost like "child's play" should in fact be part of nonalgorithmic mathematics. (The problem, as I have stated it, is actually somewhat more involved than the one treated by Berger. I am allowing polygons of general shape, whereas he used tiles based in squares, without rotations of the tiles. With the more general shapes, one would need some adequately computable way of displaying the individual tiles. One way of doing this would be to give their vertices as points in the Argand plane given as algebraic numbers. The issues raised by these distinctions are not important for us here.)

9.   *The Relevance of Complexity Theory*

The arguments that I have given above, and in the preceding chapters, concerning the nature, existence, and limitations of algorithms have been very much at the "in principle" level. I have not discussed the question of whether any algorithms which arise are likely to be in any way practical. Even for problems where it is clear that algorithms exist and how such algorithms can be constructed, it may require much ingenuity and hard work to develop the algorithms into something usable. Sometimes a little insight and ingenuity will lead to considerable reductions in the complication of an algorithm and sometimes to absolutely enormous shortening of its length of operation. These questions are often very detailed and technical, and it would not be pertinent for me to attempt to enter into a detailed discussion of them. There are various general things that are known, or conjectured, concerning certain *absolute* limitations on how much the speed of an algorithm can be increased. It turns out that even among mathematical problems that *are* algorithmic in nature, there are some classes of problems that are intrinsically vastly more difficult to solve algorithmically than others. The difficult ones can be solved only by very slow algorithms (or perhaps, with algorithms which require an inordinately large amount of storage space, etc.). The theory which is concerned with questions of this kind is called *complexity theory.*

Complexity theory is concerned not so much with the difficulty of solving *single* problems algorithmically but with infinite classes of problems where there is a general algorithm for finding answers to all the problems of the class. The different problems within such a class would have different "sizes", where the size of a problem is measured by some natural number n. The length of time – or more correctly, the number of elementary steps – that the algorithm would need for each particular problem of the class would be some natural number N which depends on n. To be a little more precise, let us say that among *all* of the problems of some particular size n, the greatest number of steps that the algorithm takes is N. Now, as n gets larger and larger, the number N is likely to get larger and larger too. Complexity theory has to do with the *rate of growth* of N, as n increases.

These questions are important for our considerations because they raise another issue, somewhat separate from that of whether or not things are algorithmic: namely, whether or not things that are known to be algorithmic are actually algorithmic in a *useful* way. I am inclined to think (though, no doubt, on quite inadequate grounds) that, unlike the basic question of computability itself, the issues of complexity theory are not quite the central ones in relation to mental phenomena. Moreover, I feel that the questions of practicality of algorithms are being only barely touched by complexity theory as it stands today.

However, I could well be wrong about the role of complexity. Indeed, the complexity theory for *actual physical objects* could perhaps be different in significant ways from the standard complexity theory of algorithms. For such a possible difference to become manifest, it seems that it would be necessary to harness some of the remarkable properties of *quantum mechanics*. According to a recent set of ideas introduced by David Deutsch (*1985*), it seems to be possible *in principle* to construct a "quantum computer" for which there are (classes of) problems which could be solved by that device in polynomial time (i.e., with N increasing not faster than some polynomial in n) yet not by any Turing machine in polynomial time. It is not all clear, as of now, how an actual physical device could be constructed which behaves (reliably) as a quantum computer – and, moreover, the particular class of problem so far considered is decidedly artificial – but the *theoretical* possibility that a quantum physical device may be able to improve on a Turing machine seems now to be definitely with us.

10. *Quantum Brains?*

Can it be that a human brain, which I am taking for this discussion to be a "physical device" – albeit one of amazing subtlety, and delicacy of design, as well as of complication – is itself taking advantage of the magic of quantum theory? Do we yet understand the ways in which quantum effects might be used for benefit in the solving of problems or the forming of judgements? Is it conceivable that we might have to go even "beyond" present-day quantum theory to make use of such possible advantages? Is it really likely that actual physical quantum devices might be able to improve on the complexity theory for Turing machines?

What about the *computability* theory for actual physical devices? According to Deutsch's analysis, the class of computable operations is the *same* for a quantum computer as it is for a Turing machine. Consequently, one would need to go beyond standard quantum theory in order to be able to describe a device that could operate in a noncomputable way. This seems to be a tall order. Most physicists and physiologists would claim that there is no evidence even that ordinary quantum effects could be of any significance in the operation of a human brain, let alone putative effects that might lie outside standard quantum theory.

To address these matters I shall turn to a philosophical issue that I believe lends some support for quantum theory being relevant to our feelings of conscious awareness. Afterwards, I shall indicate why I believe that one must, indeed, go *beyond* standard quantum theory for a full picture.

An idea sometimes discussed in relation to conscious perception is the *teleportation machine* of science fiction (cf. Hofstadter and Dennett *1981*). This is intended as a means of "transportation" from, say, one planet to another – but whether it actually would be such is what the discussion is all about. Instead of being physically transported by a spaceship in the "normal" way, the would-be traveler is scanned from head to toe, the accurate location and complete specification of every atom in his body being recorded in full detail. All this information is then beamed (at the speed of light) by an electromagnetic signal to the distant planet of intended destination. There the information is collected and used as the instructions to assemble a precise duplicate of the traveler – together with all his memories, his intentions, his hopes, and his deepest feelings. At least that is what is expected; for every detail of the state of his brain has been faithfully recorded, transmitted, and reconstructed. Assuming that the mechanism has worked, the original copy of the traveler can be "safely" destroyed. Of course the question is: is this *really* a method of traveling from one place to another or is it merely the construction of a duplicate, together with the murder of the original? Would *you* be prepared to use this method of "travel" – assuming that the method had been shown to be completely reliable, within its terms of reference? If teleportation is *not* traveling, then what is the difference *in principle* between it and just walking from one room into another? In the latter case, are not one's atoms of one moment simply providing the information for the locations of the atoms of the next moment? We have seen, after all, that there is no significance in preserving the identity of any particular atom. The question of the identity of any particular atom is not even meaningful. Does not any moving pattern of atoms simply constitute a kind of wave of information propagating from one place to another? Where is the essential difference between the propagations of waves which describes our traveler ambling in a commonplace way from one room to another and that which takes place in the teleportation device?

Suppose it is true that teleportation *does* actually "work", in the sense that the traveler's own "awareness" is actually reawakened in the copy of himself on the distant planet (assuming that this question has genuine meaning). What would happen if the *original* copy of the traveler were not, after all, destroyed as the rules of this game demand? Would his "awareness" be in two places at once? (Try to imagine your response to being told the following: "Oh dear, so the drug we gave you before placing you in the Teleporter has worn off prematurely, has it, sir? That is a little unfortunate, but no matter. Anyway, you will be pleased to hear that the other you – er, I mean the *actual* you, that is – has now arrived safely on Venus, so we can, er, dispose of you here – er, I mean of the *redundant* copy here. It will, of course, be quite painless.") The situation has the air of paradox about it. Is there anything in the laws of physics which could render teleportation *in principle* impossible? Perhaps, on the other hand, there is nothing in principle against transmitting a person – and a person's consciousness – by such means, but the "copying" process involved inevitably destroys the original? Might it then be that the preserving of *two* viable copies is what is in principle impossible?

I believe that despite the outlandish nature of these considerations, there *is* perhaps something of significance concerning the physical nature of consciousness and individuality to be gained from them. I believe that they provide one pointer, indicating a certain essential role for *quantum mechanics* in the understanding of mental phenomena. For it is a feature of standard quantum mechanics that a quantum state cannot simply be copied – unless the original state is destroyed in the process. If a mental state is like a quantum state, then the "teleportation paradox" may perhaps be resolved in this way (cf. Penrose *1987*).

Let us see how the point of view of strong AI relates to the teleportation question. We shall suppose that somewhere between the two planets is a relay station, where the information is temporarily stored before being retransmitted to its final destination. For convenience, this information is not stored in human form, but in some magnetic or electronic device. Would the traveler's "awareness" be present in association with this device? The supporters of strong AI would have us believe that this must be so. After all, they say, any question that we might choose to put to the traveler could in principle be answered by the device, by "merely" having a simulation set up for the appropriate activity of his brain. The device would contain all the necessary information; and the rest would just be a matter of computation. Since the device would reply to questions just exactly as though it were the traveler, then (Turing test!) it would *be* the traveler. This all comes back to the strong-AI assertion that the actual hardware is not important with regard to mental phenomena. This assertion seems to me to be unjustified. It is based on the presumption that the brain (or the mind) is, indeed, a digital computer. It assumes that no specific physical phenomena are being called upon, when one actually thinks, which might demand the particular physical (biological, chemical) structure that brains actually have.

No doubt it would be argued (from the strong-AI point of view) that the only assumption that is really being made is that the effects of any specific physical phenomena which need to be called upon can always be accurately *modeled* by digital calculations.

I feel fairly sure that most physicists would argue that such an assumption is actually a very natural one to make on the basis of our present physical understandings. However, it is precisely this assumption that I am disputing. If we accept this (commonly held) view that all the relevant physics *can* always be modeled by digital calculations, then the only real assumption (apart from questions of time, and calculation space) is the "operational" one that if something *acts* entirely like a consciously aware entity, then one must also maintain that it *"feels"* itself to be that entity.

11. Does Physics Allow Nonmechanisms?

But if our conscious thinking involves nonalgorithmic physical action, there would need to be scope for nonalgorithmic behavior in physical laws. This is assuming that – as I indeed believe – the actions of our brains or minds are indeed to be understood in terms of physical laws. What evidence is there that there is anything essentially nonalgorithmic that can be harnessed from the laws of physics? In discussing this question, it is necessary to make clear which laws one is regarding as relevant. We must ask what physics is likely to be actually involved in mental activity. Is Newtonian physics sufficient? Can we adequately regard our brains as functioning according to the laws governing the collisions and attractions between "Newtonian billiard balls"? Is there anything essentially new involved when we include the fields of Maxwell's electrodynamics? Does Einstein's special relativity have any significant role to play, or do the absurdly tiny space-time curvatures that arise because of the effects of his *general* relativity? What about quantum effects? Is there any conceivable reason to believe that quantum theory has relevance in the functioning of such a large and "warm" object as a human brain?

Even if so, do these more subtle physical theories provide any useful scope for nonalgorithmic action? Though there is little which is actually proved in this direction, I believe that there actually some good indication that there is *not* such scope in any useful sense. There is some interesting work which is related to these problems, by Pour-el and Richards (*1979, 1981, 1982*), which perhaps does point to the possibility of something nonalgorithmic, but I do not regard this as *useful* scope, if it is really scope at all.

Related to this is the kind of nonalgorithmic behavior that in effect arises when the gross performance of a system depends very delicately on its initial state. There are many situations when this is the case, such as in those circumstances which lead to "chaotic behavior". The large-scale behavior of a system may not be computable because it is not possible to know the initial state to anything like the precision required. This kind of nonalgorithmic behavior, as it arises in practice, is certainly not what I would mean by "useful" nonalgorithmic action. It is not something that could be harnessed by the brain in order to achieve anything that a Turing machine cannot achieve. If we require something nonalgorithmic in the laws of physics that the brain might indeed be harnessing then, it seems to me, this must go *beyond* the physical laws that we now understand!

However, it is already hard to see how the physics of relativity or quantum theory could have any relevance to the functioning of the brain. It seems an even *more* remote possibility that there could be relevance for some *hitherto undiscovered theory* that goes beyond quantum theory and our now-standard picture of space-time according to relativity? Perhaps surprisingly, I think that there is some reason to believe that such an undiscovered theory *is* playing a role – indeed, even a functional role – in the actions of our conscious thinking. I cannot elaborate greatly on what I have in mind here, but roughly my picture is based on the following.

Quantum theory has been marvelously well substantiated at the molecular level and smaller. Here, the term "level" does not refer to spatial size but to the fact that only comparatively small amounts of mass-energy are involved. The effects of quantum theory are certainly maintained over substantial distances – at least over several meters as, in particular, the experiments of Aspect and co-workers (*1976, 1986*) strikingly demonstrate. However, the rules of the theory run into profound theoretical difficulties if applied to physical objects that are very much "larger" (in the sense of much larger mass-energy distributions, not physical size) than the levels at which the theory has been experimentally verified. Quantum theory insists that all the different alternative possible ways that a system *might* behave have to be allowed to *coexist* in (linear) superposition. At some stage, however, just *one* alternative out of this array of superposed alternatives finds realization in the actuality of experience. How does this happen? Quantum theory has no adequate explanation; and the question is debated endlessly by theoreticians. My own view (which is shared by a number of respectable people, I might add, including several of the founders of the subject!) is that quantum theory is a *provisional* theory, at the macroscopic level, and must eventually be replaced by something more satisfactory (and presumably nonlinear). I believe that persuasive reasons can be provided that these changes must come about at the stage when the differing mass-energy distributions that are to be linearly superposed have quantum-mechanically significantly differing *gravitational effects* (thus affecting the space-time geometry differently). Preliminary calculations indicate that such effects will already have become relevant at the level of one ten-millionth of a gram – but, I hope, sometimes also at significantly smaller levels (cf. Penrose *1986, 1987a*).

Can this sort of thing be relevant for the workings of the brain? I think that it is likely. It is already known that some nerve cells are sensitive to a single quantum stimulus (e.g., in a frog's retina, where a single photon can trigger a macroscopic nerve signal). For such cells, the effect is to convert something which has apparently to be treated quantum mechanically into something which has apparently to be treated classically. As yet, there is no adequate theory for this. My guess would be that for some classes of cells, or perhaps of chemical transmitters, in the brain, this kind of quantum/classical to-and-fro is an essential ingredient of our thinking. If this is the case, we would not, as yet, have an adequate theory to cope with it. The normal way that physicists would try to handle the situation would be to keep invoking the probabilistic rules which quantum theory uses whenever an "observation" is deemed to have taken place. But, in my opinion, these are only approximations, valid at the extreme limits when things go from the completely quantum to the completely classical (as with the standard "Copenhagen interpretation"). Here, I am imagining that things are more delicately poised at the quantum/classical borderline, where we really have no adequate theory at all.

I am guessing that the theory we shall need is essentially nonalgorithmic in nature. I believe that there are clear indications that it must also be nonlocal: that is to say, the total behavior cannot be considered to be composed of separated independent parts, but is, in some appropriate sense, "holistic". This kind of holistic behavior is already an observed feature of standard quantum theory, and is what the experiments of the Aspect type demonstrate. As yet, there is no indication from *physical* theory (as opposed to philosophical/mathematical considerations) that we shall need nonalgorithmic laws. But, as I indicated above, we do not yet have the correct physical theory. It should be made clear that the issue of nonalgorithmic behavior is quite a separate one from determinism or mathematical well-definedness. (For example, one can easily make up deterministic, though quite artificial, "toy models" which are not algorithmic. The state of our "toy physical system" could be given by a list of pairs of words. If the two words in the final pair are "equal" by virtue of all the other pairs being deemed "equal" then the list is to be modified in some preassigned way at the "next instant of time"; but if they are "unequal" the list is to be modified in a *different* preassigned way at the "next instant".) What I would mean by a physical "nonmechanism" would be some physical object whose behavior is governed in some such well-defined, but nonalgorithmic way.

Whether or not our physical theory eventually actually moves in such a direction, it has to be admitted that nonalgorithmic behavior is a possibility that must be seriously considered. It seems to me that the philosophical issues that this raises have barely begun to be explored. Alan Turing's seminal work of fifty years ago may now be telling us that these possibilities are things that we may need to contend with in our understanding of the physical world. Perhaps, eventually, such considerations may help us to understand how it is that we can actually *be* part of that world!

References

Aspect, A.
   1976 Proposed experiment to test the nonseparability of quantum mechanics. *Phys. Rev.* **D14** (1976) 1944- 51.

Aspect, A., and P. Grangier
   1986 Experiments on Einstein-Podolsky-Rosen-type correlations with pairs of visible photons. In: *Quantum Concepts in Space and Time,* eds. R. Penrose and C.J. Isham. Oxford: Oxford University Press (1986).

Benacerraf, P.
   1967 God, the Devil and Gödel. *The Monist* **51** (1967) 9–32.

Berger, R.
   1966 The undecidability of the domino problem. *Mem. Am. Math.* **66** (1966).

Bowie, G.L.
   1982 Lucas' number is finally up. *J. Philos. Lo.* **11** (1982) 279–285.

Deutsch, D.

  1985 Quantum theory, the Church-Turing principle and the universal quantum computer. *P. Roy. Soc. A* **400** (1985) 97–117.

Geroch, R., and J.B. Hartle

  1986 Computability and physical theories. *Found. Phys.* **16** (1986) 533.

Gödel, K.

  1931 Über formal unentscheidbare Sätze der Principia Mathematica und verwandter Systeme I. *Monats. Math. Phys.* **38** (1931) 173–198.

Good, I.J.

  1969 Gödel's theorem is a red herring. *Br. J. Phil. S.* **18** (1969) 359–373.

Grünbaum, B., and G.C. Shephard

  1986 *Tilings and Patterns.* New York: W.H. Freeman and Co. (1986).

Hofstadter, D.R.

  1979 *Gödel, Escher, Bach: an Eternal Golden Braid.* Stanford Terrace, Hassocks, Sussex: the Harvester Press, Ltd. (1979).

  1981 A conversation with Einstein's brain. In: Hofstadter and Dennett *1981.*

Hofstadter, D.R., and D.C. Dennett

  1981 *The Mind's I.* Harmondsworth, Middx.: Basic Books, Inc., Penguin Books, Ltd. (1981).

Lucas, J.R.

  1961 Minds, machines and Gödel. *Philosophy* **36** (1961) 120–124; reprinted in: Alan Ross Anderson, *Minds and Machines,* Englewood Cliffs (1964).

Mandelbrot, B.B.

  1986 Fractals and the rebirth of iteration theory. In: *The Beauty of Fractals: Images of Complex Dynamical Systems,* eds. H.-O. Peitgen and P.H. Richter. Berlin: Springer-Verlag (1986) 151–160.

Paris, J., and L. Harrington

  1977 A mathematical incompleteness in Peano arithmetic. In: *Handbook of Mathematical Logic,* ed. J. Barwise. Amsterdam: North-Holland Publ. Co. (1977).

Penrose, R.

  1986 Gravity and state-vector reduction. In: *Quantum Concepts in Space and Time,* eds. R. Penrose and C.J. Isham. Oxford: Oxford University Press (1986) 126– 146.

  1987 Minds, machines and mathematics. In: *Mindwaves,* eds. C. Blakemore and S. Greenfield. Oxford: Blackwells Publ. (1987).

  1987a Newton, quantum theory and reality. In: *Three-Hundred Years of Gravity,* eds. S.W. Hawking and W. Israel. Cambridge: Cambridge University Press (1987).

Pour-El, M.B., and I. Richards

  1979 A computable ordinary differential equation which possesses no computable solution. *Ann. Math. Log.* **17** (1979) 61–90.

  1981 The wave equation with computable initial data such that its unique solution is
  not computable. *Adv. Math.* **39** (1981) 215–239.

  1982 Noncomputability in models of physical phenomena. *Int. J. Theor.* **21** 553–555.

Robinson, R.M.

  1971 Undecidability and nonperiodicity for tilings of the plane. *Invent. Math.* **12** (1971)
  177–209.

Searle, J.R.

  1980 Minds, brains, and programs. In: *The Behavioral and Brain Sciences* vol. 3. Cambridge: Cambridge University Press (1980). Reprinted in Hofstadter and Dennett
  1981.

Tseitin, G.S., and D. Scott

  1955 see *Logic Machines and Diagrams,* by M. Gardner, p. 144. Chicago: University of
  Chicago Press (1958).

Turing, A.M.

  1936-7 On computable numbers, with an application to the Entscheidungsproblem. *P.
  Lond. Math. Soc. (2)* **42** (1936-7) 230–265; A correction, ibid. **43** (1937) 544–
  546.

  1950 Computing machinery and intelligence. *Mind* **59** (1950) no. 236. Reprinted in
  Hofstadter and Dennett *1981*.

Effective Processes and Natural Law

Robert Rosen

1. Introduction

One of the most remarkable confluences of ideas in modern scientific history occurred in the few short years between the publication of Gödel's original papers on formal undecidability in 1931, and the work of McCulloch and Pitts on neural networks, which appeared in 1943. During these twelve years, fundamental interrelationships were established between logic, mathematics, the theory of the brain, and the possibilities of digital computation, which still literally takes one's breath away to contemplate in their full scope. It was believed at that time, and still is today, over half a century later, that these ideas presage a revolution as fundamental as that achieved by Newton three centuries earlier.

The name of Alan Turing is preeminent in the history of these astonishing developments. For it was Turing, in his seminal paper of 1936, who first really juxtaposed the relevant ideas through the construction of the class of "machines" which bear his name. These *Turing machines* were, on the one hand, explicitly extrapolated from the mental processes of a human being engaged in a mathematical computation; on another hand, they represented a formal embodiment of logical or algorithmic processes as manifested in mathematics; and on yet another hand, through the use of the term "machine" they suggested both the harnessing of material processes to extend our own mathematical capabilities (soon to be realized through the creation of digital computers) and, at an even deeper level, a new and powerful metaphor for exploring life itself.

At a purely mathematical/logical level it was quickly recognized that the Turing machines were one of a number of equivalent formalisms for embodying the concept of an *algorithm*. In its turn, an algorithm is regarded as the epitome of an *effective process* for solving a problem. Now an effective process connotes an idea of absolute necessity; it is an inferential chain which in every case *must* lead from appropriate initial data to the corresponding answer or solution. Moreover, an algorithm is a *rote process* which, once set in motion, requires no further intervention, no reflection, and no thought in its relentless progression from data to solution. That is why it seems, in retrospect, so natural to embody it in a "machine", as Turing did.

Since the notion of "effective process" is an informal, intuitive concept, while the notion of "algorithm" is a precisely formalized mathematical one, it was early suggested that the latter can replace the former. This is precisely the substance of Church's Thesis (cf. Kleene *1952*), which asserts that any process one would want to call "effective" can already be carried out by some properly programmed Turing machine.

Now, strictly speaking, all of the developments described so far are entirely formal; they take place in a logical and mathematical universe of propositions and production rules. In a sense, they are "all software". However, much of their interest resides in the fact that terms like "machine" or "effective" have connotations which are nonmathematical, which pertain to the external world of natural, material phenomena. Indeed, as we have already noted, Turing himself designed his mathematical machines as abstractions from a real-world phenomenon, a human being performing a computation. The suggestion is thus irresistibly conveyed that if *this* aspect of human mental activity can be "mechanized", why not others? Why not all? Likewise, if mental processes, involving what happens in a *material* brain (i.e., in hardware) can be represented entirely formally, why can there not be other kinds of material systems (i.e., other hardware) which can be made to do the same thing the brain does? Here we see, in embryo, the field of "artificial intelligence" in its widest ramifications, and much else besides.

But once we have admitted a material significance to words like "machine" or "effective", we have left the world of mathematics and entered the world of (in the broadest sense) physics. And whereas the Turing machines are, as we have noted, "all software", physics is by contrast "all hardware". We have thus, for good and ill, introduced a fundamental distinction between hardware and software which is, in itself, not part of the formal theory with which we began; nor, by the same token, is it part of physics either. This distinction, as we shall see, is vital, but it is insidious; it is camouflaged in the all-encompassing, umbrella term "machine". We can see how insidiously the distinction creeps in from the following quotation taken from Martin Davis's book *Computability and Unsolvability* (*1958*):

> For how can we ever exclude the possibility of our being presented someday (perhaps by some extraterrestrial visitor) with a (perhaps extremely complex) device or "oracle" that "computes" a noncomputable function?

This was clearly meant to be the most rhetorical of rhetorical questions, presented in the context of an entirely formal mathematical development. But we see here clearly the equivocation on the term "machine", resting on a tacit distinction between *real* hardware and logical software.

The present author has been troubled for a long time by the deep epistemological ramifications of this question. In particular, once we admit "hardware", or material systems, into our discussion (and as we have already noted, that was from the beginning the clear if tacit intention), what happens to the idea of an "effective process"? A long time ago (Rosen *1962*) I considered the question of what Church's Thesis means in this new context. Specifically: Is Church's Thesis a fundamental restriction on material nature (akin to the exclusion of the *perpetuum mobile* by the Laws of Thermodynamics), or not? And what happens to recursiveness if Church's Thesis can be violated by natural processes?

It may be useful to review the salient points of our earlier argument here. Suppose we are given a *physical* system *S* (perhaps the alien "computer" of Davis). Its behavior is governed by physical laws, which we learn about by doing experiments. A typical experiment will involve either doing something to the system (i.e, perturbing it from outside) or else letting the system do something to its environment, and then observing or measuring the result. Clearly, both the experimenter's intervention and the measured results are material events. Events are described or characterized by means of numbers, whose values are determined by the application of suitable meters (cf. Rosen *1978*). Suppose for simplicity that our experimental intervention α is characterized by such a number $r(\alpha)$, and that the resultant behavior of our system is characterized by another such number β. In this way, our experimenter can generate a table of values

$$r(\alpha) \longmapsto \beta$$

which defines a function f from numbers to numbers. The reader will recognize this procedure as a typical input-output characterization of our system *S*. The form of the function *f* clearly tells us something about the *laws* which govern the behavior of *S*. This is, after all, the whole function of experiment in science.

Now surely this experimentation process is in some sense *effective*. Indeed, as we shall argue at great length below, sequences of events in the material world (e.g., in the system *S*) are governed by *causal relations,* which bind them together quite as inexorably as implication relations bind propositions. Thus, if our experimental procedures are *repeatable* (which means that the same causal sequence in *S* can be recreated at will), then Church's Thesis must mean that *any input-output function f, generated as we have described from any material system S, must also be recursive or computable.* Otherwise, the system *S* would be precisely the "computer" which Davis assured us is excluded from possibility.

Seen in this light, Church's Thesis is an attempt to draw inferences or conclusions about hardware (physics) from premises about software (algorithms). Another well-known attempt to do the same thing is embodied in von Neumann's arguments about "self-reproduction" (Burks *1966*; Arbib, this volume; cf. Section 5 below). Here again, the intent is to learn something about the behavior of *material* systems (especially organisms) from a *formal* theory of computation. As might be expected, such enterprises are risky in the extreme; but *if* it could be successfully pulled off, the rewards would be great indeed.

At the very least, we can perhaps already see that introducing an idea of "hard-ware" into formal theory will have some peculiar ramifications. Complementary peculiarities arise from the other side, when we attempt to introduce ideas of "soft-ware" into physics. However, these very peculiarities promise to tell us something interesting about both. In the remainder of the present paper, we shall explore some of these possibilities.

2.   *Church's Thesis in Formal Systems*

The essence of Church's Thesis is that it identifies *logical inference* in any formal system with *string processing*. In its turn, string processing, or word processing, is a purely *syntactic* activity. String processing is, of course, precisely what the Turing machines do. Nevertheless, it seems on the face of it to be a very strong, perhaps excessively strong condition to require that every inference in a formal system should be expressible in syntactic terms *alone*; i.e., that every trajectory from premises to conclusion should be navigated entirely through the manipulation of the symbols in which these propositions are encoded.

Nevertheless, a rather strong case can be built to support this rather unlikely-looking Thesis. It distills a trend towards formalization which began with Euclid, be-came a matter of urgency in the confusion following the discovery of non-Euclidean geometries (i.e., geometries which, as formal systems, were consistent as God-given Euclid), and of absolute desperation when the paradoxes in naive set theory were revealed. The formalistic response to this situation, pioneered by David Hilbert, was precisely to empty mathematics of any semantic content whatsoever, arguing in ef-fect that it was a needless semantics which was at the root of the difficulties. This in effect turned all of mathematics into a kind of game in which meaningless sym-bols were manipulated according to (a finite family of) arbitrary syntactical rules. Indeed, the whole point of Hilbertian Formalism is to create systems in which there is nothing but syntax.

Perhaps the clearest statement of this kind of Formalist program was given by Kleene *1952*:

> This step (axiomatization) will not be finished until all the properties of the undefined or technical terms of the theory which matter for the deduction of theorems have been expressed by axioms. Then it should be possible to perform the deductions treating the technical terms as words in themselves without meaning. For to say that they have mean-ings necessary to the deduction of the theorems, other than what they derive from the axioms which govern them, amounts to saying that not all of their properties which matter

for the deductions have been expressed by axioms. When the meanings of the technical terms are thus left out of account, we have arrived at the standpoint of formal axiomatics ... Since we have abstracted entirely from the content matter, leaving only the form, we say that the original theory has been *formalized*. In this structure, the theory is no longer a system of meaningful propositions, but one of sentences as sequences of words, which are in turn sequences of letters. We say by reference to the form alone which combinations of words are sentences, which sentences are axioms, and which sentences follow as immediate consequences of others.

Clearly, the idea here is that it is always possible to replace *semantics* ("meanings") with syntactics, so that *logically* no information is lost; any inference involving semantics possesses a purely syntactical image in the formalization.

In such a formal system, we start with the idea that the axioms are *true*. This notion of truth is *hereditary*; if the axioms are true, then so also are the symbol sequences obtained by applying the inferential rules of the system to them; thus truth passes from axioms to theorems. So far, we never need to import a notion of "truth" into the system from outside, as it were; we simply construct true propositions (theorems) as we go along.

The troubles embodied in Gödel's celebrated theorems (Gödel *1931*) arise from trying to compare this constructive notion of internal truth with preassigned *external* truth-value in formal arithmetic; as Gödel showed, they do not match. Furthermore, one way of looking at Turing's theorem on decidability (Turing *1936-7*) is that there is *no internal inferential mechanism* for deciding whether a proposition in the given formalization is a theorem (i.e., true) or not.

Thus, we *know* that the Formalist program, in which only purely syntactic inferences are allowed, is too impoverished to even play the game of number theory. That is, we must either allow into our system some "informal" inferential procedures, which according to Gödel's Theorem *cannot* be reduced to syntactics even in principle, or else restrict ourselves forever to mere fragments of number theory. Such "informal" inferential procedures are what are disallowed by Church's Thesis; they are *ineffective*.

Let us put the above discussion into more familiar terms. In ordinary ("Platonic") mathematics, which of course has both a syntactic and a semantic aspect, we know that if we are given a set, a variety of other sets can always be built from S via canonical constructions. For example, we have the power set 2^s; we have the free algebraic structures (semigroup, group, etc.) generated by S, we have the set $H(S, S)$ of all maps from S to S; we have the Cartesian product $S \times S$, etc. These associated sets give us *inferential capabilities* which have the power of logical relations, without necessarily being expressible in terms of the formal inferential laws which govern the mathematical system from which S came. For instance, if $Q : S \rightarrow S$ is an automorphism of S (i.e., an element of $H(S, S)$), and $s \in S$, we may say that $Q(s) = s'$ establishes an implication relation between s and s'. But this "implication" need not coincide with any we can draw from the production rules governing the system; i.e., need not follow from system syntactics alone. If it does, we may say that our Q is *computable* in the system; otherwise, *not computable.* In the latter case, we would have to say that the mapping Q is not *effective,* according to Church's Thesis. But clearly, if Q has any meaning or existence at all, once it is given to us, it *is* effective.

In the "all-software" world of formal systems, we can of course restrict ourselves in any way we like. Thus, we can agree not to allow ourselves any automorphisms Q which cannot be expressed in purely syntactical terms. In such a world, and only in such a world, could Church's Thesis hold unrestrictedly. Whether such a formal world would be at all interesting is, of course, another question. And as we shall see, the situation gets even worse when we allow "hardware" into our world.

Before turning to material systems, we should say a word about the encoding of propositions in a formal system onto Turing machine tapes, and decoding tapes back into propositions in the system. It is fairly clear that we may use the word "effective" in its usual intuitive sense in connection with encoding and decoding; the familiar Gödel numbering, for example, is clearly an effective mapping from syntax to arithmetic and back. Indeed, in showing that a process in some formal system is effective, or recursive, we can clearly combine the encoding, the computation, and the decoding into a single Turing machine which does all three. However, we merely note here for future reference that the encoding and decoding are logically distinct from each other and from the actual computation; if these are in some sense "ineffective", then Church's Thesis may appear to fail, even though the computation itself is completely recursive.

3. *Implication and Causality*

As we have seen, in the realm of formal systems, Church's Thesis identifies the intuitive notion of "effective process" with the purely syntactic idea of string processing. That is, any implication which can be "effectively" performed within the system can already be carried out by means of a finite set of production rules, operating on finite strings of symbols taken from a finite alphabet.

We have also pointed out that when we deal with the *material* world (as opposed to the formal ones of mathematics and logic) ideas of implication are replaced by ideas of causality. Nevertheless, we can still retain the idea of an "effective" process. In this context, Church's Thesis means that any *causal* sequence can be represented by a corresponding recursive process; i.e., *any causal sequence can be described by purely syntactic means.* If this is true, it of course places severe limitations on what *physics* can be like. *The question is no less than whether the Laws of Nature can themselves be formulated in purely syntactical terms, or whether they can possess an inherent semantic component which cannot be finitistically formalized.*

It should be noted that the urge to formalization (i.e., to pure syntactics) in mathematics is exactly parallel to similar trends in theoretical science. Indeed, the whole thrust of atomic theory (or nowadays, the theory of "elementary particles") is to reduce all material processes to the motion of ultimate constituent units, devoid of any internal structure ("meaning"), possessing only an instantaneous position ("configuration") and the temporal derivatives of position. The forces which push these ultimate units around are the precise analogs of the production rules in a formal system. Hence the paths or trajectories traced out by a material system under the influence of given forces are the analogs of formal theorems, with initial conditions as axioms.

The idea that causal relations between events in material systems can be related to implication relations between propositions describing those events is the *sine qua non* of theoretical science. Indeed, the belief in what used to be called *Natural Law* requires (a) that the sequences of events we perceive in the external world are not arbitrary or whimsical, but are governed by definite rules (this is *Causality*), and (b) that these rules can be articulated in such a way that they can be grasped by the human mind. Taken together, this formulation of Natural Law asserts precisely that causal relations in material systems can be brought into congruence with implications in a formal (ultimately, mathematical) system of propositions about those events.

This situation can be most succinctly expressed in terms of a diagram (see Figure 1).

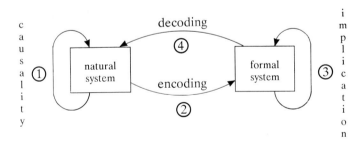

Figure 1.

We say that a *modeling relation* exists between the natural system on the left of the diagram, and the formal system on the right, when the following commutativity holds:

$$① = ② + ③ + ④ \tag{1}$$

That is, we get the same answer whether we simply sit as observers, and watch the unfolding sequence of events in the natural system, or whether we (a) encode some properties of the natural system into the formalism, (b) use the implicative structure of the formal system to derive theorems, and then (c) decode these theorems into propositions (*predictions*) about the natural system itself. When the diagram commutes, we have established a congruence between (some of) the causal features of the natural system and the implicative structure of the formal system. We can then say that the formal system is a *model* of the natural one, or alternatively, that the natural system is a *realization* of the formal one.

These little diagrams themselves possess a number of rich and important epistemological properties, which we cannot enter into here; for a fuller discussion, see e.g. Rosen *1985*.

Once we have thus constructed a formal system which is a model for some natural process, we have left the realm of science and entered that of mathematics. We can then treat a model as we would any other formal system. In particular, we can look at its purely syntactic aspects, which we can immediately identify with the "effective" processes of Church's Thesis, and ask whether these exhaust the implicative resources of the system itself.

In this way, we can construct a purely syntactic "machine" model of our original natural system, as indicated in Figure 2.

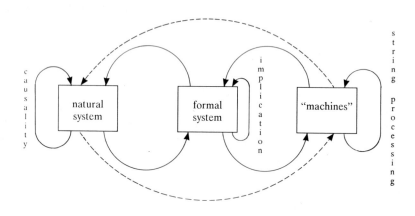

Figure 2.

This can be seen by looking only at the outer two systems, and forgetting about our original model, which now plays the role of a "transducer" between them.

However, when we do this, the following points must be explicitly noticed: (a) the *encoding* and *decoding* arrows between them (the dotted arrows in Figure 2) cannot be described as *effective* in any formal sense, and (b) these encoding and decoding arrows involve exclusively the *input* and *output* strings of the machines inhabiting the right-most box. Both of these observations are important. We shall briefly discuss them each in turn.

In applying Church's Thesis to formal systems, we noted above that the corresponding encoding and decoding arrows themselves represented *formal* processes, which could in fact be amalgamated into the Thesis itself. However, when we wish to compare a natural system, governed by causality, with a formal system, governed by implication, this is no longer the case. The encoding instruments, or transducers, are now themselves material systems; i.e., governed by causality and not by implications. As noted above, they are (in the broadest sense) *meters*. Since these meters are governed by causality, anything they do is *effective* in a material sense. But it is clear that a formalization of the encoding process would require more *models*; these would in turn require their own encoding and decoding processes, which would require more models, etc., an infinite regress. It is precisely this fact which makes "the measurement problem" so hard in physics. The question as to whether this potential infinite regress can be terminated at some finite point is a deep epistemological question about the nature of the world, with close ties to such things as reductionism. We cannot of course enter into such matters here; we will merely assume that a set of meters or other transducers from the natural world to input tapes is *given*, and that Church's Thesis can be investigated relative to *these* encodings (which are then, by hypothesis, effective in the material sense).

Our second observation is also epistemologically important. It says that all relevant features of a material system, and of the model into which it was originally encoded (cf. Figure 1), are to be expressed as *input strings* to be processed by a machine *whose structure itself encodes nothing*. That is to say, the rules governing the operation of these machines, and hence the entire inferential structure of the string-processing system themselves, have no relation at all to the material system being encoded. The *only* requirement is that the requisite commutativity hold, as expressed in Section 1 above, between the encoding on input strings and the decoding of the resultant output strings. As we shall see, this is the essence of *simulation*.

We have already noted that Church's Thesis amounts to asserting that all causal relationships can be expressed in purely syntactic terms. We can formulate the Thesis still more sharply now: relative to any given encoding of a natural system into input strings, the string-processing machinery itself must not encode any aspect of the material system. That is to say, the "hardware" of the machines *must* be totally independent of the "hardware" generating the strings to be processed. If this cannot be done, then Church's Thesis cannot be true.

Given the above, Church's Thesis asserts that the Turing machines constitute a class of *universal simulators* for all material processes. There are then two questions to be asked: (a) is it true? and (b) if so, what does it mean?

4. Is Church's Thesis Physically True?

The upshot of the argument of the preceding sections is the following. Nature provides us with a plethora of material processes which we would want to call "effective". These processes are (we suppose) governed by causality, and not by implication or production rules, as in a formal system. In these terms, Church's Thesis can be expressed as follows. Given any such process, we can encode appropriate propositions about the natural system generating it onto a set of input tapes to a Turing machine; and the corresponding output tapes can be decoded so as to perfectly *simulate* the process in question. Equivalently, Church's Thesis asserts that all "information" about material processes, and hence all of Natural Law, can be expressed in purely syntactic terms.

We know from Gödel's Theorem that sufficiently rich *formal* systems always contain inferences which cannot be obtained syntactically. More specifically, given any encoding of propositions of the system onto input tapes to Turing machines, there will be propositions of the system which are "true" but will never appear on any output tape. The processes by which the "truth" of such inferences are established are thus *ineffective*; they cannot in principle be simulated by any Turing machine in the given encoding. We can always change the encoding, of course, but this will not change Gödel's conclusion.

Thus, in formal systems, we already find that a purely syntactical encoding will in some sense *lose information*. The information lost must then pertain to an irreducible, unformalizable *semantic component* in the original inferential structure. By changing the encodings, we can shift to some extent where this semantic information resides, but we cannot eliminate it.

By itself, this result of Gödel does not bear on the *physical* truth of Church's Thesis, since it is a purely formal result. But it is in fact suggestive of how the physical form of Church's Thesis might be verified or falsified.

As we have seen above, the manner in which we compare material processes with formal ones is through the establishment of modeling relations, as diagrammed in Figure 1 above. Formal models of material systems are then perfectly good formal systems, whose inferential structures by definition reflect causal processes in the natural system being modeled. Thus, if a *model,* arising in this fashion, should fall within the purview of Gödel's argument, this would at least be strong evidence that Church's Thesis is false as a physical proposition. Stated another way, there would exist physical processes which could effectively compute nonrecursive functions. It would also mean that Natural Law cannot be expressed entirely in syntactical terms.

The obvious thing to look for, then, is a model of a material system which is rich enough as a formalism to "do arithmetic". The formalisms which physics provides, as models of purely physical systems, are unfortunately extremely impoverished, considered simply as formal systems. But, as we have argued elsewhere, these formalisms are in fact highly nongeneric and do not suffice to image material systems like organisms (cf. Rosen, in press). One way of expressing this nongenericity is precisely in terms of the way they image causal structures. When this nongenericity is lifted, a new class of (*potential*) models is obtained in which the image of causal structure is infinitely richer and more complicated. Since it is precisely the formal imaging of causal structures which is at the heart of Church's Thesis, we may perhaps find in these formalisms many processes which are *causally* effective, but *mathematically* ineffective. This would mean that the behavior of such systems must contain an irreducible semantic component, one intimately related to the *complexity* of the system.

A different approach was taken long ago by John Myhill (cf. Myhill *1966*). He was able to show that, modulo some idealizations regarding measurement and performance tolerances, there are already classical analog devices (analog computers) which could "compute" nonrecursive functions. Such systems would thus already manifest behaviors which could not be predicted by any purely syntactic encoding, and hence would also have an irreducible semantic aspect.

Finally, we have already mentioned that Newtonian particle mechanics, and more recently, the unified physical theories based on elementary particles, are in themselves an attempt to express the Laws of Nature in purely syntactic terms. Insofar as any material system is comprised of such "meaningless" (i.e., structureless) elementary subunits, these theories at heart assert that to understand any behavior of any such system it suffices to describe it in terms of these subunits and their interactions. As noted earlier, this is the essence of reductionism.

In these terms, the familiar Laplacian Spirit is a purely syntactical concept: the embodiment of Church's Thesis if he could exist. As a matter of fact, he could not exist (or at any rate, not as a material system built of particles himself), for reasons we have already indicated. But even if he could exist, he would be a very poor biologist, for example; organisms, and open systems in general, are constantly turning over their constituent particles. Thus, to even find an organism, let alone follow it in time, he would need to supplement his purely syntactical information with other (semantic) information not formalizable within his system.

Thus, for a variety of reasons, there is cause to believe that Church's Thesis fails as a physical proposition. Nevertheless, as we have seen, to state and analyze the Thesis in material terms touches on some of the deepest and most basic aspects of theoretical science.

5. The Role of Simulations

We have already alluded above to the use of the Turing machines both as metaphors for the material world and as effective descriptions of that world. In both cases, though in different ways, we seek to draw conclusions about material processes from a purely syntactic formalism. In this final section, I will very briefly consider one well-known example of this: the "self-reproducing automata" of von Neumann (Burks *1966*; Arbib, this volume).

The basis of von Neumann's argument was the inference of the existence of a "universal constructor" from Turing's argument for a universal simulator (computer). On the purely formal side, von Neumann constructed a universe ("cellular space") of intercommunicating Turing machines arranged along some regular geometric array, like the cells in a multicellular organism, or the neurons in a brain, or atoms in a crystal. Each machine communicates with its nearest neighbors in the array. In the obvious fashion, the array as a whole changes "state" in time. The question was whether some sub-array ("tessellation automaton") could induce certain interesting behaviors in its complement, which could be *interpreted* as construction, replication, growth, development, evolution, and so on.

At the same time, von Neumann clearly believed that a "universal constructor" could exist as *hardware*. He envisaged equipping a Turing machine with sensors, so that it could "read" a blueprint, and with effectors, so that it could extract physical components from its environment, and assemble them as instructed by the blueprint being read. (An "effector", in this context, is a transducer from numbers to things, i.e., an "inverse meter".) The idea was that *both* computation and construction were algorithmic processes, and therefore whatever was true of the one must be true of the other.

Von Neumann also felt that the cellular spaces were not merely formal constructs, but actually comprised *models* of real-world constructors and their activities. In this, he was tacitly assuming Church's Thesis in its strongest form.

We have argued elsewhere (cf. Rosen *1985*), on grounds of causality, that any inference regarding a *material* universal constructor from the existence of a *formal* universal computer is unjustified. In that argument, we essentially showed that there was no intrinsic way of distinguishing between those input strings which encode "real-world information" and those which do not. Thus there was no way to distinguish between a computation which could be claimed to simulate a "real-world" process, and one which has no such realization.

Conversely, we can also see that the falsity of Church's Thesis means that there are aspects of material processes which cannot be formalized with any given encoding. Thus there are (a) formal constructions without material counterpart, and conversely, (b) material constructions without formal counterpart. Therefore, on both counts, von Neumann's argument is without material content, and merely involves the familiar equivocation on the terms "automaton" (or "machine") and "construction".

These considerations show how dangerous it can be to extrapolate unrestrictedly from formal systems to material ones. The danger arises precisely from the fact that computation involves only *simulation,* which allows the establishment of no congruence between causal processes in material systems and inferential processes in the simulator. We therefore lack precisely those essential features of encoding and decoding which are required for such extrapolations. Thus, although formal simulators can be of great practical and heuristic value, their theoretical significance is very sharply circumscribed, and they must be used with the greatest caution.

There are, of course, many other ramifications of Church's Thesis which we cannot touch on in this brief space. Its main role, as we have seen, is to separate out what is syntactic in a formal system from what is not; when the formal system is also a model of a material system, Church's Thesis does the same for causal relations. The Thesis in fact raises a host of deep questions about Natural Law, about causality, about modeling, and about the material realization of formalisms. Its central feature, the Turing machines, embody the essence of syntactics or string processing in a single, conceptually rich package. Even if (as I believe) Church's Thesis fails, it does so in a most instructive way. Its implications for the material sciences, and especially for biology, have barely begun to be explored.

References

Arbib, M.A.

    1988 From universal Turing machines to self-reproduction. This volume.

Burks, A.

    1966 *The Theory of Self-Reproducing Automata.* Urbana, IL: University of Illinois Press (1966).

Davis, M.

    1958 *Computability and Unsolvability.* New York: McGraw-Hill (1958).

Gödel, K.

    1931 Über formal unentscheidbare Sätze der Principia Mathematica und verwandte Systeme. *Monats. Math. Phys.* **38** (1931) 173–198.

Kleene, S.C.

    1952 *Introduction to Metamathematics.* New York: von Nostrand (1952).

Myhill, J.

    1966 *Creative Computation Revisited.* USAF Technical Report (1966).

Rosen, R.

    1962 Church's Thesis and its relation to the concept of realizability in biology and physics. . *B. Math. Biol.* **24** (1962) 375–394.

1978 *Principles of Measurement.* New York: Elsevier (1978).

1985 *Anticipatory Systems.* New York: Pergamon Press (1985).

in press Biology and physics: An essay in natural philosophy (in press).

Turing, A.M.

1936-7 On computable numbers, with an application to the Entscheidungsproblem.*P. Lond. Math. Soc. (2)* **42** (1936-7) 230–265; A correction, ibid. **43** (1937) 544–546.

Turing Naturalized:
Von Neumann's Unfinished Project

Helmut Schnelle

1. Naturalization of Logic and of Obvious Knowledge

Turing's enterprise must be seen in its historical context: the Turing machine is the result of an attempt to establish an epistemologically absolute foundation of mathematics and, through it, of science. Since it was realized that an absolute foundation of science is impossible, a program of a relative foundation seemed to be appropriate, based on the assumption that successful parts of science are true. In this perspective we are free to naturalize logic and perception of obvious facts by specifying psycho-physical mechanisms of logic and of perception and by investigating whether science could be reconstructed in all its ramifications on the basis of such mechanisms. The obvious charge of circularity has lost its force since it was agreed upon that certain kinds of circularity are unavoidable.This naturalization of logic may take Turing's proposal as a first step and continue along the lines indicated by von Neumann and other neo-intuitionists, who consider mental mechanisms as the basis of logic, mathematics, and science. Let me explain these introductory remarks in further detail.

For more than a hundred years the basic question of epistemology was the absolute foundation of science. Science in the modern – Galilean – sense, is based on mathematics. Knowledge cannot become precise unless mathematized. It was hoped that the contents of a mathematized science could be split up into statements of mathematical form and statements which are actually true (e.g., numerical statements logically implied by obvious statements of fact). In this context it seemed to be clear that the absolute foundation of science required an absolute foundation of logic and, in particular, of the logic of mathematics.

It turned out that the attempt of providing such a foundation was much more difficult than expected. Frege's very careful undertaking failed; it could not, as Russell and others showed, provide exact criteria for the exclusion of inconsistencies. Hilbert defined a program, the success of which would have led to the goal. Gödel showed, however, that in principle it was impossible to execute the program. In order to demonstrate this, Gödel correlated the process of numerical computation and formal proof. He wrote: "It is well known that the development of mathematics in the direction of greater precision has led to the formalization of extensive mathematical domains, in the sense that proofs can be carried out according to a few mechanical rules" (Gödel *1931/1965,* p. 5). Gödel's characterization of "formal mathematical system" uses the notion of a mechanical rule or, more correctly, of a finite mechanical process instantiating the rule.

But what constitutes a finite mechanical process? Several attempts have been made to render this notion precise, among them Turing's. He wrote:

> A function is said to be "effectively calculable" if its value can be found by some purely mechanical process. We may take this statement literally, understanding by a purely mechanical process one which could be carried out by a machine. It is possible to give a mathematical description, in a certain normal form, of the structure of these machines. (Turing *1939/1965,* p. 160)

We shall later return to this very interesting statement about Turing's own approach. This approach was successful as a basis for the later development of electronic machines for universal computation. It did not and could not, however, provide an absolute foundation for the whole of mathematics as applied in the sciences. On the contrary Gödel's and Turing's results showed that this goal could not be reached.

As a consequence, a new goal was defined: the justification of science and logic plus epistemology simultaneously in contrast to the goal of justifying the former on the basis of the latter. According to Quine, the task is no longer to develop an absolutely true prima philosophia but to show that science can explain its own emergence by providing a scientific analysis of cognitive perception or at least of scientific perception.

> The old epistemology aspires to contain, in a sense, natural science; it would construct it somehow from sense data. Epistemology in its new setting, conversely, is contained in natural science, as a chapter of psychology. But the old containment remains valid too, in its way. We are studying how the human subject of our study posits bodies and projects his physics from his data, and we appreciate that our position in the world is just like his. Our very epistemological enterprise, therefore, and the psychology wherein it is a component chapter, and the whole of natural science wherein psychology is a component book – all this is our own construction or projection from stimulations like those we are meting out to our epistemological subject. There is thus reciprocal containment, though containment in different senses: epistemology in natural science and natural science in epistemology. This interplay is reminiscent again of the old threat of circularity, but it is all right now that we have stopped dreaming of deducing science from sense data. We are after an understanding of science as an institution or process in the world and we do not intend that understanding to be any better than the science which is its object. This attitude is indeed one that Neurath was already urging in Vienna Circle days, with

his parable of the mariner who has to rebuild his boat while staying afloat in it. (Quine *1968/1969* p. 83–4)

The position that logic, epistemology, physics, and scientific psychology could and should be justified simultaneously and constructively has far reaching consequences for logic which may lead to such extreme statements as von Neumann's:

> There is an equivalence between logical principles and their embodiment in a neural network, and while in the simpler cases the principles might furnish a simplified expression of the network, it is quite possible that in cases of extreme complexity the reverse is true. All of this does not alter my belief that a new, essentially logical, theory is called for in order to understand high-complication automata and, in particular, the central nervous system. It may be, however, that in this process logic will have to undergo a pseudomorphosis to neurology to a much greater extent than the reverse. (von Neumann *1951/1962*, p. 311)

In other words: our ordinary ways of stating logic by means of symbols referring to ordinary objects and their attributes in the world may be naive. What may be needed are symbols referring to simple interactive units, being in their function similar to neurons, and to build network-combinations of such symbols, representing organs for perception, activation of action, or inner representation of processes of thought and imagination. The system of subnetworks may be the basis for natural logic, whereas our symbols on paper are only secondary surrogates of certain states and processes in such networks. Let me emphasize: *certain* processes, because in complicated cases it may be too difficult for us to specify a complete calculus for the functioning of such processes.

As a case of extreme complication von Neumann mentions the problem of defining the notion of visual analogy. We might add the problem of defining linguistic structures in phonology, syntax, and semantics.

If what von Neumann says is true, a Turing machine program for defining visual analogy or phonological, syntactic, and semantic structures would be too complicated, whereas an appropriate distributive analysis into subtask units (or even hierarchies of them) may be the only practical way to come to a theoretical description and the only one that could be controlled by human beings.

In order to provide an understanding of the problem, let me turn to the discussion of a similar problem. Consider the situation of linguistic communication in a city at a given moment. Let us assume that among the 100,000 inhabitants, five percent, i.e., 5,000, are speaking, i.e., uttering words at that moment. We could code the set of words uttered at the moment as the linguistic communicative state of the city. Since there is a finite number of speakers, the sequence of time points at which at least one speaker starts a new word is discrete. The sequences of sets of words thus defines the communicative activities in the city. The class of possible communicative activities in the city is probably Turing-computable. But the definition of the system is much too complicated, since the compounding of words of different speakers and the formation of sequences of these (unnatural) units prevents each natural constructive consideration of defining the "communicative-city" machine. The same might hold for the collectivity of neurons defining visual analogy for an organism.

Thus, the demonstration that a given behavior is Turing-computable may be devoid of any practical interest when it is too complicated or too unnatural to be understood. Moreover, it is even devoid of epistemological interest, since what we need is a constructive analysis of actual processes and not the principled demonstration of rock-bottom. What we need is a naturalization of logic and linguistics, and von Neumann's statement could at least be taken as a definition of this challenge.

But do we know that the definition of a *natural* logic and linguistics is not equally complicated? What is the basis of our confidence that formally constructed logic in terms of sequences of symbols does not only serve the purpose of founding *written, scientific* discourse but is an equally good model for natural inferences (with all their swiftness and all their error)? Since logic was not invented as a model for natural inference but as a tool for judging scientific inference this confidence is not well founded at all. It may well be that what von Neumann claimed for the notion of analogy also holds for the notion of our natural faculty of drawing inferences.

The situation is even more complicated when we return from logic to epistemology, since there we must also be clear about what to accept as obvious knowledge.

As von Neumann points out, visual analogy (and perhaps linguistic-acoustic analogy) are processes providing "obvious" knowledge. Still, if considered in their details they belong to most complicated structures.

Quine tried to describe them in disposition terms, taking a description in disposition terms as "a promissory note for an eventual description in mechanical terms" (Quine *1974*, p. 14), such that "conjectures about internal [dispositional] mechanisms are laudable insofar as there is hope for their being supported by neurological findings" (ibid., p. 37). It seems that at least one source for von Neumann's similar considerations were his early discussions of the process of observation and of the observer in quantum physics. The problem had been introduced and often been discussed by Bohr. Von Neumann states,

> Let us compare the facts of nature and of their observation. We must agree that, in principle, the act of measurement, or the correlated process of subjective apperception, is a new phenomenon not reducible to the physical world. It transcends the latter, or better, it leads into the uncontrolled, viz. in each experiment presupposed, conceptual internal life of the individual. Nevertheless, there is the fundamental requirement of natural science, namely the principle of psycho-physical parallelism, that it must be possible to describe the actually extraphysical process of subjective apperception in such a way, as if it occurred in the physical world – i.e., to assign to its components physical processes in the objective world, i.e., in ordinary space. (Obviously, it is necessary to localize these processes in points inside the space region occupied by our body.) This consideration is to be applied

to an example as follows: Consider the measurement of temperature. [Follows a short description of the external physical processes up to the length of the column of mercury observed as well as the process of vision in the eye up to the retina.] Were our knowledge of physiology more precise than it is today, we could continue, considering the chemical reactions caused by the retinal picture in the nerves and in the brain and finally state: it is the chemical changes in his brain cells which cause the apperception of the observer. However far we proceed in the computation: to the column of mercury, to the retina, to the brain, there comes the point at which we must say: this is perceived by the observer. I.e., we must always split the world into two parts, one being the observed system, the other the observer. In the former we may analyze all physical processes in an arbitrarily precise way (at least in principle), in the latter this is meaningless. The border between the two is arbitrary. . . . That this border can be arbitrarily transferred into the inner parts of the body of an actual observer is the content of the principle of the psycho-physical parallelism – but it must be drawn somewhere else the description stays free wheeling i.e., does not allow comparison with an experience. Because the experience only produces statements of the following type: an observer has had a certain (subjective) observation, and never of the following type – a physical magnitude has a certain value. (von Neumann *1932*, p. 223–4; my translation)

It is obvious that these considerations dating from 1932 were a big challenge for von Neumann: How far can the border between the physical processes and the inner experiences of the observer be transferred into the inner body, viz. the brain, of the observer? Von Neumann's general and logical theory of (high complication) automata and his comparison of sequential, program-controlled computers and the brain seem to be designed to meet this challenge.

The challenge can even be increased in two ways: How are we to specify the internal experience? Which is the role of inner language in having an internal experience?

The experiments with split brain patients seem to indicate that consciousness is connected with that hemisphere of the brain which also controls the distinctive use of speech. We linguists have proceeded very far in describing the control of language and speech in formal ways which correspond closely to formal descriptions of structures and processes in the natural sciences.

This leads to the other question: Can we describe the structures and processes of the internal experiences in terms of a psychological mechanics (Herbart 1821) or, today, in terms of a psychological quantum mechanics? The program is indicated by von Weizsäcker. After having discussed von Neumann's problem of transferring the border, he asks whether it is possible to apply the quantum theory to conscious experience (cf. von Weizsäcker *1985*, § 11,2e, p. 536ff. and § 12,3). "The assumption that this application is possible is certainly hypothetical, but it does not contradict the logical structure of the abstract quantum theory. The abstract quantum theory refers to decidable alternatives without any dependency on their special nature" (ibid., p. 537). The analysis shows that, formally, "the consciousness of an observer behaves just as some arbitrary quantum object". The probability function used by some analyst Z to describe the "state of consciousness of Y is reduced each time Y has said something to Z and has been understood" (ibid.). We see how the two considerations of the increased challenge are related. In this context von Weizsäcker defines the following task: in trying a quantum theoretical description of consciousness "we shall constitute the alternatives in the self recognition of consciousness from elementary alternatives. We shall then have to *conclude* that it must be possible to describe the consciousness in three dimensional space. The conclusion is obvious that this must be the human body" (von Weizsäcker *1985,* pp. 581–2). Von Weizsäcker agrees that this is a reduction of the problem in "scientific style". Analyzing consciousness in terms of decidable alternatives means nothing else, prima facie, than applying to it the same questions and methods of analysis as to nature in natural science. "It remains to be seen what can be learned from this." Von Weizsäcker emphasizes in this context again, just as did Quine, that the obvious charge of circularity has lost its force since we have accepted a priori *this* kind of circularity.

We see very clearly how these specifications of the challenges lead straight to the core problems of modern cognitive science and formal linguistics in "Galilean style" (cf. Chomsky *1980, 1986*). We also see that the development of formal logic as an a priori of epistemology has lost its privileged status. We now need accounts of internalized languages and process structures as prerequisites of a naturalized logic. But let us return to the concept of a Turing machine as a way of specifying mental representations.

2. Turing's Abstract Concept of a Computing Machine

As we saw, there is no need of insisting on the formal-symbolic character of a Turing machine; it does not play an essential role for the empirical description of actual processes of computation in the brain. Its main role is that of providing "a mathematical description, in a certain normal form, of the structure of . . . [computational] machines" (Turing *1939/1955,* p. 160).

The remark "in a certain normal form" indicates that, though *only a special case out of the class of high-complication automata* – or, more generally, of dynamic systems – is considered, it is quite obvious that this special case is determined by Turing's original problem of defining a machine that mimics the behavior of a human computer in executing a consciously controlled task of complicated calculation. It is clear that a general and logical theory of functional processes in organisms and of psychological mechanisms must be able to demarcate the special procedural character of Turing's assumptions. Let us analyze these assumptions in detail.

The introspective analysis shows that the basic feature is a *distributed configuration of changes of state* (cp. Schnelle *1981*). In considering the human computer the following specializations may be introduced: the changes of state occur in two realms – on paper and in mind. What is changed are *states of paper* (i.e., configurations of marks on paper) and *states of mind*.

Furthermore, the states of paper are changed in virtue of and determined by the states of mind. The mind is a *control unit* for the processes of change on paper.

The notion of a Turing machine specializes even further: the states of paper can be conceived as a *spatially sequential* organization of elementary symbol units; this is motivated by our conventions of writing. The states of mind occurring during a process of computation show a *temporally sequential* order; this is motivated by our introspective experience in consciously controlled activity.

There are further specifications of the control relation existing between the changes of state in mind and the changes of state on paper: in each moment, the mind is related only to a *subsection of the configuration* on paper, which may either cause a change of state (in an act of perception) or become changed (acted upon) as determined and caused by mind. This is motivated by the fact that a human computer has locally focussing eyes and locally active writing organs (hand plus pencil or eraser).

We need a further, essential, restriction, which is motivated by the fact that the process of computation is not arbitrary but rather (in its ideal form) the model of a *regular procedure* (which can be expressed by some regularity of computation). This regular procedure is manifested in a *regular sequence of acts of control* (of perceptions and actions relative to the states of paper). This now is the important step of abstraction: we may neglect all features of the constitution and the distributed processing of the mind, of paper structure, of optics, etc., keeping only two abstractly specified features:

a) the mental space is represented by a finite set M of states of mind and the operative space (paper) is generated by a finite set S of local states of a unit section of paper (one square) such that the operative space of a sequence of n squares is represented by the Cartesian product S_n of the finite set of local states;
b) the "dynamics" of the system is represented by three functions each having the pair of a momentary mental state and of a momentarily perceived local paper state as its argument. The functions are distinguished by their values for:
c) the next mental state (in the first case);
d) the next local paper state (at the currently perceived position) (in the second case);
e) the local paper state to be perceived next (in the third case).

The dynamic functions specify

1) the change of mental state;
2) the change of paper state;
3) the change of local attention.

The examination of these restrictions and specifications clearly shows the specialized nature of these systems as considered from a more general system-theoretical point of view. The dominant feature is the separation into two fields of activity, one intended to be the active part (mind) and one intended to be a passive part (paper). The formal analysis shows, however, that the dynamic definition is totally symmetric: mind states and paper states simply interact; but the formalism does not show this intended asymmetry between active and passive part, except for the arbitrary pairing of mind states and dynamic functions.

This leads to the conclusion that it must be possible to define everything with respect to one realm only. This realm would represent the totality of states and the dynamics of change, instead of separating these factors into a range of symbolic data and a "mental" control unit. This is the basic idea of von Neumann's high complication automata: structure should be rendered by connectivities which determine the specificities of interaction and thus the specificities of the behavior of the system as a whole.

The so-called von Neumann computer, which is based on Turing's concept of a separation of data storage and control, can be analyzed as a very special case of high-complication automata. Its special character is determined by its special architecture which specifies certain subunits: one for the local change of attention, one for the change of access to parts of the store based on a management of addresses, or similarly, units implementing universally basic operations, addressing and decoding of rules or expressions of operations to be executed. We shall return to von Neumann's attempts to free computer design and conceptual design principles of organisms from these architectonic specializations which are, as we have seen, specifically motivated by manipulating symbols on paper by a locally and temporally focussing agent conceived abstractly in view of this task only. The human organism is designed for many other tasks, and its structure is much better adapted to these than the locally and sequentially operating symbol manipulator. It is thus to be expected that it has a quite different functional architecture.

In contrast to von Neumann's attempts most researchers were completely satisfied with the operations focussing locally on sequences of marks on papers. They were rather interested in further abstractions based on these assumptions instead of further analysis of the distributed design of processes.

Post suggested a reconstruction of a Turing machine in terms of a Semi-Thue System. The essential feature is that each state of mind is itself represented by a symbol which is represented on paper, at each moment, immediately left to the momentarily perceived symbol. The shift of attention of the reading-and-writing head of a Turing machine is represented by the shift of the "mind-state" symbol over the string. The "dynamics" of the system is now definable with respect to changes of state of paper only, though the acts of change now require by necessity that the local section to be operated upon is longer than a single symbol (usually a pair of symbols: q for a mind state symbol and S for paper state symbol). The processes of computation can now be defined by replacements of local symbol configurations only. Post's proposal is usually considered to be a further step of abstraction in comparison to Turing's.

A further step is Post's suggestion to abstract away from the actual sequence in which symbol configurations are changed according to rules of replacement by considering only the *generated sets* of configurations (whatever their sequence of production). This is, as is well known, Chomsky's starting point in considering linguistic structures as the generated sets determined by local symbol replacement rules applied on strings of symbols (whether on paper or in another store).

Post's proposals, and similar ones, are often characterized as *the* model of computational abstraction (e.g., by Chomsky a.o.). Given the fact that it is based on very special assumptions this claim appears as rather strange. Why should (spatially) sequential orderings of marks and (temporally) sequential local replacement rules provide a universally necessary basis for formal abstraction? These structure specifications rather seem to be determined by the accidental nature of our eyes as scanners and our hands as writing organs. Couldn't we assume instead that each local stretch of symbol units realizing a local configuration may itself instantiate the regularities of change? Instead of moving an active unit executing the replacements of state over the configuration each configuration would realize a change as soon as the conditions for such a regular change are obtained. This would not exclude sequential order of processing, where necessary: among the conditions there may be signals enforcing a sequential operation of changes; on the other hand, the changes may operate simultaneously everywhere and at every moment in which a condition for change (the argument part of the dynamical function) is obtained.

3. Axiomatization

High complication automata are dynamical systems, i.e., systems having states whose changes are determined by dynamical laws. The development of a general theory of automata as distributed dynamical systems is von Neumann's third exercise in axiomatization, after his first, the development of the foundations of quantum mechanics, and his second, the development of the theory of games and economic behavior.

These enterprises were based on von Neumann's credo as a mathematician:

> Mathematical ideas originate in empirics, although the genealogy is sometimes long and obscure. But, once they are so conceived, the subject begins to live a peculiar life of its own and is better compared to a creative one, governed by almost entirely aesthetical motivations, than to anything else and, in particular, to an empirical science. There is, however, a further point which, I believe, needs stressing. As a mathematical discipline

travels far from its empirical source, or still more, if it is a second or third generation only indirectly inspired by ideas coming from "reality", it is beset with very grave dangers. It becomes more and more and purely aestheticizing, more and more purely *l'art pour l'art* ... whenever this stage is reached, the only remedy seems to me to be the rejuvenating return to the source: the reinjection of more or less empirical ideas. (von Neumann, p. 9).

The axiomatization of an empirical framework starts with "the gradual development of a [mathematical] theory, based on a careful analysis of the ordinary ... interpretation ... of facts. This preliminary stage is necessarily *heuristic,* i.e., the phase of transition from unmathematical plausibility considerations to the formal procedure of mathematics" (von Neumann and Morgenstern *1944,* p. 7).

The development then aims at a theory that is "mathematically rigorous and conceptually general" (ibid.). This development of generality has several phases.

Its first applications are necessarily to elementary problems, where the result has never been in doubt and no theory is actually required. At this early stage the application serves to corroborate the theory. The next stage develops when the theory is applied to somewhat more complicated situations in which it may already lead to a certain extent beyond the obvious and the familiar. Here theory and application corroborate each other mutually. Beyond this lies the field of real success: genuine prediction by theory. It is well known that all mathematical sciences have gone through these successive phases of evolution. (ibid.)

In the case of quantum theory, von Neumann achieved the highest form of mathematical rigor and conceptual generality, because the theories he started with had already reached the third stage. In the case of game theory and economic behavior and of automata theory (including the theory of organisms!) von Neumann reached, perhaps, the second stage.

There are more specific precepts, which von Neumann obeyed:

... models are theoretical constructs with a precise, exhaustive and not too complicated definition; and they must be similar to reality in those respects which are essential in the investigation at hand. To recapitulate in detail: the definition must be precise and exhaustive in order to make mathematical treatment possible. The construct must not be unduly complicated, so that the mathematical treatment can be brought beyond the mere formalism to the point when it yields complete numerical results. Similarity to reality is needed to make the operation significant. And this similarity must be restricted to a few traits deemed "essential" pro tempore – since otherwise the above requirements would conflict with each other. (von Neumann and Morgenstern *1944,* p. 33)

The starting point for axiomatization is thus a field of intuitive concepts. In his third approach, von Neumann compares organisms and artificial high complication automata:

Natural organisms are, as a rule, much more complicated and subtle, and therefore much less well understood in detail, than are artificial automata. Nevertheless, some regularities which we observe in the organization of the former may be quite instructive in our thinking and planning of the latter; and conversely, a good deal of experiences and difficulties with our artificial automata can be to some extent projected on our interpretations of natural organisms. (von Neumann *1951/1963*, p. 288–9).

The real power of this comparison becomes clear only when we apply the first of two possible ways of axiomatization: the synthetic and the integral. The *integral way* is exemplified by Turing. He "started axiomatically describing what the whole automaton is supposed to be, without telling what its elements are, just by describing how it's supposed to function" (von Neumann *1949/1966*, p. 43). This is, by the way, also Chomsky's method of axiomatizing linguistics, as I understand it.

The *synthetic way* of axiomatization is the one exemplified for organisms by McCulloch and Pitts. In it,

> the organism can be viewed as made up of parts, which to a certain extent are independent, elementary units. (von Neumann *1951/1963*, p. 289)

> Axiomatizing the behavior of the elements means this: we assume that the elements have certain well-defined, outside, functional characteristics; that is they are treated as "black boxes"! They are viewed as automatisms, the inner structure of which need not be disclosed, but which are assumed to react to certain unambiguously defined stimuli, by certain unambiguously defined responses. (ibid., p. 289)

In this sense McCulloch and Pitts axiomatized an elementary unit which they called formal neuron, because it had a similarity to an actual neuron. "They believed that the extremely amputated, simplified, idealized object which they axiomatized, possessed the essential traits of the neuron, and that all else are incidental complications, which in a first analysis are better forgotten." (von Neumann *1949/1966*, p. 44)

I still believe that the notion thus introduced by McCulloch and Pitts and taken up by von Neumann was a fruitful one, because other similar axiomatizations, though slightly more complicated, can be related to them (e.g., the basic units assumed in the models of parallel distributed processing (see Rumelhart et al. *1986*, or T. Kohonen *1977, 1984*).

Here, as in the general case of axiomatization, we must make a careful choice of the constraints appropriate for explanation:

> If you chose to define as elementary objects things which are analogous to whole living organisms, then you obviously have killed the problem, because you would have to attribute to these parts just those functions of the living organism which you would like to describe or to understand. One also loses the problem by defining the parts too small, for instance, by insisting that nothing larger than a single molecule, a single atom, or single elementary particle [perhaps: a single spin, a single Ur, cf. von Weizsäcker] will rate as a part ... We are interested here in organizational questions about complicated organisms, and not in questions about the structure of matter or the quantum mechanical background of valency chemistry. So, it is clear that one has to use some common sense criteria about choosing the parts neither too large nor too small. (von Neumann *1949/1966*, p. 70)

The essential part of the synthetic method is the explanation of how the functioning of the whole is expressed in terms of the functioning of its elementary constituents and their interaction along their connections. The understanding is to be guided by architectonic partitionings into intermediate units of various levels composed from simpler units. In other words: "What principles are involved in organizing these elementary parts into functioning organisms, what are the traits of such organisms, and what are the essential quantitative characteristics of such organisms?" (ibid., p. 77)

In considering the different types of complexities involved, von Neumann assumed the following hierarchy of automata:

a) automata,
b) high complexity automata (showing at least certain kinds of intelligent behavior),
c) natural automata or organisms,
d) artificial automata,
e) computing machines,
f) analog devices,
g) digital devices,
h) digital computers
(cp. von Neumann *1949/1966*, p. 32/3).

Only the last mentioned automata are the so-called von Neumann machines, all others are not. The difference between digital computers such as von Neumann machines and most other automata lies in the architectonic simplicity of the former. An extreme case of architectonic simplicity are the Turing machines which, nevertheless, contain the essential units of digital computers, as we saw in the previous paragraph. In these machines, complexity of operation is fully expressed in the complexity of configurations fed into these devices and interpreted by them as programs.

The other automata have a much more involved architecture. There the synthetic axiomatization becomes imperative to describe them adequately; and consequently it becomes central for von Neumann's approach to such automata.

Another essential aspect connected with the axiomatization must be mentioned: the behavior defined for the elementary units and for their connectivities is given by functions

> ... under certain suitable operating conditions. ... These operating conditions are the ones under which it is normally used; they represent the functionally normal state of affairs within the large organism, of which it forms a part. Thus the important fact is not whether an organ has necessarily and under all conditions the ... character [defined] – this is probably never the case – but rather whether in its proper context, it functions primarily, and appears to be intended to function primarily, as [defined]. (von Neumann *1951/1963*, p. 298)

In the case of a human organism, it may function in a variety of ways, when asleep, waking up, being drunk, not having a sufficient supply of oxygen, etc. All of this may have an influence on the functioning of the neurons; however, the functioning to be defined in a systematic analysis of normal behavior is only the functioning of neurons for a normally attentive brain.

4. *Logical Calculi, Turing-Machines, and Brain Modules*

Though it may well be that all kinds of regular behavior could in principle be represented by a Turing machine – in particular when time and space of realization are neglected – von Neumann insists that there are essential differences between ways of descriptions.

The example of the communicative states of a city, discussed in § 1, shows that an integral description of a phenomenon may be possible in principle but that the resulting analysis may not be perspicuous or understandable. An appropriate distribution into naturally related and causally dependent parts of the phenomenon may contribute to perspicuity.

The same holds when one tries to enumerate general properties of a complicated phenomenon or to define it in terms of a logical proposition as close as possible. It may be that the phenomenon is best understood not in terms of a conjunction of properties but rather in terms of a specific analysis of cooperating units. Von Neumann believed that this requires the development of appropriate tools for the definition of distributed systems with specific connectivities.

The concepts of visual analogy are cases in point. Von Neumann argues as follows: the problem we are facing is "a general concept of identification of analogous geometrical entities" (von Neumann *1951/1963,* p. 311). Let me add that, from the point of view of the linguist, a similar, though in some sense even more complicated problem is the concept of identification in hearing, especially in the identification of speech sounds.

Von Neumann emphasizes that these problems provide only microscopic pieces of the general concept of analogy (or of similarity for the communicating organism). He assumes that "any attempt to describe it by the usual literary or formal-logical method may lead to something less manageable and more involved than the structured-design description of the connections of the visual brain" (ibid.). It is understood that an adequate understanding of the connectivities requires, in more complicated cases, that a hierarchical structure explains the organization of the net of connectivities into functional modules, submodules, sub-submodules etc. A corresponding description might be required for the structured design analysis of the identification (and production) mechanisms of speech sound configurations.

This is the challenge! Turing's original question was: Can a machine mimic a human computer in essential aspects? The concept of the Turing machine is an idealized but sufficient approximation. Turing's later question was: Can a machine mimic a human being in general? With respect to this question von Neumann's arguments must be taken seriously. His answer is: probably yes, but not if the problem is approached in terms of sequences of symbols and their manipulation, i.e., in terms of those representations uniquely applied in the usual "literary or formal-logical method". In other words: a Turing machine (or a Post-algorithm, formal calculus, or formal grammar, etc.) are structurally inadequate methods in problem areas having a high-complication structure, among them mathematical thinking and linguistic competence.

It is obvious that this relates to the basic question about the foundation of mathematics. It had been assumed by Hilbert, von Neumann's teacher, "that there is a discrepancy between the laws according to which the mind of *homo mathematicus* works, and the laws governing objective mathematical fact" (Detlefsen *1986*, p. ix). The latter is established by proof and thus is presented either, in the context of mathematical science, "by ordinary language" or, in the metamathematical science, by "formulas with mathematical and logical symbols, put in a sequence according to specific rules" (Hilbert *1925/1964*, p. 95); this is what von Neumann refers to by the term "literary or formal-logical method".

Hilbert stressed the simplifying and unifying effects of applying the method using symbols – his ideal method. In the process of simplifying our thought, ideal elements unify it. Von Neumann would agree; it is the very nature of axiomatization to introduce ideal conceptual elements which serve this purpose. When, however, Hilbert assumes furthermore that these ideal elements can always be expressed by symbols arranged in sequential order, von Neumann disagrees, as shown above: the arrangement and operation of ideal elements could be of a kind which is much more adapted to the analytic purpose of the organism than is a linear expression:

> When we talk mathematics, we may be discussing a *secondary* language, built on the *primary* language truly used by the central nervous system. Thus the outward forms of *our* mathematics are not absolutely relevant from the point of view of evaluating what the mathematical or logical language truly used by the central nervous system is . . . Whatever the system is, it cannot fail to differ considerably from what we consciously and explicitly consider as mathematics. (von Neumann *1958*, p. 82)

The contrast to Turing (or Post, etc.) is obvious: the question is not what we could compute or generate in principle on the basis of symbol sequence representations. The question is whether these representations are adapted to their purpose (and simple enough for treating complex problem areas) and whether they have a real basis. In an evolutionary sense the former is related to the latter: the units of the brain and the structure of their connectivities and interaction are those representations which are best adapted to the purpose of organizing understanding and controlled activity in the world. We must be interested in the structure of the former; otherwise understanding of complicated problems areas will be impossible in fact, in spite of being, perhaps, possible in principle.

We may now understand better von Neumann's statement that, in the process of trying to understand high-complication automata and, in particular, the nervous system in terms of an essentially logical theory, logic will have to undergo a pseu-domorphosis to neurology to a much greater extent than the reverse.

A first step in this direction was the conception of cellular growing automata developed by von Neumann (von Neumann *1949/1966*; see Arbib, this volume). They are certainly dissimilar to cellular networks in brain since they are strictly regular. But this initial idealization permitted to arrive at a number of very important conclusions, namely that a growing cellular automaton has the same recursive power as a Turing machine (Burks and Wang *1957*). This result shows that the notion of symbol sequences is not essential for computational theory.

A second step was the discussion of machines composed of unreliable units (von Neumann *1956*). Von Neumann considered this step as very important in clarifying the basis of a naturalized logic.

5. *The Importance of Architecture*

The importance of architectonic considerations has often been emphasized by von Neumann's pupil A.W. Burks. He wrote:

> The number of states assumed by any actual computer in even a short time is much too large for a state-by-state analysis to be of much use in viewing the whole computer, though such a mode of analysis is often valuable in designing small portions of the computer. In planning and operating a computer we must, therefore, organize it into interrelated units or compounds at various levels: half-adders, flip-flops etc. at the lowest level; registers, memory switches, counters etc. at the next higher level; arithmetic units, memories, input-output units, controls etc. at a still higher level. The particular organization of current computers reflect strongly the present state of technology and our current problem interests, and I feel sure that radically new organizations will appear in the future. But in any case, we must organize computers into some such hierarchy of units simply because they are too complicated for us to understand otherwise. (Burks *1963*, p. 108)

Burks points out that a Turing machine is extreme in having almost no archi-tectonic structure, this architecture being tuned very specifically to the introspective analysis of a human computer (as indicated above and studied in Schnelle *1981*). It has just two parts: reading and writing head and tape reflecting the difference between control and data. This is the minimal architecture for a data processing machine (in contrast to a data flow network, see below); it is thus extreme in its neglect of architectonic considerations.

In a recent article Burks (*1986*) summarizes the present situation in computer architecture in which we have to distinguish

a) the centralized structure of the von Neumann single-bus architecture, and
b) evolutionary von Neumann architectures relating to higher organizations of memories, arithmetic logic units, control, input-output, parallelism "in space" and "in instruction".

The common characteristics of the former systems is brought out by the syntax and semantics of von Neumann's program language: syntactically, each instruction has an address linking it to one or more instructions to be executed next. Semantically, each instruction has an address or addresses linking it to the data variables it is to operate on.

Radically non-von-Neumann architectures are those computer organizations which have a different type of programming language. Examples are data flow and demand-driven machines, for they package instructions and their datas together. Even more so are special purpose signal flow systems in which a distinction between instruction and data is impossible.

Burks generalized von Neumann's consideration discussed above. He stated in 1963 that since a general purpose computer is "designed to solve all problems in a very wide class ... [it] does not take advantage of the special properties of a particular problem. This limitation is inherent in the idea of a general-purpose computer" (Burks *1963*,p. 114). He suggested that for many problems it may be easier to think of the problem in terms of a special-purpose computer designed to solve the problem. In a sense the solution of the problem is then a kind of analog treatment, i.e., the structure of the processes to be analyzed is "paralleled by the structure of the special-purpose computer that solves the problem. In such a case the mathematical equation describes the behavior of a physical model" (Burks *1963*, p. 117). An example is a special purpose system for solving partial differential equations. Burks suggested that it may be convenient for computing the value of a function at all grid points simultaneously.

This example is a simple case. Further developments of this idea would require exercises in more complicated problem areas. I have studied systems of this type for language perception and production. The basic architectonic ideas are as follows:

The adequate representation of "internalized language" (cf. Chomsky *1986*, § 2.3) is a network of interactive units building the module of language competence. The network can be decomposed into two cascades of subnetworks, one for production and one for reception, with correlator networks between them (cf. Schnelle *1983*). The subunits of the cascade may be made to correspond roughly to linguistic levels (cf. Chomsky *1986*,p. 46, with reference to Chomsky *1975*, written in 1955-56).

In contrast to Chomsky's proposal to represent the levels in terms of strings of passive symbols and their manipulation as specified by rules, I assume each level as decomposed further into sub-sublevels each of which is represented as a set of simple data-flow shift registers together with configurations of arrays of information-flow structure detecting units (realizable by programmed logic arrays PLA's). In terms of such configurations it is possible to specify all possible phonological rule processors as well as constituent rule processors for production. (cf., Schnelle and Job *1983,* Schnelle and Rothacker *1986,* and Schnelle *1986*). The parallel processing of such systems may be easily simulated by programming them on an electronic worksheet. The parallel processing becomes very perspicuous and observable, step by step, on the screen of a personal computer (cf. Schnelle *1987*).

I shall not go into the details in the present context. What I want to emphasize is the following: in addition to von Neumann's proposal of regular cellular nets for the specification of behavior (forcefully taken up again recently, see D. Farmer et al. *1984*), we need architectonically guided constructions of nets whose structure of composition and of processing is the best representation of the inherent regularities of behavior. What we are looking for is a mathematically definable analogue of the perceptual and productive mechanisms in terms of architectonically determined interactive connectivities instead of in terms of rule determined manipulations of strings of passive symbols. This is what would finally complete von Neumann's unfinished project of specifying the logical properties of processing in complicated organisms – and of natural logic a fortiori – in terms of processing structure.

Turing's work provided an important starting point for von Neumann. Turing seemed to believe that his type of models were sufficient for a logical and practical understanding of human behavior. In contrast to this, von Neumann thought that insight into architecture and composition of organisms is essential, at least where proofs of feasibility in principle are not sufficient but practical and constructive understanding is required.

References

(Many citations are from reprints; for these, the last mentioned year is relevant)

Arbib, M.A.

    1988 From universal Turing machines to self-reproduction. This volume.

Burks, A.W.

    1963 Programming and the theory of automata. In: *Computer Programming and Formal Systems,* eds. P. Braffort and D. Hirschberg. Amsterdam: North-Holland Publ. Co. (1963).

    1986 A radically non-von-Neumann architecture for learning and discovery. In: *Conpar 86,* eds. W. Händler, D. Haupt, R. Jeltsch, W. Juling, and D. Lang. Berlin: Springer-Verlag (1986).

Burks, A.W., and H. Wang

    1957 The logic of automata I–II. *J. ACM* (1957) 193–218, 279–297, reprinted in: *A Survey of Mathematical Logic,* ed. H. Wang, Chap. VIII, pp. 145–223. Peking: Science Press (1962).

Chomsky, N.

    1980 *Rules and Representations.* New York: Columbia University Press (1980).

    1986 *Knowledge of Language. Its Nature, Origin and Use.* New York: Praeger (1986).

Detlefsen, M.

    1986 *Hilbert's Program. An Essay in Mathematical Instrumentalism.* Dordrecht: Reidel (1986).

Farmer, D., T. Toffoli, and S. Wolfram (eds.)

    1984 *Cellular Automata.* Amsterdam: North Holland Physics Publ. (1984) (= *Physica* **10D**).

Gödel, K.

    1931 On formally undecidable propositions of Principia Mathematica and related systems I. Reprinted in: *The Undecidable,* ed. M. Davis, pp. 4–38. New York: Raven Press (1965).

Hilbert, D.

    1925 Über das Unendliche (from *Math. Annal.* **95** (1925)) in: *Hilbertiana.* Darmstadt: Wissenschaftliche Buchgesellschaft (1964).

Kohonen, T.

    1984 *Self-Organization and Associative Memory.* Berlin: Springer-Verlag (1984).

Quine, W.V.

    1968 Epistemology naturalized. In: *W.V. Quine, Ontological Relativity and Other Essays,* pp. 69–90. New York: Columbia University Press (1969).

    1974 *The Roots of Reference.* La Salle, IL: Open Court (1974).

Rumelhart, D.E., G.E. Hinton, and J.L. McClelland

    1986 A general framework for parallel distributed processing. In: *Parallel Distributed Processing,* vol. 1, eds. D.E. Rumelhart and J.L. McClelland. Cambridge, MA: MIT Press (1986).

Schnelle, H.

    1981 Introspection and the description of language use. In: *A Festschrift for Native Speaker,* ed. F. Coulmas, pp. 105–126. The Hague: Mouton (1981).

    1983 Some preliminary remarks on net-linguistic semantics. In: *Psycholinguistic Studies in Language Processing,* eds. G. Rickheit and M. Bock, pp. 82–98. Berlin: de Gruyter (1983).

    1986 Array logic for syntactic production processors. In: *Language and Discourse: Text and Protext,* ed. J. Mey, pp. 477–511. Amsterdam: John Benjamins (1986).

1987 Ansätze zur prozessualen Linguistik. In: *Sprache in Mensch und Computer,* ed. H. Schnelle. Wiesbaden: Westdeutscher Verlag (1987).

Schnelle, H., and D.M. Job

1983 Elements of theoretical net-linguistics Pt. 2: Phonological nets. *Theor. Ling.* **10** (1983) 179–203.

Schnelle, H., and E. Rothacker, E.

1984 Elements of theoretical net-linguistics, Pt 3: Principles and fundamentals of dynamic nets for language processing. *Theor. Ling.* **11** (1984) 87-116.

Turing, A.M.

1936-7 On computable numbers, with an application to the Entscheidungsproblem. *P. Lond. Math. Soc. (2)* **42** (1936-7) 230–265; A correction, ibid **43** (1947) 544–546. Reprinted in: *The Undecidable,* ed. M. Davis, pp. 115–151. New York: Raven Press (1965).

1939 Systems of logic based on ordinals. Reprinted in: *The Undecidable,* ed. M. Davis, pp. 155–222. New York: Raven Press (1965).

von Neumann, J.

1932 *Mathematische Grundlagen der Quantenmechanik.* Berlin: Springer-Verlag (1932).

1947 The mathematician. In: *Collected Works,* vol. I, ed. A.H. Taub. New York: Pergamon Press (1963).

1949 *Theory of Self-Reproducing Automata,* edited and completed by A.W. Burks. Urbana, IL: University of Illinois Press (1966).

1951 The general and logical theory of automata. In: *Collected Works,* vol. V, ed. A.H. Taub. New York: Pergamon Press (1963).

1956 Probabilistic logics and the synthesis of reliable organisms from unreliable components. In: *Collected Works,* vol. V, ed. A.H. Taub. New York: Pergamon Press (1963).

1958 *The Computer and the Brain.* New Haven: Yale University Press (1958).

von Neumann, J., and O. Morgenstern

1944 *Theory of Games and Economic Behavior.* Princeton, NY: Princeton University Press (1944).

von Weizsäcker, C.F.

1985 *Aufbau der Physik.* München: Hanser Verlag (1985)

Complexity Theory and Interaction

Uwe Schöning

1. Introduction

Complexity theory is dealing with resource-bounded universal computing devices, and the most common device used is the Turing machine. The boundedness in resources affects either the space (the number of tape squares allowed) or the time (the number of elementary steps allowed). This kind of standard complexity theory has developed further in various ways. One particular interesting extension is the structural one, and one major issue in structural complexity theory is to develop (Turing machine) models which allow some kind of interaction with their environment, and to study these new models and corresponding complexity classes. These particular kinds of investigations which will be sketched in this overview have arisen from the still unsolved P-NP-problem: P is the class of sets (or rather: problems) that can be "efficiently" decided (or solved) by deterministic Turing machines, whereas the class NP consists of those sets that can be "efficiently" solved by nondeterministic Turing machines, i.e., those which, in each computational situation, are allowed to have several legal following actions. Furthermore, this model assumes that the machine is always "lucky" and chooses the best action, the one which leads to decide the status of the input quickly. Many interesting practical problems are located within NP and are not known to be in P, that is, are not known to be efficiently solvable by standard algorithms. Moreover, for these problems it can be shown that either none of them is in P, or all of them are, and the latter case happens if and only if $P = NP$ (which is the open question). This "if and only if" statement arises from the notion of NP-completeness which originated from the papers of Cook *1971* and Karp *1972,* and independently by Levin *1973.*

In the context of the P-NP-problem the question came up: How much can the computational behavior of a Turing machine be improved if some kind of interaction with the outside world is allowed? This can be formalized in several ways. First, we will consider "oracles" which allow certain parametrizations of the underlying complexity classes, like P and NP. We will present Baker, Gill, and Solovay's result from *1975* (independently obtained by Dekhtyar *1976*) that there are oracles A and B such that $P(A) = NP(A)$ and $P(B) \neq NP(B)$. This somewhat confusing result has led to a couple of interesting interpretations and investigations. The question can be considered: What happens if the oracle information available is restricted by some kind of bottleneck. We will see that this leads to another characterization of "polynomial-size circuits" or "nonuniform complexity". A third approach is the notion of "helping" which leads to a different characterization (and understanding) of the class $NP \cap$ co-NP, i.e., those problems in NP whose complementary problems are also located in NP.

A one-sided and probabilistic version of helping finally leads to the notion of interactive proof systems, an intriguing notion which has just come up very recently.

2. *Preliminaries*

The standard model of computation in complexity theory is the *multitape Turing machine*. It has a separate (read-only) input tape, an arbitrary, but fixed, number of work tapes, and operates in the well-known way. Two versions of such Turing machines have to be considered: the Turing machine *acceptor* has a special *accepting* final state, and it is intended for defining the *language* accepted by this Turing machine M. Let $L(M)$ be the set of all strings x over the input alphabet Σ (which should contain 0 and 1; in most cases it is sufficient to assume $\Sigma = \{0, 1\}$) such that M started with x on the input tape has a computation that leads to the accepting final state. This definition of $L(M)$ can be applied to both deterministic and nondeterministic Turing machines.

The second Turing machine version to be considered is the *transducer,* which is always deterministic and additionally has a (write-only) output tape. The final states need not be separated into accepting and nonaccepting states. A Turing machine transducer M simply is intended to define a function $f_M : \Sigma^* \to \Sigma^*$, such that $f_M(x)$ is the content of the output tape when M started on input x reaches a final state.

In complexity theory, the Turing machines considered have to obey certain restrictions, either on the number of computation steps allowed, or on the number of tape squares used, or both. Define a Turing machine M to be $t(n)$-time bounded where t is a function on the natural numbers if M on inputs of size n stops always after at most $t(n)$ many computation steps. Similarly, M is $s(n)$-space bounded if no computation of M on inputs of size n uses more than $s(n)$ many tape squares (where only the tape squares on the work tapes are counted).

Define

$DTIME(t(n)) = \{L(M)|M$ is deterministic and $t(n)$-time bounded$\}$,

$NTIME(t(n)) = \{L(M)|M$ is nondeterministic and $t(n)$-time bounded$\}$,

$DSPACE(s(n)) = \{L(M)|M$ is deterministic and $s(n)$-space bounded$\}$,

$NSPACE(s(n)) - \{L(M)|M$ is nondeterministic and $s(n)$-space bounded$\}$.

Using these notions, the most prominent complexity classes considered in the literature can be defined:

$$DLOGSPACE = \cup_{c>0}DSPACE(c \cdot \log n)$$
$$NLOGSPACE = \cup_{c>0}NSPACE(c \cdot \log n)$$
$$P = \cup_{k>0}DTIME(n^k)$$
$$NP = \cup_{k>0}DSPACE(n^k)$$
$$DPSPACE = \cup_{k>0}DSPACE(n^k)$$
$$NPSPACE = \cup_{k>0}NSPACE(n^k)$$

Since it can be shown (see Hopcroft and Ullman *1979*) that $DPSPACE = NPSPACE$, we can abuse the notation and call this class $PSPACE$. Furthermore, the following inclusions are not hard to see:

$$DLOGSPACE \subseteq NLOGSPACE \subseteq P \subseteq NP \subseteq PSPACE$$

but none of these inclusions is known to be proper, it can only be shown yet that $NLOGSPACE \neq PSPACE$.

In the following, we will focus attention on the most famous open problem in this context: the *P-NP*-problem, that is, the question whether $P = NP$ or $P \neq NP$. This question is so important because the class P can be considered as the class of languages (or problems) that can be solved by efficient algorithms, but on the other hand, for many problems no such efficient algorithms are known – but are most desirable from the practical standpoint. For many of these problems it turns out that they are a member of NP. Hence, if $P = NP$, all these problems do have efficient algorithms. The most outstanding result in theoretical computer science during the last decades is that the converse is also true for these problems: if any of these problems (which now are called NP-complete) has an efficient algorithm, then $P = NP$, and hence all of them have.

What does NP-completeness mean? First we need to define the concept of polynomial reducibility. Call a set $A \subseteq \Sigma^*$ polynomial *reducible* to set $B \subseteq \Sigma^*$ if there exists a $p(n)$-time bounded Turing machine transducer M where p is a polynomial such that for all strings $x \in \Sigma^*$, $x \in A$ if and only if $f_M(x) \in B$. We write symbolically $A \leq_p B$ if this definition holds.

In a sense, the transducer M is able to efficiently transform membership questions for A into membership questions for B. Hence, it is not hard to see that $A \leq_p B$ and $B \in P$ ($B \in NP$) implies $A \in P$ ($A \in NP$, resp.). Now, call a set $A \in NP$ to be *NP-complete* if for all sets $B \in NP$, $B \leq_p A$.

Assume there exists a NP-complete set, call it A. Using the above observations it follows immediately that if $A \in P$ (i.e., A is efficiently solvable) then for any $B \in NP$ we get $B \in P$, hence $P = NP$.

Today, hundreds of problems from all areas of computer science, logic, game theory, operations research, and number theory have been shown to be NP-complete (see Garey and Johnson *1979*. But this would not be possible if there were not a first problem shown to be NP-complete. This is due to Cook *1971*, and the problem was the satisfiability problem in propositional logic (cf. also Levin *1973*). Stated as a language, this is

$$SAT = \{x \in \Sigma^* | x \text{ is encoding of some propositional formula}$$
$$\text{which has a satisfying assignment}\}$$

Furthermore, it is the merit of Karp *1972* to show how to build upon Cook's result to demonstrate NP-completeness of many other combinatorial problems.

Now we start to leave the "standard" way of doing complexity theory and turn to more "structural" issues. One such important concept is the *polynomial-time hierarchy* (cf. Stockmeyer *1977*). First observe that the class NP can be equivalently characterized as follows. A set A is in NP if and only if there is a set $B \in P$ and polynomial p such that

$$A = \{x \mid (\exists y, |y| = p(|x|)) \, (x, y) \in B\}.$$

Here $|y|$ denotes the size of string y, and (x, y) denotes some standard way of pairing strings, i.e., an efficiently computable bijection from $\Sigma^* \times \Sigma^*$ to Σ^*.

This characterization of NP resembles very much a similar characterization of the recursively enumerable sets, and also suggests to consider the analogue of the arithmetical hierarchy (see Rogers *1967*). Define Σ_k^p to be the class of sets A for which there exists a set $B \in P$ and a polynomial p such that

$$A = \{x | (\exists y_1, |y_1| = p(|x|)) \quad (\forall y_2, |y_2| = p(|x|)) \dots$$
$$(Q y_k, |y_k| = p(|x|)) \quad (x, y_1, \dots, y_k) \in B\}$$

where $Q = \forall$ if k is even, and $Q = \exists$ otherwise. Let Π_k^p be the set of complements of the sets in Σ_k^p, in symbols: $\Pi_k^p = \text{co}-\Sigma_k^p$. Then we get

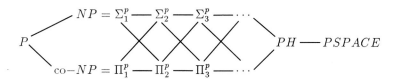

where a line means inclusion (but none of them is known to be proper), and PH is defined as the union over all Σ_k^p classes, $k \geq 1$. We summarize some facts about the polynomial-time hierarchy.

1. Every level contains complete sets.
2. If PH has a complete set, then the hierarchy "collapses", i.e., for some k, $\Sigma_k^p = \Sigma_{k+1}^p = \cdots = PH$.
3. If $PH = PSPACE$, then the polynomial hierarchy collapses (reason: $PSPACE$ has a complete set).
4. If $\Sigma_k^p \neq \Sigma_{k+1}^p$ for some k, then $P \neq \Sigma_1^p \neq \Sigma_2^p \neq \cdots \neq \Sigma_k^p$. (In particular, it follows $P \neq NP$.)
5. $\Sigma_k^p = \Sigma_{k+1}^p$ if and only if $\Sigma_k^p = \Pi_k^p$. (In particular, $\Sigma_1^p = \Sigma_2^p$ if and only if NP is closed under complementation.)

We consider the polynomial-time hierarchy more to be a structural notion, because from the standpoint of existence or nonexistence of efficient algorithms, the best known algorithms for the sets in the polynomial hierarchy use an exponential number of computation steps (i.e., $2^{p(n)}$ for some polynomial p), which is just the same bound as for NP and $PSPACE$.

3. Oracles

When studying the relative computability (or complexity) of problems, we want to prove statements like: "Problem A is (efficiently) computable if any desired information about problem B is given for free". Note that B might even be noncomputable (in this case, A probably will be so, too). In a sense, the idea behind the above statement is the same as in the notion $\leq_p$ used to define NP-completeness: "Every problem in NP can be efficiently solved if (enough) information about any of the NP-complete sets is given for free."

Indeed, the notion we want to study can be considered as some kind of reducibility, and has been named Turing reducibility after its inventor. The machine model used to define Turing reducibility is the oracle Turing machine: in addition to the usual devices an oracle Turing machine has a (write-only) *oracle tape* and three additional states, the *query-state,* the *yes-state,* and the *no-state.* At any time during the computation the machine is allowed to transfer into the query state. After this action, the next state can be either the yes-state or the no-state, but this decision does not depend on the machine, rather it is given from "outside": if the string present on the query tape at that moment is element of the set B (now called the *oracle set),* then the next state is the yes-state, otherwise the no-state. After this "oracle query" the oracle tape is "magically" (i.e., simply by definition) erased, and the computation can proceed as normal. But observe that the machine computation after this oracle query now has this one bit of information available – whether the string on the oracle tape is in the oracle set or not – and is free to use this information in the following course of the computation.

We denote by $L(M, B)$ the set accepted by oracle Turing machine M when using the oracle set B. Note that the accepted set is not uniquely determined by M alone, it also depends on the oracle set B. That is, we may consider M to be an *operator* on the set of languages. Given (oracle) language B, M maps it to the language $A = L(M, B)$, i.e., the set of strings which can be computed by M provided information about membership in B is given "for free".

Resource-bounded oracle Turing machines can be defined in just the same way as we did it for usual Turing machines. Call $P(C)$ (or $NP(C)$) the class of languages that can be computed by a deterministic (or nondeterministic, resp.) and polynomial-time bounded oracle Turing machine when using the set C as the oracle.

As an application of this notion, it can be shown that for each $k \geq 1$,

$$\Sigma_k^p = \cup \{NP(C) \mid C \in \Sigma_{k-1}^p\} \quad \text{(letting } \Sigma_0^p = P).$$

In recursive function theory, a certain "relativization principle" is known (which is not a theorem, but a general principle, i.e., no counterexample to it is known): if some statement about Turing machine computations is true, then, switching to oracle Turing machines, the analogue statement is true in the presence of any oracle set. For example, it can be proved that there exist sets which are recursively enumerable, but not recursive (the halting problem for Turing machines is such a set). Virtually the same proof shows that for each oracle set A there exist sets which are recursively enumerable in A but not recursive in A. The notions "recursive in A" and "recursively enumerable in A" have as their counterparts in complexity theory the notions $P(A)$ and $NP(A)$, resp. Therefore, the following unexpected result surprised the experts quite a bit. It shows that the relativization principle definitely does not hold in complexity theory.

Theorem (Baker, Gill, and Solovay *1975*; Dekhtyar *1976*):
a) There exists an oracle set A with $P(A) = NP(A)$.
b) There exists an oracle set B with $P(B) \neq NP(B)$.

Proof. For a) it is sufficient to take any *PSPACE*-complete set A. Then, $P(A) = P(PSPACE) = PSPACE$, and $NP(A) = NP(PSPACE) = NPSPACE = PSPACE$. This is a consequence of Savitch's theorem (see e.g., Hopcroft and Ullman *1979*).

b) The set B is constructed by diagonalization. First note that for any set B, the set

$$L_B = \{0^n | \text{ there exists a string of size } n \text{ in } B\}$$

is obviously in $NP(B)$. Now a specific set B is constructed with $L_B \notin P(B)$ which proves the theorem. Let $M_1, M_2, \ldots$ be an enumeration of all polynomial-time bounded deterministic oracle Turing machines, and let $p_1, p_2, \ldots$ be their corresponding polynomial time bounds. The desired set B is constructed in stages as follows:

Stage 0: Let $B := \emptyset$ and $n := 0$;

Stage $k > 0$: Let m be the smallest integer satisfying
   (1) m is greater than the size of the longest string queried by $M_1, \ldots, M_{k-1}$ on inputs of size $\leq n$,
   (2) $2^m > p_k(m)$.
Give n the new value $n := m$;
If $0^n \in L(M, B)$ then goto the next stage
   else let w be some string of size n which is not queried by M on input 0^n using oracle B (such a string exists because of (2)), and let $B := B \cup \{w\}$ and **goto** the next stage.

The so constructed set B has the property $L_B \notin P(B)$. Suppose to the contrary that $L_B \in P(B)$, then there is an index k such that $L_B = L(M_k, B)$. But now, look at the construction of B at stage k and let n' be the value of n at stage k. If $0^{n'} \in L(M_k, B)$ then there is no string of size n' put into B. On the other hand, if $0^{n'} \notin L(M_k, B)$ then there is a string of size n' in B (namely w). Also observe that no stage interferes with other stages because of the choice of n done in (1). This argument shows that $0^{n'} \in L_B$ if and only if $0^{n'} \notin L(M_k, B)$, which is a contradiction to the assumption $L_B = L(M_k, B)$. □

This result has led to a lot of different interpretations and various directions of further research. Note that the structure of the two statements a) and b) looks very much like an independence result: to show that some assertion is independent of a set of axioms, construct a model for the set of axioms and the assertion; and construct another model for the set of axioms and the complement of the assertion. An exploitation of this argument has started with Hartmanis and Hopcroft's *1976* article. The general leading idea is to show that the statement "$P = NP$" is independent of the formal system of reasoning in mathematics, like Zermelo-Fraenkel's set theory. But the results obtained along these lines are very far away from any strong statement like this.

A less extreme standpoint is to say that if for an assertion (like "$P = NP$") oracles are known which can make the relativized assertion both true and false, then the assertion seems to have a difficult proof in the unrelativized case. And certainly, a potential proof must contain aspects which do not carry over when introducing oracles. So the standard proof techniques known from recursive function theory will not suffice alone to prove a statement like "$P = NP$" or "$P \neq NP$".

A third argument says: When relativizing a statement like "$P = NP$" then the underlying problem disappears. Because, examining Baker, Gill, and Solovay's proof we see that the diagonalization which makes $L_B \notin P(B)$ has nothing to do with the nature of the P-NP-problem, but rather the "unfair" situation is exploited that a deterministic and polynomial-time bounded oracle Turing machine can only search through a polynomial-size oracle space. Hence such a machine can be "fooled" by hiding crucial information in the oracle where the machine cannot "see" it. So, the idea came up to make the access to the oracle information "fair", in the sense that the nondeterministic oracle machine has to obey the same bounds as the deterministic machine naturally has. Let $NP_b(C)$ be the class of languages that can be recognized by nondeterministic and polynomial-time bounded oracle Turing machines when using the oracle set C, where for each input there must be a uniform polynomial bound on the number of different strings to be queried in the computation tree induced by M, the input, and C.

Theorem (Book, Long, and Selman *1984*):
$P = NP$ if and only if for each oracle set A, $P(A) = NP_b(A)$.

(For a more formal definition of $NP_b(A)$ and the proofs see the above reference). The theorem says that a potential proof of "$P = NP$" does relativize with respect to arbitrary oracles when the $NP_b(\ )$-operator is used. An analogue statement is not known to hold for "$P \neq NP$" instead "$P = NP$".

Instead of restricting the access to the oracle information one might consider restricting the oracles themselves. Call a language A *sparse* if there is a polynomial function p such that for each n, $p(n)$ is an upper bound for the number of strings in A of size n.

Theorem (Long and Selman *1986*; Balcázar, Book, and Schöning *1986*):
a) The polynomial-time hierarchy has only finitely many distinct levels if and only if the polynomial-time hierarchy is finite with respect to any sparse oracle.
b) The polynomial-time hierarchy has infinitely many distinct levels if and only if the polynomial-time hierarchy is infinite with respect to any sparse oracle.

So, regarding the question whether the polynomial-time hierarchy "collapses" or whether it is infinite, it is enough to settle this question with respect to any sparse oracle set.

If such an assertion would also hold for the *P-NP*-question then $P \neq NP$ would be proved because the oracle B constructed in Baker, Gill, and Solovay's theorem is sparse!

4. *Advice*

Karp and Lipton *1980* define and investigate a Turing machine model which permits the insertion of outside information into the computation. This model is similar to the oracle machine model – certain links exist which we will later explore – but the models are not equivalent. Actually, the motivation for introducing Turing machine models with "advice" stems from the context of nonuniform or circuit-size complexity theory. We will present Karp and Lipton's definition in the following, as well as the interrelationships that exist to the complexity notions we have studied before.

For a class of languages C and a class of functions F (from N to Σ^*), let C/F be the class of languages such that $A \in C/F$ if there exist $B \in C$ and $f \in F$ with

$$A = \{x \mid (x, f(|x|)) \in B\}.$$

That is, for each size n, $f(n)$ gives some kind of "advice" which helps to decide membership questions for A with the resources sufficient to recognize B (which might be much less than those for A). Therefore, the function f is called the *advice function*. Karp and Lipton were particularly interested in the case $C = P$ and $F = $ Poly, where Poly denotes the *polynomially bounded* functions $f : N \rightarrow \Sigma^*$, i.e., $f \in$ Poly if and only if there is a polynomial p such that for all n, $|f(n)| \leq p(n)$.

A collection of Boolean circuits (with "and", "or", "not"-gates) can be used to describe a language A as follows. For each size n we fix a circuit c having n inputs and one output where c on input $x = x_1 x_2 \ldots x_n$ outputs 1 if and only if $x \in A$. A set A has *polynomial-size circuits* if there is a polynomial p such that for each n there is a circuit for A with n inputs as above having at most $p(n)$ many gates.

Theorem (Karp and Lipton *1980*; Berman and Hartmanis *1977*):
The following are equivalent:
(a) A has polynomial-size circuits,
(b) $A \in P/Poly$,
(c) $A \in P(S)$ for some sparse oracle set S.

Proof. $(a) \to (b)$: Suppose that A has polynomial-size circuits. For each n let $f(n)$ be some encoding (as a string) of the smallest circuit with n inputs that describes A in the above sense. Then, clearly, $f \in$ Poly, and also there is a set $B \in P$ that evaluates the circuit on the given input x of size n. Thus, f and B witness that $A \in P/\text{Poly}$.

$(b) \to (c)$: Let $B \in P$ and $f \in$ Poly be given such that $A = \{x | (x, f(|x|)) \in B\}$. Encode f into a language S as follows: $S = \{(0^n, w) | w$ is a prefix of $f(n)\}$. Then it can be seen that S is sparse, and one can design an oracle Turing machine which, on input x of size n, queries the oracle for strings of the form $(0^n, w)$, starting with the empty string w. By this, the machine obtains the information about $f(n)$ in a bit by bit fashion. Then the machine evaluates B on input $(x, f(n))$. This argument shows that $A \in P(S)$.

$(c) \to (a)$: Turing machine computations can be encoded as Boolean circuits such that the size of the circuit (the number of gates) is proportional to the product of the time and the space used by the machine. Also note that, by direct constructions, sparse sets have polynomial-size circuits. Both constructions can be combined to yield polynomial-size circuits for the set $A \in P(S)$. $\square$

So we have seen that adding the advice capability to a Turing machine essentially yields the nonuniform or circuit-size complexity model. This is called nonuniform since for each size a new computational model (circuit) is provided, whereas uniform complexity indicates that there is just one model (the standard Turing machine) which can do all computations, for all sizes.

What are the interrelationships between uniform and nonuniform complexity theory? There are some obvious ones. E.g., if there is some NP-complete set which can be shown not to have polynomial-size circuits, then $P \neq NP$. Karp and Lipton *1980* found an astonishing answer to the inverse question: What happens if each set in NP does have polynomial-size circuits?

Theorem: *If each set in NP has polynomial-size circuits (or equivalently, $NP \subseteq P/Poly$), then the polynomial hierarchy collapses to its second level, i.e., $\Sigma_2^p = \Pi_2^p = \Sigma_3^p = \Pi_3^p = \ldots$*

Proof (following Hopcroft *1981*): It suffices to show that the hypothesis implies $\Sigma_2^p = \Pi_2^p$. Let $A \in \Pi_2^p$, then A has the representation:

$$A = \{x \mid \forall y \exists z \, (x, y, z) \in B\}$$

for some set $B \in P$ and polynomially length-bounded quantifiers.

Define

$$L = \{(x, y) \mid \exists z \, (x, y, z) \in B\}.$$

Clearly, $L \in NP$. By the hypothesis, L has polynomial-size circuits. It can be shown that in this case L has also polynomial-size circuits which not only output 1 or 0 depending on whether $(x, y) \in L$ or not, but in case of membership in L, they output the "witness" string z. (Actually, to make this argument really formal, one has to consider the "self-reducibility" properties of NP sets; see Balcázar *1987*). Now, A can be expressed in a different way:

$$A = \{x | \exists \text{ circuit } c \, \forall y [c \text{ on input } (x, y) \text{ outputs } z \text{ and } (x, y, z) \in B]\}.$$

Since the expression in brackets is a predicate in P, this representation of A shows that $A \in \Sigma_2^p$, and hence the polynomial hierarchy collapses. $\square$

5. *Helping*

In both, the oracle machine model and the advice model, the accepted language can change once another oracle or advice is "plugged in". Exactly this is forbidden in the following definition. Call an oracle Turing machine M *robust*, if for all oracle sets A the language accepted by the machine M when using oracle A stays the same, e.g., the same as with the empty oracle.

But, one might ask what effect the oracle now can have if the machine is not allowed to "rely" on the oracle information in the strong sense that the accepted language can be affected. Still, the oracle can influence the running time of the machine. This looks very much like the situation in a potential man-machine interaction: the human problem-solver and the machine try to solve some computational problem. From time to time the machine stops and asks the user how to proceed, which strategy to use, based on the partial results obtained so far. Now the human problem-solver uses his/her intuition or just luck to give his/her answers. But even under bad decisions the problem should finally be able to be solved. If the problem-solver gives wise (or lucky) answers then we can hope for an efficient computation.

Let's say that an oracle (or problem-solver) A *helps* the robust oracle Turing machine M if M, when using A as oracle set, becomes polynomial-time bounded. (In general, using other oracles, M can be exponential-time or even worse.) Now, what problems can be solved using lucky (or wise) problem-solvers? More formally, we want to know what the following class of languages is.

$$P_{help} = \{L | \text{ there is a robust oracle machine } M$$
$$\text{accepting } L \text{ and there is an oracle } A$$
$$\text{that helps } M\}.$$

The answer is given in the following theorem.

Theorem (Schöning *1985*):

$$P_{help} = NP \cap \text{co}-NP.$$

Proof. Let $L \in P_{help}$. Then there is a robust oracle machine M accepting L which is being helped by some "wise" oracle A. Let p be the polynomial-time bound which M achieves when using A as oracle. Construct a new machine M_1 as follows: M_1 simulates the behavior of M but treats each oracle query as a nondeterministic guess (thus M_1 becomes a nonoracle and nondeterministic machine). Further, a "clock" is attached to M_1 that shuts off each computation after $p(n)$ steps (and in this case M_1 rejects). Then it can be seen, using the robustness of M, that M_1 accepts L, and hence $L \in NP$. On the other hand, considering a dual machine M_2 (which accepts whenever M rejects within $p(n)$ steps) gives $\overline{L} \in NP$ where $\overline{L}$ is the complement of L. Hence $L \in$ co-NP.

Conversely, suppose $L \in NP \cap$ co-NP. Then there are sets $B_1, B_2 \in P$ and a polynomial p such that

$$L = \{x \mid (\exists y, |y| = p(|x|)) \ (x, y) \in B_1\} \text{ and,}$$
$$\overline{L} = \{x \mid (\exists y, |y| = p(|x|)) \ (x, y) \in B_2\}.$$

Further, since nondeterministic, polynomial-time computations can be simulated deterministically in exponential time, there exists an exponential-time bounded deterministic Turing machine M' that accepts L. Now the following oracle Turing machine M accepts L and is robust: M on input x, $|x| = n$, asks the oracle for the string $(x, 1)$. In case of answer yes, M continues querying for $(x, 11)$, otherwise for $(x, 01)$. Continuing in this fashion (inserting a 1 in case of a yes-answer, and 0 otherwise), M finally comes up with a string y of size $p(n)$. If $(x, y) \in B_1$ then M accepts; if $(x, y) \in B_2$ then M rejects. If none of them hold, then M starts a "slow" computation simulating M' and by this finally recognizes L correctly. Now, M is clearly robust. No matter what the oracle is, the set L will be correctly recognized (possibly in exponential time). But when using the following oracle A, M runs in polynomial time. This shows that $L \in P_{help}$.

$$A = \{(x, y) \mid w \text{ is a prefix of a string } y \text{ such that}$$
$$\text{either } (x, y) \in B_1 \text{ or } (x, y) \in B_2\}. \qquad \square$$

This result shows that "helping" (in the above sense) can make sets in $NP \cap$ co-NP efficiently solvable, although there is the potential possibility that $P \neq NP \cap$ co-NP. (This is in interesting contrast to recursive function theory, since there it is obvious that a set is recursive if and only if the set and its complement are recursively enumerable).

6. Interactive Proof Systems

The class NP can be considered to consist exactly of those languages for which membership is succinctly provable. Let $A = \{x | \exists y (x, y) \in B\}$ be a typical set in NP. A "proof" of the assertion "$x \in A$" consists of a presentation of y. The correctness of the "proof" can be efficiently decided using the predicate B. Of course, the fact that the proof y is short (i.e., polynomial in the length of x) does not speak to the question how difficult it is to find y among the exponential number of such potential y's. (This is the very nature of the N-NP-problem!)

The situation can be viewed in terms of an interaction between a *prover* and a *verifier,* where the prover's aim is to convince the verifier that the given string x is in the language A. The protocol of this interaction is quite simple: the given string x is shown to both, the prover and the verifier, then the prover starts by "magically" finding a suitable y and sending it to the verifier. The verifier's task then is to check y (i.e., test whether $(x, y) \in B$) and finally accept or reject. This protocol for NP sets is so simple: why not allow more extensive interactions between the prover and the verifier? In particular, why not including the possibility that the verifier sends something to the prover? These ideas led Goldwasser, Micali, and Rackoff (*1985*) to the invention of *interactive proof systems* (see also Goldreich, this volume). Their definition (which we will present formally later) leads to a class of languages being possibly a proper superset of NP. Furthermore, the authors allow the verifier (possibly also the prover) to act randomly. This will be modeled by probabilistic Turing machines.

Let us start with an example. The graph isomorphism problem is, given two graphs G_1 and G_2, determine whether they are isomorphic. This problem (encoded as a language) clearly belongs to NP. Stated in the prover-verifier terminology, given G_1 and G_2, the prover sends a permutation of the vertices of G_1 to the verifier who checks whether the permutation applied to G_1 gives G_2. If so, the verifier accepts (the fact that the graphs are isomorphic).

It is not known whether graph isomorphism is NP-complete, neither is it known whether the problem is in P. For NP-complete problems it is very unlikely that short proofs of *non*-membership in the language exist (this would imply $NP = $ co-NP). It is not even likely that there are short *interactive* proof-systems for nonmembership. But Goldreich, Micali, and Wigderson *1986* found such an interactive proof protocol for graph *non*-isomorphism. Given G_1 and G_2, the verifier starts by randomly selecting $i \in \{1, 2\}$ and randomly selecting a permutation π of the vertices of G_i. The verifier then sends the graph $H = \pi(G_i)$ to the prover (while keeping the random choices secret). The prover then has to decide whether the permuted graph was obtained from G_1 or from G_2. That is, the prover picks $j \in \{1, 2\}$ and sends j to the verifier. The verifier checks whether $i = j$, and if so, he/she accepts. (Even better: prover and verifier begins playing this game n times, and the verifier finally accepts only if the prover was correct each time.) Now, if the graphs were isomorphic, then, under which strategy whatsoever, the prover has no better chance than 50% to find the correct j. For nonisomorphic graphs, the prover can always answer correctly. Hence, if the prover answers correctly n times, then the verifier accepts, and makes a mistake with probability 2^{-n} (in case the prover was just lucky). This argument does not prove that graph isomorphism is in $NP \cap$ co-NP but it is very close to it, and it makes it very unlikely that this problem ever turns out to be NP-complete.

Another point should be noted. In the simple protocol for graph isomorphism stated first, the verifier is also able to convince a third person of the fact that the graphs G_1, G_2 are isomorphic, since now he/she knows the isomorphism π.

The situation seems quite different in the second probabilistic protocol for graph nonisomorphism. Even if the verifier is convinced that the graphs are nonisomorphic, he/she will probably not be able to convince anyone else of this fact. The prover apparently hasn't given away any additional information to the verifier apart from the fact that the graphs are nonisomorphic. That is to say, interactive proof systems do or do not have the so-called *zero knowledge* property. The formal definition of zero knowledge is not easy to grasp, it speaks about the probability spaces involved (before and after the interaction). Furthermore, several definitions exist which have quite subtle differences in their strength. E.g., under one definition every NP set has zero knowledge interactive proof systems, but apparently not under another stronger definition (cf. Fortnow 1987). Also, to make the above protocol for nonisomorphism really zero knowledge according to the formal definitions, one has to allow the prover also to act randomly. We will not go deeper into these issues, and instead, consider the notion of interactive proof system as such – as an interesting generalization of the class NP.

We will now define formally what an interactive proof system is. First, we need to define probabilistic Turing machines. A probabilistic Turing machine is just a nondeterministic Turing machine where each nondeterministic branch is assigned a probability. For simplicity (and without loss of generality) we may assume that in each step the machine has exactly two choices, each of them has probability $\frac{1}{2}$. This implies that a specific computation of length k occurs with probability 2^{-k}. Let Prob[$M(x) = 1$] be the probability for the event that M on input x reaches an accepting final state.

A probabilistic Turing machine can additionally be equipped with the oracle facility. Let M^A denote that M uses the oracle set A. Hence, Prob $[M^A(x) = 1]$ is the probability that M on input x, when using the oracle set A, accepts. Now we have all tools available to define formally what an interactive proof system is. The definition can be understood as a one-sided and probabilistic version of helping.

Definition: *A set L has an interactive proof system (in symbols: $L \in IP$) if there exists a probabilistic and polynomial-time bounded oracle Turing machine M such that (1) and (2) hold.*

$$(1) \ \exists A \ \forall x \in L \quad \text{Prob}\,[M^A(x) = 1] \geq \sqrt[3]{4},$$

$$(2) \ \forall A \ \forall x \notin L \quad \text{Prob}\,[M^A(x) = 1] \leq \sqrt[1]{4}.$$

The definition says that there exists a "prover strategy" (the oracle A in (1)) to convince the verifier (the machine M) with high probability of the fact that $x \in L$. On the other hand, for strings x not being in L, no prover whatsoever can convince the verifier of their membership in L. This is what (2) says. Note that Goldwasser, Micali, and Rackoff *1985* define interactive proof systems in terms of *two* Turing machines (one for the prover, one for the verifier) which communicate by a common tape.

There is nothing special about the constants ¾ and ¼ occurring in the definition. Without changing the definition, the constants can be changed as long as the first is greater then ½ and the second is smaller than ½. It is even possible to substitute the constants by $1 - 2^{-n}$ and 2^{-n}, resp.

It is easy to see that $NP \subseteq IP \subseteq PSPACE$, but none of these inclusions is known to be proper. The class IP is not even known to be included in the polynomial hierarchy. Now we are in the situation to give a more formal proof of the above mentioned protocol for graph nonisomorphism.

Theorem (Goldreich, Micali, and Wigderson, *1986*):
Graph Non-Isomorphism $\in IP$.

Proof. We have to describe a suitable probabilistic oracle machine M. On input $x = (G_1, G_2)$, M randomly selects $i, j \in \{1, 2\}$ and also random permutations π, σ. Then M asks its oracle for the two strings $(G_1, \pi(G_i))$ and $(G_1, \sigma(G_j))$. If it is true that the first oracle answer is yes if and only if $i = 1$, and also the second oracle answer is yes if and only if $j = 1$, then M accepts, otherwise M rejects. Now, on input $x = (G_1, G_2)$ being two nonisomorphic graphs, using $A = $ Graph Isomorphism as the oracle set, M always accepts. This proves (1). On the other hand, if the input $x = (G_1, G_2)$ is a pair of isomorphic graphs, then for each oracle A, Prob $[M^A(x) = 1] = 1/4$. This proves (2).   □

In the graph isomorphism example it turns out that only a constant number of interactions between prover and verifier is needed. Therefore, one can define separately IP_{const} to be the class of sets having an interactive proof system with only a constant number of interactions. Then, obviously $NP \subseteq IP_{\text{const}} \subseteq IP$. But it turns out that the subtle difference between IP and IP_{const} seems more significant that it looks at first glance: IP_{const} is a member of the polynomial hierarchy, $IP_{\text{const}} \subseteq \Pi_2^p$, and also $IP_{\text{const}} \subseteq NP/\text{Poly}$ (cf. Goldwasser and Sipser *1986*). These both inclusions are not known to hold for the general class IP.

Theorem (Boppana, Hastad, and Zachos *1987*; cf. also Schöning *1987*):
If every set in co-NP has a constant-round interactive proof system (in symbols: co-NP $\subseteq IP_{\text{const}}$) then the polynomial hierarchy collapses to $\Sigma_2^p = \Pi_2^p$.

We will not prove this theorem here, but just point out that by the fact that $IP_{\text{const}} \subseteq NP/\text{poly}$, the assumption of the theorem implies co-$NP \subseteq NP/\text{Poly}$ (or equivalently, $NP \subseteq$ co-NP/Poly) which looks very similar to the second theorem proved in Section 4.

The theorem has some interesting consequences. If the complement of any NP-complete set would have a constant-round interactive proof system, then the polynomial hierarchy collapses. Further, by the fact that graph nonisomorphism does have an interactive proof system, it follows that the graph isomorphism problem cannot be NP-complete unless the polynomial hierarchy collapses.

We finish the overview about results concerning different forms of interactions with Turing machines at this point. Hopefully, this paper has demonstrated that the study of structural issues in complexity theory is a worth-while and interesting field of research. Also, it should be seen which important role the Turing machine model plays in all these investigations.

References

Baker, T, J. Gill, and R. Solovay
  1975 Relativizations of the P=NP question. *SIAM J. Comp.* **4** (1975) 431–442.

Balcázar, J.L.
  1987 Self-reducibility. *STACS 87,* Lecture Notes in Computer Science 247. New York: Springer-Verlag (1987) 136–147.

Balcázar, J.L., R.V. Book, and U. Schöning
  1986 The polynomial-time hierarchy and sparse oracles. *J. ACM* **33** (1986) 603–617.

Berman, L., and J. Hartmanis
  1977 On isomorphism and density of NP and other complete sets. *SIAM J. Comp.* **6** (1977) 305–327.

Book, R.V., T.J. Long, and A.L. Selman
  1984 Quantitative relativizations of complexity classes. *SIAM J. Comp.* **13** (1984) 461–487.

Boppana, R.B., J. Hastad, and S. Zachos
  1987 Does co-NP have short interactive proofs? *Inf. Proc. L.* **25** (1987) 127–132.

Cook, S.A.

1971 The complexity of theorem proving procedures. In: *Proceedings of the 3rd Annual ACM Symposium on Theory of Computing* , pp. 151–158. New York: ACM (1971).

Dekhtyar, M.I.

1976 On the relativization of deterministic and nondeterministic complexity classes. In: *Mathematical Foundations of Computer Science 76,* Lecture Notes in Computer Science 45, pp. 282–287. New York: Springer-Verlag (1976).

Garey, M.R., and D.S. Johnson

1979 *Computers and Intractability – A Guide to the Theory of NP-Completeness.* San Francisco: Freeman (1979).

Fortnow, L

1987 The complexity of perfect zero-knowledge. In: *Proceedings of the 19th Annual ACM Symposium on the Theory of Computing.* New York: ACM (1987).

Goldreich, O.

1988 Randomness, interactive proofs, and zero-knowledge. This volume.

Goldreich, O., S. Micali, and A. Widgerson

1986 Proofs that release minimum knowledge. In: *Mathematical Foundations of Computer Science 86,* Lecture Notes in Computer Science 233. New York: Springer-Verlag (1986) 639–650.

Goldwasser, S., S. Micali, and C. Rackoff

1985 The knowledge complexity of interactive proof systems. In: *Proceedings of the 17th Annual ACM Symposium on Theory of Computing,* pp. 291–304. New York: ACM (1985).

Goldwasser, S, and M. Sipser

1986 Private coins versus public coins in interactive proof systems. In: *Proceedings of the 18th Annual ACM Symposium on the Theory of Computing, pp. 59–68. New York: ACM (1986).*

Hartmanis, J., and J.E. Hopcroft

1976 Independence results in computer science, *SIGACT News* **8/4** (1976) 13–23.

Hopcroft, J.E.

1981 Recent directions in algorithmic research. In: *Theoretical Computer Science,* Lecture Notes in Computer Science 104, ed. P. Deussen, pp. 123–134. New York: Springer-Verlag (1981).

Hopcroft, J.E., and J.D. Ullman

1979 *Introduction to Automata Theory, Languages, and Computation.* Reading, MA: Addison-Wesley (1979).

Karp, R.M.

    1972 Reducibility among combinatorial problems In: *Complexity of Computer Computations,* eds. Miller/Thatcher, pp. 85–103. New York: Plenum Press (1972).

Karp, R.M., and R.J. Lipton

    1980 Some connections between uniform and non-uniform complexity classes. In: *Proceedings of the 12th Annual ACM Symposium on the Theory of Computing,* pp. 302–309. New York: ACM (1980).

    1982 Turing machines that take advice. In: *Logic and Algorithmic,* Monographie No. 30 de l'Enseignement Mathematique, pp. 255–274. Genève: Université de Genève (1982).

Levin, L.A.

    1973 Universal search problems (in Russian). *Probl. Pered. Inf.* **9** (1973) 115–116. English translation in *Probl. Inf. Tr.* **9** 265–266.

Long, T.J., and A.L. Selman

    1986 Relativizations of complexity classes with sparse oracles. *J. ACM* **33** (1986) 618–627.

Rogers, H

    1967 *Theory of Recursive Functions and Effective Computability.* New York: McGraw-Hill (1967).

Schöning, U.

    1985 Robust algorithms: A different approach to oracles. *Theor. Comp.* **40** (1985) 57–66.

    1987 Graph isomorphism is in the low hierarchy. In: *STACS 87,* Lecture Notes in Computer Science 247, pp. 114–124. New York: Springer-Verlag (1987).

Stockmeyer, L.J.

    1977 The polynomial-time hierarchy. *Theor. Comp.* **3** (1977) 1–22.

Mechanisms for Computing
Over Arbitrary Structures

John C. Shepherdson

Abstract. Gandy *1980* has given four "Principles for Mechanisms", which appear to be satisfied by any discrete mechanical device and are sufficient to ensure that the function computed by such a mechanism can also be computed by a Turing machine. Friedman *1971* has generalized the theory of computation to the case where the data objects are not necessarily natural numbers or finite strings, but may be elements of any relational structure, and where the operations and relations of the structure are given as basic computational operations. These approaches are combined here; Gandy's principles are extended to deal with machines which incorporate processors for computing the operations and relations of a structure, and it is shown that functions over total structures computed by machines satisfying these principles can also be computed by a "generalized Turing algorithm" in Friedman's sense. Structures with partially defined operations and relations are also considered. These illuminate the differences between serial and parallel procedures discussed in Shepherdson *1975*. To preserve its universality over partial structures the generalized Turing algorithm needs to be given some kind of parallelism.

1. Introduction

Gandy *1980* (hereafter referred to as [G]) has argued that the analysis in Turing *1936-7* gives a convincing proof of:

Theorem T: *What can be calculated by a human being working in a routine way is computable by a Turing machine,*

but not of:

Thesis M: *What can be calculated by a machine is computable by a Turing machine.*

Certainly what Turing actually concentrated on in that paper – written before the existence of electronic computers – was human calculation. And, as Gandy points out, Turing assumed that the computation was serial, proceeding as a sequence of elementary steps, whereas a machine may do several steps simultaneously. (Actually one of Turing's reductions, to a one-dimensional tape, is not immediately obvious even for human calculation. He says: "I think it will be agreed that the two-dimensional character of paper is no essential of calculation". It would probably "be agreed", but a direct simulation of a two-dimensional Turing machine by a one-dimensional one is quite tedious.)

I think the most convincing argument for this Thesis M, and possibly for Theorem T as well, is the simple proof given by Church *1936* that if a step by step procedure is such that its states can be coded by natural numbers so that the next state function, input/output coding, and property of being an output state, are recursive, then so is the function calculated by the procedure. Having proved the equivalence of "recursive" and "computable by a Turing machine" and acquired the basic techniques of elementary recursion theory this is usually a very quick way of showing that a particular concrete device or abstract procedure satisfies Thesis M. But the target notion of recursive or Turing computable is essentially involved in this demonstration; to show that the function calculated is recursive one has to show that the atomic acts of the machine are recursive. It would be more satisfactory to have a general description of machine which did not involve the notion of recursion but was sufficient to show that Thesis M was satisfied. This is what Gandy did very convincingly in [G] with his four "Principles for Mechanisms". As Friedman *1971* (hereafter referred to as [F]) said:

> ... Turing's analysis could probably be significantly improved in relation to configurational computability; it seems instead to be an analysis of linear computability. The problem is to fill in the blank in: Turing's operations are to finite linear configurations as are − − to arbitrary finite configurations.

I think Gandy has solved the problem by replacing the blank with "mechanisms satisfying Principles I–IV".

[F] is the source of the other stream which the present article merges with [G]. It was a very comprehensive answer to the question:

> *What becomes of the concepts and results of elementary recursion theory if instead of considering only computations on natural numbers, we consider computations on data objects from any relational structure?*

An example would be computation over real numbers, given the ability to add, multiply, subtract, and divide them and test for zero. Another would be querying a database. If one is thinking of programs the obvious thing to do is to allow basic instructions

$$y := f(x_1, \ldots, x_n),$$

$$\text{if } R(x_1, \ldots, x_m) \text{ go to } i, \text{ else go to } j,$$

where f is one of the functions of the structure and R one of its relations, and the variables $y, x_1, x_2, \ldots$ take values in the structure. Such program schemes had been considered earlier by Luckham and Park (*1964*) and Paterson (*1967*) (Luckham, Park, and Paterson *1970* is the best source), but from a somewhat different point of view, the schemes being thought of as applicable not to one particular relational structure but to any structure of the right similarity type (i.e., having functions and relations of the right arity). Whether you get any more computational power by adding number variables and instructions such as $x := x+1$, involving them, depends on the particular structure. Friedman gives very complete answers to such questions. There was also an independent line of development by computer scientists in which they investigated the effect of adding counters (essentially number variables with the obvious instructions), stacks, push down stores, queues, arrays, etc. Chandra and Manna *1973* is a delightful summary; for a detailed analysis and comparison of Friedman's work and the parallel computer science work see Shepherdson *1985*.

In particular Friedman gives the appropriate generalization of the Turing machine. This is the *generalized Turing algorithm* which is allowed to store data elements on its tape as well as auxiliary symbols from the usual finite alphabet of tape symbols. In addition to the usual Turing machine moves it can switch the content of the scanned square with that of its left or right neighbor, can replace the content of the scanned square by $f(x_{i_1}, \ldots, x_{i_n})$, where $1 \le i_1, \ldots, i_n \le n$ and $x_1, \ldots, x_n$ are the contents of the squares to the immediate right of the scanned square, and f is one of the functions of the structure, and similarly if R is one of the relations of the structure it can test $R(x_{i_1}, \ldots, x_{i_m})$. When reading the symbol on the scanned square a data element looks the same as a blank square. The computation is started by placing the argument values $x_1, \ldots, x_n$ on adjacent squares, immediately to the right of the scanning head, with all other squares blank. The output is the data element (or Boolean value if a relation is being computed) on the scanned square when a halt state is reached. This differs from [F] in inessential ways and also in allowing computation of $f(x_{i_1}, \ldots, x_{i_n})$ with $1 \le i_1, \ldots, i_n \le n$ whereas [F] only allows $i_1 = 1, i_1 = 2, \ldots, i_n = n$. That would seem to be inadequate to allow e.g. the computation of $f(x, x)$, since copying of data elements is not allowed. Also [F] tests all of the relations of the structure on $x_1, \ldots, x_n$ in each move, which is harmless on total structures, but on structures with partial relations this means that the machine cannot even be a universal *serial* procedure. We shall show below how to modify our version so as to make it also a universal parallel procedure.

Like ordinary Turing machines, generalized Turing algorithms are tedious to work with. Friedman showed they were equivalent in power to his *effective definitional schemes*. That is a very convenient and easy to handle logico-combinatorial notion. It is simply a recursively enumerable set of *clauses*

$$E_1 \& E_2 \ldots \& E_k \to t$$

with intended meaning:

> "*If E_1 and $E_2 \ldots$ and E_k hold then the value of the function computed by this procedure is t.*"

Here t is a term built up from the input data variables $x_1, \ldots, x_n$ and the function symbols of the structure, and $E_1, \ldots, E_k$ are either atomic formulae $R(t_1, \ldots, t_m)$ or their negations, where R is a relation symbol of the structure and $t_1, \ldots, t_m$ are terms. Any two distinct clauses are required to have incompatible antecedents, i.e., one of the conjuncts of one must be the negation of a conjunct of the other.

Shepherdson *1975* (hereafter referred to as [S]) attacked the analogue of Thesis M for computation over an arbitrary structure. The basic idea is to supplement the machines by processing units which can compute the functions and relations of the structure. But what is proved about Thesis M in that paper is only the analogue of the Church argument referred to above, that if the atomic acts of the machine are recursive then so is the function computed. The fundamental notion of a *synchronous parallel procedure* required the next control state to be a recursive function of the present one. What we do in the present article is to provide the analogue of Gandy's argument for computation over arbitrary structures, i.e., to give natural modifications of Gandy's four principles which are sufficient to ensure that any function calculated by such a device can be calculated by a *generalized Turing algorithm* or a parallelized version of it.

Clearly if one wants to include all conceivable machines one must allow parallel operation. In the case usually considered, of computation over total structures, i.e., ones whose functions and relations are defined for all arguments, parallel procedures are no more powerful than serial ones, for one can obviously serialize a parallel procedure by subdividing the time scale. This is no longer true if there are partial functions, e.g., the function f defined by

$$f(x) \;=\; x, \text{ if } f_1(x) \text{ is defined or } f_2(x) \text{ is defined}$$
$$=\; \text{undefined otherwise,}$$

obviously cannot be computed by any serial procedure because it might choose the wrong one of f_1, f_2 to evaluate first. And there is no bound on the number of processing units which may need to be simultaneously active as is shown by the example

$$f(x) = x, \text{ if any of } f_1(x) \text{ or } f_1(f_2(x)) \text{ or } f_1(f_2(f_2(x))) \text{ or } \ldots \text{ ad inf, is defined.}$$

(Our assumption about the processing unit for $f_1(x)$ is that if $f_1(x)$ is defined it will return its value, but it will not tell you in finite time if $f_1(x)$ is undefined. For even if f_1 is a partial recursive number theoretic function that may not be computable.)

In order to bring out the differences between parallel and serial processes we shall follow [S] in also considering partial structures. In that case we shall find that

Thesis M needs to be modified. The generalized Turing algorithm, being a serial procedure, is no longer universal and must be replaced by a parallel device.

For simplicity we consider only single-sorted structures, but the extensions to many-sorted structures with several different datatypes (e.g., *integer, real, complex,* or a vector space over the reals) are obvious.

We also discuss only the calculation of data-valued functions. The extensions to the computation of relations over the structure, i.e., Boolean-valued functions, are straightforward.

Although we only discuss synchronous devices here, where actions take place only at clock times $1, 2, 3, \ldots$, the results of Section 9 of [S] show that the results also hold for asynchronous behavior, where it is assumed that the times taken by processing units are real numbers and that the instants at which changes take place are those at which processors report their results or the state of part of the control unit changes. It is sufficient to assume that each unit has a next "firing time" (possibly $= \infty$ if it is quiescent) at which it will change state except possibly if some other unit changes state in the meantime. That is to avoid the possibility of firings occurring at a descending sequence of times $\ldots, \frac{1}{4}, \frac{1}{3}, \frac{1}{2}, 1$, when it would appear to be impossible to describe the action in finite terms. A natural assumption which implies the existence of such a next firing time is the existence of a lower bound $\varepsilon > 0$ such that if a unit fires at time t it cannot fire again before $t + \varepsilon$.

2. *Gandy's Principles for Mechanisms*

Although Gandy's principles were obtained by a very natural analysis of Turing's argument they turned out to be rather complicated, involving many subsidiary definitions in their statement. In following Gandy's argument, however, one is led to the conclusion that that is in the nature of the situation – though it might be worth looking for stronger conditions which were simpler to state, and Dahlhaus and Makowsky *1986* show a close relation between Gandy's four principles and their *computable directory transformations* which might lead to a simpler treatment. Rather than repeat here a large part of [G] we shall only give his principles in sufficient detail to enable the reader to understand our modifications and proofs.

He wanted a form of description sufficiently abstract to apply to mechanical, electrical, or merely notional devices (e.g., programming languages) and chose to use hereditarily finite sets. That is to say starting from a potentially infinite set L of labels you form finite sets, finite sets of finite sets and so on. Formally

$HF_0 =_{df} \emptyset$
$HF_{n+1} =_{df}$ the set of nonempty finite subsets of $(L \cup HF_n)$
$HF =_{df} \cup \{HF_n : n \in \omega\} \cup \{\emptyset\}.$

It is elements of *HF* which are used to describe mechanisms and parts of mechanisms. The reason for excluding the empty set from the construction is that it is an abstract object and the idea is to use completely concrete structures. However the empty set is put into *HF* at the end because the empty structure is to be allowed. Labels are used for positions in space (e.g., squares on a Turing machine tape), for physical attributes (e.g., the symbol on a square, the state of a transistor) and for parts of a machine. However the actual labels used are irrelevant; only the structure is important. That is embodied in

Principle I: The form of description. *Any machine M can be described by giving a structural set $S \subseteq HF$ of state-descriptions together with a structural function $F : S \rightarrow S$. If $x_0 \in S$ describes an initial state then $F(x_0), F(F(x_0)), \ldots$ describe the subsequent states of M.*

To say that a set is structural means that it is closed under permutations of the set *L* of labels. To say that the function *F* is structural means that there is a permutation which sends Fx^π into $(Fx)^\pi$ and leaves all labels involved in x^π fixed.

Principle II: Limitation of Hierarchy. *The set-theoretic rank of the states is bounded, i.e., $S \subseteq HF_k$ for some k.*

For all the principles Gandy gives simple counter-examples showing that if the principles are weakened they allow the calculation of *any* number-theoretic function or relation. Here the counter-example is

$$S = \{\{a\}, \{\{a\}\}, \ldots\}$$
$$F(\{\ldots n \text{ times } \{a\} \ldots\}) = \{a\} \text{ if } \alpha(n)$$
$$\{\{a\}\} \text{ otherwise,}$$

where α is an arbitrary number-theoretic property.

A useful example to bear in mind when considering the principles is a cellular automaton like Conway's game "life" where the next state of each cell is determined by that of its neighbors.

Principle III: Unique reassembly. *Each state x in S can be uniquely reassembled from parts of bounded size.*

This one needs a few definitions for its precise statement, but the idea is simply that the parts are labeled, like the parts of a model construction kit, so that there is a unique way of putting them together. The parts are allowed to overlap.

Thus the tape of a Turing machine can be uniquely reassembled from the collection of all pairs of consecutive squares with their symbols: two such pairs are glued together with overlap if they both contain the same label for some square.

The size of a part here is the number of labels used in its construction.

Principle IV: Local causation. *The next state $F(x)$ of a machine can be uniquely reassembled from its restrictions to overlapping "regions" and these restrictions are locally caused. That is for each region s of $F(x)$ there is a "causal neighborhood" t of bounded size such that $F(x|s)$ depends only on $x|t$.*

Here $x|t$ denotes the restriction of x to t; if t is a set of labels it is simply the structure whose construction follows that of x omitting all labels not in t. For more general t the definition is more complicated.

What we have just given is only the preliminary version of Principle IV; what we intend to refer to is the final version ([G], p. 14) which includes conditions determining overlaps.

Gandy's main Theorem is:

If F and S satisfy I–IV then, to within isomorphism, F is computable by a Turing machine.

This means that the stereotype, the ε-isomorphism type of $F(x)$, is computable from the stereotype of x.

3. Principles for Mechanisms
Computing Over Arbitrary Structures

Our basic intuition is that the mechanism of the kind discussed in the last section acts as a control which has memory locations which can store data objects (elements of the structure) and which sends out processors for computing the functions and relations of the structure applied to the data objects in specific named locations. For example part of the control might be a factory for assembling these processors. Since the functions and relations of the structure are considered to be atomic, the processors are, from the point of view of the control, black boxes. One example would be analogue devices for adding and multiplying real numbers. Since the control is a discrete mechanism it could do nothing itself with these real number data objects but would simply direct the adders and multipliers to attach their inputs and outputs to the correct locations. All it would know about the locations which could store data objects would be whether, at a given time, they were actually storing them.

If we were only interested in total structures it would be simplest to assume that all the processors completed their evaluation of a functional or relational value in one unit of time. But that would rule out some interesting concrete examples and is an unnecessary restriction. And for the case of partial structures it is obviously the wrong assumption. If there is any bound on the time taken by a processor, then by waiting long enough you can find out whether it is going to produce a value, i.e., whether the function in question is defined for the given arguments. And that is something you are not supposed to know. For example, one of the cases covered is that of relative recursiveness, where the structure in question is the natural numbers, but the functions and relations are not necessarily computable functions. If the structure had one partial computable function f, then allowing evaluations of that should not allow the computation of noncomputable functions, but it would if you were allowed to know whether $f(x)$ was defined. You should not even assume that processors computing $f(x)$ for a given value of x always take the same time, for if a processor evaluating $f(y)$ takes a different time you have discovered $y \neq x$, but we do not want to assume that the equality relation is one of the given computable relations of the structure.

So the only assumption we want to make about the processors is that if the functional or relational value they are trying to calculate is defined they will eventually produce it. And the only thing which is visible to the control is whether or not they have yet produced a value, and in the case of a relation, what that Boolean value is. To avoid having to discuss relations separately it is convenient to replace them by their characteristic functions, i.e., allow Boolean valued functions of data objects. Constants are treated as 0-ary functions.

Of course, under the weak assumption that all we know about the processors is that if the value they are trying to compute is defined, then they will eventually produce it, most parallel mechanisms or procedures will be indeterminate, i.e., the value they produce will depend on the times taken by the processors. So we shall always have a proviso that the mechanism is determinate, for the initial argument values in question.

The easiest way of incorporating the storage of data objects and the initiation of black box processors into Gandy's description would appear to be to regard a state as an ordered pair consisting of two parts, one internal, the other external.

The *internal state* can be thought of as an overall *control state* together with a *processor despatcher,* a *memory map,* and a *processor record.* The processor despatcher describes the processors which are to be despatched in the next move. The processor record records which processors are still calculating and which have returned values; in the case they have returned Boolean values these are recorded; in the case of a data value all that is recorded here is "data-value". The memory map describes the location of data elements. Since the mechanism cannot read or write data elements it refers to these symbolically, e.g., "the second of the given arguments", "the result of processor so and so". The *external state* completes this description; it is a *data record* saying which actual data elements correspond to these symbolic descriptions. This part is "invisible" to the mechanism.

The internal state will therefore be a quadruple

$$< x, p_d, m, p_r >$$

where x is the control state, p_d the processor despatcher, m the memory map and p_r the processor record. The *control state* x is simply an element of HF as in [G]. The *processor despatcher* is a finite set of tuples

$$< y, f, x_1, \ldots, x_n > .$$

This tuple is an instruction to send off a processor with label y to compute $f(\delta_1, \ldots, \delta_n)$, where $\delta_1, \ldots, \delta_n$ are the data elements (if any) in "locations" $x_1, \ldots, x_n$. Here $x_1, \ldots, x_n, y$ all belong to HF and f is a new symbol. These new symbols $f_1, \ldots, f_r$ referring to the functions of the structure are to be distinct from the original label set L, and will remain fixed when we consider permutations of L. The *memory map* is a finite set of ordered pairs of the form $< l, d >$ where $l \in HF$, meaning that the data element described by d is "in" "location" l. Here d is a label of one of two kinds. The first is A_i, where $1 \le i \le u$, and u is the number of arguments of the function the device is being used to calculate (we shall only be concerned with using a mechanism to compute a function of a given number of arguments). This refers to the ith of the u given argument values. The symbols $A_1, \ldots, A_u$ are to be new.

The second kind of d is a member y of HF. This refers to the value (if any) calculated by the processor (if such exists) with label y. For each l there will be at most one such pair $< l, d >$, i.e., each location holds at most one data element.

The *processor record* describes, as far as it can without using data values, the current state of each of the processing units which has been despatched. As above each such processor is described by a label d. The current state of this unit is described by an ordered pair

$$< d, D > \quad \text{or} \quad < d, T > \quad \text{or} \quad < d, F > \quad \text{or} \quad < d, W >$$

meaning respectively "has already returned a data value", "has returned the Boolean value T", "has returned the Boolean value F", "waiting – has not yet returned a value". Here D, T, F, W are new symbols. (It is possible that a given kind of mechanism might not be able to store or use the memory map or processor record, but since we aim to show that a wide class of mechanisms can compute no more than a generalized Turing machine it behooves us to give such mechanisms all the apparent power we can imagine.)

The *external state* or *data record* complements the memory map by showing which actual data elements are referred to by the labels. It is a finite set

$$\{< d_1, \delta_1 >, \ldots, < d_r, \delta_r >\}$$

where $\delta_1, \ldots, \delta_r$ are the data values corresponding to those labels $d_1, \ldots, d_r$ which currently have data values associated with them, i.e., all those d of the form A_i describing the given arguments, and those y such that the processor which was despatched with label y has already calculated a data value. This part of the state is invisible to the mechanism and is updated by some external agency so that it can be consulted at the end of the computation to output the actual data value corresponding to a given label.

There is to be one exceptional kind of state, a halting state, whose internal state is of the form $< H, y >$ where H is a new symbol and y is the location of the result to be output. Gandy does not specify how the output is to be obtained. He says,

> It is of little importance whether we designate certain states as "halt" states encoding the output, or whether we consider, as Turing did, an infinite sequence which enumerates a (possibly empty) set whose members are encoded by certain of the S.

The second alternative is not meaningful for mechanisms computing arbitrary data objects as values so we have adopted the first. By making the machine signal so clearly that it is in a halting state and where to find the output we make the next-state function F do all the work and so avoid the need to state conditions similar to Principle IV on the output function and the property of being a halting state.

The execution sequence of the machine is a little different from [G] where the succession of states was

$$x_0, F(x_0), F(F(x_0)), \ldots$$

where F was the next-state function. Now the execution sequence depends on the behavior of the processors, i.e., the times they take to report back. If the current state is a halting state with internal state $< H, y >$ the machine halts and outputs the value δ (if any) such that $< y, d >$ belongs to the memory map and $< d, \delta >$ to the data record.

Otherwise the current state will be of the form

$$<< x, p_d, m, p_r >, \xi >$$

where $< x, p_d, m, p_r >$ is the internal state and ξ the external state, x being the control state, p_d the processor despatcher, m the memory map, p_r the processor record, and ξ the data record. The next state is determined in 3 stages:

1) *Processors are despatched.* For each tuple $< y, f, x_1, \ldots, x_n >$ in the processor despatcher p_d, if there are pairs $< x_1, d_1 >, \ldots, < x_n, d_n >$ in the memory map m and pairs $< d_1, \delta_1 >, \ldots, < d_n, \delta_n >$ in the data record ξ of the current state, then a processor with label y is sent off to compute $f(\delta_1, \ldots, \delta_n)$. If any of $\delta_1, \ldots, \delta_n$ is missing, e.g., if there is no pair $< x_1, d_1 >$ in m and corresponding pair $< d_1, \delta_1 >$ in ξ then we have to choose between several different models.

We can either send off the processor knowing it is destined never to produce a result, or we can say that the whole procedure terminates without result (or with an error message "undefined argument"). If the machine has enough resources it is perfectly capable of avoiding this situation by checking in the memory map and processor record that the locations do hold data values. If one wanted to be really kind to a feeble machine one could use a third model where you gave it the error message but did not terminate the procedure. Yet another model would allow the processor to wait until its arguments were defined, taking each as soon as it became defined, or waiting until all were simultaneously defined. Which model to use depends on the particular realization of "processor" one has in mind; the last one for example calls for a smarter processor than the others. The choice does not affect the results below, since these relate various types of computation, provided we make the same choice in all of them. And although the choice may affect the behavior of individual mechanisms, it is fairly obvious that it does not affect the computing power of the whole class of mechanisms.

If a processor is given a label y which has already been used the computation terminates without result

2) *The processor and data records are updated.* For each processor $< y, f, x_1$ $, \ldots, x_n >$ which has just been despatched, a pair

$$< y, W >$$

is added to the processor record p_r. Then for each processor with label d which has just returned a value, a pair $< d, D >, < d, \mathbf{T} >, < d, \mathbf{F} >$ describing that value (as either data, $\mathbf{T}$ or $\mathbf{F}$) is added to p_r and the pair $< d, W >$ is deleted from it. If the value is a data value δ then $< d, \delta >$ is also added to the data record ξ.

3) *The new internal state is calculated.* If p'_r is the new processor record then the new internal state is

$$< x', p'_d, m', p'_r >= F(< x, p_d, m, p'_r >)$$

where F is the next-state function. This move can thus effect a change of control state to x', a change of processor despatcher to p'_d and a change of memory map to m'. This latter might involve a physical copying of data from one location to another, inserting a pointer from one location to another, or just stamping another name on a processor which is storing its own result. Note that it does not involve a further change of processor record; that is controlled by the behavior of the processing units and stays at its updated value p'_r.

To obtain appropriate versions of Gandy's principles we first enlarge the label set L to include the new symbols $A_1, \ldots, A_u, f_1, \ldots, f_r, D, W, \mathbf{T}, \mathbf{F}$, but we regard these as symbols which are to be fixed when we consider permutations of L. We then modify Gandy's four principles as follows:

Principle I': *The set of internal states and the next-state function F are to be structural. The sequence of states is determined as above.*

Principle II': *The set-theoretic rank of the internal states is bounded.*

Principle III': *The internal states can be uniquely reassembled from parts of bounded size.*

Principle IV': *The next internal state $F(< x, p_d, m, p'_r >)$ can be uniquely re-assembled from its restrictions to "overlapping regions" s and these restrictions are locally caused. That is for each region s of $F(< x, p_d, m, p'_r >)$ there is a causal neighborhood t of bounded size such that $F(< x, p_d, m, p'_r > \restriction s)$ depends only on $< x, p_d, m, p'_r > \restriction t$.*

As above this is only a crude statement of Principle IV'; what we intend is to be obtained from the full version of IV by replacing "state" by "internal state", i.e., x by $< x, p_d, m, p'_r >$.

So the changes made to Gandy's principles are: a) replace "state" by "internal state", which is to be of the form $< x, p_d, m, p'_r >$ described above, and b) change the definition of the sequence of states given in Principle I to incorporate the influence of the returning times of the processors. (One might get a simpler treatment by adding the data objects as fixed nonpermutable elements to the label set L. This would allow one to treat the state as a single element of HF. But it would appear that one would then need Principle IV not only for the next-state function but also for the functions giving the processor despatcher, the memory map, and the processor record. By exhibiting the graphs of these functions explicitly in the internal state we have made the next-state function do all the encoding work.)

As observed above we cannot hope to get the exact analogue of Gandy's main Theorem in the case of partial structures, because the generalized Turing algorithm, being a serial procedure, cannot be universal. So we proceed in two stages and show first that the function computed by a machine satisfying Principles I'–IV' can be computed by a *synchronous parallel procedure* as defined in [S]. The definition given there may be simplified and brought into line with the present treatment as follows. In the definition just given replace elements of HF by natural numbers (or strings on a finite alphabet) and require the next state function F to be recursive. (Remember that this was intended as the model corresponding to the machine of Church's argument whose basic act is required to be recursive.) It is easy to check that this definition is equivalent to the one in [S].

Main Theorem: *A mechanism satisfying Principles I'–IV' is equivalent to a synchronous parallel procedure.*

Proof. The equivalence here is in the strongest possible sense. The two procedures produce the same result over any partial structure and for any *execution* *sequence,* i.e., any given set of times taken by those processing units which do eventually produce a result. In fact we can assert that, with a suitable coding of visible states, a mechanism satisfying Principles I'–IV' *is* a synchronous parallel procedure. The first step is to show that, to within isomorphism, $F(< x, p_d, m, p'_r >)$ is a recursive function of the internal state $< x, p_d, m, p'_r >$. This follows by Gandy's argument because we have required the internal states to satisfy Principles I'–III'. All that remains to be done then is to use the set of natural numbers as the label set L and to code up the elements of HF by natural numbers. □

4. Computation Over Total Structures

In this section we make the following abbreviations:

> *eds* for effective definitional scheme; see introduction for definition;
> *gTa* generalized Turing algorithm; see introduction for definition;
> *spp* for synchronous parallel procedure; see last section for definition.

When discussing computation over structures we shall talk about the behavior of the devices on a given *interpretation,* i.e., a particular structure $\mathcal{M}$ of the correct similarity type together with a tuple $x_1, \ldots, x_u$ of argument values from this structure. If the structure is total, i.e., all the functions $f_1, \ldots, f_r$ of the structure are defined for all argument values we say the *interpretation is total.* In this case all the processing units which are despatched in the course of a computation will report back after a finite time. But since we have assumed nothing else about the times they take, it is likely that the function value output by a mechanism or synchronous parallel procedure will depend on the times taken by these various processors. E.g., send off processors to compute $y := f_1(x), y := f_2(x)$, and as soon as one reports back halt and output y. When the output is the same for all possible timings we say the device is *determinate on the given interpretation*; when it is determinate on all interpretations with a given structure (i.e., for all choices of argument values) we say it is *determinate on the structure.* When talking about the equivalence of devices we shall regard them as computing functions of a given number u of arguments and say they are *equivalent over an interpretation* $(\mathcal{M}, x_1, \ldots, x_u)$ if they both output the same value or both output no value. (A stronger notion is sometimes of interest where the second case is subdivided into "stop without result" and "does not halt".) They are *equivalent over a structure* if they are equivalent over all interpretations $(\mathcal{M}, x_1, \ldots, x_u)$.

Theorem: *Over a total structure any function calculable by a mechanism satisfying Principles I'–IV' is computable by a generalized Turing algorithm.*

Proof. By the Main Theorem above such a mechanism is equivalent to an *spp*. By [S] Theorem 4.3, this is equivalent, over all total interpretations for which it is determinate to an *eds*, and by [F] Theorem 1.2 to a *gTa*.   □

What the *gTa* does is to carry out a symbolic calculation first and then attach data values to the variables (for more detail see the corresponding argument for parallel Turing machines below).

Here we are taking "*f* is calculable by a mechanism *M* over the structure $\mathcal{M}$" to mean that *M* is determinate over all interpretations ($\mathcal{M}, x_1, \ldots, x_u$), outputs the value $f(x_1, \ldots, x_u)$ if that is defined, otherwise outputs no value. It is a priori conceivable that more functions might be calculable if one could assume that all processors reported back in the next move. But this is not the case:

Corollary: *The theorem still holds under the assumption that all processors report back in the next move.*

Proof. In [S] Theorem 4.3 the *eds* was constructed to behave like the *spp* behaves when this timing assumption is made. Clearly we could get the same result for any other assumption about the timings of the processors as long as the time taken is a recursive function of the internal state at the time of despatch.

The 'converse' of the above theorem, that every generalized Turing algorithm can be represented as a mechanism satisfying Principles I'–IV', follows easily from the representation of an ordinary Turing machine as a mechanism satisfying Principles I–IV. The operation of replacing the content of the scanned square by $f(x_{i_1}, \ldots, x_{i_n})$ amounts to despatching a processor which, to ensure that processor labels are not used more than once, could be labeled $< \ell, t^* >$, where ℓ is the label of the scanned square and t^* is a structure which encodes the 'time' (number of steps) *t*.

5. Computation Over Partial Structures

As we saw in the introduction, over partial structures there may be functions computable by parallel procedures which cannot be computed by any serial procedure. But let us first examine what procedures acting serially can do. What do we mean by "acting serially"? Certainly that at most one processor is in action at any time. But also that no control changes are taking place while the processor is calculating its result. ([S] Theorem 6.18 shows that this is a stronger condition.) In terms of the definition of next state given above for mechanisms and synchronous parallel procedures we may define this formally by saying that such a device *acts serially on an interpretation* when for all possible timings of the processors, the processor despatcher p_d contains at most one tuple, and if p'_r contains any tuple $< d, W >$ indicating that the result of a processor is awaited,

$$F(< x, p_d, m, p'_r >) = < x, \emptyset, m, p'_r >$$

(If these conditions are satisfied for all states then the device is guaranteed to act serially on all interpretations and is effectively a serial procedure.)

Theorem: *For every mechanism satisfying principles I'–IV' there is a generalized Turing algorithm which is equivalent to it on all interpretations on which it acts serially and on all total interpretations on which it is determinate.*

   Proof. By the Main Theorem and Theorem 6.12 and p. 493 of [S].   □

Lemma 6.2. of [S] confirms the obvious fact that an *spp* acting serially on an interpretation is determinate on that interpretation.

   In order to make it into a universal parallel procedure we must give the *gTa* the ability to send off an unbounded number of processing units. The easiest way to do this would appear to be to reinterpret a *gTa* instruction "replace the content of the scanned square by $f(x_{i_1}, \ldots, x_{i_n})$" as "send off a processor to compute $f(x_{i_1}, \ldots, x_{i_n})$ and, if and when it reports back, put the result in the presently scanned square". It can do this quietly, just replacing the original content of the scanned square by this result, without informing control, since it can be arranged in the proof of the theorem below that this square is cleared of data before initiating the processor, so that by continually coming back to check whether the square contains a data element it can be discovered when the processor has returned a value (and it is not necessary to have more than one processor in operation at a time which will put its result on this square). There is one exception to this: when the processor is due to put its result on the square currently under scan. In that case the control must be able to ask the processor to wait for one step, or to divert its output to a neighboring square. This appears to be needed when the currently scanned square contains a data element one wants to preserve, e.g., one of the original arguments which (see proof below) is being taken to another square.

   Partial relations would appear to be best dealt with in the way we have suggested above, as partial Boolean-valued functions.

   Let us call such a machine a *parallel Turing machine (pTm)*. We then have:

Theorem: *For every mechanism satisfying Principles I'–IV' there is a parallel Turing machine which is equivalent to it on all interpretations on which it is determinate.*

   Proof. By the Main Theorem and [S] Theorem 5.5 we get a *reds* (recursively enumerable definitional scheme) which is equivalent to the mechanism on all interpretations on which it is determinate. A *reds* is like an *eds* except that as well as atomic formulae and their negations conjuncts of the form $t \downarrow$ (t is defined), where *t* is any term, are allowed in the antecedents of the clauses, and the condition the distinct clauses have incompatible antecedents is dropped, i.e., any r.e. set of clauses is allowed. So it usually gives more than one value; it is said to be determinate on a interpretation if it gives at most one value.

To get a *pTm* equivalent to the *reds* we enumerate the clauses of the *reds* and as each one is generated we test its antecedent to see whether it is satisfied and, if so, whether the consequent term *t* is defined; if so we output that as the value. Since we have to wait indefinitely for processors to report, we start testing the next clause while we are waiting for the result of this one. The testing involves calculating some Boolean-valued functions, or discovering whether a data valued function is defined. This is done starting with innermost subterms, marking the squares on which these results will be put and continually coming back to check them, proceeding with a new stage in the evaluation when all the arguments of a term have been evaluated. The basic procedure for initiating processors is to "drive" their arguments by "switch left" or "switch right" operations to the square immediately to the right of the square on which you want the result put, initiate the processor, and then drive the arguments back to their parking places so that they can be found next time they are wanted.□

Note that this confirms the fairly obvious fact that although an unbounded number of processors may need to be simultaneously active, they can be sent off one at a time by a serial control.

6.   *Other Universal Devices*

For total structures the *eds* and the *spp* are equivalent in power to the *gTa* and to mechanisms satisfying Principles I'–IV', as follows from the results of [F] and [S]. There are some other program-type definitions given in [F] and [S] which are also equivalent to these. Most of them derive from Friedman's *fap – formalized algorithmic procedure* – which is a finite program with instructions of the forms

$$y := f(x_1, \ldots, x_n)$$
$$\text{if } R(x_1, \ldots, x_m) \text{ go to } i \text{ else go to } j.$$

In general, as he shows, these are not universal, but they are if supplemented with two stacks or one queue or one stack and one counter or one array and an unlimited number of counters (this is essentially the same as the *fapir* or *fap* with index registers of [S]). (See also Chandra and Manna *1973* or Shepherdson *1985* for a general survey.) The *countable algorithmic procedure* of [S] is universal. That is simply a countable program of the above kind where the instruction is a recursive function of the line number. Similar universal devices are the *esk – effective schemes of Kfoury* (Kfoury *1973, 1974*). These are infinite binary trees each node of which, apart from the input node, which is labeled with the list of input variables, is either labeled with an assignment instruction $y := x$ or $y := f(x_1, \ldots, x_n)$ and has one successor, or is labeled with a test $R(x_1, \ldots, x_m)$ and has two branches labeled "true", "false", or is an output node labeled with an output variable y. The scheme must be effective, i.e., the label of the node must be a recursive function of its number where the nodes are numbered from top down and left to right.

There are also equivalent definitions which are the analogue of the recursive functions; see Fenstad *1975, 1980,* Moschovakis *1969, 1971,* and Moldestad et al. *1980.*

Over partial structures there are three levels of universality. The lowest is for procedures acting serially; for this case all the devices just listed are still universal. The highest level is for procedures which can do anything any parallel procedure can do. The results mentioned in the last section show that the parallel Turing machine, the synchronous parallel procedure, and the recursively enumerable definitional scheme are universal in this strongest sense. So are the suitably augmented formalized algorithmic procedures, and countable algorithmic procedures of the last paragraph if instructions of the form $y := f(x_1, \ldots, x_m)$ are interpreted as initiating processors which start working in parallel while the program goes on to the next instruction. If there are partial predicates these are best dealt with by replacing them by their Boolean-valued characteristic functions.

There is also an intermediate level, for procedures which are determinate on all interpretations. The function

$$f(x) = x, \text{ if } f_1(x) \downarrow \text{ or } f_2(x) \downarrow$$
$$= \text{ undefined otherwise,}$$

is calculable by such a device but not by any serial procedure. Clearly all the devices just mentioned restricted by the condition that they be determinate on all interpretations will be universal in this intermediate sense. In the case of the *reds* this can be put in a more testable form; just impose the condition that any two clauses with compatible antecedents must have the same consequence. This notion was introduced by Strong *1971* and called an *effective functional.* For example the function just defined is computed by the effective functional:

$$\{f_1(x) \downarrow \to x, f_2(x) \downarrow \to x\}.$$

There does not appear to be any simple way of modifying mechanisms, synchronous parallel procedures or parallel Turing machines so that they are determinate on all interpretations.

These three notions of universality correspond to the three different notions of "g is recursive in f" which arise in classical recursion theory when f is partial. The narrowest notion "g is Turing-reducible to f", i.e., computable by an oracle Turing machine (which must wait for an answer before proceeding), amounts to allowing only serial procedures. The broadest notion "g is partial recursive in f", or equivalently, "there is a partial recursive operator Φ such that $g = \Phi(f)$", corresponds to allowing any parallel procedure which is determinate on f. The intermediate notion of Davis *1958* or Rogers *1967* "there is a recursive operator Φ such that $g = \Phi(f)$", amounts to allowing only parallel procedures which are determinate on *all* partial functions.

References

Chandra, A., and Z. Manna

    1973 On the power of programming features. *Stanford AI Memo* 185.

Church, A.

    1936 An unsolvable problem of elementary number theory. *Am. J. Math.* **58** (1936) 345–363.

Dahlhaus, E., and J.S. Makowsky

    1986 Computable directory queries. In: *Proceedings of CAAP '86,* Lecture Notes in Computer Science, vol. 214, pp. 254–265. Berlin: Springer-Verlag (1986).

Davis, M.

    1958 *Computability and Unsolvability.* New York: McGraw Hill (1958).

Fenstad, J.E.

    1975 Computation theories: an axiomatic approach to recursion on general structures. In: *Logic Conference, Kiel 1974,* eds. G. Müller, A. Oberschelp, and K. Potthoff, pp. 143–168. Berlin: Springer-Verlag (1975).

    1980 *General Recursion Theory. An Axiomatic Approach.* Berlin: Springer-Verlag (1980).

[F] Friedman, H.

    1971 Algorithmic procedures, generalized Turing algorithms, and elementary recursion theory. In: *Logic Colloquium '69* eds. R.O. Gandy and C.M.E. Yates, pp. 361–389. Amsterdam: North-Holland Publ. Co. (1971)

[G] Gandy, R.O.

    1980 Church's thesis and principles for mechanisms. In: *The Kleene Symposium,* eds. J. Barwise, H.J. Keisler, and K. Kunen, pp. 123–148. Amsterdam: North-Holland Publ. Co. (1980).

Kfoury, D.

    1973 Comparing algebraic structures up to algorithmic equivalence. In: *Automata, Languages and Programming,* ed. M. Nivat, pp. 253–263. Amsterdam: North-Holland Publ. Co. (1973).

    1974 Translatability of schemas over restricted interpretations. *J. Comput. Sy.* **8** (1974) 387–408.

Luckham, D., and D. Park
1964 The undecidability of the equivalence problem for program schemata. Bolt, Beranek and Newman Inc. Report No. 1141 (1964).

Luckham, D., D. Park, and M. Paterson
1970 Formalized computer programs. *J. Comput. Sy.* **4** (1970) 220–249.

Moldestad, J., V. Stoltenberg-Hansen, and J.V. Tucker
1980 Finite algorithmic procedures and computation theories. *Math. Scand.* **46** (1980) 77–94.

Moschovakis, Y.N.
1969 Abstract first-order computability I, II, *T. Am. Math. S.* **138** (1969) 427–464, 465–504.
1971 Axioms for computation theories – first draft. In: *Logic Colloquium '69,* eds. R.O. Gandy and C.E.M. Yates, pp. 199–255. Amsterdam: North-Holland Publ. Co. (1971).

Paterson, M.
1967 Equivalence Problems in a Model of Computation. Doctoral Dissertation, Cambridge.

Rogers, H.
1967 *Theory of Recursive Functions and Effective Computability.* New York: McGraw Hill (1967).

[S] Shepherdson, J.C.
1975 Computation over abstract structures: serial and parallel procedures and Friedman's effective definitional schemes. in: *Logic Colloquium '73,* ed. H.E. Rose and J.C. Shepherdson, pp. 445–513. Amsterdam: North-Holland Publ. Co. (1975).
1985 Algorithmic procedures, generalized Turing algorithms, and elementary recursion theory. In: *Harvey Friedman's Research on the Foundations of Mathematics,* eds. L.A. Harrington et al., pp. 285–308. Amsterdam: North-Holland Publ. Co. (1985).

Strong, H.R. Jr.
1971 High level languages of maximum power. Proceedings of the I.E.E.E. Conference on Switching and Automata Theory (1971).

Turing, A.M.
1936-7 On computable numbers, with an application to the Entscheidungsproblem. *P. Lond. Math. Soc. (2)* **42** (1936-7) 230–265; A correction, ibid. **43** (1937) 544-46.

Comparing the Church and Turing Approaches: Two Prophetical Messages

Boris A. Trakhtenbrot

1. Introduction

The search for a precise mathematical characterization of what "algorithm" and "computable function" should mean resulted half a century ago in the discovery of three well-known equivalent approaches. Their chronological order is as follows: λ-definability (Church-Kleene, 1932–34); general recursiveness (Gödel-Herbrand, 1934) and Turing machines (1936).

The type-free λ-calculus was conceived by A. Church as a foundation for logic and mathematics, but this aim failed. In spite of this failure Church realized that a consistent part of this calculus is a paradigm for computation in the same way as predicate calculus is a paradigm for deduction. In 1934 he proclaimed his famous Thesis, which identifies the intuitive notion of computable function with the formal notion of λ-definable function, but there still was some lack of consensus about this Thesis. We learn from Davis *1982* that it was only after Turing's work that Gödel accepted Church's Thesis which had then become the Church-Turing Thesis. This is the way the miracle occurred: the essence of a process that can be carried out by purely mechanical means was understood and incarnated in precise mathematical definitions.

In retrospect it is not hard to understand why, unlike the previous approaches, only Turing's succeeded to convince once and for all that the genuine formulation had been achieved. The point is that Turing worked with the machine concept. Relying on this concept, he was able to give a direct analysis of computing processes and to provide clear arguments that all possible algorithms for computing functions can be embodied in such machines.

In modern jargon we can characterize Turing's approach as computer- (or hardware-) oriented; hence, also the expectation of significant practical conse-quences. Indeed, a Turing machine is like an actual computer except that it is error-free and has access to unlimited external memory in the form of a (potentially) infinite tape. Turing's theorem about the existence of a universal machine of this kind was a prophetical message which anticipated the era of universal digital computers; it exhibits in a rudimentary form the main ingredients of such computers. Turing's prophecy is in no essential way affected by the development of modern computing techniques which rely on more sophisticated technology and design principles. It is known that Turing himself went into computing at the National Physical Laboratory in 1945–48 and from 1948 on at the Computing Machine Laboratory in Manchester (England).

Let us also recall that, independently of Turing, E. Post also elaborated a similar analysis of processes which are performable in a purely mechanical way. However, Post's formulation is in terms of "combinatorial systems" and does not explicitly use the attractive machine paradigm.

Due to the logical interchangeability of hardware and software, one can develop programming languages based on Turing machines; indeed in the early period of computing some people developed such languages. This is an awkward programming style, but it should not be a hindrance in those situations in which one has to prove the existence of computable functions with specific properties because such proofs can be produced without having to explicitly write down the programs.

Now, returning to the earlier Church-Kleene and Gödel-Herbrand approaches, the idea occurs to characterize them (in contrast to the Turing-Post approach) as programming or software-oriented. With respect to Gödel-Herbrand recursiveness such a characterization is quite clear. Not only does recursiveness come closer to traditional mathematics, but as a matter of fact some versions of recursion and recursive schemes occur in most programming languages. Less evident is the relevance of λ-definability and in general of λ-calculus for programming languages. At first sight λ-definitions look very unusual and far from traditional mathematics and modern programming. S.C. Kleene (*1979*) remembers the rather chilly reception of audiences around 1933–35 to disquisitions on λ-definability. That is why after general recursiveness had appeared he had chosen to put his work into this format, which is more familiar to mathematicians:

> In retrospect, I now feel it was too bad I did not keep active in λ-definability as well. So I am glad that interest in λ-definability has revived, as illustrated by Dana Scott's 1963 communication. (Kleene *1979*).

Paradoxically, whatever the first impressions of λ-definitions may be, the objective truth is that the λ-calculus does implicitly incorporate some of the most important features of modern high-level programming languages. In this sense Church's approach is not just a companion to Turing's, but it also contains its own prophetical message, namely about future programming languages. Subsequent theoretical research and development of programming confirmed this message.

It took about ten years after the mathematical birth of Turing machines for the real computers to appear. The road from the λ-calculus to the study of existing programming languages and the design of new ones was considerably longer. In many respects this is astonishing, and the following are some preliminary suggestions as to why it is so.

First of all, the syntax of the λ-calculus is very simple. Therefore it is not immediately apparent how it can contain – even in an implicit and rudimentary form – features of high level programming languages, which usually rely on a very elaborate syntax.

In addition, there are the intricate semantical problems of the type-free λ-calculus. In order to produce λ-definitions for all the "computable" functions, Church was forced to put aside the semantically easier typed λ-calculus. In this way self-application became inevitable and hence also the nightmare of the set-theoretical paradoxes.

Thirdly, there is the dilemma between functional languages (and that is really what lambda-calculus may be pretending to be) and the languages with imperative features (like Fortran, Algol, Pascal, etc.) which actually dominated from the very beginning of the development of high-level programming languages.

Finally, let us mention one more restrictive factor: the λ-calculus is intrinsically sequential, whereas parallelism features are highly desirable in a modern language. (Sequentiality was also a primary restriction of Turing machines).

All these circumstances suggest that the realization of Church's vision and the rooting of the λ-calculus into the theory and practice of programming might not be free of evasiveness and obscurity. Before appropriate research was done by prominent computer scientists and logicians, it was impossible to guess the full impact of Church's approach on programming.

Of course, by now these ideas and facts are well known to people who are engaged in this specific area. Hopefully, putting some of these ideas and facts in a certain order may be useful for a broader audience.

That is the main goal of this essay which is organized as follows: After having sketched in Section 2 the syntax of the λ-calculus, in Sections 3 and 4 the focus is on the languages LISP and ISWIM, which absorbed and promoted ideas of the λ-calculus. Landin's ISWIM was especially instrumental for standards of forthcoming functional programming languages such as Edinburgh ML and others; it paved the way to a consensus about what in Section 4 is called the Church-Landin Thesis.

Section 5 is about denotational semantics for λ-calculus and through it for programming languages in general.

The main goal of Sections 6 and 7 is to put into the right perspective the relevance of the λ-calculus for languages which allow such features (like assignments or parallelism) that to all appearances are alien to the spirit of the λ-calculus.

The style of exposition remains eclectic throughout; more or less rigorous definitions and claims alternate with informal speculations. I do not claim to have painted a complete picture. On the other hand numerous quotations from various sources have been included in order to keep the historical flavor and to reflect personal views of investigators in the area.

2. *Lambda Calculus*

λ-calculus amounts to λ-notations plus rules of manipulating them. The λ-notation introduced by Church is very natural and efficient; it is really a pity that it is not in general use. In daily life one has sometimes to guess whether an expression like x^y denotes a function or, alternatively, a number. Even when limited to functions one may still be confused about what function is intended. λ-notations avoid such ambiguity through the *abstraction* operation, that is a binding mechanism which explicitly points out that:

1. $\lambda x.x^y$ denotes the function, which for each x returns the value x^y;
2. $\lambda y.x^y$ denotes the exponential function.

Hence, the respective derivative functions are $\lambda x.yx^{y-1}$ and $\lambda y.x^y lny$, whereas x^y denotes a number and it does not make any sense to ask about its derivative.

Naturally, one expects that $\lambda x.x^y$ should have the same meaning as $\lambda z.z.^y$, i.e., renaming of the bounded (dummy) variable x is allowed (unlike the renaming of the free variable y) so long as there is no "collision" with the free variable.

$\lambda x.E$ represents a function with formal parameter x and body E. According to more common programming language syntax, one would prefer to assign a name to this function – say f – via a declaration like

define f: function (x);
return E end.

The other fundamental notation is for *application*. Namely, (E_1, E_2) – a slight deviation from the more familiar $E_1(E_2)$ – denotes the application of the function (operator) E_1 to the argument (operand) E_2.

Formally, the expressions (terms) of the (pure) λ-calculus are introduced by induction; at the same time one defines the set Free(E) of free variable occurrences in E and bindings in E.

(0) A variable x is a term; x is its only free occurrence.

(I) *Application.* If E, F are λ-terms, so is $(E F)$ and the free occurrence and the bindings are those in E and in F.

(II) *Abstraction.* If E is a λ-term and x a variable, then $\lambda x.E$ is a term. Free $(\lambda x.E) =$ Free $(E) - \{$free occurrences of x in $E\}$. The bindings in $(\lambda x.E)$ contain all those of E and in addition the free occurrences of x in E are bound by the abstractor λx.

To make λ-terms more readable, save brackets as suggested by the examples:
Instead of $((E_1 E_2)E_3)$ write $E_1 E_2 E_3$ (applications associate to the left).
Instead of $\lambda x_1.(\lambda x_2.(\lambda x_3.E))$ write $\lambda x_1 x_2 x_3.E$ (abstractors associate to the right).

Note that only single-argument (monadic) functions are formalized in this way. This is not an essential restriction because there is a straightforward method (currying) which reduces the use of a function f of several arguments to the use of a related monadic function $\tilde{f}$. For example, suppose we wish the binary function f to apply to x and y and to produce x^y. Then the monadic $\tilde{f}$, applied to x alone, produces the monadic exponential function $\lambda y.x^y$, whose value is just x^y when applied to y. In his *1984* paper J.B. Rosser comments in connection with this:

> This is the way computers function. A program in a computer is a function of a single argument. People who have not considered the matter carefully may think, when they write a subroutine to add two numbers, that they have produced a program that is a function of two arguments. But what happens when the program begins to run, to produce the sum $A + B$? First A is brought from memory. Suppose that at that instant the computer is completely halted. What remains in the computer is a program to be applied to any B that might be forthcoming, to produce the sum of the given A and the forthcoming B. It is a function of one argument, depending on the given A, to be applied to any B, to produce the sum $A + B$.

λ-notations are consistent with the use of higher-order functions which allow as arguments and/or return as values other functions. As a matter of fact, this underlies the idea of "currying". Using more programming jargon, one might say that in the λ-calculus programs themselves may be dealt with as data. Such a flexibility is a direct consequence from the lack of any type constraints. The price one has to pay for this is that self-applicable functions are allowed as well. For example, in the expression

$$\phi = def \; \lambda x.f(xx) \tag{1}$$

x is applied to itself.

Many programming languages share this feature of the type-free λ-calculus and allow procedures which can take themselves as arguments. But it is well known that self-application leads to contradictions, as is evident from the following definition:

$$P(f) = def \; \text{if } f(f) \neq 0 \text{ then } 0 \text{ else } 1. \tag{2}$$

According to (2) $P(f) \neq f(f)$ for all functions f, hence applying P to itself one gets $P(P) \neq P(P)$.

Clearly, this points to serious semantical problems we have to confront when trying to interpret λ-terms as definitions of functions in the set-theoretical sense (mappings from sets to sets). We shall return to this topic in Section 5 but meanwhile let us recall the operational semantics of the λ-calculus. It relies on three operations which intuitively are expected to preserve the meaning of terms:

1. *α-reduction* viz, renaming of bounded variables, as explained above. This corresponds to the *static scope* discipline in programming terminology.
2. *β-reduction* viz evaluation by substitution (or *call by name*):

$$(\lambda x.M)N \text{ reduces to } M[x := N]$$

in which $M[x := N]$ means the result of replacing each free occurrence of x in M by N (after appropriate renaming of bound variables in order to avoid collision of free and bound variables).
3. *β-expansion* – the operation inverse to *β*-reduction.

A *red B* means that A may be transformed to B by 0 or many operations of these sorts.

If no *β*-reductions are possible on B either immediately or after some α-reductions, B is said to be in normal form and then if A *red B*, B is a normal form of A.

Consider the term ϕ as in (1). $\phi\phi$ reduces via one *β*-reduction to $f(\phi\phi)$; in this sense $\phi\phi$ is a fixed point of the function f. The term $Y =_{\text{def}} \lambda f.\phi\phi$ being applied to arbitrary F produces a fixed point of F, i.e.,

$$YF \text{ red } F(YF). \tag{3}$$

Y is the so-called paradoxical combinator (fixed point combinator). The λ-calculus mimics the computation of a program by reductions. Church identified the positive integers $1, 2, 3, \ldots$ with λ-terms (in normal form) $\hat{1}, \hat{2}, \hat{3}, \ldots$ defined as follows:

$$\lambda fx.fx, \quad \lambda fx.f(fx), \quad \lambda fx.f(f(fx)), \ldots$$

Now, given a closed (i.e., without free variables) λ-formula F it expresses (λ-defines) the partial function f such that, for each positive integer n, $f(n) = m$ or $f(n)$ is undefined, according to whether $F\hat{n}$ *red* $\hat{m}$ or not. This is a correct definition since for a given n there can be at most one m such that $F\hat{n}$ *red* $\hat{m}$ (consequence of the Church-Rosser theorem). On the other hand, starting from $F\hat{n}$ the reduction process may be performed in a determinate and calculable way. In this sense a λ-definable function is "effectively calculable".

Finally, in 1936 Church published his definite proposal, which due to Kleene is well known as

Church's Thesis: *The λ-definable functions are* all *the effectively calculable functions.*

Note: Due to the relation between recursion and fixed points, the use of self-application is very useful in producing λ-definitions for functions. In order to mimic recursion Turing suggested the use of the fixed point combinator to provide λ-definitions that are less complicated than those originally elaborated by Kleene.

For readability we use mixed expressions in the sequel with many argument functions and other familiar mathematical notations, though in all cases we could keep the standard formal syntax of λ-terms.

Let us now return to the semantical difficulties caused by self-application. A simple way to avoid them would be to use the typed λ-calculus which does not allow self-application at all. In fact we shall consider a family of languages parametrized by Ω – the set of primitive types (e.g., integers, booleans, etc.). Moreover, we shall also parametrize with respect to C – the set of constants of the language, abandoning in this way the commitment to consider "pure" λ-calculus. The formal definitions are as follows:

Type expressions (or simply-types):
Each element of Ω (each primitive type) is a type.
If α, β are types, so is $\alpha \to \beta$.
Now, $C = \underset{\alpha}{U} C^{\alpha}$ where C^{α} is the set of constants of type α. For each type α an infinite set X^{α} of variables of type α is considered as well; $X = \underset{\alpha}{U} X^{\alpha}$.

The languages $L(\Omega, C)$ consists of typed terms; $L^{\alpha}(\Omega, C)$ is the set of terms of type α. Below we omit the parameters Ω, C supposing them fixed:

Basis: $C^{\alpha} \subset L^{\alpha}, X^{\alpha} \subset L^{\alpha}$

Application: $u \in L^{\alpha \to \beta}, v \in L^{\alpha}$ implies $(u\,v) \in L^{\beta}$

Abstraction: $x \in X^{\alpha}, u \in L^{\beta}$ implies $\lambda x.u \in L^{\alpha \to \beta}$

Binding and reductions are as in the untyped case. Clearly, $L(\Omega, C)$ is a proper part of the untyped language with constant form C. In particular no self-application may occur in a typed term.

3. LISP

LISP was designed by J. McCarthy in the late fifties–early sixties. It is the second (after FORTRAN) oldest programming language which is still widely used, especially for work on artificial intelligence. The design of LISP relies on several fresh ideas which became quite popular, e.g., computing with symbolic expressions (rather than numbers) and their appropriate representation by list structures. However, for our consideration, those features that are inherited from the lambda calculus, are more relevant.

1. The most striking feature is the explicit use of the LAMBDA operator for naming functions:

> To use functions as arguments, one needs a notation for functions, and it seemed natural to use the λ-notation of Church *1941*. I didn't understand the rest of the book, so I wasn't tempted to try to implement his more general mechanism for defining functions. Church used higher order functionals instead of using conditional expressions. Conditional expressions are much more readily implemented on computers. (Quotation from McCarthy in Wexelblat *1981*, p. 173–197)

2. LISP promoted the idea of treating basic operations (*car, cdr,...* and ultimately even conditionals) as functions. Hence, they could be composed and dealt with as constants of the λ-calculus. Some slight deviations from this standard were recognized in the sequel. So, as D. Park pointed out,

> The LABEL notation invented by N. Rochester was logically unnecessary since the result could be achieved using only LAMBDA – by a construction analogous to Church's (paradoxical) *Y*-operator, albeit in a more complicated way. (Wexelblat *1981*).

3. LISP also inherited "currying" from Church's Lambda Calculus. McCarthy recognized procedures as functions of one argument; it is possible to apply one procedure to another and to return a third one as a result.

4. The representation of LISP programs as LISP data is in full accordance with the type free lambda calculus. McCarthy's universal function EVALUATE strongly relies on this idea.

5. The clear vision that the functional style of programming is the most appropriate way to assure referential transparency. At first McCarthy considered it important to express programs as applicative expressions built up from variables and constants using functions, i.e., to pursue a functional applicative style of programs in which side effects are avoided (Pure LISP).

Let us now mention two essential points in which LISP deviated from the λ-calculus ideal.

The first one concerns the introduction of "dirty" features such as assignments and goto's. Considering that tricks with side effects may be a source of computational efficiency, in the sequel McCarthy gave up the Pure LISP. (Much later, for similar reasons, other languages (for example, ML) that are generally recognized as functional languages resorted to some of these "dirty" features.)

But the most striking deviation from the λ-calculus legacy is the dynamic scope discipline of LISP as opposed to the static scope of the λ-calculus. Thus, (up to specific LISP notations) the expression

$$\lambda x.(\lambda p.((\lambda x.px)0)(\lambda u.(x+2))1$$

would be evaluated to 2 in LISP whereas its value according to the static scope is 3.

That this was not a conscious decision, but rather an unfortunate oversight of McCarthy, is testified to in his survey (Wexelblat *1981*). It is a very instructive story, and it is worth recalling that

> James R. Slagle programmed ... LISP function definition and complained when it did not work right ... In modern terminology, lexical (static) scoping was wanted but dynamic scoping was obtained ...
>
> Firstly, ... I regarded this as a bug and expressed confidence that it would soon be fixed. Unfortunately, the devices invented to fix it were not able to manage definitely with the problem ...

After all, modernized and sanitized versions of LISP such as SASL (Turner) and LIPSKIT (Henderson) abandoned this unfortunate dynamic scope and restored the original scoping of the λ-calculus.

4. ISWIM

In 1966 Peter Landin published his famous paper (Landin *1966*) under the intriguing title, "The Next 700 Programming Languages", where he introduced the language ISWIM (If you See What I Mean). In comparison with the λ-calculus the major innovation is in an additional binding mechanism through the *let* and *letrec* constructs. These constructs significantly improve the binding by abstraction permitting the statement of declarations (definitions) in a convenient way. At the same time ISWIM was also high order: hence functions which have other functions as arguments could be declared. We limit ourselves here to a few examples which will hopefully illustrate this point:

$$let \ (x = y + 1) \ in \ (x^2 - x + 1) \tag{1}$$

may be considered as a notational version ("syntactical sugar" – according to Landin) of the standard λ-expression

$$(\lambda x.x^2 - x + 1)(y + 1). \tag{2}$$

What is gained here is only a more articulate and more suggestive exhibition of the relevant subexpressions: the definition $x = y + 1$ with "definiendum" x and "definiens" $y + 1$, and the "main part" (i.e., the user of the definition) $x^2 - x + 1$.

On the other hand, *letrec* introduces recursive definitions and provides a more substantial improvement: note that

$$\text{\textit{letrec }} f = \lambda n.(\textit{if } \text{equal}(n, 0) \textit{ then } 1 \textit{ else } n \cdot f(n - 1)) \textit{ in } f(3) \qquad (3)$$

is quite different from

$$\text{\textit{let }} f = \lambda n.(\textit{if } \text{equal}(n, 0) \textit{ then } 1 \textit{ else } n \cdot f(n - 1)) \textit{ in } f(3). \qquad (4)$$

In (3) the defined f binds the occurrences of f in both the definiens and the main part $f(3)$; hence the intended value of the whole ISWIM expression is $3! = 6$

In (4) unlike (3) only the occurrence of f in $f(3)$ is bound, whereas the occurrence in $f(n - 1)$ is free and hence refers to a global function supplied by the environment (say, by an external declaration).

Usually, instead of (3) and (4) ISWIM would use the sugared forms, say, (3') below instead of (3):

$$\text{\textit{letrec }} f(n) = \textit{ if } \text{equal}(n, 0) \textit{ then } \ldots \qquad (3')$$

The *letrec* construct was conceived by P. Landin as syntactical sugar for specific λ-terms which use the (paradoxical) fixed-point operator Y. For example (3) might be desugared to (the nonrecursive declaration)

$$\text{\textit{let }} f = Y(\lambda f.\lambda n(\textit{if } \text{equal}(n, 0) \ldots)) \textit{ in } f(3)$$

which might be further desugared, like (1) to (2).

In general, ISWIM allows arbitrary mutual declarations and nested declarations as illustrated in (5):

$$\text{\textit{letrec }} (f(n) = (\textit{if } \text{equal}(n, 0) \textit{ then } 1 \textit{ else } f(n - 1) + g(n - 1)) \textit{ and } g(n) = 2n + 3)$$
$$\text{\textit{in }} \quad (\text{\textit{let }} f(n) = (\textit{if } \text{equal}(n, 0) \textit{ then } 1 \textit{ else } n \cdot f(n - 1)) \textit{ in } f(3)) \qquad (5)$$

Consider the defined f in the mutual recursive declaration for f and g; clearly by this declaration it is specified as the function $\lambda n.(n + 1)^2$. On the other hand it binds the occurrence of f in $n \cdot f(n - 1)$; hence the value of the whole expression will be $3 \cdot (3 + 1 - 1)^2 = 27$.

The *let* and *letrec* constructs reflect the down-top style of definitions and their use, and this is the common style in programming. In more common programming notations instead of, say

$$\text{\textit{let }} f(n) = n + y \textit{ in } f(5) \qquad (6)$$

one might use something like

$$define \; f : function \; (n);$$
$$return \; n + y$$
$$end; \tag{6'}$$
$$f(5).$$

In fact, ISWIM also allows as synonymous for *let* and *letrec* the *where* and *whererec* formats which exhibit a top-down approach. For example, in the simple case of (1) the "where" format looks like:

$$x^2 - x + 1$$
$$where \tag{1'}$$
$$x = y + 1$$

Often, when the λ-formulas of LISP programs are cumbersome and deeply nested, the use of the "*let*" and "*letrec*" constructs provide more readable notations which come closer to the conventional mathematical style of formulating and referring to definitions.

In addition to the reduction rules of the λ-calculus, the operational (reduction) semantics of ISWIM also includes specific rules for manipulation with declarations. For example, in the case of a block with a simple (nonmutual) declaration, expansions of calls are performed according to the

Expansion Rule:

    letrec $x = N$ in M reduces to *letrec* $x = N$ in $M[x := N]$

Note that for the nonrecursive *let,* this rule in fact yields a slightly improved sugaring of the β-reduction of the λ-calculus (compare (1) and (2) above).

We considered above the untyped ISWIM. In order to get its typed version ISWIM (Ω, C) we have only to extend the syntax for $L(\lambda, C)$ with declaration clauses. For example, using I^α as a shorthand for ISWIM$^\alpha(\Omega, C)$ we have:

Block with mutual recursive declarations:
$$x_1 \in X^{\alpha_1}, \ldots, x_k \in X^{\alpha_k}$$
$$u_1 \in I^{\alpha_1}, \ldots, u_k \in I^{\alpha_k}, \; v \in i^\beta \; imply$$

$$letrec \; (x_1 = u_1 \; and \; \ldots x_k = u_k) \; in \; v \in I^\beta.$$

ISWIM was originally a purely functional language, and here we consider it only as such, although Landin later added an imperative feature – the "program point". ISWIM (unlike LISP) obeys the static scoping of the λ-calculus with respect to both binding mechanisms, hence renaming bounded variables will in no way affect the meaning of an ISWIM program.

One other deviation from the λ-calculus is that ISWIM evaluates "by value", whereas the λ-calculus evaluates "by name". For example:

$$(\lambda x.6)(5/0) \tag{7}$$

returns the value 6, whereas *let* $x = 5/0$ *in* 6 requires $5/0$ to be evaluated first and hence is not defined. This is not a dangerous deviation for two reasons:

1. Call by name and call by value cannot produce two different values (as the alternative scoping rules may do). The worst that can happen (see the example (7)) is that evaluation by name produces a result, whereas call by value is not defined.
2. Call by value can be modeled through call by name plus appropriate primitive constants.

Later, the ISWIM innovations became widely used in all programming languages which developed the λ-calculus paradigm.

We mention here only two such developments: The first one is Edinburgh ML (Gordon et al. *1979*), with a rich type structure and many facilities which in particular facilitate interactive proofs of programs (hence the initials ML for Meta Language).

The second one is LUCID (Wadge and Ashcroft *1985*), conceived by its authors as a Data Flow Language, with the intention of a parallel implementation. In fact LUCID may be desugared to ISWIM with an appropriate set of constants. (The authors deal mainly with untyped ISWIM).

But beyond the important technical aspects of his contribution, Landin was the first to fully grasp and promulgate what we would like to call

The Church-Landin Thesis: *Programming Languages are λ-calculus sweetened with specific sugar.*

Remarkably, from the very beginning Landin had in mind not only functional languages but also imperative ones such as Algol (Landin *1965*). On one hand, he defined the so-called SECD machine, on which λ-formulas may be manipulated in the same way as common imperative programs in real computers. On the other hand – and that was the heart of the thesis – he had the clear vision of the λ-calculus underlying both functional and imperative languages.

5. Semantics

It is not the purpose here to advocate the importance of having formal (mathematical) means for the specification of the semantics of programming languages; clearly, relaxing the requirement of formality may lead to inconsistencies (Knuth *1967*).

The need of precise formal syntax was recognized much earlier, and FORTRAN and later ALGOL already established high standards of syntax specification. The problems with semantics were and still are considerably more difficult.

In Sections 3 and 4 semantics of λ-calculus and of ISWIM was specified by the computational behavior of symbolic manipulation (say – reductions of the λ-calculus). This sort of specifying semantics is called *operational,* and it is quite different from the *denotational* one, which is generally accepted in mathematics and logic. In contrast to operational semantics, the denotational meaning of a piece of syntax is an object belonging to a (hopefully) well-understood semantical domain. Moreover, the meaning of an expression is defined by induction through the meanings of its subexpression. Hence the denotational approach provides a referentially transparent semantics which obeys the *compositionality principle* of the kind usual in mathematics and logic.

Scott and Strachey (*1974*) took on the challenge of elaborating a denotational semantics (they used the term "mathematical semantics") for programming languages. Scott promoted the fundamental idea that models of the λ-calculus might be an appropriate tool for providing programming languages with formal semantics. Moreover, it was clear that the type-free λ-calculus had to be dealt with. As in general for other algebraic calculi, it is not difficult to consider a superficial model by taking appropriate equivalent classes of syntactical object. For the λ-calculus, it would roughly amount to factorizing the class of all λ-terms with respect to convertibility; but such a model would be of little help. The famous solution presented by Scott in 1969 was the construction of λ-models based on *continuous lattices.* After that Scott developed the theory of *domains,* as the basis for mathematical semantics of programming languages.

For the typed λ-calculus and typed ISWIM, where no self-application may occur, models for denotational semantics are easy.

To define the semantics of $L(\Omega, C)$ we need a *type-frame* $D(\Omega, C)$, that is a family of sets D^α called domains, one for each type α such that:

1. $D^{\beta \to \gamma}$ consists of *some total* functions from D^β into D^γ.
2. The constants from C^α are interpreted by some elements from D^α.

An environment for a type-frame $D(\Omega, C)$ is a mapping $e : X \to D$ which respects types. $e[d_1/x_1, \ldots, d_k/x_k]$ is the environment that differs from e only on the variables $x_1, \ldots, x_n$ for which the values $d_1, \ldots, d_n$ are assumed. Env denotes the set of all environments.

Establishing a semantics for L amounts to specifying a "nice" mapping

$$[[\quad]] : L \times \text{Env} \to D$$

which respects types; $[[u]]e$ is the meaning of the term u under the environment e.

We expect that these meanings are consistent with the operational semantics; in particular reductions should preserve the meanings of terms. This is indeed the case if we assume that D is a *full frame,* that is for each β, γ the domain $D^{\beta \to \gamma}$ consists of *all total* functions from D^β into D^γ. It is very easy to prove:

Theorem: *Assume D is a full frame. Then there exists a (unique) mapping* $[\![\ ]\!]$ *which correctly interprets the constants from C and such that:*

0) *for* $x \in X^\alpha$ $[\![x]\!]e = e(x)$
1) *Application:* $[\![uv]\!]e = [\![u]\!]e \, ([\![v]\!]e)$
2) *Abstraction:* $\forall d \in D^\alpha \ \forall x \in X^\alpha \ [\![\lambda x.u]\!]e \, (d) = [\![u]\!] \, e[d/x]$

In fact, in addition to the full frame, there may be other frames for which the theorems holds as well. All such frames are called λ-models of $L(\Omega, C)$.

Call λ-terms u, v denotationally equivalent (notation $u \approx v$) iff in each model:

$$[\![u]\!]e = [\![v]\!]e \text{ whatever the environment } e \text{ may be} \tag{$*$}$$

The claims we formulate below point to fundamental properties of denotational semantics in general; their analogues hold for other languages as well:

Claim 1 (Replacement Rule): *Assume that $u \approx v$ and that w_2 is the result of literally replacing (without renaming of bound variables) a subterm u of w_1 by v. Then $w_1 \approx w_2$.*

Claim 2 (Consistency of the denotational semantics with the operational one): *Assume the terms u, v are convertible (i.e., one can reduce u to v using α-reductions, β-reductions and expansions); then they are denotationally equivalent.*

Denotational semantics for ISWIM (Ω, C) is handled very similarly to, though a bit more complicated than, $L(\Omega, C)$. As in the case of $L(\Omega, C)$ one needs a type frame $D(\Omega, C)$, and again $D^{\beta \to \sigma}$ consists of some *total* functions from D^β into D^σ.

The main point is that one has to assure in each domain $D^{(\alpha \to \alpha) \to \alpha}$ the existence of a fixed-point operator A_α such that for all $f \in D^{\alpha \to \alpha}$ both (Yf) and $f(Yf)$ are the same element in D^α. It is also necessary that the fixed-point operators be chosen consistently with the structure of the frame. A well-known approach is to use *continuous models*; this amounts to considering frames whose domains are *complete partial orders (cpo's)* with continuous functions from D^α into D^β as elements of $D^{\alpha \to \beta}$. We won't go into detail here on this subject, but the following point should be emphasized. Since (according to the definition of a cpo) in each domain D^α there exists the specific element $\perp_\alpha$ – the undefined element of type α – considering only *total* functions is not a restrictive requirement (roughly – mimic a partial function by a total function which sometimes returns $\perp$).

Two ISWIM expressions E_1, E_2 are said to be equivalent (notation: $E_1 \approx E_2$) if they have the same meaning in all continuous models and for each environment.

As expected, the analogue of Claim 1 holds for ISWIM as well. Consistency with operational semantics is manifested by a lot of important equivalences. Here are some illustrations:

Expansion
  (*letrec* $x = N$ *in* M) $\approx$ (*letrec* $x = N$ *in* $M[x := N]$)

Denesting :
  let $x_1 = ($*let* $x_2 = u_2$ *in* $u_1)$ *in* $v \approx$
  let $(x_1 = u_1$ *and* $x_2 = u_2)$ *in* v
  assuming x_2 is not free in v

Explicit parametrization:
  (*let* $f = u$ *in* v) $\approx$ *let* $h\,x = u[f := hx]$ *in* $v[f := hx]$.

There are a lot of other important equivalences, concerning expansions of recursive calls, transformations of simultaneous declarations into iterated ones. We shall return to this topic later in Section 6.

Unlike the typed languages where things are relatively easy, for the type-free λ-calculus, the crux is to give a consistent meaning of self-application. Scott developed a rich theory of models of self-applicable functions for the semantics of type free λ-calculus as well as other programming languages. This theory relies on a novel understanding of the notion of function and application of a function to an argument because ordinary mathematical functions are not self-applicable.

Scott's deep and elegant theory was too overloaded with subtle mathematical details to be accessible to the broad programming community. That is why he returned to this topic over and over again, looking for a more conventional approach.

Even prominent experts in programming used to complain on the cumbersome machinery of denotational specifications. In his *1978* Backus estimated the state of affairs as follows:

> Why did the good mathematics of the Scott-Strachey approach not work? ... It results in a bewildering collection of productions, domains, functions and equations that is slightly more helpful ... than the reference manual of the language.

And here is a quotation from Scott *1982*:

> When I remember how much headaches I have caused to people in Computer Science who have tried to figure out the mathematical details of the theory of domains I have to cringe ...
>
> The difficulty in the presentation of the subject is in justifying the level of abstraction in comparison with the payoff; often the effort needed for understanding ... does not seem worth the trouble – especially if the notions are unfamiliar or excessively general ...

It takes some time to learn the notations and terminology and to become comfortable with them, to gain sufficient intuition ...

However, it may be that, for the first time since Church started his programming exercises on the λ-calculus, λ-programs obtained a precise and robust model-theoretic semantics. Moreover, from now on mathematical semantics of constructs in diverse programming languages might be specified via translations into appropriate forms of λ-calculus.

6. Imperative Features

Two prominent features – assignments and goto's – are directly inherited from the von Neumann computer architecture. Both are completely alien to the spirit of λ-calculus and are nicknamed "dirty features" by the adepts of pure functional programming. The goto's were long ago recognized as a troublesome control mechanism by the pioneers of "structured programming" (Dijkstra: goto's are harmful). Assignments are the main vehicle through which the computer memory (or store) is altered by program execution. Due to them (even without goto's) one can count to achieve tricky side effects and computational efficiency. Imperative languages were intensively developed beginning with the early FORTRAN. Notably John Backus, the designer of FORTRAN, later joined the critics of the "dirty features"; in his Turing lecture (*1978*) he called for the liberation of programming from the von Neumann style and outlined a new functional language.

Instead of commenting on the controversy between functional and imperative programming we shall confine ourselves here to the following claim: λ-calculus is not solely the core of functional languages but it is also of exceptional relevance to imperative languages as well.

In order to give evidence to this claim we focus here on fully typed algol-like languages (Pascal comes close enough to them). These languages allow an extremely broad repertoire of program constructions: assignments, conditionals, command sequencing, recursion, high-order procedures with value parameters (call by value), variable parameters (call by reference) and procedure parameters. No restrictions are assumed on the nesting of blocks and of mutually recursive procedure declarations. Nevertheless, what we are going to explain is that despite the multifariousness of these features, the "true" syntax of Algol-like languages is nothing but ISWIM(Ω,C) with appropriately chosen parameters Ω, C.

Concretely, we consider an illustrative toy language PROG, whose syntax reflects familiar programming languages except for one essential point: an explicit distinction is made between the type *loc* of memory locations and the type *val* of storable values.

(In common programming terminology one refers implicitly to this distinction via "left" and "right" values of expressions.)

For example in the assignment

$$x := cont(y) + a$$

x is of type *loc*; on the other hand in the value expression $cont(y) + a$, y is of type *loc*, a is of type *val*, and *cont* is intended to perform the explicit dereferencing from locations to their value contents. Note also the difference between the Boolean expressions

$$x = y \qquad cont(x) = cont(y);$$

the first one expresses sharing of locations, whereas the second one expresses equality of values.

Below, after having sketched some features of PROG we proceed to the description of a translation *Tr* from PROG into ISWIM. This translation is the first step in a two-step process of formalizing the semantics of programs π in PROG. The second one amounts to assign meanings to $Tr(\pi)$ in the standard ISWIM way. Programs simply inherit their semantics directly from the ISWIM terms in which they translate. Moreover, the abstract syntax, viz. parse tree of $Tr(\pi)$ is actually *identical* to that of π. The translation serves mainly to make explicit the binding conventions and the implicit type coercions of PROG. Having in mind these circumstances we argue that IWSIM (Ω, C) provides the *genuine* syntax of PROG – a worthy alternative for the ALGOL-jargon which came down through history.

Types of PROG. As primitive types consider

$$\Omega = \text{def}\{loc, val, bool, stat\}$$

where *stat* is intended to be the type of statements and parameterless procedures.

Blocks and bindings in PROG. Procedures of all higher finite types derived from Ω may be declared, passed as parameters, and returned as values.

Procedure identifiers are bound in PROG via declarations occurring at the head of a procedure block, e.g.,

$$\begin{aligned}
&\textit{begin} \\
&\qquad \textit{proc } P_1(formal1, formal2) \Longleftarrow Body1, \\
&\qquad \textit{proc } P_2(formal) \Longleftarrow Body2 \qquad end; \qquad\qquad (*) \\
&\qquad \textit{Blockbody} \\
&\textit{end}
\end{aligned}$$

Blocks with variable declaration have the format

$$\textit{begin} \quad \textit{var } x; St \quad \textit{end}$$

where x is of type *loc* and *St* is a statement, i.e., a phrase of type *stat*.

The syntax preserving translations Tr. We choose the parameters Ω, C of ISWIM as follows:

1. Ω – coincides with the set of primitive types in PROG.
2. C consists of two parts:

   Σ – the signature of PROG, i.e., the set of function symbols and predicate symbols it uses (say, *plus, equal, less*$\cdots$)

   $\Delta = \{cont, seq, var, assign, \cdots\}$ contains symbols that correspond to the program constructors.

Procedure blocks are translated, using *letrec*. For example the block ($*$) above is translated into

$$letrec\ P_1(formal1, formal2)Tr(Body1)\ and$$
$$P_2(formal) = Tr(Body2)\quad in\quad Tr(Blockbody)$$

By translating procedure declarations in this way it follows from the definition of ISWIM that static scope applies.

Blocks with variable declarations are handled with λ-abstraction and with constant *var* of type $(loc \to stat) \to stat$:

$$Tr(begin\ var\ x;\ St) = var(\lambda x.Tr(St)).$$

Hence, the binding effect of *var x* in the block is reflected in the binding effect of λx on $Tr(x)$.

Other expressions and statements are translated directly by introducing suitable constants from Δ, but no binding operators.

Examples: Consider the value expression $cont(y) + a$. The intended meaning of $cont(y)$ is not a value, but a function from stores to values. (In Algol jargon such functions are called *thunks*.) Below we use *valexp* as a notation for its type (a shorthand for $(loc \to val) \to val$). But then we expect also the coercion to *valexp* for the type of a. Hence

$Tr(cont(y) + a) = plus(cont(y), Ka)$
with *cont* of type $loc \to valexp$
K of type $val \to valexp$
plus of type $(valexp, valexp) \to valexp$.

The translation of assignments is straightforward, e.g.,

$$Tr(x := cont(y) + a) = assign(x, Tr(cont(y) + a))$$

with *assign* of type $(loc, valexp \to stat)$.

In this way the syntax preserving translation Tr may be accomplished and hence most of the semantics of PROG. All that remains is the appropriate choice of domains and checking that the constants in Δ are continuous functions and also that they are consistent with the underlying intuition.

All this is almost a routine exercise with the significant exception of *var*. Here is where one of the crucial points of the whole enterprise concerning the connection of local variables with global procedures and the semantics of local storage (Trakhtenbrot et al. *1984*) is hidden.

What are the advantages of presenting PROG in the ISWIM format?

An immediate benefit is that it reveals the fundamental difference between the binding mechanisms in blocks with variable declarations and in blocks with procedure declarations. This is made explicit by their contrasting translations into ISWIM; the first are translated using λ-abstraction and the constant *var*, whereas the second are translated using the *letrec* binding mechanism.

A principal consequence of the syntax preserving translation is that the basic properties of the procedural mechanism in programs can be recognized as direct consequences of more elementary and well-understood properties of ISWIM. Thus many familiar and relatively simple equivalent transformations of ISWIM correspond to significant transformations of programs. The soundness of these transformations is therefore independent of even the meaning of the primitive constructs of the programming language (i.e., ultimately – of the interpretation of the constants in Δ). This observation brings us to the clear distinction between two levels of abstraction for algol-like programs: on one hand there are the *program schemes* with uninterpreted signature Σ but still relying on the meaning of program constructs; they are just the objects studied in classical *comparative schematology*. On the other hand there are the λ-schemes (following the terminology of Damm and Fehr) which abstract from both Σ and Δ.

7. *Parallelism Features*

The λ-calculus is generally recognized as a language which specifies sequential (or serial) computations. In modern programming language design this feature had been deemed too restrictive; as a remedy various facilities of parallelism were suggested. Remarkably, the λ-calculus shares this drawback with the Turing approach to computability. To preserve its universality the generalized Turing algorithm needs to be given some kind of parallelism. Shepherdson (this volume) deals with this issue; the following two illuminating remarks are quoted from there:

[1] Certainly what Turing actually concentrated on in his paper – written before the existence of electric computers – was human calculation. And ... Turing assumed that

the computation was serial, proceeding as a sequence of elementary steps, whereas a machine may do several steps simultaneously.

[2] Clearly, if one wants to include all conceivable machines one must allow parallel operation. In the case *usually considered* [my italics – B.T.] of computation over total structures, i.e., ones whose functions and relations are defined for all arguments, parallel procedures are no more powerful than serial ones, for one can obviously serialize a parallel procedure by subdividing the time scale. This is no longer true if there are partial functions, e.g., the function f defined by

$$f(x) \;=\; x, \text{ if } f_1(x) \text{ is defined or } f_2(x) \text{ is defined}$$
$$= \text{ undefined otherwise,}$$

obviously cannot be computed by any serial procedure because it might choose the wrong one of f_1, f_2 to evaluate first. And there is no bound on the number of processing units which may need to be simultaneously active as is shown by the example

$$f(x) = x \text{ if any of } f_1(x) \text{ or } f_1(f_2(x)) \text{ or } f_1(f_2(f_2(x))) \text{ or } \dots \text{ ad inf, is defined.}$$

The remark about serializing parallel procedures points out a possible source of confusion arising when one tries to formalize the notion of sequential function and to contrast it with the notion of parallel function. This distinction makes sense when the domain and the range of the function are appropriately structured and do not reduce to (what Shepherdson refers to as being) a "total structure".

Church and von Neumann, independently of each other, formulated and investigated a model of growing automata with cells working concurrently and in a synchronized way. But the computations performed by Church-Neumann automata can be serialized just for "total structure" reasons. Therefore this model does not compute any parallel functions, despite the explicit exhibition of an unbounded number of processing units which are simultaneously active.

On the other hand, recall that in semantical models of ISWIM whose domains have a cpo structure there occur de facto partial mappings, albeit disguised as total functions. Hence, one can expect the distinction between sequential and parallel functions to make sense. A typical example is the parallel OR, which unlike the sequential one, returns the value *true* when one of its arguments is true, even if the other one is $\perp$ (undefined). OR behaves essentially as the function $f(x)$ in Shepherdson's example. Similarly one can contrast the parallel and the sequential versions of the conditional "if... then... else".

Precise definitions of the notion "sequential function" (and by negation of the notion "parallel function") for different well-structured domains were formulated and investigated in Kahn and Plotkin *1978*, Plotkin *1977*, and Sazonov *1976*. In general these definitions are consistent with each other and support the hope that indeed they capture the essence of the phenomenon.

Now returning to the λ-calculus as a programming language, the questions arise: (i) in what precise sense is it sequential (serial)? and (ii) what might be the ways to enrich it with parallelism?

As to the first question, one of the possible explanations is the following: Assume that in a model of ISWIM (Ω, C) all constants from C are interpreted as sequential functions; then each function definable in ISWIM (Ω, C) is also sequential. In other words this phenomenon may be explained as follows: application (as a binary mapping) and fixed points operators (as unary mappings) are sequential functions.

One way of enriching ISWIM with parallelism might be to impose on some of the constants in C specific interpretations requiring them to be parallel functions (say, parallel conditionals). This approach is reminiscent of the one used earlier in Section 6 to enrich ISWIM with imperative features; it was realized by Plotkin *1977* and Sazonov *1976,* where the comparative power at parallel functions (constructs) was also analyzed.

As an alternative attractive approach it would be nice to incorporate concurrent computations directly into the calculus. Milner *1984* and Hoare *1985* created special calculi for concurrent computation, conducted by communication among independent agents: CCS (Milner) and CSP (Hoare). This trend is developing very successfully, both in theory and in applications to programming and communication. Unfortunately, its relationship to the λ-calculus and to the notion of parallel function as explained above is not clear.

In his *1984* Milner testifies:

> Sequential computation has a well-established (model) theory, due to the λ-calculus, which existed long before any notion of implementing a programming language. The λ-calculus was (and is) a paradigm for evaluation, in the same way that the predicate calculus is a paradigm for deduction. More recently and largely due to Dana Scott, the model theory of the λ-calculus has grown and has been harmonized with the evaluation theory.
>
> CCS is an attempt to provide an analogous paradigm for concurrent computation, conducted by communication among independent agents. It arose after several unsuccessful attempts by the author to find a satisfactory generalization of the λ-calculus, to admit concurrent computation.
>
> The relationship between a calculus for communication and the lambda calculus is far from clear.
>
> The notion of higher-order function which fit so well with the λ-calculus, seems to find no obvious generalization in the setting of concurrent communicating systems.

8. Concluding Remarks

The previous sections hopefully gave evidence to the claim that in many essential aspects high-level programming languages are inherited from the λ-calculus. This claim was referred to as the Church-Landin Thesis.

Dealing with lambda definitions for numerical functions, Church pursued the sole object of giving a precise characterization of computability. This was before the existence of computers and programming languages. Certainly, it would be an anachronism to discern in it direct and explicit indications on the future development of high-level programming languages. All this was realized only later, especially due to Landin and Scott's investigations.

Let us shortly recall the main features of the λ-calculus and its ISWIM extension, considered as functional programming languages:

1. Abstraction as a fundamental binding mechanism, to which other bindings may be adequately reduced.
2. Declaration mechanism via the *letrec* construct (ISWIM)
3. Static scoping rule for bindings; hence bound variables play the role of place holders and may be renamed.
4. Call by name (by substitution) as an evaluation rule. Call by value can be mimicked.
5. Clear distinction between the flexible type-free language and the more restrictive, but more amenable, typed language.
6. Currying, as a way to deal only with one-argument operators.

And, in addition to these syntactical features,

7. An illuminating denotational semantics, which serves as a robust guide in reasoning about programs.

On a high schematological level even imperative and/or parallel programming languages behave as the λ-calculus and may inherit most of the features listed above.

Less clear is the relation between the λ-calculus and the calculi for communications that were developed to support parallel programming.

But actually the post-ISWIM development of the λ-calculus was and still is impetuous, and so is its impact on programming language design. This comes to light basically through the elaboration of more developed type structure and type discipline – an area which is beyond the scope of this essay. Reynolds *1985* and Burstall and Lampson *1984* give a good idea of the activities in the area. The following illustrations are borrowed from these papers:

(i) Burstall and Lampson designed PEBBLE – a core language to support the writing of large programs in a modular way, which takes advantage of type checking. In order to achieve the goal of "programming in large" designers usually invent various features and sometimes make ad hoc decisions. As to PEBBLE, "it provides a precise model for those features, being a functional language based upon the λ-calculus with a peculiar type structure in which types are values. It is addressed to the problems of data types, abstract data types and modules."

(ii) Reynolds *1984* discusses, among other things, the polymorphic λ-calculus which was defined independently by Girard and Reynolds. Unlike the common "first-order λ-calculus" the polymorphic one (called also second-order λ-calculus) allows the definition of polymorphic functions by abstraction on type variables. For example, in the second order lambda-term

$$\Lambda\alpha.\lambda f^{\alpha\to\alpha}. \ \ \lambda x^{\alpha}.f(fx) \tag{*}$$

the abstraction $\Lambda \ \alpha$ is on the type variable α, whereas $\lambda f^{\alpha\to\alpha}$ and λx^{α} abstract on "common" variables f and x of types $\alpha \to \alpha$ and α respectively. The term $(*)$ specifies the polymorphic "doubling" function that can be applied to any type (say *integer*) to obtain a doubling function for that type.

Certain forms of polymorphism are allowed in some widely used programming languages (e.g., ADA). But the polymorphic λ-calculus suggests a novel and powerful programming style with the following peculiarities:

1. all programs terminate;
2. a wide spectrum of data (integers, booleans, lists, ...) are represented as polymorphic functions.

In fact (2) is inspired by the way in which Church encoded natural numbers, truth values, ... in his λ-definitions. For example the term $(*)$ is nothing but a polymorphic variant of Church's untyped code for the number two.

And again we are faced with a wonderful anticipation: a seemingly ad hoc technicality in Church's λ-definitions is reincarnated after half a century in a novel approach to data representation.

References

Backus, J.

  1978 Can programming be liberated from the von Neumann style? *Comm. ACM* **21** (1978) 613–641.

Barendregt, H.

  1984 *The Lambda Calculus: Its Syntax and Semantics,* revised edition. Studies in Logic 103. Amsterdam: North-Holland Publ. Co. (1984).

Böhm, C.

  1966 The CUCH as a formal and description language for computer programming. In: *Formal Language Description for Computer Programming,* ed. Steel. Amsterdam: North-Holland Publ. Co. (1966).

Burstall, R., and B. Lampson

  1984 A kernel language for modules and abstract data types. In: *International Symposium on Semantics of Data Types,* Lecture Notes on Computer Science 173. Berlin: Springer-Verlag (1984).

Church, A.

    1941 *The Calculi of Lambda-Conversion.* Princeton, NJ: Princeton University Press (1941).

Damm W., and E. Fehr

    1980 A schematological approach to the procedure concept of Algol-like languages. In: *Proc. 5-iem Colloque Sur Les Arbres,* Lille, pp. 130–134 (1980).

Davis, M.

    1982 Why Gödel didn't have Church's Thesis. *Inf. Contr.* **54** (1982) 3–24.

Fortune, S., D. Leivant, and M. O'Donnell

    1983 The expressiveness of simple and second-order type structure. *J. ACM* **30/1** (1983) 1451–185.

Gordon, M., R. Milner, and C. Wadsworth

    1979 *Edinburgh LCF,* Lecture Notes on Computer Science 78. Berlin: Springer-Verlag (1979).

Halpern, Y., A. Meyer, and B. Trakhtenbrot

    1984 From Denotational to Operational and Axiomatic semantics for Algol-like Languages, Lecture Notes in Computer Science 164, pp. 474–500. Berlin: Springer-Verlag (1984).

Hoare, C.A.

    1985 *Communicating Sequential Processes.* London: Prentice Hall (1985).

Kahn, G., and G. Plotkin

    1978 Structures de donnees concretes. IRIA-LABORIA Report 336 (1978).

Kleene, S.C.

    1979 Origins of recursive function theory. *Ann. Hist. Comp.* **3** (1979) 52–67; see also Conf. Rec. 20th Amm. IEE Symp. on FOCS, 1979, pp. 371-382.

Knuth, D.

    1967 The remaining trouble spots in ALGOL-60. *Comm. ACM* **10/10** (1967).

Landin P.

    1965 A correspondence between Algol-60 and Church's Lambda-Notation. *Comm. ACM* **8** (1965) 89–101 and 158–165.

    1966 The next 700 programming languages. *Comm. ACM* **9** (1966) 157–166.

Meyer, A.

    1982 What is a model of the Lambda calculus? *Inf. Contr.* **52** (1982) 87–122.

Milner, R.

    1984 Lectures on a calculus for communicating systems. In: *Seminar on Concurrency CMU, Pittsburg.* Lecture Notes on Computer Science 197, pp. 197–220. Berlin: Springer-Verlag (1984).

Morris, J.

1968 *Lambda Calculus Models of Programming Languages.* Dissertation MIT (1968).

Plotkin, G.

1977 LCF considered as a programming language. *Theor. Comp.* **5** (1977) 223–257.

Reynolds, J.C.

1981 The essence of Algol. In: *Proceedings of the International Symposium on Algorithmic Languages,* eds. de Bakker and van Vliet. Amsterdam: North-Holland Publ. Co. (1981).

1985 Three approaches to type structure. In: *TAPSOFT Advanced Seminar on the Role of Semantics in Software Development.* Lecture Notes in Computer Science. Berlin: Springer-Verlag (1985).

Rosser, J.B.

1984 Highlights of the history of the lambda calculus. *Ann. Hist. C.* **6/4** (1984) 337–349.

Sazonov, V.

1976 Expressibility of functions in D. Scott's LCF language. *Alg. Log.* **15** (1976) 308–330 (in Russian).

Scott D.

1972 Continuous lattices. In: *Topics in Algebraic Geometry and Logic,* Lecture Notes in Mathematics 274, pp. 97–136. Berlin: Springer-Verlag.

1982 *Domains for Denotational Semantics,* Lecture Notes in Computer Science 140, ICSLP. Berlin: Springer-Verlag (1982).

Scott, D., and C. Strachey

1971 Towards a mathematical semantics of computer Languages. In: *Proceedings of a Symposium on Computer and Automata, New York* (1971).

Shepherdson, J.C.

1988 Mechanisms for computing over arbitrary structures. This volume.

Trakhtenbrot, B.

1976 Recursive program schemes and computable functionals. In: *MFCS, Proceedings 1976,* Lecture Notes in Computer Science 45, pp. 137–151. Berlin: Springer-Verlag (1976).

Trakhtenbrot, B., Y. Halpern, and A. Meyer

1984 *From Denotational to Operational and Axiomatic Semantics for Algol-like Languages,* Lecture Notes in Computer Science 164, pp. 474–500. Berlin: Springer-Verlag (1984).

Wadge, W. and Ashcroft, E.

1985 *LUCID, The Data-flow Programming Language.* London: Academic Press (1985).

Wexelblat, R. (ed.)

    1981 *History of Programming Languages,* pp. 173–197. New York: Academic Press (1981).

Form and Content in Thinking Turing Machines

Oswald Wiener

1.

For purposes of epistemology, the crowning idea in Turing's work on computability is the conjecture known as the Church/Turing thesis: that every "working" procedure can be formulated in terms of a Turing machine (TM). Does this mean that a mind's knowledge of a "working" procedure takes the form of a TM or of some equivalent? Introspection clearly supports the hypothesis that understanding natural processes amounts to being capable of simulating them by some "internal" TMs – the clearest indication might perhaps be seen in our arrival at the TM concept itself. But even before the strict concept (Turing *1936-7*) was available, physicists like Hertz and other scientists had described this notion of structural simulation in less stringent terminologies. Compare Schrödinger's version:

> In trying to grasp the observed behavior of natural objects one forms a notion (Vorstellung), relying upon the experimental data in one's possession, but without obstructing intuitive imagination. This notion is precisely elaborated in every detail – *much* more so than any limited experience could warrant. The notion in its absolute determinateness is like a mathematical entity or like a geometrical figure that can in every respect be computed on the basis of a number of *defining features*; like for instance in a triangle one side and the two adjoining angles as defining features determine the third angle, both other sides, the three altitudes, the radius of the inscribed circle, etc. The sole important distinction between a notion and a geometrical figure is the circumstance that in addition to the latter's determinateness in the three dimensions of space it is likewise determined in time, as in a fourth dimension. That is to say (which is self-evident) that it always is an entity that changes in time, taking on various *states*. And if a state is known by the required number of defining features, then not only all the other features are given for this particular moment (as explained above in the case of the triangle), but also all

the features, the exact state, for every particular point in time thereafter – quite as the situation at the base line of a triangle determines the situation at the opposite apex. Part of the internal law of this entity is that it changes in a definite manner – left on its own in a certain initial state it will continuously run through a determined series of states and attain each single one at a definite point in time. This is its nature, this is the hypothesis that is posed on the grounds of intuitive imagination, as I said before.

Of course, one is not simple-minded enough to suppose that one might in this manner find out how nature actually is run. To indicate that one does not think so, one prefers to call the precise thinking aid one has created an *image* or a *model* ... (Schrödinger *1935*, Schrödinger's emphasis).

Today the indicated hypothesis is widely taken at face value and even has been given the form of computer programs cum explanations (Simon and Hayes *1974*). While at higher levels – explicit knowledge as programs – the issue is not really controversial today, it appears that we still have hardly come to realize the extent of complexity needed for the mechanisms that *generate* the simple conscious thoughts of humans – the "surfaces" of those TM models that become "visible" in introspection and behavior. Our knowledge in this deeper domain is limited to the degree that the crucial question for today's attempts at scientific epistemology still seems to be: *Are beliefs justifiable that these construction mechanisms* – presumably including Schrödinger's "intuitive imagination" and the closely related notion of Turing's "intuition" (Turing *1939*) – *are TMs (or something equivalent) in their own turn?*

Searle's criticism (Searle *1980*) marks an open question with regard to the Church/Turing thesis: if indeed this apparatus cannot be explained in terms of TMs then our hypothesis implies that components are involved that we may not hope to understand (and Searle will then be justified in employing unexplained notions of "understand" and all the rest, "intentional states" and "causal powers of the brain"). The moment this is realized to be a fact will be the moment to seriously start looking out for physical processes that are not describable by TMs (for possible incorporation into a thinking machine without our understanding their internal workings).

However, while search for processes of this kind has merits of its own, we may, as Turing put it, never say that we have looked hard enough for a conventional – a Turing machine – solution. In this spirit, though fully aware of the warnings of Husserl and neurology, I believe that attempts will not be entirely futile to sort out instances where this stronger hypothesis – that explicit knowledge is constructed by TMs – appears to make sense after all. In what follows I derive from introspection and speculation a catalogue of capabilities that seem necessary for a machine in order for me to believe it to be thinking. I present the catalogue as a little dictionary of TM metaphors for epistemological concepts, and proceed to note some of the difficulties that are likely to arise in implementations of these "definitions"; I ask whether difficulties with philosophical transparence might be hoped to be reducible to mere technical obstacles – an ambivalent question because in proportion to growth of complexity technical problems merge into philosophical difficulties. My main point is reconcilement of introspective experience and technical metaphor.

I will not here try to justify my pervading references to introspection (I do this in Wiener *199x*) but just point out that, although I do subscribe to the "polite convention that everyone thinks" (Turing *1950*), I am not prepared to extend this compliance to thinking machines brought about in nontraditional ways. I shall not as an experiment in epistemology play Turing's famous game, but will have to be shown in detail how the machine arrives at its performance. Notwithstanding stories of birds and airplanes, I think that the only way to construct a thinking system will be to imitate the general ways of the human mind.

The choice of TMs as building blocks for definitions (with or without quotation marks) in epistemology is justified by the concept's elementarity and generality, by its natural embodiment of elementary kinetics, by the fact that it is used in research areas such as learning theory or complexity theory, by its conceptual nearness to parts of organic chemistry, and not least by the persistent, ever helpful presence of the results of the theory of computation.

With further regard to epistemology it is encouraging to notice that the TM not only furnishes an elementary description of the behavior of a human computer, but also an elementary description of the facts of human observation: with natural processes we always observe some changes that impinge upon our sense organs (strings of "signs" in a time/space series), but most often we have to actively account for some "invisible" part of the process (strings of signs of some "internal alphabet") in order to perceive regularity.

For the metaphors below I prefer TMs (in place of, say, programming languages or formal systems) mainly for psychological and didactic reasons. In textbook definitions of the TM, atomic components – "tapes", "signs", "scanners/printers", "states" – are introduced to present a clear-cut concept. However, the elementarity of this concept is characterized by the fact that those components are TMs in their own right. A *sign*, for instance, may be viewed as a TM computing the Boolean function of some set and having its affirmative output coupled to some constant, or alternatively as the constant itself, produced in the manner described. (A *constant* is a trivial TM – see below – regarded as computing a zero argument function; a *string* S is a trivial TM capable of interacting with some other TM "in the manner S": "printing the string S" then denotes constructing the trivial TM "PRINT S".) In what seems to be more than a figure of speech, the TM is a sign or a symbol in space/time, a sign endowed with internal structure.

Thus the TM concept appears to be just right for modeling purely dynamic or "organic" entities that function as a sign in one place and as a process in another. To specify the table of a definitely given machine is to maintain that the relations described will hold no matter which way the supporting "layers" arrive at the specified outputs – appearance of the expressions "sign" or "string" in a text signals that some system believed entirely decomposable shall be analyzed down to a certain level only. To name some part of the dynamics a "state" is nothing but a sometimes useful shorthand (further complicated by the introduction of the Universal TM: do we by "state" mean a state of the UTM or a state of the TM it is executing?). In order to usefully employ the TM concept in descriptions of complex systems, however, one will want to allow for ongoing changes in the supporting layers also, for changes in the boundaries provisionally delimiting the "components" as well as the "layers" – e.g., parts of the "TM proper" (described in the table) may become part of the scanner or vice versa, or, some state may have to be replaced by an entire TM. Introspection seems to demonstrate that we most often proceed to analyze underlying structure whenever we find ourselves baffled by difficulties on the level that temporarily "represents" – the TM model offers every desirable malleability in that any part of a TM supporting some representation may immediately become a part of the representation itself (what constitutes a specified TM within a complex system of TMs will depend on another TM specifying it).

In short, what recommends the use of the TM as a building block for epistemological theories is its triple-facedness qua description, process, and physical object. While in a sufficiently powerful programming language there is nothing to prevent similar uses, programming languages and the von Neumann computer certainly do not facilitate them (von Neumann's cell automaton certainly does, but on the one hand it poses technical difficulties for a general computational theory, and on the other it appears that one has to fall back on TMs in describing complex cell automaton behavior).

With logic things appear a trifle more awkward still. Conversion of a formal system into a TM system proceeds by taking expressions (signs, terms and formulas) for tape inscriptions and setting up the transformation rules as atomic TMs (or rather, as states of "the" TM of the system). To create the flexibility presumably needed for the description of cognitive processes one would have to regard proofs and derivations as entities of the system: the various ways of arriving at an expression would have to constitute the various structures of that string. But logic as we know it is not designed to preserve tracks and study regularities of processes if they do not show in the regularities of inscriptions – all dynamics is trusted to the mind of the logician.

Consequently there is a risk of employing all too simple notions of what TMs are and what they can do, if in philosophical arguments the TM model is regarded as just another way of presenting formal systems. As an example take Putnam's TM concept (see various papers in Putnam *1975*). While identifying a "psychological state" with some TM state has the air of a pun, and while it certainly seems unfeasible, as Putnam notes in a refutation of his former views, to define it extensionally as a "disjunction of TM states" (p. 298 loc. cit.) or as a "disjunction of conjunctions of a machine state and a tape (i.e., a total description of the memory bank)" (p. 299), such a "state" – if one bothered at all to meddle with the vague notion – might jolly well be conceived of as a trait common to various internal actions, and this trait will have to be pictured as a full-fledged TM proper (a TM that "embodies" the perceived regularities characterizing the considered set of actions – the "intension"). I take an enigmatic statement, "... memory and learning are not represented in the Turing machine model as acquisition of new states, but as acquisition of new information printed on the machine's tape" (p. 298), to indicate that Putnam did not think of a more flexible notion of the TM. Conscious learning of new material is achieved precisely by the construction of some TM – regardless, in principle, of whether it is given by a "tape" inscription describing its machine table or by some actually working machine.

2. *A Budget of Metaphors*

1. ABBREVIATIONS
A *trivial* TM is a TM that functions in the mode of a finite automaton only: a TM that by its layout is unable to utilize the tape as an intermediate memory for its action. To distinguish trivial TMs verbally from other TMs I name the latter *folded* TMs.

By a *trace* (of a computation of some TM T) I mean the succession of the "instantaneous descriptions" of Davis *1958,* state symbol removed. A trace can be regarded as one single string. By *output* of T I mean some part of T's trace that serves as an input to other TMs.

A *structure* of a string is a TM that, with or without input, is capable of printing the string (a string, then, viewed as a trivial TM, is a structure of itself); alternatively: a (part of a) TM that "accepts" the string (that is, the function it computes is defined on the string); alternatively: a TM computing a Boolean function and affirming the string. This concept of "structure" is amenable to considerable refinement (see e.g. Koppel, this volume), but the above will do for my purposes.

If some set of strings can be produced by some TM by varying the input then I will call this TM a structure of the set. The input will be called *data;* more specifically, a datum is a trivial TM that, if joined to or folded into a given structure of a given set of strings, will augment the structure by at least one state that will be activated only in the single instance of the structure's producing one particular element of the set. (A datum may be viewed as a program for a structure to retrieve a particular element of the set it structures.) The conceptual boundary between a structure and its possible data will be termed a *partition.*

A *projector* is a TM computing a characteristic function for a set of strings, its affirming output being coupled to a constant. The "inverse" of a given projector is a trivial TM that prints some element of the projector's affirmation subset. At times I will by "projector" also indicate some collection of projectors appropriate for the given context.

By a *prototype* I mean a TM viewed as a modularization of another more elaborated TM. A prototype might be regarded as a TM coupled to projectors on the input side (that is, accepting a set of constants as its alphabet) and to their "inverses" or to other trivial TMs on the output side (the set of projectors and trivial TMs may be taken as its "read/print head"). Alternatively it may be regarded as a program core, where the "calling" outlets can be coupled to "dummies" or to other appropriate TMs. A prototype of T might print some subset of T's trace (on one or various inputs). The simplest prototype is a trivial TM (a "name" – see below) regarded as a (actual or potential) pointer to some other TM.

An *organism* is a set of TMs that are capable of constructing, modifying and coupling (concatenating, fusing, inserting, composing, indexing) TMs as well as of activating ("calling") one another and the newly constructed ones. The *operating system* (OS) of an organism contains the basic construction instruments; it is supposed to be unchangeable relative to those tools. The original set-up of the OS is supposed to contain some basic set of projectors.

Outside (a given organism) designates the rest of the universe (including the organism's hardware) in the state of interacting with the organism.

A *screen* is a device functioning as a tape that is jointly used by some subset of an organism. The elements of the alphabet common to these TMs are the *screen primitives,* or "signs" proper.

In saying that a organism *matches* two strings I shall imply that the strings are found to be "the same" modulo some projector.

An *effector* is an element of an organism that prints (parts of) its traces onto the outside.

An *observer* is a human observing the entire internal and external action of an organism embedded in its outside.

2.

(a) If strings can be printed onto the screen of some organism also from the outside and be marked *external,* and if the organism or its OS is coupled to some effector capable of causing changes in the outside that will, at least partially, in their turn appear as strings or changes of strings on the screen;

(b) and if the organism or its OS can store strings up to a certain length (for instance by converting them into trivial TMs to be incorporated into the organism);

(c) and if the organism is capable of constructing, modifying, and coupling TMs so that they become folded structures of strings that appear on the screen;

(d) and if the organism is capable of using TMs resulting from (b) and (c), or their output, or parts thereof, in the control of the effector;

(e) and if it is capable of insulating these structures from the external strings as well as from the effector, and of "running" them in this *detached* mode and of marking the strings hereby appearing on the screen as *internal* –

then I will call these structures *models* (of the strings) and say that this organism (or its OS) has such models. The organism's acquisition of new models will be called *learning* (or *induction*). I assume that the organism is capable of adding appropriate versions of its models to its stock of projectors. A model actually run to print onto the screen will be called *in focus*.

Strings that are caused by the organism but printed onto the screen without mediation of a model will be classified external. Structures that do not meet conditions (c) or (e) will be considered elements of the OS proper; one might term such TMs "implicit models" or *pre-models*.

The construction, modification, etc., of a model M will depend on a set of TMs that possibly includes other models; this set is the *construction environment* of M. The running of M will also depend on other TMs, among them possibly models that provide input for M or otherwise govern its momentary action (select a state as the initial state, etc.); this set of TMs is a *running environment* of M. Those parts of an environment consisting of TMs that in their turn can be made to print internal strings will collectively be called the *traceable environment* of M. M is understood to be a component of its running environment. A structure run by an environment without other model components is an *isolated* structure.

Generally I will talk about models without paying too much attention to their "depth". What to the observer might appear to be a model of a square that is divided into smaller squares could, inserted into a suitable running environment, appear to be a model of a chess board; depending on circumstances he might also call this whole running environment (or its traceable part) a model of a chess board.

3.

Belief: all that an organism can summon in its interactions with the outside are its models and pre-models (reflexes, ...). For the observer to say that an organism believes so-and-so is to say that it will run such-and-such models under certain conditions, and perhaps also that it has such-and-such preferences among the strings that those models are capable of producing.

If the organism is capable of retrieving a model in stock to match parts of this model's trace to a given external string, or to an internal string produced by a different model, then it is *oriented* with respect to this string. If various external strings (that may appear at different times) can be matched to various parts of traces of the same model, I will call the set of these external strings an *object* (this extends to the matching of internal strings, so I introduce "real" and "mental" objects).

To indicate an object of the observer, single quotation marks will be used. At times I will permit myself the sloppy – with regard to the metaphor – usage "M is a model of O", where O is an 'object', for further brevity of diction.

While – to the observer – an object is constructed as an organism's model of certain strings, people act as if they perceived automata situated "on the other end of the sense organs". There seems to be nothing wrong with this very economical way of handling the problem! The "objectivity" of objects, of course, with people depends on mainly two factors: "consistency" within the individual, and possibility of communication about the object with other people. Since I wish to maintain that the fundamental mechanisms of cognition are not dependent on natural language and communication, I will say little about the unquestionable, supportive role of language in thinking (but see Wiener *1988*), and nothing about "correct" thinking. As to the objects of organisms, I assume that to model a string is to create a device that will, by reproducing the screen string, trip (part of) the "feel" of the original "sensory experience". Whether relative to an object there is consistency within the organism will depend on the respective model's environment as well as on the organism's modeling capacity and on economizing forces that might be operating on it. In particular, for the organism to "have a real object" will in the observer's view mean that the model can become part of a running environment that to the observer will represent (parts of) the 'outside environment' of the 'object'.

I assume that orientation behavior in an organism will manifest itself in four modes:

– the organism finds itself oriented;
– it has lost orientation, albeit having available additional models that as components of the running environment can aid in a search for reconnection (for instance by providing missing parts of the target string or by guiding the effector);
– the organism switches to the learning mode;
– it ignores the state of affairs – which on a low level of course is trivial but on a higher level might be a feature of sophistication.

4.

A string is *possible* in a strong sense if it is part of a trace of a model run in the detached mode. Just for the record (I do not wish to go into the concept of consistency – it can be done in the metaphor): it is possible in a weak sense (relative to some model M) if it is produced by a different model or an isolated structure, and the organism detects no inconsistency within the running environment of M after incorporation of the model or structure in question.

If the organism is capable of identifying certain possible strings as "desirable" (i.e., there are some preference functions defined on the strings) and of utilizing for control of the effector the traceable environments of the models that produce the desirable strings, so that eventually external strings may be caused that can be matched to the desirable strings, then the organism is characterizable as *purposive.* Since success of the behavior is not a criterion for the presence of purpose, purposiveness will show only within the frame of the observer's theory of the organism. On the other hand, humans, while trying, now and then know that they are trying – whether an organism will have the potential to "know" something similar about certain configurations of its own models will depend on its sophistication (see below, "intentionality").

In the given formulation, incidentally, arrangements of the outside in the manner of drawings or other "external representations" are likewise admitted, provided that they survive the matching process (for a discussion of "representation" see the last part of the paper).

Let me try a distinction between "purpose" and "goal-seeking". As the observer would see it, for a purposive organism the traceable environment of any purposively appropriate model will have to contain models of the 'effector' (as far as it enters the screen in the guise of external strings) and of ways of employing it; it will have to contain models of how to gain access to the 'object' causing the external string (that gave rise to the model) by activation of the 'effector' etc.; this metaphor of purposiveness will have to be expanded to instances of replacing OS preference functions by models, of abolishing the restriction to external strings, of establishing levels of representation, etc. A goal-seeking entity, in spite of having been termed a "purposive mechanism" in authoritative explanations (Rosenblueth et al. *1943*), will make do with some "flat" feedback set-up, and perform accordingly.

5.

An organism *understands* a string *clearly* if it has an adequate model for it.

I do not think that it would be useful (or sensible) to define lower limits for the "depth" of understanding, but surely one line is drawn by the requirements of the apparatus that enables an organism to build models. Conditions for the ascription of understanding will be quite severe if we require the organism to be purposive, and even more so if we require it to be referential.

As to "clearly": there might be other modes of understanding – which I will not here venture to explore. "Adequacy" will depend on "consistency" in some of the model's running environments, on the demands on effectiveness of the organism's purposive behavior, etc.

6.

A *designation* is a string (not necessarily appearing on the screen) viewed as a program for the OS to retrieve and activate a model M together with a running environment for it. The retrieval process may or may not utilize other models. A designation S is a *name* if the OS succeeds, on input S, in directly localizing some M among the stock of its models. S is a *description* if it is a program for the OS to compose and run M from, and with the help of, other available models or their components. Normally a description will consist of names of models to be run, and of names of other models to be added to the actual running environment (the latter providing input to the OS for its control of the running mode). What I have in mind here derives with a little editing from Minsky's statement:

> There are a variety of ways to assign names. The simplest schemes use what we will call *conventional* (or *proper*) names; here, arbitrary symbols are assigned to classes. But we will also want to use complex *descriptions* or *computed names*; these are constructed for classes by processes which *depend on the class definitions.* The notion of description merges smoothly into the more complex notion of *model*; as we think of it, a model is a sort of active description. It is a thing whose form reflects some of the structure of the thing represented, but which also has some of the character of a working machine. (Minsky *1963,* Minsky's emphasis).

To give an example, the sentence "Shall I compare thee to a summer's day" in a human addressee's organism will usually initiate a complicated process that involves orientation in the surroundings, some variant of its model of the speaker (mainly models of the speaker's actual running environment), its models of summer and of day, and divers models of coupling those models or parts thereof, derived in part through the sentence's other designations and partly from models of general language use; how exactly modification and composition of all the models, and their outputs, will unfold will depend on the actual running environment of the addressee.

On the other hand, the described notion of designation presupposes no inter-organism communication language. What the observer may take to be strings caused by a 'tree', showing in external strings on the organism's screen, may function as a name in a more direct mode than the word "tree" – in case the organism should happen to possess an English language module (such cases, and the question of feature detection in strings, will be addressed briefly later on).

Hesitantly following imprecise but ubiquitous uses in everyday language, I will take the expressions "name" and "description" also to refer to the internal strings generated by M, and even further to some of the external strings that might be caused by the organism's effector under direction of M, or to strings with respect to which the organism is oriented under M.

7.

In his famous paper, "Über Sinn und Bedeutung" (reprinted in Frege *1967*), Frege established a seminal distinction of "denotation" (Bedeutung) from "meaning" (Sinn) of expressions to resolve the ambiguities alluded to. In our metaphor, this distinction makes *meaning* (of a designation S) refer to a model M retrieved or composed by the OS under control of S (or more precisely to some particular part or "position" of M with respect to some running environment), in contrast to *denotation,* which term will refer to the output of M.

Thus, the denotation of the string $\sum_{k=1}^{\infty} 1/2^k$ is the sign "1", while its meaning is the structure described by "½ + ¼ + ...", or a prototype thereof, as realized in the organism. The meaning of "1" will vary with the different structures (all, possibly among other designations, bearing the name "1") that the organism will have available for producing the string "1". Again, assuming some previous mild exposure of the organism to mathematics, the denotation of every single element of an appropriate set of strings (like for instance "1", "1^k", "the first element of the series of perfect squares", etc.) will invariably be the internal string "1"; whereas, depending on the model called or composed by virtue of the actual designation, and depending on the running environment of that model, the meaning will be the trivial structure of "1", the TM computing 1^k with some input datum provided by the running environment, and so on. In the eye of the observer, with an organism commanding an English-language module the sign "tree" will produce the denotations of the descriptions, respectively, "perennial plant with single woody self-supporting stem", "a piece of wood or metal to be used for stretching boots", "a diagram of branching lines", and what not – i.e., internal signs that for the observer will be representations or even images of a 'tree', of a 'shoe-stretcher', of a 'tree graph', etc.; the meanings of "tree" will be the respective TMs that the organism will have retrieved.

The distinction between denotation and meaning is by no means superficial, because the models might be used by the organism in ways very different from the uses of their outputs.

8.

We have defined "structure of a string" to designate a TM capable of printing the string unto its tape, etc. Contemporary authors tend to use the expressions "structure" and "form" synonymously, or regard "structure" as heir to romanticist "form". Didactic reasons, however, recommend a distinction. In the context of the TM metaphor, I understand *form* (of a string) to designate a prototype of some structure of the string that may be coupled to various projectors in order to also accept a respective variety of other strings – thus creating an equivalence class (under the prototype) of strings that generally will not be accepted by the original structure. Form, from this point of view, is a generalization of some collection of structures. TMs (and prototypes) being processes, I will also use the expression "form(s) of processes" (for an example take the form of multiplication: some prototype doing multiplication proper that may be coupled to divers projectors, which not only "decode" number (or other) representations but also will have to "call" the additionally needed bookkeeping units).

Judging from introspection, the detection of forms depends to a considerable degree on the choice or construction of appropriate projectors, which choice quite frequently will constitute, or at least prepare for, the creative aspect of the process. I surmise that exchange of projectors is the metaphor equivalent for "casting the problem in the appropriate form", the "changes of representations" of Amarel *1966* and others; but for those changes to become effective for discovery, work at the creation of "content" (see below) is indispensable.

Instead of accepting as simple a suggestion as the propounded one, analytical philosophers, bound to the trajectories of their metaphysical extraction, mostly abstain from offering constructive criticism of the notion. Kambartel, for instance (in Kambartel *1968*, pp. 176f. and 197f.), stresses the circumstance that "the form of a sentence, the structure of a concrete relational inventory" may be given only "seemingly with the help of an appropriate symbolism as the *form-in-itself*".

> Our considerations on the symbolizing of structures ... have made it clear that even so-called "formal" representations always are but other exceptionally transparent examples of a "form", but never the – as it were – "form-in-itself". The *intended* structure ... may be underlined in an "example" of a certain figure, because the figure is an *example* of this structure ... And this implies: we do not *directly* perceive the intended "form". Something appears to be "thus", because we are shown a so-called "form". But to cause this effect, already the form will have had to appear to be "thus". In this manner even the formal is a phenomenon of interpretation. (Kambartel's emphasis).

While this certainly hits upon a fact, Kambartel produces no tangible notion onto which one could stick "form".

This seems to be the place to point out that one specific feature of the TM metaphor is that the organism "sees" strings only – but never TMs. The organism might construct a model to produce strings that the observer might take to be an image of an automaton, but this will always remain an image produced by another different automaton. In writing this paper I construct a model of my own modeling process. However, the thought that somehow my "screen images" could become identical with the apparatus that produces them is simply erroneous within the frame of my approach (it cannot even be formulated within the metaphor).

9.

As with "denotation" versus "meaning", and as in the case of "form" and "structure", I think that important distinctions are lost if one simply choses to let "content" refer to certain strings, or even to some generalization of a form of strings (which is attempted for instance in Hunt et al. *1966*). In our metaphor, a generalization of a form will be a form again, no matter what. While the meaning of "content" is most indefinite in philosophical deliberations and in everyday discourse, I will use the name to refer to those aspects of a structure that are not directly involved in producing its traces. These are the aspects of the structure's genesis (its construction environment), its dependence on other models, its particular functions in governing the effector, its relations to other structures that are equivalent under a form, and its possible decompositions into parts of other available models. Concisely, by a *content* (of a string S) I mean a running environment of a model capable of printing S.

Of course I do hold that in the last analysis content, like form, is revealed to be nothing but structure. The didactically important tag arises with regards to the ways a organism learns, which I now proceed to consider.

3. *Clues from Introspection*

> ... je jugeai que je pouvais prendre pour règle générale, que les choses que nous concevons fort clairement et fort distinctement sont tous vraies, mais qu'il y a seulement quelque difficulté a bien remarquer quelles sont celles que nous concevons distinctement.
>
> Descartes, *Discours de la Méthode, quatrième partie*

1. SCREENS AND EFFECTORS

Some of the more obvious difficulties and vaguenesses within the preceding descriptions evolve around the notion of the screen.

It is supposed to transform signals of the outside into strings that affect the TM organism, thus defining the latter's level of "sign". This poses the question of screen primitives, but we will not be begging the question of model building as long as the screen strictly remains a part of the OS, unchangeable relative to the modeling process and unable to model strings. It is a pre-model of the outside, "created by natural evolution".

Marr *1982* demonstrates that the autonomous performance of the human visual system in transforming intensity arrays into 3D-shapes ready for recognition is truly stupendous. What is more, it is reassuring to now perceive feats that only a short time ago were inconceivable as ranking in the vicinity of algorithms (I even believe that Marr's highest processing level already belongs to the realm of cognition proper or otherwise emerges in collaboration with the orienting process).

Aside from having to stipulate limits for the screen's holding capacity, it seems therefore that we are to a certain extent free to define the correspondences of our general screen primitives to 'objects'. Whether we, for instance in a purely visual screen, have pixels for primitives (values of their attributes – intensities, colors, ...), or rather some topological (closures, convexities, discontinuities ...) or geometrical entities (lines, surfaces ...), or even ready-made representations of physical shapes:

in principle the modeling problem will be the same for every level, although the complexity of the model world (unless one postulates "freezing" of lower-order models into the OS) and the time factor in the evolution of "mesocosmic" models will vary tremendously.

I fully agree with Marr that recognition – not to speak of the ensuing mental elaborations – cannot be adequately modeled by feature detection and feature processing (in the sense that features were the primitives of recognition and ideation); I think, however, that a feature recognition module in sensory perception, controlled by the running environments, is necessary in order to quickly sort out candidate models (where existent) for the strings that eventually issue from the channel.

For the effector I ask nothing that has not been on the schedule of robotics research for a long time. I suppose the effector to be some sort of a servomechanism with "slots" to accept high level control input. Efficient control of the effector may be pictured as arising from the organism's experimentation in modeling the changes that the effector will cause in external strings and concurrent monitoring of its own effector control strings. It goes without saying that this, as in the case of the screen, amounts to serving the organism with a highly structured device from the outset. One might go one step further and also pose that the effector will cause a special characteristic in the external strings that follow its appearance in the "field of perception" of the screen; further stipulations seem possible without danger of unduly straining contemporary notions of technical feasibility.

From the postulate of monitoring effector control input alone it will have become obvious that the organism's screen is conceived to be a general purpose module "behind" and apart from the visual and other sensory channels. While arguments (for instance proposed in Gazzaniga and Smylie *1985*) in favor of a total severance of the "mental image" screen from the visual system are not yet really conclusive, I assume that many model outputs do not have visual character at all. Whether the screen should be thought of as being accessible from the model side via some "internal effector" servomechanism that interprets model output as well as input of the same general design as the effector input proper is a question that I do not here have space to discuss.

However, it is important to my approach that the notion of mental images as discussed anew during the last decade or so is *not* co-extensive with my more general notion of internal strings which is meant to comprise all the internal signals in the organism that may be compared to others via models, to provide feedback detectable by the modeling process, etc. I am convinced, for instance, that pseudo-innervations of the limbs – mainly the hands, of course, but at times even of the whole body – are very important and ubiquitous ingredients in thinking, especially, but by no means solely, in geometrical domains and in the exploration of dynamics. It seems a safe guess that our notions of space, of resistance, of continuity, etc., derive in the greater part from kinesthetic experience. On top of this the quasi-autonomous operation capability of the effector's servomechanism may distinctly be introspected to aid in parallel screen processing, helping to expand the limited focus facilities.

Finally, for this brief sketch, the visual system seems to constitute, perhaps even developmentally, the nucleus of conscious experience by its curious property of offering clear and distinct images only in a small area in the center of the visual field, opening up the surrounding capacity to different, ancillary functions. I am assuming that the screen has inherited a comparable property.

2. FOLDING AND PROTOTYPES

A TM T producing a given string can be folded (or refolded) if similarities are detected in parts of the string that are produced by different components of T; those parts will in the newly folded TM be replaced by one single component. The replacement will more often than not require additional modifications in the internal configuration of T, necessitated by execution technicalities, or, more important, by pre-treatment of inputs (that is, by projector variations of the alphabet); those changes will as a rule lead to a new partition, to changes in the extension of the admissible input set or, in the folding of trivial TMs, to the creation of an input set from the unfolded "waste".

Folding will ultimately have to be regarded as an economizing measure, but the most important side-effect of this parsimony is that it structures the way an organism will generally conceive of the outside: a regularity *is* a TM, an object *is* a model. Viewed in this way, meaning is but a trick to overcome limitations of formal capacity; this insight extends to the guidance of formal manipulations by content (see the next paragraph).

While the problem of constructing a trivial TM is trivial, effective methods are known (Biermann *1972*) for construction of folded TMs if traces complemented by indications of the scanner movements are given. However, the organism's folding of structures is supposed to proceed otherwise. For one thing, it is rarely the case that such traces are available; for another, Biermann's recursive procedure is cumbersome and there is no indication in introspection that its formalism is normally used in human model construction. Likewise limitations on time and other computing resources, and introspectional observations as well, bar considerations of the organism's engaging in recursive enumerations of models (to say nothing of the noncomputability of models with certain strongly defined properties).

Only in special cases, on the other hand, will "to build a model" mean to find any old TM that will just satisfy a prescribed input-output relation. At least some intermittent parts of the model's traces will most often likewise have a role in the model's overall fitting into the construction or running environments: the model or its prototype will have to "move" in certain prescribed ways in order to adapt to strings other than the ones to be structured, that is, to interact properly with other models in the environment. Some of these constraints will be imposed by the projector apparatus used by the environment for detection of relevant similarities in strings.

In addition, by virtue of introspection one may know of numerous cases where the construction environment supplies extra ("virtual") strings to complement the given ones or where it conceals parts of the screen strings, in order to facilitate construction and to smooth out performance of the developing model – to create virtual symmetry, as it were. The virtual strings might be derived from the "more complete" trivial TMs among the set to be folded, or stem from other models in the environment. They are, of course, what Occam advises to avoid; but the organism's "primary concern" will be to do which way ever, and the question of economy will – if at all – arise later on in the overall embedding of the model into its various running environments; and there are instances where the virtual parts suddenly do become invested with an 'object'.

How, then, shall one picture to oneself the actual construction procedure?

I assume that normally construction will start from a set of trivial structures that are somehow "associated" to an existing model nucleus or pre-model. Such appending might come about through chance appearances of external strings in some context involving the prototype nucleus. In excess to just having them appended to a prototype, the OS might at one time detect some similarity among them, perhaps tipped off by some momentary configuration of screen strings or some chance presence of some model in the running environment of the prototype (no regress is caused by this assumptions since pre-models and projectors will be part of the OS' original endowment). It may from other models even "know" that some regularity can be found.

Introspective experience abounds with instances of awkward and incomplete folding (for instance application of a learned strategy in a game may be tripped by routine recognition of a string's membership in a certain cluster of configurations, while other seemingly eremite strings or apparent singularities will require fresh creative effort). Graceful degrading of the performance of poor foldings and a chance for resipiscence will be warranted by the running environment that wherever possible will "try" to furnish bridges over untread ground, by "jumping" from a known part in a string to the next known section, or by provisionally filling in missing links in a model through systematic or chance runs of other models. Generally a set of trivial structures appended to a grossly oriented prototype may perform excellently under known conditions (the paradigm is mental topographical maps). I think that indeed much human behavior is governed by this kind of representation – in spite of the human organism's capability to start conscious efforts with a good average chance of improving on the situation.

Experience indicates that folding in general is a slow process, the single folding acts (and their corrections) being spread out in time and widely left to chance.

One even experiences occasional stepwise "amalgamation" of single pairs of strings. However, folding attempts may become routine in an individual, and greatly supported by knowledge of how to assign existing models to the task. Even cases of over-folding do exist, as for instance with a hacker who condenses structures to a measure that is not warranted by the content of the string.

I assume that folding will normally proceed from a prototype – some folded or trivial machine accepting and producing coarse summaries of the strings to be folded – "downward" to fine models that cover many primitives of those strings. A worked-out model might later on function as a substrate for a different modeling process that will create a different form. I assume furthermore that the organism in many cases can immediately bring into focus the modules of the elaborate model corresponding to the respective states of a running prototype, and likewise contract such modules into states or smaller modules of prototypes. Observations in one's own model world however show that under certain circumstances – adequate performance on some frequently recurring inputs, among others – detailed models can become compacted and streamlined to a degree that recall for inspection becomes difficult; we speak then of *skills* acquired.

3. CONSCIOUSNESS, CONTENT, AND FORMAL SYSTEMS

An organism with the properties described in this paper is within the metaphor termed *conscious*. More specifically, consciousness is the orientation mechanism; at the core of consciousness is the construction or modification of models.

I assume that model construction proceeds with "fitting" TM parts into a "mould" constituted by the actual construction environment. The environment not only provides input (taken from an elaboration of external strings or from other models) and "expects" output of a definite sort, but also trims the "shape" of the growing TM by comparing interconnections of its parts with interconnections of the states of a given form.

Every stage of the construction process may in the guise of its output appear on the screen as one particular string. Nevertheless it cannot simply be rendered as a symbol because this symbol is only a temporary result of the ongoing structural dynamics – its function is, in a kind of feedback, to trip hitherto inactive parts of the running environment, to incite regulation of configurations of the active parts, to initiate changes in projectors etc., and eventually to activate success or failure mechanisms.

In the interaction of several processes, this output is composed from the outputs of TMs that keep their individuality all the time. It may be seen to "symbolize" this circumstance only by an observer in possession of comparable processes to supply the dynamics. Since the "mould" processes have to be kept active all the time, necessity arises for some sort of parallelism.

There is nothing, however, to suggest that changes in the configuration will have to unfold in a strictly parallel fashion. On the other hand, if the screen surface were inert in comparison with the models that play on it, the slow fading of strings would assist the OS in selecting model parts with a dynamics to match, but this will necessitate a refreshing of the strings by the constellation that produced it, if the process cannot move on swiftly. As to the question of the seemingly severe limitation in the screen's focus, it may be argued that this limitation is in accordance with the functions of the model building mechanism. Model construction is a local process; apparent globality derives from intermittent prototype (and environment) jumps. If the scope were extended then more of the components of the running environment competing for screen presence would show, and again some higher level filter would have to be installed to admit the actually relevant model features exclusively.

Composition and modification of models will start by composition or modification of their prototypes. I submit that intermittent runs of this prototype compositions or modifications, showing in some of their effects on the screen, will concurrently be accompanied by the running of the more elaborate versions in progress. Should the devolutions run into trouble, as compared to the prototypes, a signal will be generated that will cause the OS to bring the respective parts of underlying structures into screen focus for inspection. The same will be true of completed fine models. I think this mechanism roughly accounts for my orientation "feeling": in conversations about well-known subjects, for instance, I rarely experience the necessity of concretely picturing to myself the detailed situation "meant". I think that prototypes produce the anonymous blobs and lines that may appear as mental images to accompany general orientation behavior. On the other hand, I believe that "verifying runs" of more elaborate versions of the prototype do appear all the time "on the fringe of the screen" in the shape of transient or even instantly "flashing" pictures of details relevant to the current aspects of my actual understanding.

Thus, we are primarily concerned with the notion of one or several concurrent "invisible" processes assuming a transient heterarchy, and operating on a single process that prints into the focus. Again and again introspection corroborates the notion of construction, in the course of understanding, of mechanisms that in their turn become the substrate of the same construction process in a different content setting. The rules of chess are transformed into a program that will be operated on by strategies or strategy construction; similarities in strategies may be detected and transformed into a device of generalizing to other games of blockade and blockbusting. In every transition of level, the old program subsides into the traceable environment.

I think that formal systems are originally constructed in this way. While content gives "depth" to signs, regularities may be perceived on the level of the environment's screen manifestation that may be embodied in some new "flat" mechanism F to reproduce screen string behavior. In a human organism, access from screen strings to content is warranted most of the time; a situation subsists that has much similarity to the notion of "interpretation" in mathematical model theory. In a computer program for F, however, content is restricted to the depth of F. A program of this kind cannot access content "below" F for guidance of string transformations like a human usually can. To give a banal example, if in calculus a content of the differential quotient is traceable as a local measure of elastic expansion or contraction of the dependent variable as compared to the independent variable, many relations – e.g., substitution in integrals, graph length computations, etc. – immediately become "intuitively" clear from experience acquired in the outside, and will straightforwardly guide manipulations in the formalism. In writing a computer program, one likewise "peels off" string behavior from content and transfers this surface to the intimately known mechanisms of the programming language.

For these reasons the well-known arguments that a machine will not be able to detect "truth" in propositions not syntactically derivable in formal systems loose much of their force. (I would like to stress that by introducing the notion of "flat formalisms" I do not want to convey an opinion that formalisms will always *have* to be flat – I just think that thinking cannot be modeled in flat formalisms.)

4. LANGUAGE

In contrast to many contemporary scholars, I believe that language – the object of linguistic study – is not involved in thinking processes in the character of a necessary condition. I am assuming that there is a distinct language module in humans, however complex, to translate focus conditions of actual running environments into the flat grammar system, and to extract cues from actual discourse for the mobilization of models. On the other hand there is no doubt concerning the advantages of using words, with their constant shape and obviously very straightforward connection to the models they name, as thinking aids for model retrieval and relief from screen capacity pressure.

I think that in analogy to other formal systems of the day natural language expressions are "flat" projections of content. In serious and careful conversation, words to be chosen are tried out in the intended running environment to compute the effect they have on one's own models. They are then "skimmed off", as it were, from content and transmitted with the tacit conjecture that the listener will have similar models to be called by the designations and compiler cues of the sentences, and thereafter will experience a similar effect.

5. INTENTIONALITY AND REFERENCE

In dreams and in the labor of trying to examine some object that is sense-perceptible in principle but at the given time not observationally present, the question of the mental images "representing" something else never arises. If for instance I wish to know how the kick starter of my friend's little motorcycle folds, a mental image of the starter appears. I "turn" some part of the "starter" in order to "see" where it will fold ... Although preposterously coarse and unstable (compared in afterthoughts to direct sensory experience), it "feels" and "looks" like turning the real thing with my hand – probably mirroring the fact that even amidst the wealth of possible sensory experience at any one time I can pick up only a few features that are filtered out by the actual context (running environment). As long as I am concentrating on what happens, *this is reality for me* – there is nothing besides it. When "concentration" falters a little, some flag seems to appear, indicating that those images are "internal". But if this should happen then this will be my reality, a reality of the same quality as before but with a different content.

The next stage that I wish to consider consists in my stepping back from the mental image in a certain direction. The "starter" "contracts" into a line or indefinite blob in some coarse image of the entire motorcycle, the "motorcycle" shrinks to a blob in the place where I usually see it, the wall and the tree near my friend's house appear ...

If I choose to try another step back into a different direction then something extraordinary happens. I "see" (some blob that is) the starter, surrounded by (some other blob that is) the motorcycle next to the wall-blob and underneath the tree-blob, and I "see" – one image becoming more distinct as the previous one(s) fade(s) out of focus – "myself" standing in front of all this, looking at the starter: I "see" that the visual impression – some kinesthetically felt transmission from the "starter" to "me" – matches the mental image (another blob) in "my" head. The extraordinary thing is that everything that happens is internal, although some of the strings, the "starter" etc., are "felt to be external".

It is at this stage that I feel ready to say that the image in "my" head *represents* the visual impression "I" am having. A rather solid argument here is that, by turning in still another direction, I "see" that I could just walk there and have a look at the starter: I expect to see something that will match the "starter".

(I can accomplish yet another step: I can "see" "me" as "I" sit at the table imagining the scene near the wall. I do not know how far I can go on stepping back, since the scene "sitting at the table and imagining ..." will repeat itself from this point onward, and eventually I will loose count of the levels. To be more precise: I do not count the levels, and there is nothing to suggest that something inside me does (there is not space here to go into arguments that preclude stacking in the OS). I just realize the step back – in afterthoughts. On the other hand the images will never have to become void through abstraction, because apparently I can always fill them up with all the knowledge that is available to the most concrete level; that is, any level's image of "I" can be supplied with everything I know about myself – ultimately, in the case of "I", with every single model in stock).

(Things can become a little more complicated still. For a variety of reasons – for instance if the screen's focus cannot simultaneously hold all relevant detail – I will draw some sketch of X (that is, a sketch of my mental image of X: using my effector I will produce external strings that match the internal strings produced by my model or by some of its prototypes) and in front of this sketch I will again feel almost exactly like in front of X: I will be able to step back, etc.)

Apparently the fact of having models by itself cannot account for the notion of "representation *for* the organism".

Whenever conditions for the organism's running of some model are right, the observer might note that the model represents some feature of the 'organism's out-side'. By contrast, in adopting the point of view of the organism he will judge that the organism "is in" the model – that it, as it were, temporarily *is* the model. Only the organism's model of the model situation will make the notion of "representing *for* the organism" possible.

The process referred to by saying that "this blob is the starter" is complicated indeed. So long as I "manipulate" the "starter", all that happens is the appearing of internal strings that change in various ways. It may then happen that entirely different string arrangements appear, where certain strings are felt to "stand for" some of the strings in the previous arrangement – by some process that is not evident from the screen arrangement alone because the connections are established by the running environment. And it may even happen that some strings stand for other strings in the same arrangement by virtue of some process that *is* indicated on the screen, for instance when I let the blob in "my" "head" and the blob in the "motorcycle" move concurrently in an identical fashion.

To establish a handy formula: a string *represents* another one if they are of the same form, i.e., if the models having produced them as components of the current running environment are in the organism reducible to a common prototype, and if this prototype is in focus. A simpler instance is incurred if the representation relation is one of strings produced by a model and of strings produced by one of its prototypes. Most generally, the relation of input and output of a model in possession of the organism is a representation, if a prototype of this process is in focus of the running environment. And to connect to the notions expounded in the last paragraph: the strings produced by a prototype in focus represent the "marginal" strings of its concurrently running fine version(s).

In the literature of epistemology sometimes the opinion is found that having a representation must be accompanied by the knowledge of the subject that it is having the representation. Similarly, in definitions of intentionality as directedness it is sometimes held that "in experiencing an act of consciousness we find ourselves directed to something – ... in perceiving we are directed to the thing perceived, in remembering we are directed to the event recalled, ... " (Gurwitsch *1982*). Put thusly, everything becomes as disturbing as the possibility of speaking of centaurs was to Brentano. But we are "directed" to something that is present, namely to strings appearing on the screen. And then, this "finding oneself ..." is not a component of immediate experience, but a theoretical assertion after the fact. The assertion is possible because screen work is memorized and can be replayed and modeled in turn – the actual model will always reside on the level "behind" the images to be produced, and description of this model will require another model one additional step "back" (or "ahead", depending on the point of view). Like all the other properties of models, "directedness" is always detected in a different environment applied to the replay of the previous experience. (By these scanty remarks I do not intend to say that introspection amounts to nothing else but replay.)

In the TM metaphor one might think of constructing levels of sophistication in intentionality – of the capability to "step back", to use models as a substrate to modeling, culminating in the modeling of the modeling mechanism. But since directedness basically is orientedness in the sense of the metaphor, the obvious candidate for the term *intentionality* is the capability of an organism to be oriented with respect to external and internal strings. Certainly higher levels will be necessary for the organism in order to achieve greater sophistication in model building, but this seems to be a matter of degrees. For instance, unless the organism is expected to be able to model its own modeling process, it is not required to "know" that its internal strings (or, in loose talk, its models) represent. In fact I believe that in the collective and individual history of humans a knowledge of this kind is a rather late achievement.

Generally, then, I think there is no enigma of intentionality. For one thing, the model construction and modifying process seems to work in an identical manner on any level whatsoever, once the basic conditions for its existence are established; and the working conditions too will have to be the same on every level, the same restrictions on immediate complexity, the same manner of screen representation, the same ways of jumping from detail to prototype, etc. Second, for ascriptions of intentionality there is no condition that the string the organism "is directed to" will have to be a 'representation' of the "correct" 'object' in the view of the observer, and there simply is no problem of object identity. The organism will be in the "intentional state" all the time during being "on", viz., it will be in one of the modes of orientation behavior. Error is possible only with respect to the adequacy of the actually running models. How, then, can the organism be sure that its internal strings represent a certain 'object', or that the object it is directed to is identically the same every time it comes up? It cannot; neither can I: things may turn out one way or other, or not at all.

6. Concluding Remarks

This paper's theses have been coarsely sketched and certainly need elaboration. Many of its decisive notions still remain in a state of impractical generality and will have to be substituted by clearer perspectives, but others just suffer the constraints of available space. Notwithstanding obvious deficiencies, I think that the overall picture contains nothing to suggest that existing notions of computation will have to be expanded radically in order to explain human cognition performance, and indeed scruples seem in place with implementation problems only – among those, however, problems of complexity.

A wholly different question, of course, is possible criticism under the image of the self as still entertained by the humanities. I believe that here radical reorientation is inevitable. Introspection in the purpose of establishing spontaneity and irregularity of the mind is the main source of the anti-mechanist; but this is a far cry from observation intended to track recurrence of events, to assist in generalization and heuristic modeling. Like every other kind of observation, sound introspection will never tell what things are, but just what they do in terms of preformed notions. Therefore consciousness will have to be described in terms of process: it is what makes introspected things behave the way they do.

Among philosophical concepts, the notion of consciousness will probably be the one to loose the most. Of what? To support its constituting the beachhead of the nonmechanical in this world of the conceivable, we possess the luminosity metaphor derived from visual experience – "a kind of illumination, the self-awareness ..." (Eccles *1970*, p. 46), and, it is true, a set of experiences that cannot yet be explained with the help of mechanical devices. But I have said nothing of sense qualities, modalities, and emotions, simply because I do not believe that these entities have a function in clear understanding other than to hinder, further, or direct it. As to philosophical uneasiness concerning the pervasive introspectional experience of a subject causing content, I think the immediate target is *performance* of the cognitive apparatus; in its insistence on this approach, Turing's *1950* has its main thrust.

In order to deposit one final, though obvious, caution, I would like to emphasize that I do not take the TM metaphor as literally describing brain events. It might well turn out that some sort of equilibrium system describable by hydrodynamics will come closer to actual reality – a system, perhaps, that has its units (here described in terms of TMs) act upon one another much more directly, in the manner of corporeal machines. In my opinion, however, the important point will remain that the mechanisms central to understanding can be cast in the image of the Turing machine.

References

Amarel, S.

    1966 On the mechanization of creative processes. In: *IEEE Spectrum* (April 1966).

Biermann, A.W.

    1972 On the inference of Turing machines from sample computations. *Artif. Intell.* **3** (1972).

Davis, M.

    1958 *Computability and Unsolvability.* New York, Toronto, London: McGraw-Hill (1958).

Eccles, John C.

    1970 *Facing Reality.* New York: Springer-Verlag (1970).

Frege, G.

    1967 *Kleinere Schriften.* Hildesheim: Georg Olms Verlagsbuchhandlung (1967).

Gazzaniga, M.S., C.S. Smylie

    1985 What does language do for a right hemisphere? In: *Handbook of Cognitive Neuroscience,* ed. M.S. Gazzaniga. New York (1985).

Gurwitsch, A.

    1982 In: *Husserl, Intentionality, and Cognitive Science,* eds. H.L. Dreyfus and H. Hall. Cambridge, MA (1982).

Hunt E.B., J. Marin, and P.S. Stone

    1966 *Experiments in Induction.* New York (1966).

Kambartel, F.

    1968 *Erfahrung und Struktur.* Frankfurt am Main: Suhrkamp (1968).

Marr, D.

    1982 *Vision.* San Francisco: Freeman (1982).

Minsky, M.L.

    1961 Steps toward artificial intelligence. In: *Proceedings of the Institute of Radio Engineers 49* (1961); reprinted in: *Computers and Thought,* eds. E.A. Feigenbaum and J. Feldman. New York: McGraw-Hill (1963).

Putnam, H.

    1975 *Mind, Language and Reality.* Cambridge: Cambridge University Press (1975).

Rosenblueth A., N. Wiener, and J. Bigelow

    1943 Behavior, purpose and teleology. in: Philos. Sci. **10/Nr.1** (1943).

Schrödinger, E.

    1935 Die gegenwärtige Situation in der Quantenmechanik. *Naturwissen.* **23/48–50** (1935).

Searle, J.R.

    1980 Minds, brains, and programs. *Behav. Brain* **3** (1980).

Simon, H.A., and J.R. Hayes

    1974 Understanding written problem instructions. In: *Knowledge and Cognition,* ed. L.W. Gregg. Hillsdale, NJ (1974); reprinted in: Simon, *Models of Thought.* New Haven, CN (1979).

Turing, A.M.

    1936-7 On computable numbers, with an application to the Entscheidungsproblem. *P. Lond. Math. Soc. (2)* **42** (1936-7) 230–265; A correction, ibid. . 43 (1937) 544–546; reprinted in Davis, M., *The Undecidable.* New York: Raven Press (1965).

    1939 Systems of logic based on ordinals. *P. Lond. Math. Soc. (2)* **45** (1939) 161–228; reprinted in Davis, M., *The Undecidable.* New York: Raven Press (1965).

    1950 Computing machinery and intelligence. *Mind* **59** (1950) 433–460.

Wiener, O.

    199x Poetik im Zeitalter naturwissenschaftlicher Erkenntnistheorien. (forthcoming).

List of Journal Abbreviations

Adv. App. Math.	Advances in Applied Mathematics
Adv. Math.	Advances in Mathematics
Alg. Log.	Algebra i Logica
Am. J. Math.	American Journal of Mathematics
Am. Math. Mo.	American Mathematical Monthly
AMS Transl.	AMS Translations
Ann. Hist. C.	Annals of the History of Computing
Ann. Math. Log.	Annals of Mathematical Logic
Ann. Math. S.	Annals of Mathematics Studies
Ann. Math.	Annals of Mathematics
Ann. U. Paris	Annales de L'Université de Paris
Arch. Math. Log. Gr.	Archiv für Mathematische Logik und Grundlagenforschung
Arch. Math. Physik	Archiv der Mathematik und Physik
Artif. Intell.	Artificial Intelligence
B. Am. Math. S.	Bulletin of the American Mathematical Society
B. Math. Biol.	Bulletin of Mathematical Biology
B. Math. Biophy.	Bulletin of Mathematical Biophysics
Behav. Brain	Behavioral and Brain Sciences
Bell Sys. T.	Bell Systems Technical Journal
Biometrika	Biometrika
Biosystems	Biosystems
Br. J. Phil. S.	British Journal for the Philosophy of Science
Br. J. Psycho.	British Journal of Psychology
Com. Sci.	Computer Science
Comm. ACM	Communications of the ACM
Comp. Biol. M.	Computers in Biology and Medicine
Comp. J.	Computer Journal
Comp. Sur.	Computing Surveys
Comp. Sys.	Complex Systems
Cont. Math	Contemporary Mathematics
Curr. M. Bio.	Currents in Modern Biology
Discr. Math.	Discrete Mathematics
Duke Math. J.	Duke Mathematical Journal
Eur. J. Oper.	European Journal of Operational Research
Found. Phys.	Foundations of Physics

Fund. Math.	Fundamenta Mathematicae
Hist. Phil. Log.	History and Philosophy of Logic
IBM J. Res.	IBM Journal of Research and Development
IEEE Trans. Inf. Th.	IEEE Transactions on Information Theory
Ind. Math.	Indagationes Mathematicae
Inf. Contr.	Information and Control
Inf. Proc. L.	Information Processing Letters
Int. J. Theor.	International Journal of Theoretical Physics
Invent. Math.	Inventiones Mathematicae
J. ACM	Journal of the Association for Computing Machinery
J. Algebra	Journal of Algebra
J. Am. Chem. S.	Journal of the American Chemical Society
J. Autom. Reas.	Journal of Automated Reasoning
J. Comput. Sy.	Journal of Computing and Systems Sciences
J. Math. Phys.	Journal of Mathematical Physics
J. Neurosci.	Journal of Neuroscience Research
J. Philos. Lo.	Journal of Philosophical Logic
J. rein. Math.	Journal für die reine und angewandte Mathematik
J. Stat. Phys.	Journal of Statistical Physics
J. Symb. Log.	Journal of Symbolic Logic
J. Theor. Bio.	Journal of Theoretical Biology
Mach. Intell.	Machine Intelligence
Math. Annal.	Mathematische Annalen
Math. Cent. T.	Mathematical Centre Tracts
Math. Comput.	Mathematics of Computation
Math. Intell.	Mathematical Intelligencer
Math. Scand.	Mathematica Scandinavica
Math. Z.	Mathematische Zeitschrift
Mem. Am. Math.	Memoirs of the American Mathematical Society
Mich. Math. J.	Michigan Mathematical Journal
Mind	Mind
Monats. Math. Ph.	Monatshefte für Mathematik und Physik
Nature	Nature
Naturwissen.	Naturwissenschaften
New Sci.	New Scientist
Notre Dame J. Form. Log.	Notre Dame Journal of Formal Logic
P. Am. Ac. Arts Sci.	Proceedings of the American Academy of Arts and Sciences
P. Kon. Ned. A	Proceedings of the Koninklijke Nederlandse Akademie van Wetenschappen Series A. Mathematical Sciences
P. Lond. Math. Soc.	Proceedings of the London Mathematical Society
P. Nat. Ac. Sci.	Proceedings of the National Academy of Sciences
P. Roy. Soc. (A)	Proceedings of the Royal Society of London, Series A. Mathematical and Physical Sciences
Peng. Sci.	Penguin Science News
Phi. T. Roy. B	Philosophical Transactions of the Royal Society of London Series B. Biological Sciences
Phil. Nat.	Philosophia Naturalis
Philos. Sci.	Philosophy of Science
Philosophy	Philosophy
Phys. Report.	Physics Reports
Phys. Rev. L.	Physical Review Letters

Phys. Rev.	Physical Review
Physica	Physica
Probl. Inf. Tr.	Problems of Information Transmission
Probl. Pered. Inf.	Problemy Peredači Informacij
Proc. Cont. Aut.	Process Control and Automation
Rev. M. Phys	Reviews of Modern Physics
Russ. Math. S.	Russian Mathematical Surveys
Sci. Am.	Scientific American
Science	Science
SIAM J. Comp.	SIAM Journal of Computing
Sov. Math. Dokl.	Soviet Mathematics – Doklady –
Sym. P. Math.	Symposia in Pure Mathematics
T. Am. Math. S.	Transactions of the American Mathematical Society
Theor. Comp.	Theoretical Computer Science
Theor. Ling.	Theoretical Linguistics
Tijd. Wijsb.	Tijdschrift voor Wijsbegeerte
Trans. AIEE	Transactions of the American Institute of Electrical Engineeers
Trans. AMS	Transactions of the American Mathematical Society
Usp. Mat. Nau.	Uspechi Matematičeskich Nauk
Z. Warsch. V.	Zeitschrift für Wahrscheinlichkeitstheorie und verwandte Gebiete

Reihe *Computerkultur*

Herausgegeben von Rolf Herken

Die Reihe „Computerkultur" beschäftigt sich mit philosophischen, kulturellen, histori-schen, mathematischen und technischen Aspekten der Computerwissenschaft, insbeson-dere der „Künstlichen Intelligenz".
Neben der Veröffentlichung von grundlegenden Werken in deutscher Übersetzung erfolgt auch eine Auseinandersetzung mit diesem Themenbereich in Form von Original-veröffentlichungen (in englischer und deutscher Sprache).

Andrew Hodges

Alan Turing, Enigma

Aus dem Englischen übersetzt von Rolf Herken und Eva Lack
Zweite Auflage. 1994. 22 Abbildungen. VI, 662 Seiten.
Gebunden DM 58,–, öS 398,–, sFr 58,–
ISBN 3-211-82627-0
(Computerkultur, Band 1)

Herbert A. Simon

Die Wissenschaften vom Künstlichen

Aus dem Englischen übersetzt von Oswald Wiener unter Mitwirkung von Una Wiener
Mit einem Nachwort des Übersetzers
Zweite Auflage. 1994. 7 Abbildungen. XII, 241 Seiten.
Gebunden DM 48,–, öS 336,–, sFr 48,–
ISBN 3-211-82629-7
(Computerkultur, Band 3)

Giorgio de Santillana, Hertha von Dechend

Die Mühle des Hamlet

Ein Essay über Mythos und das Gerüst der Zeit

Von der Autorin durchgesehene Übersetzung aus dem Englischen von Beate Ziegs, Berlin
Zweite Auflage. 1994. 56 Abbildungen. X, 522 Seiten.
Gebunden DM 68,–, öS 476,–, sFr 68,–
ISBN 3-211-82630-0
(Computerkultur, Band 8)

Springer-Verlag Wien New York

Sachsenplatz 4–6, P.O.Box 89, A-1201 Wien · 175 Fifth Avenue, New York, NY 10010, USA
Heidelberger Platz 3, D-14197 Berlin · 3-13, Hongo 3-chome, Bunkyo-ku, Tokyo 113, Japan

Weitere Bände in Vorbereitung:

Bernhard Dotzler (Hrsg.)

Babbage's Rechen-Automate

Ausgewählte Schriften von Charles Babbage

1995. Zahlreiche Abbildungen. Etwa 500 Seiten.
Franz. Broschur etwa DM 100,–, öS 700,–, sFr 100,–
ISBN 3-211-82640-8
(Computerkultur, Band 6)

Stephen Graubard (Hrsg.)

Probleme der Künstlichen Intelligenz

Eine Grundlagendiskussion

Aus dem Englischen übersetzt von Rike Felka, Berlin
1995. Zahlreiche Abbildungen. Etwa 320 Seiten.
Franz. Broschur etwa DM 70,–, öS 490,–, sFr 70,–
ISBN 3-211-82641-6
(Computerkultur, Band 9)

Oswald Wiener

Schriften zur Erkenntnistheorie

(Computerkultur, Band 10)

Sachsenplatz 4–6, P.O.Box 89, A-1201 Wien · 175 Fifth Avenue, New York, NY 10010, USA
Heidelberger Platz 3, D-14197 Berlin · 3-13, Hongo 3-chome, Bunkyo-ku, Tokyo 113, Japan

Springer-Verlag
and the Environment

WE AT SPRINGER-VERLAG FIRMLY BELIEVE THAT AN international science publisher has a special obligation to the environment, and our corporate policies consistently reflect this conviction.

WE ALSO EXPECT OUR BUSINESS PARTNERS – PRINTERS, paper mills, packaging manufacturers, etc. – to commit themselves to using environmentally friendly materials and production processes.

THE PAPER IN THIS BOOK IS MADE FROM NO-CHLORINE pulp and is acid free, in conformance with international standards for paper permanency.